The Evolution of Life

This book is dedicated to my incomparable students, in gratitude.

The Evolution of Life

Graham Bell

McGill University

OXFORD
UNIVERSITY PRESS

OXFORD
UNIVERSITY PRESS

Great Clarendon Street, Oxford OX2 6DP,
United Kingdom

Oxford University Press is a department of the University of Oxford.
It furthers the University's objective of excellence in research, scholarship,
and education by publishing worldwide. Oxford is a registered trade mark of
Oxford University Press in the UK and in certain other countries

Published in the United States of America by Oxford University Press
198 Madison Avenue, New York, NY 10016, United States of America

British Library Cataloguing in Publication Data
Data available
Library of Congress Control Number: 2014950233

ISBN 978–0–19–871257–2

Printed in Great Britain by Bell & Bain Ltd., Glasgow

FOREWORD

This book was written after a decade's experience of teaching the Evolution course to a class of about 150 second-year students at McGill University. Students take the course because evolution is an item in their curriculum, together with courses about genetics, cell biology, ecology, and so forth. I try to explain why it is more than that. Most students are already familiar with the biology of individuals, and with the principles that govern, say, photosynthesis or DNA replication. They are often not as familiar with the biology of variable populations, and therefore with the principles that govern evolutionary change. In developing the course, I found myself emphasizing these principles, even to the extent of excluding many topics that are often taught in the context of evolution. The focus of the course sharpened as it became dedicated exclusively to the distinctive properties of evolving populations.

In particular, the course is about adaptive evolution. The intricate design of living bodies, and how exquisitely they are tailored for a myriad ways of life, has always aroused wonder. The purpose of evolutionary biology is to account for this remarkable observation. The main goal of a course on evolution must be, then, to explain how very complex and highly integrated organisms can have arisen through a natural process, with no guiding hand. This is a genuinely difficult problem that students struggle to understand, I believe, unless the mechanism of Darwinian evolution and its surprising consequences form the framework of the course and inform all its aspects.

The outcome of evolution on Earth has been the millions of species of animals and plants and other organisms that live in the sea and on the land. The underlying theme of a course on evolution is thus the radiation of living organisms during the course of Earth history. We are only one of these millions of species, but we are of course particularly interested in our own place in the world. It seems appropriate, then, to orient an evolutionary account of diversity by showing how we are related to other organisms and by tracing the human lineage deeper and deeper into the past.

The focus of the course on adaptation and radiation made it necessary to discard or downplay many familiar topics, such as the Hardy–Weinberg law, the "shifting balance", and the voyage of the *Beagle*. I regret their omission, all the more because it has been accompanied by the loss of all but the very simplest mathematical propositions, and by the restriction of theory to statements that can be supported directly by evidence. This book is not a substitute for more specialized or more extensive treatments. Its sole purpose is to provide a grounding in the main features of evolution on Earth.

A book can assist the delivery of a course in two ways. It may simply present the topics as they are taught in the course, so that the lectures follow the text quite closely and the text encapsulates the material that the students should learn. On the other hand, it may extend the material in both directions, on the one hand by reinforcing areas such as basic genetics, and on the other hand by introducing more advanced topics in areas such as population genetics, genomics or developmental biology. There is much to be said for both models, but this book belongs in the former category. I have set out to write a brief text that follows very closely the sequence of lectures in my own course. Its purpose is to provide a full course for university students who have not previously taken a course dedicated to evolution, with as little extraneous material as possible. More advanced students may benefit from a more comprehensive treatment that covers a wider range of topics in related fields. I hope, however, that this book will serve to introduce students at any level to the main features of evolution and the main principles of evolutionary biology. The Further Reading that is suggested at the end of each chapter has been chosen mainly from the secondary literature, which consists of reviews and syntheses written to provide a fairly non-technical introduction to a broad field. They are in turn based on the primary literature that reports original research findings, and on which this book is ultimately based. The reference lists of the Further Reading provide a convenient entry to the primary literature.

ONLINE RESOURCE CENTRE

online resource centre

www.oxfordtextbooks.co.uk/orc/bell_evolution/

The Online Resource Centre to accompany *The Evolution of Life* features the following resources:

For registered adopters

- Figures from the book in electronic format, for use in lecture slides
- Additional questions to accompany each chapter, to augment those in the printed book
- Journal Clubs: discussion questions that guide students through research papers related to each chapter.

To register as an adopter of the book, go to **www.oxfordtextbooks.co.uk/orc/bell_evolution** and follow the on-screen instructions.

For students

- Programs and interactive spreadsheets related to specific topics covered in the book
- Updated further reading suggestions to include those released since the book's publication.

SUMMARY OF CONTENTS

FULL CONTENTS

PART 1
Basics

The first part of this book describes some of the characteristics of evolving populations and the main features of the process of evolution. It is intended to introduce evolution to students who have had little or no systematic instruction in the subject, and to remind others of the outline of the field.

The Evidence for Evolution

Why do you need to learn about evolution? The natural world is very complicated—think of the enormous variety of animals and plants that we take for granted. It would be very difficult to learn about every one of them separately. Science is how we understand complicated situations in terms of a few simple principles that can be very widely applied. The motion of a bird in the air, for example, is very complicated, but it is governed by a few simple laws of dynamics that also govern the flight of an arrow or an airplane. The science of evolution explains how the vast diversity of living things is likewise governed by a few simple natural principles. The object of this book is to explain these principles and to show you how all living things—including ourselves—are the products of evolution.

1.1 The uniformity of life. *Living organisms share a narrow range of materials and processes.*

Almost everything is a pebble, a watch, or a beetle. Think of an object. The chances are that it will belong to one of three categories, which I shall call pebble, watch, and beetle.

Pebbles have very simple structures. They may be a mixture of substances and they may have complicated outlines, but they do not consist of separate parts that must interact with one another in order to perform some function or activity. Snowflakes, crystals, and mountains are pebbles, in this sense. Pebbles come into being by some simple natural process that does not require conscious design.

A *watch* is a complex integrated structure whose parts must interact in a particular way in order to produce a characteristic outcome. A hammer, a building, and the computer on which I am writing this sentence are all watches, in this sense. All are prefigured in the mind of a designer and constructed by an artisan, both almost always human.

A *beetle* is also a complex integrated structure whose parts must interact in a particular way in order to produce a characteristic outcome, but it has been neither designed nor constructed: instead, it has evolved and developed. A dog, a daisy, and the organism who is typing this sentence are all beetles, in this sense.

A commonplace and meaningful distinction is that beetles are alive, whereas pebbles and watches are not. This is not a profound philosophical insight, just a rough guide for thinking about the world. There are some objects it would be difficult to classify in this way. A dead snail shell, or a fragment of the shell, for example, might appear to be a pebble but had once been a beetle. At the other extreme, the whole Earth (shorn of its living inhabitants) would carry out cycles of erosion, subduction, and mountain forming without either design or evolution. So is the Earth a pebble, a watch, or a beetle? You can probably think of many other debatable cases.

The distinction between living and non-living things is useful, though, because it focuses the mind on the problem of beetles. If they neither arise spontaneously through simple physical laws, like pebbles, nor are consciously designed and constructed, like watches, how do they come into being? The answer is that they evolve: they are the descendants of previous generations of beetles, modified by natural selection acting on variation through environmental change.

Part 1 of this book describes in broad terms why we know that the diversity of life has evolved, and explains how the mechanism of evolution works. Subsequent parts discuss the main themes of evolution in more detail. This book will show you how all the physiology, biochemistry, and genetics that you have learned in other biology courses are the consequences of a few natural mechanisms. If you have not learned about these mechanisms before, this will be the most interesting course you will ever take.

1.1.1 All living organisms share the same basic features.

The outward diversity of animals and plants is based on a very restricted range of materials, body plans, power sources, and instructions, relative to those we use in everyday life. Figure 1.1 lists the features that

distinguish beetles from watches. This is not because animals and plants are better designed than our own tools; on the contrary, they are in many ways inferior. It is because they have been neither designed nor constructed, but have instead evolved.

Suppose that you visit the hardware store, then the pet store. The hardware store has a wide range of "watches" (tools) and the pet store has a wide range of "beetles" (animals). At the hardware store you could buy a screwdriver, a barbecue, or a light bulb. At the pet store, you could buy a goldfish, a parrot, or a puppy. The variety available at these stores might seem to be comparable, but this is misleading: in many ways, tools are more diverse than animals.

- Tools are made of a great variety of different materials: metals, ceramics, plastics, concrete, silicon, and glass, besides organic materials like wood or rubber. Animals and plants are made of only a few basic kinds of substance—amino acids, nucleotides, and sugars—all based on carbon.

Figure 1.1 Features that distinguish beetles from watches.

Watch courtesy of Pierre EmD. Beetle courtesy of AlbertHerring. These files are respectively licensed under the Creative Commons Attribution-Share Alike 3.0 Unported and 2.0 Generic licenses.

Watch		Beetle
Very diverse: metals, ceramics, plastics, concrete, silicon and glass, besides organic materials like wood or rubber	Material	A few basic kinds of substance — amino acids, nucleotides, and sugars — all based on carbon
Many: lever, wedge, crank, pulley, screw, wheel, etc.	Mechanism	Few, e.g. lever; screw and wheel unknown
Many: clockwork, gravity, electricity, internal or external combustion engines, nuclear power, etc.	Energy source	Weaker and more complex processes such as fermentation, respiration, and photosynthesis
Wide: temperature of liquid nitrogen to the melting point of iron or beyond	Working range	Narrow: freezing to boiling point of water
Many kinds: an oral description, a drawing, a recipe, a computer file, or a model	Instructions	One kind: sequence of nucleotides in a nucleic acid

- Tools vary because the material used for each is the best suited for its purpose: metal is best for edge tools, ceramic for tableware, glass for windows, and so on. Organisms use a much more restricted set of materials that are neither as strong nor as stiff as those we can make: calcium carbonate, cellulose, chitin, and bone.

- Tools use a variety of mechanisms: lever, wedge, crank, pulley, screw, wheel, and so on. Organisms use only a few of these: levers are used by animals with jointed limbs, for example, but screws and wheels are almost unknown.

- Again, machine tools make use of many sources of energy: clockwork, gravity, electricity, internal or external combustion engines, and so forth. None of these are used by organisms as prime movers; the living cell is instead powered by weaker and more complex processes such as fermentation, respiration and photosynthesis.

- Tools can be devised to operate over a very wide range of conditions, from the temperature of liquid nitrogen to the melting point of iron or beyond; organisms operate only between the freezing and boiling points of water.

- The instructions for making a tool can be presented in many ways: an oral description, a drawing, a recipe, a computer file, or a model. The instructions for making an organism of whatever kind are encoded in only a single way, as a sequence of nucleotides in a nucleic acid.

Evolution is modification through descent. All organisms are similar, not because they have been perfectly engineered, but rather because they all descend from the same common ancestor. The mechanism that produces modification is described in the next chapter. This chapter is concerned mainly with descent—how we know that the diversity of living forms springs from a single root.

1.1.2 Similarity implies common descent.

Children resemble their parents. Everyone knows that children tend to look like their parents. In some organisms, in fact, they look exactly like their parents. Dandelions, duckweed, water fleas, and many other animals and plants reproduce in a very simple manner by using cells from a single parent to develop into new individuals. This asexual reproduction is very common among microbes and microscopic animals, but it also occurs in some fish, salamanders, and lizards. In these species, offspring are almost exact copies of their parents and resemble them very closely. In most animals and plants, of course, reproduction is more complicated because it involves two parents, and the cell from which a new individual develops bears material from both of them. Offspring that are produced sexually therefore resemble both parents, although the resemblance is incomplete.

Children of the same parents resemble one another. Because children in the same family resemble their parents, they necessarily also resemble one another. In sexual organisms they differ because they receive different combinations of material from their mother and father. Nevertheless, they will often share attributes such as stature and facial features, and on average they will be much more similar than two randomly chosen individuals from the same population. In exceptional cases they may be very similar indeed. If the fertilized egg divides in two, and each daughter cell then develops into a new individual, the result is identical twins, children who have received the same genetic material from their parents. Identical twins may be so similar in appearance and behavior that strangers cannot tell them apart.

Relatives are similar because they share ancestral genes. The physical basis of inheritance is the gene, which is a short length of DNA that can be translated into a protein by the biosynthesis machinery of the cell. Genes do two things. In the first place, they encode the proteins that are responsible for the underlying physical make-up of an individual. Individuals that have the same genes are therefore likely to be very similar in most ways, although they will always differ to a greater or lesser extent because development is influenced by the environment as well as by the genes. Secondly, genes are copied in the process of forming the reproductive cells, so that offspring bear copies of the genes borne by their parents. Consequently, offspring will tend to resemble their parents because they bear copies of the same genes. The same principle applies to all kinds of relatives: great-grandchildren or second cousins, for example, are more similar than randomly chosen individuals because they share some fraction of their genes through descent from a common ancestor.

Sister species are very similar because they are closely related. At a broader scale, all individuals belonging to a given species are more closely related to one another than they are to members of any other species. This is because the most recent common ancestor of any two individuals belonging to the same species will normally itself have been a member of that species.

Species that have recently evolved from the same common ancestor (sister species) will also be closely related. The whitefish of northern lakes provide a good example of species that are currently emerging. Many lakes contain two types, a large type that feeds on worms and shrimps in the mud and weeds, and a small type that feeds on plankton in mid-water, as illustrated in Figure 1.2. Although they are different in appearance and behavior they separated very recently (since the retreat of the glaciers about 10,000 years ago), are still capable of interbreeding, and share most of their genes. At this very early stage, in fact, they are somewhere between two ecotypes of the same species and two completely separate species. In situations like this, we can actually witness the emergence of new species—Figure 1.2 also shows some of the many kinds of whitefish that have recently evolved in Swiss lakes.

Species share attributes of their common ancestor. As time goes on, newly formed species will become steadily more different because they have separate and independent histories, each with a unique pattern of adaptation to environmental change and variation. The physical basis of adaptation to some new environment is altered versions of genes. Hence, two species will tend to become less closely related over time as they become adapted to different ways of life. At first (about 100,000 years after diverging from their common ancestor), they are as closely related as the two kinds of lake whitefish; later (more than 1 million years after) only as closely related as different kinds of whitefish such as lake whitefish and mountain whitefish; then (more than 10 million years after) as different kinds of salmonid fishes, such as whitefish, grayling and trout; then (more than 100 million years after) as salmonid and centrarchid fishes, such as bass and pumpkinseed; and so forth. We will explain how these events are dated in Chapter 5.

Clearly, species become steadily more different and less closely related as time goes on. Nevertheless, they continue to resemble one another in many ways. Whitefish and bass both have fins and jaws, for example. This is not because fins and jaws have evolved independently in each new species of fish, but rather because the most recent common ancestor of whitefish and bass (a member of a group called the Pholidophoroformes living in the upper Triassic period, about 220 million years ago) possessed both characteristics and transmitted the genes responsible to its modern descendants. Whitefish and rabbits are still more different, but both possess a backbone and a thyroid gland, for example, which they have inherited from a still more distant common ancestor.

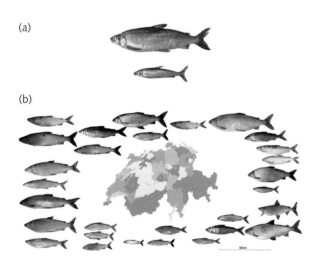

Figure 1.2 Sister species of whitefish, *Coregonus*. (a) Normal (benthic) and dwarf (limnetic) forms from a Quebec lake. (b) A cloud of recently evolved species of whitefish from lakes in Switzerland.

(a) Reproduced with permission from Jeukens, J. and Bernatchez, L. 2012. Regulatory versus coding signatures of natural selection in a candidate gene involved in the adaptive divergence of whitefish species pairs (Coregonus spp.) *Ecology and Evolution* 2 (1): 258–271 doi: 10.1002/ece3.52. (b) From Pascal Vonlanthen, Eawag, and kindly provided by Bänz Lundsgaard-Hansen.

Universal characters are inherited from a distant common ancestor. The characters involved in the formation of new species are usually adaptive responses to some ecological challenge or opportunity, such as avoiding a predator or being able to grow in salty soil. They may be caused by alterations in only a few genes, or even a single gene. More fundamental changes, such as the ability to form bone and to use it to develop a skeleton, happen less often because they require alterations in many genes.

The most fundamental features of the biochemical machinery of the cell have scarcely changed at all since life began, since they are common to all known organisms. The simple explanation is that they have been inherited, with minimal alteration, from a very remote common ancestor. The uniformity of basic processes and the molecules responsible for them shows how all living organisms have evolved in a single unbroken line of descent from a universal common ancestor that lived between three and four billion years ago.

1.2 Extinction. *Fossils of extinct organisms are the physical remains of past evolution.*

Living organisms provide very strong circumstantial evidence of evolution. It would be even more convincing to see the ancestors themselves. This is seldom possible, because most individuals decay and disappear soon after they die. In exceptional circumstances, however, their remains may be preserved in sediments that have hardened into rocks under heat and pressure. These are fossils. They are the direct evidence of how ancient organisms have evolved into modern animals and plants.

1.2.1 The majority of species are extinct.

Fossils are extremely abundant and diverse. Wherever you live, there are likely to be outcrops of fossil-bearing rocks not far away. The local natural history museum will know where they are and may have a representative collection of fossils on display. Many fossils are immediately recognizable because they clearly resemble some familiar living organism such as a clam or a fish. Others are more difficult to identify because they are only part of an organism, or may be difficult to see because they are very small.

Even if you searched a site very carefully and found every fossil, however, you would still not have found every species that once lived there, because whether a particular species is likely ever to be fossilized depends on how it is built and where it lives. Small soft-bodied animals such as worms and jellyfish quickly decompose leaving no remains, so they are preserved as fossils only in exceptional circumstances. The remains of animals and plants that live in forests or grassland are likely to be consumed or simply weather away. The remains of animals that have hard parts such as shells and skeletons and that live in the sea or freshwater, on the other hand, are likely to sink into sediments that later turn into rock, where they are preserved as

fossils. Marine invertebrates, for example, are very abundant as fossils and have been very thoroughly described. About 300,000 living species of sponges, corals, crabs and shrimps, barnacles, starfish, and similar animals have been catalogued, grouped into about 60,000 genera. Only about one-third of these species and genera, however, have hard parts that would fossilize easily. Another 30,000 genera are known only from fossils, and almost all of these can be confidently assumed to be completely extinct. If only one-third of extinct forms were fossilized, then about 90,000 genera have previously existed, but are now extinct. This is about 50% more genera, and hence about 50% more species, than are currently alive. This is an underestimate of the true number of extinct species, because fossils are much more difficult to find than living organisms, and because the fossils of many species are likely to have been destroyed by geological processes in the past. Hence, more species have become extinct than are currently living.

A species has a limited lifespan. The frequency of extinction emphasizes the limited lifespan of a species. This can be estimated from the first and last occurrences of a species in the fossil record. For marine invertebrates the average lifespan of a species is about 4 My ("My" stands for "million years," and "Mya" stands for "million years ago"), and other groups give rather similar values. Horses provide a very well-studied example with abundant fossils. At present there is only a single species of horse, which is native to Asia but has been domesticated and spread around the world by humans. In the past there were many species of horse, living in different parts of the world, including North America. Their relationships are shown in Figure 1.3; each species lived for about 3 My on average. There are some species in other

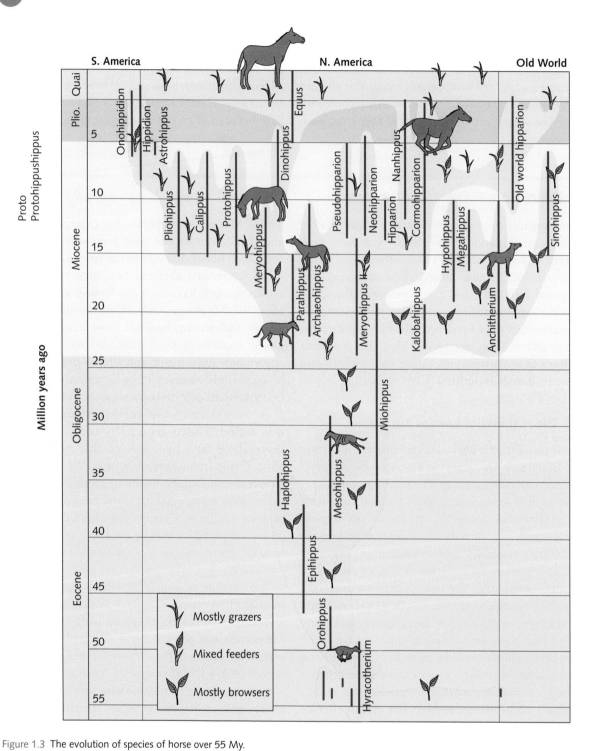

Figure 1.3 The evolution of species of horse over 55 My.

From MacFadden, Bruce J. 2005. Fossil horses – evidence for evolution. *Science* 307: 1728–1730. Reprinted with permission from AAAS.

groups which are much longer lived, including those known as "living fossils." Species very similar to the gingko tree, the horseshoe crab and the tadpole shrimp lived as long as 100 Mya, so the lineages leading to the modern species have changed very little over this vast span of time. Others are much shorter lived, becoming extinct in less than 1 My. As a broad general rule, most species last between 1 and 10 My, and 5 My is a reasonable representative value for many groups of animals and plants.

Consequently, there are not just more extinct species than living species—there are many times more. Given that a diverse marine invertebrate fauna first evolved about 500 Mya and that the average longevity of a species is roughly 5 My, the species now living make up only about 1% of all the species that have ever existed. This emphasizes how dynamic the world biota is. On the timescale of human history, a few thousand years at best, it appears to be static. On the longer scales of geological time, however, the biota is in continual flux and has been completely replaced many times.

1.2.2 Extinction reveals the long history of life.

Older rocks have stranger fossils. If you are collecting fossils from the youngest deposits, such as glacial sands and gravels, most of them will belong to species still living, or will be closely related to living species. Moving back in time, more and more unfamiliar species will be found. If you are collecting mollusks, for example, only about half the species you find from Pliocene rocks (6–7 Mya) will still be living, and Miocene faunas (20 Mya) will contain few if any modern species. Dig deeper and you will find whole groups that have utterly vanished.

The dinosaurs are the most familiar example of an extinct group, but there are dozens of major groups of animals and plants that may have once been very abundant but are now completely extinct. Trilobites, for example, were common marine animals for over 100 My. There are no trilobites alive today, but they were clearly arthropods, with a distant resemblance to crustaceans or sow bugs (also called pill bugs or woodlice). Dig deeper still, and there are groups that are not only extinct, but do not look at all like anything now living. The earliest echinoderms, for example, do not resemble any living group, although they are easily identified as echinoderms from the form of their skeleton. Later on, forms recognizably similar to modern groups appear in the rocks (Figure 1.4; for more details, see Section 5.13). This is easily understood when we recollect that modern species descend from ancient species through an unbroken chain of intermediate species. Older fossils become steadily less familiar as we move backwards through time because modifications accumulate over time.

Earth history involves a long succession of faunas and floras. Not only species but major groups of organisms succeed one another through time. This gives each geological period a characteristic flavor, so to speak, according to the new groups that flourish and the old ones that dwindle. In the present Cenozoic period, for example, placental mammals, birds and flowering plants have become abundant and diverse whereas cycads and sea-lilies have dwindled, and dinosaurs and ammonites have disappeared. This is not a very tidy process; one fauna is seldom if ever abruptly and completely replaced by another. Sometimes the end of a period is marked by the final extinction of a once flourishing group, like trilobites at the end of the Permian or ammonites at the end of the Cretaceous. More often, the rise and fall of distinctive kinds of organisms results in a gradual but continuous shift from one kind of community to another.

Life has evolved over about 3500 My. Very long periods of time (by the standards of human history) have been occupied by the long succession of faunas and floras that have left their remains in the rocks. The methods that have been used to measure geological time are described later in this book (Chapter 6), but a few milestones may be helpful at this point. The continental ice sheet of the last glacial period was in full retreat about 10,000 years ago. Our own species emerged in Africa about 100,000 years ago. The asteroid impact, which ended the period when dinosaurs flourished, happened about 65 Mya. A diverse fauna of animals with fossilizable hard parts first appeared about 500 Mya. There are abundant remains of organisms resembling modern bacteria from 1000–2000 Mya. In deepest time the fossils become scarce and often debatable; there

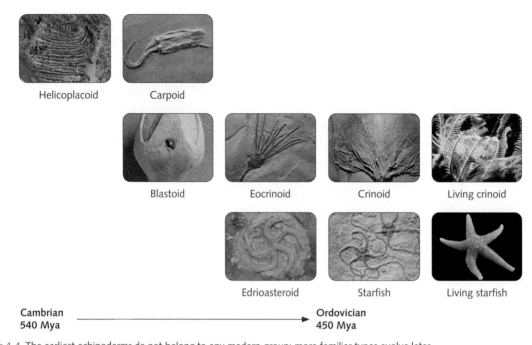

Figure 1.4 The earliest echinoderms do not belong to any modern group; more familiar types evolve later.

Carpoid image courtesy of The Virtual Fossil Museum. Blastoid image courtesy of Richard Schrantz. Eocrinoid image courtesy of John P. Adamek, Tethys Fossils.com. Edrioasteroid image courtesy of Joseph Koniecki. Image of starfish fossils courtesy: Richie Kurkewicz Pangaea Fossils <http://www.fossilmall.com>. Living starfish image © Hans Hillewaert / CC-BY-SA-3.0.

are unequivocal remains of cellular organisms from about 2800 Mya, and indirect evidence that life first appeared about 3500 Mya or a little more. The age of the Earth is about 4500 My.

All of these figures will reappear later in this book when the history of life on Earth is described in more detail. The crucial point is that the succession of faunas preserved in the fossil record has taken place over very long periods of time. This is sometimes thought to be difficult to imagine, but it is not. We are all familiar with an atlas that may condense a continent onto a single page. This is a spatial scale of about 1: 10 million; on a temporal map of the same scale the history of animal life would occupy about 50 years, and would be completed within a single human lifespan.

1.2.3 Extinction implies the continual evolution of new kinds of organism.

Marine diversity has changed little over the last 400 My. The relentless course of extinction has not depleted the world biota. When the global diversity

of marine invertebrates is assessed by compiling all the described species of fossils, it shows a rapid increase after the appearance of species with readily fossilized hard parts about 500 Mya, followed by a long period during which overall diversity does not change very much, despite the continual turnover of species and major groups. After the great mass extinction at the end of the Permian period, about 240 Mya, diversity rapidly recovered, then exceeded previous levels and is still increasing. This may be in part attributable to the better preservation of more recent faunas, as the most recent analyses have shown that marine diversity has probably not changed much over the last 400 My of Earth history. Meanwhile, novel kinds of organisms such as flowering plants, insects, and tetrapods, have added to the diversity of recent times. At all events, it is clear that diversity has not tended to decrease over geological time.

Speciation balances extinction in the long term. Since overall diversity remains more or less the same, or tends to increase over time, new species must be continually arising to replace those that become extinct. We can sometimes witness the early stages of this process, as in the lake whitefish discussed above. In other cases we can detect a recent burst of speciation by finding large numbers of closely related species in isolated lakes or on isolated islands. For example, the 38 species of the cricket *Laupala* occur only on the Hawaiian Islands. The remarkably rapid divergence of these animals is illustrated in Figure 1.5. We know the age of the islands (from geological information) and can work out the relationships among the species (from genomic information). We can calculate the age of a focal species by first calculating the genetic distance to its closest relative as a fraction of the distance to the most deeply branching species. The most deeply branching species cannot be older than the islands themselves. Hence, multiplying the scaled genetic distance by the age of the islands gives an estimate of the maximum age of the focal species. This gives an average rate of speciation of about 4 new species/My. Other well-known examples of rapid recent speciation include the cichlid fishes of Lake Victoria in Africa and the *Partula* land snails of the Society Islands in the South Pacific.

There are also examples of recent rapid extinction, unfortunately. At least 50 of the 60 or so species of *Partula* have been completely wiped out by an introduced predatory snail, which was intended to control the giant African land snail, which was itself introduced as a source of food.

In most cases, however, speciation proceeds much more slowly. The average rate of speciation for arthropods is about 0.1–0.2 species/My, as we would expect from the average species lifespan of 1–10 My. What this rate means is that the probability that an average species will give rise to a new species during a million-year period is between one in five and one in ten. There are probably about 10 million species of multicellular animals and plants living at any given time, so we expect one or two new species to appear each year. Hence, the total number of species that have ever existed is about 500–1000 million, only about 1–2% of which are alive today. This gives a clear idea of the vast diversity produced by evolution.

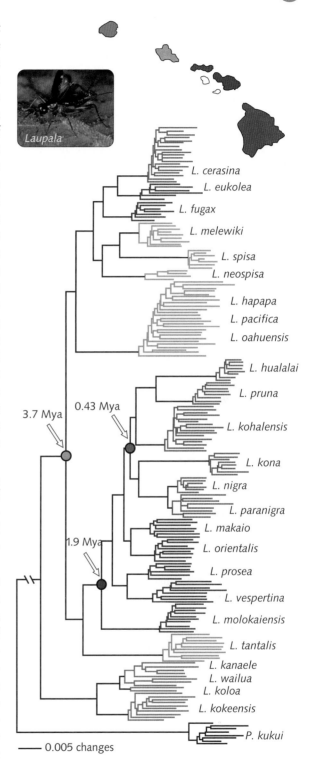

Figure 1.5 The divergence of lineages of a cricket, *Laupala*, on the Hawaiian Islands.

Reprinted by permission from Macmillan Publishers Ltd: *Nature* 433 (7024): Mendelson and Shaw, Sexual behaviour: rapid speciation in an arthropod. © 2005.

1.3 **Adaptation.** *Evolution explains how organisms are constructed.*

The central problem of biology is to explain how organisms are well adapted to their ways of life. One answer is that they were designed and constructed like a watch. We know that this is wrong, because they follow a common plan that reveals their common ancestry and we can find the remains of their ancestors in the rocks. When we look more closely at how animals are constructed, it becomes clear that they must have evolved, because they retain many of the features of their ancestors, even when this leads to imperfect or inefficient design.

1.3.1 **Differences between organisms reflect different ways of life.**

Organisms with similar ways of life share many features. Many large marine animals such as sharks, tuna, and whales cruise at several meters per second and can reach a top speed of 10–20 meters per second. These groups are not closely related. Tuna and similar fish such as swordfish and albacore are teleosts, a group of bony fish that last shared a common ancestor with sharks in the Silurian period, a little more than 400 Mya. Sharks and bony fish are primary aquatic animals, because all their ancestors lived in the sea. Whales are mammals, a group of mainly terrestrial animals whose ancestors belonged to a different group of bony fish, the crossopterygians or lobe-finned fishes. Whales are thus secondary aquatic animals because their ancestors were terrestrial.

Because of their distant relationship and contrasting histories, these three groups are different in many ways—if you were shown a skull, a liver, or a blood cell, to take three features at random, it would be easy to tell which of the three it came from. Nevertheless, their external appearance is very similar (Figure 1.6). All have slim, fusiform bodies that are widest in the middle and taper towards each end, and are about three to five times as long as their greatest width.

Ichthyosaurs provide another example. They were large predatory marine reptiles that became extinct at the end of the Cretaceous period, about 65 Mya. They are unrelated to the other three groups, but they are remarkably similar in external appearance.

Bony fish, sharks, ichthyosaurs, and whales show that there is a general body form characteristic of all large fast aquatic animals. The ideal shape of such animals is determined by the principles of hydrodynamics. Any swimming animal must overcome drag, the resistance of the water to movement. At high speed the main source of drag is the difference in pressure at the leading (head) and trailing (tail) edges of the body. A fusiform body minimizes this pressure difference and so makes swimming more efficient. The best shape (least drag for greatest volume) has a ratio of length to breadth of 4.5. This is called the "fineness ratio," and shapes that have fineness ratios close to the optimum are said to be streamlined. A streamlined shape also reduces turbulence: the pressure drag is least when the water passes smoothly along the sides of a swimming animal, and is increased if the animal

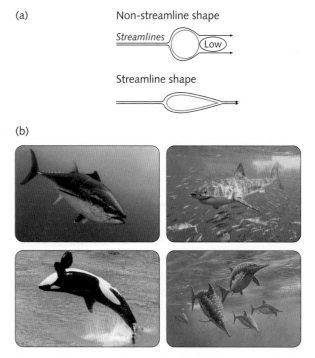

Figure 1.6 The convergent evolution of body shape in large fast swimmers. (a) Reduction of drag by a streamlined shape. (b) Independent evolution of streamlined bodies in four groups of marine vertebrates: teleosts, sharks, ichthyosaurs, and whales.

Tuna ©istock/Whitepointer. Shark courtesy of Terry Goss. This file is licensed under the Creative Commons Attribution-Share Alike 3.0 Unported license. Orca courtesy of Shamu @ One Ocean Show. This file is licensed under the Creative Commons Attribution 2.0 Generic license. Ichthyosaurs courtesy of Jørn Harald Hurum.

leaves a broad turbulent wake. The ideal body shape is therefore smooth, with no projecting flanges. The engineering solutions to the movement of a neutrally buoyant object through a fluid are submarines and airships. The evolutionary solutions are large fish, sharks, ichthyosaurs, and whales. At the level of overall body shape, engineering and evolution both lead to a characteristic smooth, streamlined body form.

Organisms with different ways of life often have different features. Conversely, if pressure drag is not the main constraint on locomotion there is no reason for streamlining to evolve. Within a single group of fish, a variety of body shapes may evolve that are adapted to different ways of life and have nothing in common with the stereotyped streamlining of large, fast-swimming animals. The fish of your local lake or river, for example, will have a wide variety of body shapes. Some are streamlined, such as pike or lake trout, because they are large predators that need to move fast at least some of the time. Others are deep-bodied, like sunfish, or big-headed, like catfish, or long-bodied, like eels. These shapes evolve in different ecological contexts: sunfish sidle through weeds picking off small prey like snails and shrimps, catfish grope through the sediment for larger prey like clams and crayfish, and eels wriggle into holes and cracks. These kinds of fish are also only distantly related, so they may be so different in part because they have different ancestry. Even rather closely-related fish may look very different, however. The Gasterosteiformes is a small group of rather bizarre-looking fish that includes seahorses, pipefishes, trumpetfishes, and sticklebacks. They are mostly small animals a few centimeters in length that move relatively slowly, so streamlining would bring them no benefits. Seahorses, indeed, must be the least streamlined of all fish; they have inflexible bodies and move slowly through weed or coral by using their fins. Sticklebacks have a more "normal" shape, but are hampered by an armor of bony plates and spines that protects them against predators.

1.3.2 Organisms are often exquisitely adapted.

Whales have evolved from terrestrial ancestors. Whales soon die when stranded on the shore; nevertheless, they evolved from terrestrial ancestors. Some of the stages in this process, known from

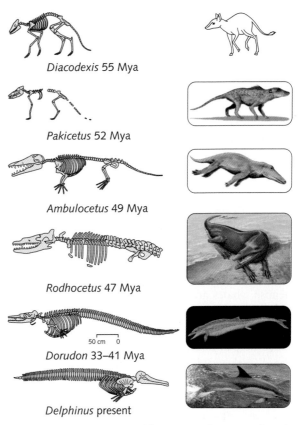

Diacodexis 55 Mya

Pakicetus 52 Mya

Ambulocetus 49 Mya

Rodhocetus 47 Mya

50 cm   0

Dorudon 33–41 Mya

Delphinus present

Figure 1.7 Whales have evolved from terrestrial ancestors through a series of intermediate forms.

Diacodexis skeleton is reproduced from De Muizon and Cifelli, 2001. A new basal "didelphoid" (Marsupialia, Mammalia) from the Early Paleocene of Tiupampa (Bolivia). *Journal of Vertebrate Paleontology* 21 (1): 87–97. With permission from Taylor & Francis Ltd. Images of Pakicetus and Dorudon skeletons kindly provided by Thewissen lab, NEOMED. Pakicetus and Ambulocetus are courtesy of Nobu Tamura. These files are licensed under the Creative Commons Attribution 3.0 Unported license. Ambulocetus skeleton adapted from National Geographic's *The evolution of whales* by Douglas H. Chadwick, Shawn Gould, and Robert Clark. Re-illustrated for public access distribution by Sharon Mooney ©2006. Open source license CC ASA 2.5. Rodhocetus courtesy of John Klausmeyer/University of Michigan Museum of Natural History. Dorudon courtesy of Nobu Tamura. This file is licensed under the Creative Commons Attribution-Share Alike 3.0 Unported license. Delphinus courtesy of Mmo iwdg. This file is licensed under the Creative Commons Attribution-Share Alike 3.0 Unported license.

the fossil record, are illustrated in Figure 1.7. The terrestrial ancestors of whales were the pakicetids, swift carnivores about the size of a wolf with long, narrow heads and slender limbs. They lived about 52 Mya in a region that is now Pakistan. The fossils of pakicetids suggest that they were related to the ancestors of modern artiodactyls, the group of even-toed ungulates that includes cattle, sheep and deer, and this has been confirmed by molecular analyses showing that the closest living relatives of modern whales are artiodactyls such as hippos. The artiodactyl line of descent led to thoroughly terrestrial animals, including the llamas and camels, which are at home in arid mountains and deserts. The cetacean (whale) line of descent, by contrast, became progressively more specialized to life in the sea.

Amphibious cetaceans with large powerful hindlimbs lived about 45 Mya, and ocean-going forms with reduced hindlimbs, such as the famous 20 m-long *Basilosaurus*, had evolved by about 40 Mya. Modern whales had appeared by about 25 Mya and are astonishingly well-adapted to a completely aquatic way of life.

Experienced human skin divers can swim about 100 m underwater, or descend to about 100 m depth, holding their breath for two or three minutes. Bottlenose whales are not much larger than skin divers but can descend to 1000 m and stay under water for a half-hour or more. Sperm whales occasionally dive much deeper, down to 2500 m, where they hunt for squid in the cold, lightless depths of the sea. These feats have been made possible by extensive modification of the ancestral body plan. The body is streamlined and smooth, with formerly projecting structures such as external ears and male genitalia being internalized; it is protected by an insulating layer of blubber; the muscles are modified for use as an oxygen store for deep dives; the forelimbs have been modified as fins for steering, while the hindlimbs are almost completely lost; a tail fin has evolved for propulsion; the skull has been shortened and the nostrils moved further back; the neck vertebrae have often become fused because the neck does not need to be flexed; and there is a long list of other adaptations for reproducing, feeding and communicating in the water. Through this long sequence of changes a long-extinct group of terrestrial carnivores gave rise to these huge creatures that are so completely at home in the deep seas.

Adaptation can also occur at much smaller scales. Some classic examples of adaptation are provided by cryptic and mimetic coloration that conceals or disguises animals that would otherwise be at risk from predators. There are some general themes that can be found in many species. Fish that swim in open water, for example, are almost always countershaded, with dark backs and pale bellies, to hide them from predators looking from either above or below, since a dark back is difficult to see against the bottom and a pale belly difficult to see against the sky. A few fish habitually swim upside down, usually in order to feed more effectively at the surface, and these species show reverse countershading, with pale backs and dark bellies. In other cases, animals hide themselves through a specific resemblance to some object that is less edible than they are. One favorite example is the leafy sea dragon, *Phycodurus*, a large seahorse living in shallow waters off the southern coast of Australia. Its body bears protuberances with flattened green lobes that give it the appearance of the seaweed amongst which it lives. Like most seahorses it swims by undulating its pectoral and dorsal fins while holding its body rigid, so that while moving it appears to be just another clump of floating weed.

1.3.3 Adaptation is constrained by ancestry.

High-speed tail fins look similar but have evolved independently. Pelagic fish normally swim by lateral undulations of the body that move it forward by pressing against the water like a series of inclined planes. High cruising speeds are best achieved, however, by moving the tail fin rapidly from side to side while holding the fore part of the body relatively still. This requires a tail with a large span (distance from tip to tip), which both tuna and sharks possess. Although similar in appearance, however, the tail fins of tuna and sharks are quite different in construction. The vertebral column of a shark passes into the upper lobe of the tail, which is larger than the lower lobe and is responsible for producing most of the thrust. The tail fin of a tuna consists of two equal lobes extending from the end of the vertebral column, which does not pass into the fin. This shows how a similar result can be obtained from quite different modes of development. Ichthyosaurs, by the way, had unequal-lobed tails in which the vertebral column passed into the lower lobe. The three types of fin are illustrated in Figure 1.8. All three groups have similar tail fins, whose shape reflects their hydrodynamic properties, but have evolved them independently.

Whales have a horizontal tail fin because they are mammals. Sharks, tuna, and ichthyosaurs all undulate from side to side, and their tail fins are therefore vertical. Whales, on the other hand, have

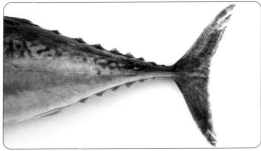

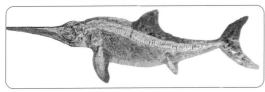

Figure 1.8 Convergent evolution of high-aspect tail fins in large fast aquatic animals by different routes. From top to bottom: shark, teleost, ichthyosaur.

Shark tail courtesy of Jean-Lou Justine. This file is licensed under the Creative Commons Attribution-Share Alike 3.0 Unported license; Tuna tail ©istock/ carlosdelacalle; Ichthyosaur skeleton provided by Professor Ryosuke Motani.

Function and ancestry both contribute to the evolution of form. For purely functional reasons, we expect to find organisms with similar ways of life to have similar forms. If you only saw the body outline of a large, fast marine animal, for example, you might not be able to say with any confidence whether you were looking at a fish, a shark, a whale, or even an ichthyosaur. All have evolved towards similar hydrodynamic shapes. If you saw only the shape of their tail fin you would be equally unable to tell one from another (unless you were an expert zoologist).

Evolution is not merely a process of modification, however, but rather a process of modification by descent. Hence, organisms may differ, even though they follow the same way of life, because they have different ancestors. Thus, if you could see the underlying structure of the fin, and saw whether it moved up-and-down or side-to-side, you would know at once what kind of animal it belonged to. This is because evolved structures do not appear from nothing but are modifications of previously existing features that may have been used for quite different activities. The signature of evolved structures is the combination of function with ancestry.

1.3.4 Adaptation is often far from perfect.

Evolved bodies have many imperfections. Whales are very impressive animals, but they are far from perfect. Their most obvious flaw is that they cannot breathe water like a fish. Lacking gills, they must return to the surface frequently in order to breathe, and although they can hold their breath for a long time this severely restricts their ability to operate at depth. They have more fundamental limitations, too. For example, they have to deform their bodies by muscular contraction in order to move, rather than having rotational devices, like propellers, that a good engineer would install.

There are more subtle flaws, too, that we can note in our own bodies. We have no third set of replacement teeth, for example, although without modern dental treatment our adult teeth often rot or break, whereas other animals (such as sharks) can renew their teeth perpetually. Our respiratory and digestive tracts cross, creating a chronic risk of choking, and mixing genital with urinary functions seems a false economy. When the testes descend in boys, the

horizontal flukes. This is not for any functional reason, but rather because whales are mammals, in which the neural spines of the vertebrae do not permit the body to be flexed laterally. If you watch a lizard running, you will see that it has sprawling limbs that move its body from side to side. If you watch horse or a deer running, you will notice that its back flexes up and down with the movement of its upright limbs. Whales, horses, and deer share the same common ancestor and have inherited the same kind of backbone. This is why fish and sharks swim, but whales gallop through the sea. Their backbone can only flex up and down, and their tail fin must therefore be horizontal. Vertical and horizontal fins provide a striking difference in design; this does not arise from any dynamic considerations, however, but solely from ancestry.

sperm ducts simply elongate, so that instead of passing directly from the testis to the seminal vesicle they make a long and unnecessary loop around the bladder. Most parts of our body are difficult to replace or repair: if a salamander loses a finger or toe it will soon grow another one, but if we suffer a similar injury we are permanently maimed.

There are no good engineering reasons for flaws like these; we inherited most of them from distant ancestors (where they may have served some useful purpose) and they are difficult to modify by evolutionary processes, however desirable this might be.

The eye is a good but flawed instrument. Whales use their eyes to capture squid, and the squid use theirs to escape. We are visual animals, and the structure of our primary sense organ has always impressed philosophers and biologists as demonstrating perfection in organic design, an optical instrument on which human ingenuity could scarcely improve. The production of such a complex structure as the human eye is indeed a remarkable testament to the power of evolutionary processes, but the eye also illustrates how evolved structures are likely to be flawed. The detailed structure of the vertebrate eye does not suggest an optical instrument engineered for optimal performance, but makes perfect sense as an evolved structure that has inherited flaws that can be palliated but not eradicated.

The basic elements of eye structure are familiar to everybody. Light enters the eye through a transparent cornea, passes through the transparent aqueous humor and is focused on the retina by a crystalline lens, whose shape can be adjusted by delicate muscles. The retina contains cells whose light-sensitive ends are directed inwards, and which are connected to other retinal cells and ultimately to ganglion cells whose axons trail over the surface of the retina. These axons pass through the retina at the optic disc, where they are bundled together to form the optic nerve, which communicates with the brain.

This arrangement works tolerably well, but it is scarcely ideal. Because the light-sensitive ends of the retinal cells are directed inward—away from the source of light—it is the inner surface of the retina that must be innervated, creating a tangle of neurons that straggles across the retina, not only interfering with light reception, but also making it necessary to bore a hole through the retina in order to get to the brain, thereby creating a permanent blind spot (at the optic disc) in the field of vision. The image on the retina is continually shifted backwards and forwards by about 1°, partly to avoid fatigue of the receptors and probably also to mitigate the shadow cast by the neurons. The resulting noise must then be eliminated from the signal by the brain. This can scarcely be defended as good engineering. These flaws are not made necessary by some unanticipated quirk of optical design in living organisms, because squid, with a lifestyle similar to that of pelagic fish, have eyes of comparable acuity but more logical design in which the photoreceptors are pointed directly towards the light source and the optic nerve passes directly to the brain. It cannot even be argued that it is made necessary because eyes must, for some unexplained reason, be designed differently in vertebrates and mollusks, because other mollusks have eyes with the same inverted design as that of vertebrate eyes.

The structure of the eye in different animals underlines the point made in the previous section. The eye of a fish or a squid would give a similar picture of the world around them, but they have evolved independently, and in consequence they are constructed differently, have different kinds of photoreceptor and develop by different pathways (Figure 1.9). Ancestry and function both contribute to evolved structures and the retention of ancestral features means that evolved structures are often clearly imperfect.

1.4 Development. *Development shows how organisms are related.*

Every structure of the adult, such as a fin or an eye, is the product of a long process of development. To understand how organisms function, we need to understand the whole series of stages leading from the egg to the adult, rather than just the final adult state. This is another important difference between beetles and watches. A watch is put together from components, and it will "grow" as these components

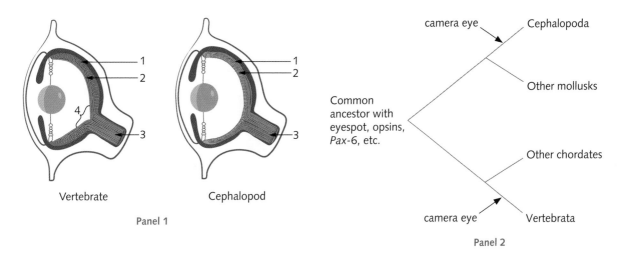

Panel 1

Panel 2

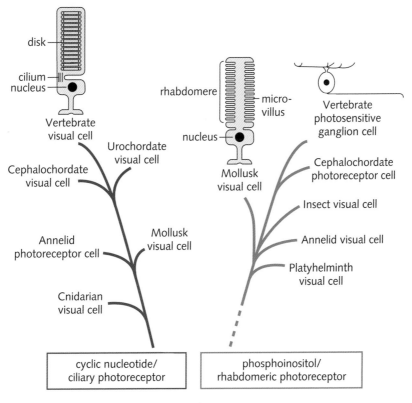

Panel 3

Figure 1.9 Panel 1: Vertebrate and cephalopods both have camera eyes of similar structure, involving photoreceptor cells, a neural retina, optic nerve, iris, lens, and cornea. Some features are quite different, however. In the cephalopod eye the neurons (2) underlie the retina (1) and pass directly to the optic nerve (3), whereas in vertebrates the neurons (2) overlie the retina (1) and pass through the retina at the 'blind spot' (4) to the optic nerve (3). Panel 2: The camera eye has evolved independently on cephalopods and vertebrates. Panel 3: The photoreceptor cells of vertebrates and cephalopods are different: vertebrates have ciliary photoreceptors (rods and cones, Section 8.2.1) whereas cephalopods have rhabdomeric photoreceptors using a different signaling system.

Eye formation in

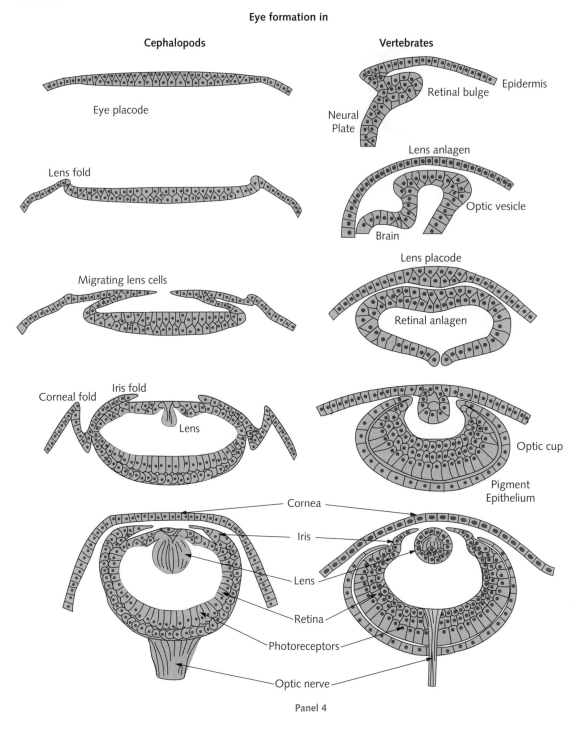

Cephalopods

Eye placode

Lens fold

Migrating lens cells

Corneal fold
Iris fold

Lens

Cornea

Iris

Lens

Retina

Photoreceptors

Optic nerve

Vertebrates

Epidermis
Retinal bulge

Neural Plate

Lens anlagen

Optic vesicle

Brain

Lens placode

Retinal anlagen

Optic cup

Pigment Epithelium

Panel 4

Figure 1.9 (*Continued*)   Panel 4: The development of the eye is different in cephalopods and vertebrates: in cephalopods the eye develops as an invagination of the surface of the head, whereas in vertebrates it develops as an extension of the surface of the brain. (a) In cephalopods the retina develops as a thickened sheet of ectoderm cells, which then undergoes a succession of folds to form iris, cornea, and an acellular lens. (b) In vertebrates, the retina develops as an optic vesicle from the forebrain, which induces the overlying tissue to form a cellular lens, which in turn induces the overlying tissue to become a clear cornea. The optic vesicle then folds inward to form an optic cup, with the iris developing at its margin around the lens.

are added to one another by the artificer, but it is only the final state of the watch that matters. It will be put together in the most efficient manner possible, and all watches of similar design will be put together in the same way. Beetles are quite different because they are not put together by anyone: instead, they develop, from egg to larva to adult, and every stage must be capable of surviving and growing. The lessons that we learned from adults can be extended to the whole of the life history. Every stage in development will be adapted to its conditions of life, so we expect that larvae and embryos with similar ways of life will share many features. Nevertheless, the sequence of stages that any particular species passes through will bear the stamp of its ancestry, just like the adult stage alone, so organisms that look very similar as adults may develop in quite different ways. Figure 1.9 has already made this point with respect to the camera eyes of vertebrates and cephalopods. Hence, we can expect that embryos, like adults, will often be imperfectly constructed because they retain inappropriate structures from remote ancestors. In these ways, the development of evolved organisms is quite different from the manufacture of designed artifacts.

1.4.1 Development depends on ancestry, not on final form.

Embryos reflect common ancestry. A rich community of animals lives on the rocks of the seashore between high and low tide, including barnacles, mussels, and shrimps. What kind of creature is a barnacle? At first glance it looks rather like a mussel. It has a hard shell that shuts tight when the tide is out and is firmly fixed to the rock. It develops from a very different kind of juvenile, however—a small creature with three pairs of legs and a single eye in the middle of its head. This is remarkably similar to the juvenile stages of copepods (small free-swimming zooplankton), ostracods (small bean-shaped creeping aquatic animals), brine shrimps, krill (the small shrimps that baleen whales feed on), tiger prawns, and many similar animals. All of these are crustaceans, like crabs and lobsters, and this characteristic juvenile stage is called a nauplius larva. Because barnacles have a nauplius larva, we know that they are crustaceans related to shrimps, rather than mollusks related to mussels. Even

crustaceans whose eggs hatch into different kinds of larvae go through a nauplius-like stage during their development. Although adult copepods, ostracods, barnacles, crabs, and shrimps look very different indeed, the resemblance of their young stages shows that they initially develop in similar ways.

Earlier alterations have larger effects. If you wish to manufacture something, you can work out the most efficient way of doing this, without regard to how any other product is manufactured. If you were making objects as different as a crab and a copepod, it is most unlikely that you would decide to make them in the same way. However, evolution can only modify the pathway that already exists. Crabs and copepods are not independent objects that have been designed, but related animals that have evolved. The nauplius is a very ancient feature of crustacean development; fossil nauplii have been found in rocks 500 My old. The development of the common ancestor of crabs and copepods, including its nauplius larva, cannot be simply discarded in order to set up two new pathways configured to produce two very different kinds of adult.

The development of an individual is somewhat analogous to the evolution of a species, in that it proceeds by a series of events, each of which builds on the outcome of the previous event. If any event is altered the final outcome of development, the adult stage, will necessarily be altered too. An alteration very late in development may give rise to a similar kind of adult that is slightly differently adapted than the original type and so exists side-by-side with it. An alteration early in development is likely to have much more serious consequences, because any alteration of an early stage will affect the next stage, which being altered will then affect the following stage, and so on. Even a minor alteration will be magnified into a large effect by the time the adult stage is reached. The most likely result of altering early development is, of course, that the embryo dies before the adult stage is reached. Even if the adult is viable, it is very unlikely to be an improvement on the original. Early stages of development are therefore likely to be more resistant to evolutionary change, and embryos of related organisms will resemble one another even when the adults are as dissimilar as barnacles and shrimps.

Embryos of different organisms resemble one another more than adults do. It is not necessarily the very earliest stages that will be the most similar. There are two reasons for this. The first is that the starting point of development, the fertilized egg, often varies quite a lot among related species. One may produce large yolky eggs that fuel development until a late stage, whereas another produces smaller eggs without much yolk that hatch quickly into actively foraging larvae. Secondly, it is not necessarily the very earliest stages that are the most sensitive, since in many organisms all the cells of very young embryos are capable of developing into a complete and normal adult. At some point early in development, however, the embryonic cells become differentiated in preparation for laying down the major organ systems; from this point on, any alteration of developmental pathways is likely to modify the adult more or less extensively. Embryos will then tend to resemble one another most closely at some intermediate stage in development. A series of examples is shown in Figure 1.10.

Completely novel developmental pathways seldom evolve, even when the adults are as different as people and fish, and instead pre-existing pathways are modified for new uses, while retaining the stamp of their ancestry. The most famous example is the "gill slits" of the human tailbud embryo. These are pharyngeal pouches that develop into gill slits in fish; human embryos do not have gills at any stage, of course, and our pharyngeal pouches develop into glands and ducts. At this stage a fish embryo has six aortic arches, which direct the flow of blood from the ventral aorta (major blood vessel) into the gill capillaries. The human embryo also has six aortic arches, three of which later develop into the systemic and pulmonary arteries while the other three disappear.

Ancestral features that have lost their utility can be preserved in development. Other structures come and go during development without contributing anything to the function of either embryo or adult. The tail of the human embryo at the tailbud stage is an obvious example (see Figure 1.11). Baleen whales develop teeth that are never used. Limb buds develop in the embryos of snakes and cetaceans and then disappear. These and many other examples show how the remnants of a developmental pathway can persist long after it has ceased to be functional.

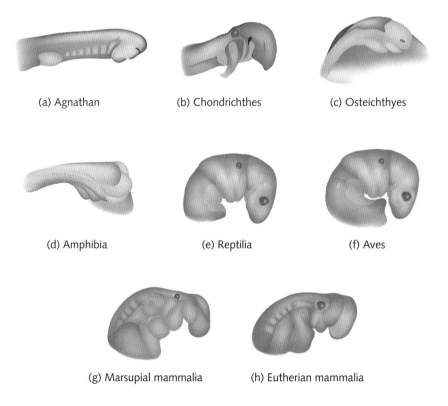

(a) Agnathan

(b) Chondrichthes

(c) Osteichthyes

(d) Amphibia

(e) Reptilia

(f) Aves

(g) Marsupial mammalia

(h) Eutherian mammalia

Figure 1.10 Representative vertebrate embryos at the tailbud stage. All possess somites, neural tube, optic anlagen, notochord, and pharyngeal pouches.

Reproduced from Richardson, M.K. 1997. There is no highly conserved embryonic stage in the vertebrates: implications for current theories of evolution and development. *Anatomy and Embryology* 196 (2): 91–106. With kind permission from Springer Science and Business Media.

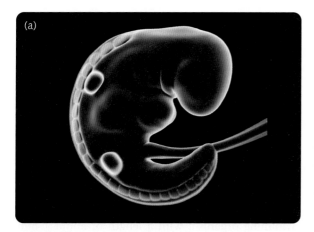

Figure 1.12 Head of a caecilian. The tentacle is the small papilla in front of the mouth; the rudiment of the eye is just visible beneath the skin.

Image from The Biodiversity Group provided by Paul Hamilton, BiodiversityGroup.org

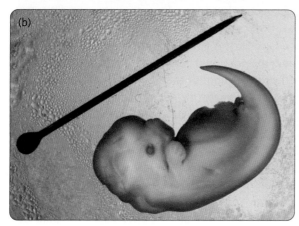

Figure 1.11 Some features appear transiently during the course of development and later disappear. (a) Human embryo at four weeks, showing tailbud. (b) Dolphin embryo showing hindlimb buds.

Human image ©Eraxion/istock. Dolphin image provided by Thewissen lab, NEOMED.

In some cases, functionless structures develop and are preserved in the adult. We have the remnants of a tail, for example, and even traces of the muscles to move it. Structures like this deteriorate generation after generation when they are no longer used because the genes that direct their development no longer function properly (they become "pseudogenes," described below). The genes continue to be copied from parent to offspring, however, so the structures themselves may persist in a vestigial form long after they have ceased to be useful. For example, caecilians are a group of amphibians that live in burrows in the soil in tropical countries. They seldom need vision, and instead have evolved a unique chemosensory tentacle organ for finding food. Consequently, their eyes are covered in skin and have become more or less reduced, as shown in Figure 1.12. They still have most of the characteristic features of vertebrate eyes, including cornea, lens, and optic nerve. Most of the muscles used to move the lens have disappeared; however, the cornea is fused to the overlying skin, the lens is often fused to the cornea, and the optic nerve is attenuated.

This process of reduction and deterioration proceeds independently in each species, and consequently the species vary considerably in the stage they have reached. Some species have aquatic larvae with a normal-sized eye that has all the usual features and is probably a reasonably good working photoreceptor. At the other extreme, one African species has a very small eye covered by bone, with no lens, an amorphous retina, and only the faintest trace of an optic nerve. At this point the eye has become functionless, and its place is taken by the tentacle. Now, the muscle that retracts the tentacle corresponds to one of the external eye muscles, the tear gland develops into a gland for moistening the sensory channel, and the muscle compressing this gland develops from another external eye muscle (the one that makes frog eyes bulge outward). Thus, the new organ is assembled from materials and developmental pathways used by a pre-existing but now non-functional organ. The long time required to shed useless structures, the unique history of each lineage, and the co-option of elements from one structure to re-use for a completely different structure are all features characteristic of evolutionary processes.

1.4.2 Development depends on embryonic rather than adult environment.

Embryos may be similar because they have similar ways of life. Think of a crocodile, or a turkey or a geranium. It is very likely that you thought of an adult reptile or bird or plant, and not of an embryo, an egg or a seed. We tend to give priority to the adult form, even though we know that every organism only becomes an adult after a long process of development from a single cell. Every step along the way is adapted to its conditions of life, just like the adult. This is most obvious when the growing embryo moves and feeds independently of its parent.

Many marine animals, for example, produce eggs that develop into small larvae that swim around and consume plankton before metamorphosing into the adult stage. Some representative forms from groups whose adults are very different in appearance are illustrated in Figure 1.13. There are two striking facts about these larvae. The first is that they are all broadly similar, in the sense that all use bands of cilia to move and feed. The second is that they are different in detail, because each main group of animals has one or more distinctive kinds of larva that differ in the arrangement of the bands of cilia. This shows how the principles applying to adults apply equally to their offspring, at any stage of life. Larvae that are a fraction of a millimeter in size are not streamlined because they do not have to be; most of the drag they must overcome is created by the friction of the

water, not by the inertia of a large body, and this is done most effectively by a low-power, continuously operating motor like a band of cilia. The particular configuration of the cilia is not as important from a hydrodynamic perspective, and in different groups it is independently inherited from different ancestors. Larvae teach us the same lesson as their parents: form evolves through both function and ancestry. The whole development of an individual from egg to adult, however, tells us more about evolution than the adult alone can do.

The nauplius has an appropriate body plan for a swimming organism with a length of about 1 mm. This is too large for cilia to generate enough power to move swiftly, but too small for streamlining to be necessary. A nauplius must overcome both inertia and the frictional resistance of the water to move effectively, and it does so by rowing. As its limbs are moved forwards the setae (hair-like structures) on them collapse, and are then extended as the limbs move backwards. This generates enough thrust to move the animal quite swiftly through the water. Rowing is an effective means of locomotion over a wide range of body sizes—for a good example in larger organisms, watch the water bugs (back-swimmers) in any small pond. The body plan of a nauplius is thus adapted to its own environment, rather than to the environment of the adult. Adult barnacles do not row themselves around! (They do move their limbs through the water, however, in order to strain out food.) Now, nauplii usually live in the open water, whereas adult crustaceans live in many different habitats: the water column (copepods and krill), plant surfaces (ostracods), seagrass

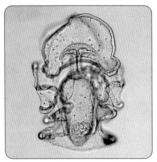
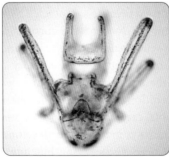
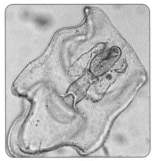

| Sea cucumber | Sea biscuit | Starfish |

Figure 1.13 Ciliated-band larvae of marine invertebrates.

beds (prawns), rocks in the intertidal zone (barnacles), and so forth. Moreover, nauplii are all about the same size, whereas adults may be very different in size even if they live in the same habitat (krill are typically about ten times longer and, hence, a thousand times as large as copepods). The embryos and larvae of related organisms are thus likely to be similar because they are adapted to similar environments, while the adults are more diversified.

Related embryos may be different when they have different ways of life. Conversely, embryos and larvae may be quite different, despite being related, if they live in different environments. Marine snails, for example, usually have a larva about the same size as a nauplius which uses wing-like lobes bearing cilia for locomotion. Some species, however, have abandoned the pelagic way of life and instead develop inside a capsule glued to the substrate. They are provided with a large yolk store, since they are unable to feed for themselves, and develop as a non-motile larva. This switch can evolve quite rapidly: some species of slipper limpet follow one lifestyle, for example, and some the other. Differences in larval environment, then, readily lead to differences in development, even in closely related species.

Function and ancestry both contribute to the development of form. Taking the arguments of the last two subsections together, we expect to find adult organisms with different ways of life to have different forms, for purely functional reasons. Nevertheless, if they are related they may develop along similar paths, to a certain point, so that their embryonic stages are similar. Development provides further evidence of how ancestry contributes to the evolution of organisms. Embryos and larvae must also be able to function normally, however, and will tend to be similar if they follow similar ways of life. There are good hydrodynamic reasons to expect small free-swimming larvae to be propelled by cilia, for example, and the ciliated-band body plan of such larvae has good hydrodynamic justification. The particular disposition of the bands is characteristic of particular groups, however, and shows the contribution of ancestry. In the complete life cycle of an organism, function and ancestry both contribute to the evolution of each stage.

1.4.3 The control of development is similar in most animals.

Development is controlled by genes. The cells that are produced by the division of the fertilized egg are all genetically identical, and yet they can somehow produce completely different kinds of tissue. A brain cell and a blood cell look as different as a caterpillar and a cow, but they share the same genome. They are not different because they have different genes (like the caterpillar and the cow) but because some genes are active in the brain cell and not in the blood cell, and vice versa. Such genes are switched on or off by other genes, which produce proteins that suppress or enhance transcription by binding to DNA.

These regulatory genes are inherited in the normal way, which is why related species usually have similar pathways of development. It also provides the basis for the evolution of life histories through the modification of the genes that control cell fate. If these genes do not work properly, grotesque individuals with altered or misplaced body parts may develop. The best-known examples come from insects, whose characteristic body plan is defined by the type of appendage borne by each segment of the body. In the head region the appendages develop as antennae or mouthparts, and in the thorax as walking legs or wings, while in the abdomen there are no appendages. Occasionally, an appendage of normal appearance will grow on the wrong segment: a leg may grow from the head of a fly instead of an antenna, for example.

Animals have a common set of development genes. The normal development of insect appendages is controlled by regulatory genes called homeobox genes, which encode proteins that control transcription. The development of a leg instead of an antenna is caused by a defective copy of a particular kind of homeobox gene called a *Hox* gene. There is a set of *Hox* genes, each of which is active in a particular region of the body of the developing fly, so they ensure the development of the correct series of appendages from head to tail. The *Hox* cluster consists of nine genes, lined up on a chromosome in the same order as the segments they control: genes active in the head region at one end and those active in the abdomen at the other. Other arthropods have slightly different *Hox* genes that lead to a different pattern of development:

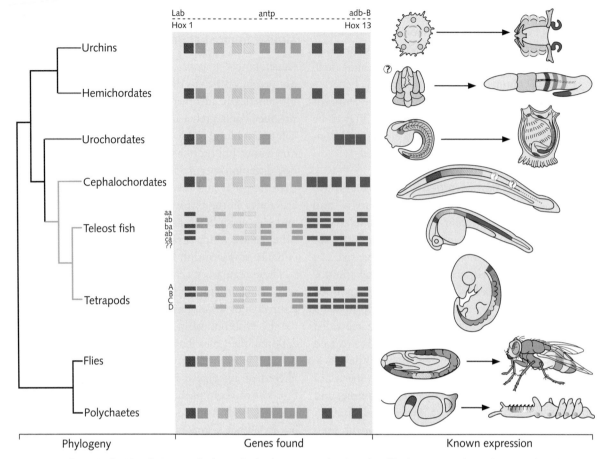

Figure 1.14 Genes in the *Hox* cluster encode diverse body plans in animals. The colored body regions indicate the sites where the corresponding genes are expressed. The phylogeny shows how the evolution of body plans can be traced by identifying the underlying modification of the genes that govern the crucial features of development.

Swalla, B.J. 2006. Building divergent body plans with similar genetic pathways Heredity 97: 235-243.

in centipedes, for example, all the abdominal segments bear legs, whereas in spiders there are no antennae and the thorax bears four pairs of legs. In each case, the basic body plan is specified by the *Hox* cluster, although in any given species some genes may be absent and others duplicated.

More surprisingly, the *Hox* genes regulate development in other kinds of animal, including vertebrates. Humans have four *Hox* clusters, all arranged front to back, and each with some genes dropped and others deleted. *Hox* genes seem to regulate development even in jellyfish, so it is likely that they appeared in the common ancestor of all animals and are responsible for much of the diversity of animal body plans. The common genetic basis of development in flies and mice, despite their very different structure, is a striking witness of their shared ancestry. The evolution of

animal body plans in relation to the *Hox* cluster is illustrated in Figure 1.14. *Hox* genes evolved from similar homeobox genes that occur even in unicellular organisms, such as yeast, where they control gene transcription but are not involved in development. Groups that have independently evolved multicellularity have independently modified homeobox genes, with a different outcome in each case.

The development of plants is also regulated by homeobox genes, but they have little in common with *Hox* genes. This is because plants and animals have no multicellular common ancestor, so their developmental control systems have evolved along different paths.

The rapid expansion of molecular developmental biology in recent years has revealed the genetic basis of the body plans of animals and plants. This new knowledge has confirmed our evolutionary interpretation of development, and enables us to find out precisely how one kind of body plan can evolve into another.

1.5 Sex. *Sex is a uniquely biological process.*

Sex is a puzzling phenomenon because it is not obvious what its function is. You may at first respond that it is necessary for reproduction, but, as we shall see, this is not the case. It is a uniquely biological process: beetles use sex to produce more beetles, but sex is unknown among pebbles, and engineers do not use sex to design watches. However, it does not have a well-defined role like fins or eyes, which are clearly adapted to improve individual performance in activities like locomotion or prey capture. Moreover, sex is potentially dangerous because it exposes organisms to a wide variety of sexually transmitted diseases. It also leads to the evolution of all kinds of contrivances to facilitate mating that can seriously interfere with locomotion, prey capture, or similar activities. The explanation is that sex not only evolves—it also strongly influences the process of evolution itself.

1.5.1 Most organisms can recombine genetic information.

Sex is not required for reproduction. In north-eastern North America you might come across a dark gray salamander with blue spots; it is quite abundant but spends most of the time hiding underground in the burrows of mice and voles. It reproduces very early in the year, moving into ponds as soon as the ice melts and laying small clumps of eggs in the reeds. It has one very unusual feature: every individual is a female. There are males of other similar species around, indeed, and they mate with the females. The sperm of these males is necessary to initiate the development of the egg, but the sperm nucleus is rejected and the offspring carry only their mother's genes. In this species, mating is clearly a rudimentary activity. Males are not really necessary, but they continue to play a vestigial role in the life cycle because the ancestor of the species mated in the normal way. The next stage, obviously, is to get rid of them altogether. This has happened in the whiptail lizards (fence lizards) of the southwestern United States. Many species of these lizards consist entirely of females who produce offspring without either males or mating. These offspring are exact genetic copies of their mother, so each family is a clone.

Sex and reproduction are different kinds of process. At first glance, this is difficult to understand. In human beings, sex and reproduction are so closely associated that we think of them as being different aspects of the same process. The truth is exactly the opposite: they are completely different kinds of process. Reproduction is an increase in number: at the most basic level, in unicellular organisms, it is one cell dividing to give two cells. Sex involves a reduction in number: two cells fuse to form one.

In all multicellular organisms sex occurs in close association with reproduction because an individual must start life as a fertilized egg (rather than having its body cells mate with one another as it develops). But there is no need for it to occur at all. Every large group of animals and plants contains species that reproduce without sex. There are two ways of doing this. One is to produce eggs that do not need to be fertilized, like whiptail lizards. This is rare in vertebrates but very common in some other groups. The small plankton animals that live in lakes and ponds, for example, often reproduce all year long without sex, before producing males and fertilized eggs in the fall. The asexual and sexual eggs, illustrated in Figure 1.15, develop into equivalent adults; sex is not necessary for the development of a complex multicellular organism. The second way is simply to grow an offspring in the same way as any other body structure. This is very common in plants, where shoots readily arise on the roots of established plants and grow up into full-sized individuals. They are still connected to their parent, of course, so the two form a sort of composite organism. Beech trees do this very readily, so when you see a beech forest you may, in fact, be looking at a single broadly distributed tree whose aerial stems are connected underground.

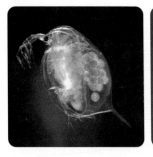

Figure 1.15 Sexual and asexual reproduction in a water flea. Both individuals are females of *Daphnia magna*. The individual on the left has produced a clutch of several diploid eggs by mitosis, in a clear brood pouch, without any sexual process. The individual on the right has produced a clutch of two haploid eggs by meiosis, in a dark brood pouch, which must be fertilized by a male. Asexual and sexual eggs develop into normal adults of identical appearance.

Left image courtesy of Hajime Watanabe. This file is licensed under the Creative Commons Attribution 2.5 Generic license. Right image courtesy of Professor Dieter Ebert.

Males are necessary for genetic recombination. If sex is not needed for reproduction, what use are males? The answer is that they are necessary to ensure genetic recombination. Gametes are produced by meiosis, during which new cells are formed that have only half as many chromosomes as normal body cells. During this process the maternal and paternal chromosomes (the chromosomes that were received from the mother and father of the individual) are broken and rejoined more or less at random to create new combinations of maternal and paternal genes. Hence, all the gametes produced by an individual will be genetically different.

Gametes fuse only if they have different gender, which will usually mean that they have been produced by different male and female individuals. Every fertilized egg then contains two recombined genomes, each of which is unique because of genetic recombination. Consequently, all sexually produced offspring are different from one another (with rare exceptions like identical twins), unlike the genetically identical offspring produced by clonal reproduction. This is most obvious when individuals produce only one kind of gamete and mate only with strangers. This is not necessarily the case: many animals and plants are hermaphrodites, producing both male and female gametes, and others mate with close relatives or even (if they are hermaphrodites) with

themselves. Moreover, I have tacitly assumed that there are only two genders, and that these are male and female, neither of which is always true. These variations of the sexual theme will be discussed later in the book.

Sex and genetic recombination facilitate adaptation. Blue-spotted salamanders and whiptail lizards originated as hybrids between closely-related sexual species. For reasons that are not very well understood, the hybrid species are able to reproduce without mating although both parent species must mate in order to reproduce. It is likely that all asexual species of vertebrates originated as hybrids. Whatever the physiological reason for this, it means that they appeared very recently, as the short-lived offshoots of much more ancient sexual lineages, and have not themselves given rise to further asexual types. Indeed, most species of asexual vertebrates arose after the last glacial period. Since these asexual groups have been derived recently from sexual ancestors, there must be a strong general tendency for asexuality to result in unusually rapid extinction.

In other groups, asexual animals and plants are not necessarily hybrids, and may arise in a number of ways. Very few of these are ancient. There is one good example of a wholly asexual group that has persisted for many millions of years, the minute bdelloid rotifers that live in transient bodies of water. This is a rare exception, however, and almost all other obligately asexual animals and plants originated very recently.

The reason that strictly asexual lineages tend to die out quickly is not that asexual progeny are necessarily inferior. Animals such as aphids and water fleas usually produce asexual offspring during the summer and sexual offspring at the end of the growing season. The asexual offspring are completely normal and just as vigorous as the sexual offspring. Sex is not an adaptation like countershading or a streamlined body form, which contribute directly to the survival of individuals. Instead, sex facilitates the survival of the lineage.

Recall that each member of a sexually produced family is unique because of genetic recombination and random gamete fusion. By contrast, asexual families consist of genetically identical offspring each bearing an unrecombined copy of its mother's genome. In the short term, asexual lineages may be very successful, because any individual that survives to reproduce is likely to have a well-adapted genome. In the long term, however, the environment will change and adaptation can be maintained only if there is sufficient variation to allow the lineage to evolve. This is where sex has an advantage. The genetic variation that is produced by recombination allows sexual lineages to respond to environmental change, whereas asexual lineages sooner or later die out because they are unable to evolve fast enough.

Most features of organisms evolve within constraints set by physical principles, such as the hydrodynamic laws governing locomotion through fluids. Sex is puzzling because it does not involve physical principles and so it cannot be explained in these terms. From the point of view of purely evolutionary principles, however, it makes sense as a process that has evolved in order to facilitate adaptation in changing environments.

1.5.2 Sexual genomes are infested by genetic parasites.

The genome is much larger than it needs to be. Our functional genome is surprisingly small and simple. The human genome consists of about 3 billion nucleotides that comprise about 20,000–25,000 genes. These collectively encode all the proteins that make up our bodies. These numbers have two very odd features.

The first is that there are so few genes. Since we seem to be so much more complicated than microbes, you would expect that we would have many times more genes, but this is not the case. We have only about four times as many genes as the common gut bacterium *E. coli* (about 4800 genes) or baker's yeast (about 6200 genes). You would also expect that a human being would have many more genes than a tiny nematode (roundworm), a puffer fish, or an annual cress (wildflower). In fact, all of these organisms have about the same number of genes as we do. We are certainly not distinguished by having a particularly large and complicated genome.

The second odd feature of these two numbers— 3 billion nucleotides and about 25,000 genes—is that they seem to be inconsistent. The average protein consists of about 500 amino acids, each of which is encoded by three nucleotides. The total number of nucleotides in the genome should therefore be about $3 \times 500 \times 25,000 = 37,500,000$. This is about 100 times too small! The quantity of genetic material we carry seems too large for the modest number of genes it represents. Some of the discrepancy is accounted for by noncoding DNA that acts to regulate the expression of nearby genes. The bulk of it is more difficult to explain.

Much of our genome has no clear function. Suppose that we were able to magnify our chromosomes so that each nucleotide was 1 cm (about a half-inch) in length. All our chromosomes laid end to end would then stretch all the way round the Earth at the equator, and we could just walk along a representative human genome (yours, for example), reading off the sequence nucleotide by nucleotide. As we did so, we could use a table of the genetic code to translate the nucleotide triplets (codons) into amino acids, and collect these until we come to a stop codon that tells us that we have reached the end of the protein. At this scale, the average protein will be about 20 paces long. Once we have read out its structure, we can discover (after a good deal of genetics and biochemistry) what it does in the cell.

If we set out from some arbitrary point, we soon find that this simple procedure does not seem to work. We can read the sequences between stop codons easily enough, but it does not correspond to a functional protein. This is because the coding sequence of most genes is interrupted by noncoding sequences called introns, which have to be cut out of the nucleotide chain in order to produce a functional protein. About a quarter of the human genome consists of introns.

As we continue our walk, we may then find a coding sequence for a complete protein, and work out that it operates in the following way. The DNA nucleotide sequence is transcribed into RNA, which moves out of the nucleus and is translated into protein, in the usual way. But the protein then attaches to its RNA template, which carries it back into the nucleus. Here, the protein nicks the chromosome and inserts a DNA copy of the RNA template. What is the point of this? It certainly does not seem to contribute to any vital cell function. What it does do, however, is to multiply the number of copies of the DNA sequence itself: in other words, the sequence is a genetic parasite that reproduces by subverting the replication machinery of the cell. Astonishingly, more than 30% of our genome consists of parasitic DNA.

Walking on, we come to a sequence that looks more like a useful protein, but right in the middle of it there is a stop codon that will prevent it from being transcribed as an intact molecule. This sequence is a pseudogene. We know that it was once a normal gene because it resembles a particular functional gene in other species, but is now merely a wreck that is passively replicated by the cell. There are about 8000 pseudogenes in our genome.

We walk past a number of other sequences that turn out to be different kinds of parasite or just gibberish—a long meaningless alternation of the same two or three nucleotides.

At last we come to a perfectly normal, intact gene that performs some vital function in the cell, but by now we have walked nearly a mile.

As the numbers suggested, normal genes make up only about 1.5% of the genome, most of the rest being junk or genetic parasites, as indicated in Figure 1.16. This is not the tightly engineered device that we might have expected. Instead, it is a rather Rube Goldberg-like contrivance that manages to work by tolerating a heavy burden of functionless DNA and continually patching the damage caused by parasitic genetic elements.

1.5.3 Sexual organisms often evolve bizarre structures and behaviors.

Males and females often look different. Eggs and sperm are very different kinds of cell because they are specialized for different activities: the egg to provide a store of material to fuel the development of the embryo, and the sperm to move swiftly around in a search for an egg to fertilize. These differences are often reflected by the female and male individuals that carry them. In most cases, females are larger and move around less, because they must mature a bulky crop of eggs or young. Males are typically smaller and swifter because they transport less bulky testes that can produce many crops of sperm.

In some cases this difference in size can be taken to extremes. Some species of angler fish seem to consist entirely of females, but most individuals bear a small organism resembling a parasite somewhere on their body. This "parasite" is the male. Young males have no digestive system and are unable to feed. Instead, they quickly seek out a female and fuse onto her body, obtaining their nutrition from her blood circulation. Their muscles degenerate and eventually they become little more than a pair of testes attached to the female, able to fertilize her eggs but incapable of an independent existence. In other fish and invertebrates there is a less extreme difference between females and males, with small size being a clear adaptation to a male way of life. In a few species, such as

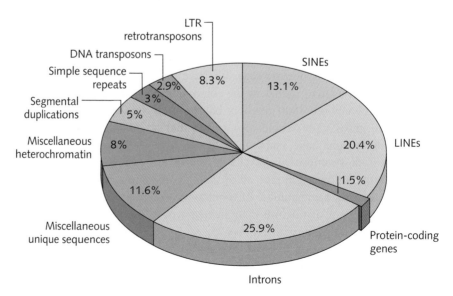

Figure 1.16 A large fraction of the human genome does not encode functional proteins. LINEs are long interspersed nuclear elements: retrotransposons encoding a reverse transcriptase, or (usually) the remains of such an element; SINEs are short versions that rely on reverse transcriptase produced by fully functional elements.

Reprinted by permission from Macmillan Publishers Ltd: *Nature Reviews Genetics* 6 (9): T. Ryan Gregory, Synergy between sequence and size in large-scale genomics. © 2005.

Figure 1.17 Female (left) and male (right) pheasants. The male is extravagantly modified in many species of birds and other animals.

Courtesy of ChrisO at the English language Wikipedia.

seahorses and sticklebacks, it is the males that protect and rear the eggs and young, and as you would expect the males are as large as or larger than the females in these species.

Males are sometimes bizarre. In other species, males have become modified in ways that are more difficult to understand. Male guppies are smaller than females, but they are often also brilliantly colored with red, orange, and blue spots. You will see other examples in the garden or the farmyard. The female cardinal is mostly green and brown, with a red bill, while the male is a uniform vivid red. The female pheasant is a rather drab, brownish bird, while the male has a brilliantly colored head and neck, and elongated tail feathers (Figure 1.17). The female turkey is likewise neutrally colored, whereas the male has red wattles hanging from its neck and is able to fan out its tail feathers to make a conspicuous display—a habit taken much further by the male peacock, with its brilliant and intricately patterned tail. The Raggiana bird of paradise (the national bird of Papua New Guinea) has extraordinary males that sport an emerald green throat encircled with a bright yellow collar and crown, huge orange or crimson plumes on its flanks and two long black tail wires that trail behind it. The female, by contrast, is a rather plain, unadorned bird, brownish in color.

The bright colors and elaborate ornaments of males are clearly connected with mating, as they are expressed only in males and are often associated with complex courtship behavior. But they do not reflect some common functional theme, like streamlining or countershading. The males of related species may have quite different kinds of ornament, and in any case there is no obvious reason for females to be attracted by wattles, for example.

Males are endangered by their sexual ornaments. It is obvious that we cannot understand the sexual ornaments of male peacocks or birds of paradise in terms of engineering principles. This difficulty extends far beyond a few species of bird, however. Male fish, for example, often sport bright colors during the breeding season, like the red throat of sticklebacks. Even among plants, male flowers are usually larger and showier than female flowers in species where the sexes are separate.

Wherever we look, we are likely to find that the coloration or structure or behavior of males has been modified in peculiar ways that are unlikely to help them survive better. Indeed, the sexual ornaments of bizarre males often endanger their lives. Brightly colored animals are more likely to be seen and captured by predators. Bulky plumes and long clumsy tails impede flight and make it more difficult to feed and escape. Features like these show how the evolution in sexual organisms can actually lead to weaker, slower and more vulnerable individuals. This does not seem to make any sense at first, because it is easy to think of less harmful ways of reproducing. It can only be understood in terms of an evolutionary process unique to sexual organisms, the mechanism of which we will explain later in this book.

1.6 Diversity. *There is variation at all levels of organization.*

Enthusiastic naturalists have spent the last three centuries describing all the different kinds of animals, plants, fungi, algae, and protists that have been collected from all parts of the world. About two million species have been described so far, and it is thought that many more remain to be found. Some groups are particularly diverse. For example, there are over 350,000 species of beetles, over 250,000 species of flowering plants, and over 50,000 species of fungi. In more recent years it has become possible to assess the amount of variation within populations of a single species. Far from being genetically uniform, most populations have a wide variety of different genotypes. Extensive variation is a key feature of biological systems at all levels from populations to ecosystems. This is because evolution tends to lead to almost indefinite diversification.

1.6.1 Most populations contain a great deal of genetic variation.

All individuals are different. The genome that we hiked along in the previous section belonged to a certain individual. If we had chosen a different individual of the same species then its genome would have been similar but not identical. We speak of "the human genome," but this is not a single genome, rather a collection of genomes that are more similar to one another than any is to a genome from another species.

In the first sections of this chapter, we emphasized that close relatives are more similar to one another than they are to unrelated individuals. Now is the time to emphasize that although they are similar they are never identical. Even identical twins are not literally identical, although they are as similar to one another as it is possible to be.

The fact of variation is obvious to anyone who looks at the crowd in a sports stadium, but nowadays it can be studied more precisely by comparing the genomes of different individuals. Instead of a single genome, we lay out a hundred, say, side by side, and as we walk along them we compare them nucleotide by nucleotide. In most cases the nucleotide at any given position will be the same in every individual, but every few hundred nucleotides we will notice that not all the genomes are the same. Sometimes about half of them will have one nucleotide and half another; more often, most individuals will be the same but a few will be different. This is not because some individuals are different from the others for all of these variable nucleotides, but rather because everyone is in the minority for some of them. Since the human genome consists of about 3 billion nucleotides, we

shall discover at least 10 million nucleotides that differ among individuals. Another way of putting this is to say that if we choose two individuals at random their genomes are likely to be different at more than 100,000 places. There is no reason to suppose that human beings are exceptionally variable animals. It is true that asexual or highly inbred species are likely to be less variable, but most sexual species probably have comparable levels of variation.

Most genetic variation has little effect. Our imaginary walk along the genome, however, showed that normal coding genes occupied only about 1.5% of its length, so only a small fraction of variable nucleotides will occur in these sequences. Moreover, the redundancy of the genetic code implies that only a fraction of nucleotide changes in genes will cause a different amino acid to be used. Even when this occurs, the altered protein may be so similar to the original that it has almost exactly the same properties. Taking all of this into account, two random individuals probably differ by no more than a few tens or a few hundreds of nucleotides that may alter their development or physiology. This is much less than the total number of differences, but it still represents a great deal of variation.

In this respect, living things are quite unlike devices. Individual specimens of any particular kind of device will be almost completely identical, especially when the device has been constructed by a sophisticated engineering process. Individuals of most species of living things, on the other hand, display a good deal of variation, some of which can be traced to variation in their underlying plan, or genome.

1.6.2 Most communities contain a very large number of species.

Ecological communities are food webs. The organisms living together in a particular place are linked together by how they feed. Green plants and algae always form the basis of the community because they can use the energy of sunlight to make sugar. They are consumed by herbivores such as caterpillars and deer, which are in turn consumed by predators such as birds and wolves. The predators may be consumed in turn by other predators, and even the top predators can fall prey to internal parasites. Other organisms specialize in consuming dead bodies, like vultures, or processing detritus, like earthworms. These interlinking relationships among organisms are often conceptualized as a food web, which describes how energy and material flow among species. Hence, the metabolism of the community as a whole depends on a range of ecologically specialized types. This is reminiscent of how the economy of a human community depends on a range of specialized trades and professions, and you might imagine that natural communities are somehow engineered to operate efficiently. But this is not the case.

Many species have very similar ways of life. A jarful of pond water will contain thousands of the minute green algae that form the basis of the food web. They are all single-celled organisms bearing a chloroplast where photosynthesis takes place, they mostly look rather similar and they have similar ways of life, floating in the sunlit surface waters of the pond until they are eaten by some small animal or die and sink to the bottom.

It comes as a surprise that there are hundreds of different kinds that can only be told apart under the microscope or by using genetic techniques. Diatoms are single-celled planktonic organisms that live inside a box made of silica; the boxes made by different species are often different in appearance, providing a visual impression of diversity illustrated in Figure 1.18. Some species are very numerous; most are rather rare. It is very unlikely that each is irreplaceable and that there are hundreds of distinct ways of life in this part of the community. Each species may in some degree be adapted to somewhat different conditions, but not all are vital for the normal life of the community, whose composition changes through time and differs from pond to pond.

Nor is each community uniquely well suited to its particular situation. We all know that foreign animals and plants often invade local communities and succeed in establishing themselves permanently. Adding or removing species can have very severe consequences, of course. Introducing the zebra mussel or the mountain pine beetle, for example, may modify the food web of an entire watershed or forest. Similarly, removing codfish from the Atlantic and sea otters from the Pacific has caused widespread ecological change.

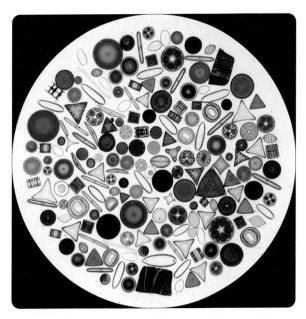

Figure 1.18 Diversity: a sample of the innumerable forms of marine diatoms.

In most cases, however, gaining and losing species has little effect on the bulk properties of the community. The food web is not like a device in which each part is carefully designed to perform a particular function and cannot be removed or replaced without a drastic reduction in functionality. It is an assemblage of evolved organisms that collectively exploit the resources that they can acquire but it is not highly coordinated or very efficient.

Closely related species are often ecologically similar. The signature of evolution is descent with modification, so we expect relatives to be alike. This applies to species as well as to individuals. If you search a woodlot or a river for a few hours you are likely to find a dozen species of violets, or minnows, or warblers. They all look similar, and they are following similar ways of life. They are not identical: anyone can see that different species of violet have white or yellow or blue flowers, and an expert will also point out the difference between long-stemmed and stemless species, or between species of wet and dry places. They are nevertheless much more similar to one another than to another random kind of wildflower (or fish, or bird). This is because they share a recent common ancestor, from which they have inherited many of their characteristics. More remote relatives tend to become increasingly more highly modified for different ways of life. The rich diversity of ecological communities springs from the universal tendency of life to diversify. This is not in the least mysterious: it is caused by a simple mechanism involving variation and selection. This is the subject of the next chapter.

● CHAPTER SUMMARY

Living organisms share a narrow range of materials and processes.

- All living organisms share the same basic features.
- Similarity implies common descent.
  - *Children resemble their parents.*
  - *Children of the same parents resemble one another.*
  - *Relatives are similar because they share ancestral genes.*
  - *Sister species are very similar because they are closely related.*
  - *Species share attributes of their common ancestor.*
  - *Universal characters are inherited from a distant common ancestor.*

Fossils of extinct organisms are the physical remains of past evolution.

- The majority of species are extinct.
  - *Fossils are extremely abundant and diverse.*
  - *A species has a limited lifespan.*
- Extinction reveals the long history of life.
  - *Older rocks have stranger fossils.*
  - *Earth history involves a long succession of faunas and floras.*
  - *Life has evolved over about 3500 My.*
- Extinction implies the continual evolution of new kinds of organism.
  - *Marine diversity has changed little over the last 400 My.*
  - *Speciation balances extinction in the long term.*

Evolution explains how organisms are constructed.

- Differences between organisms reflect different ways of life.
  - *Organisms with similar ways of life share many features.*
  - *Organisms with different ways of life often have different features.*
- Organisms are often exquisitely adapted.
  - *Whales have evolved from terrestrial ancestors.*
  - *Adaptation can also occur at much smaller scales.*
- Adaptation is constrained by ancestry.
  - *High-speed tail fins look similar but have evolved independently.*
  - *Whales have a horizontal tail fin because they are mammals.*
  - *Function and ancestry both contribute to the evolution of form.*
- Adaptation is often far from perfect.
  - *Evolved bodies have many imperfections.*
  - *The eye is a good but flawed instrument.*

Development shows how organisms are related.

- Development depends on ancestry, not on final form.
  - *Embryos reflect common ancestry.*
  - *Earlier alterations have larger effects.*
  - *Embryos of different organisms resemble one another more than adults do.*
  - *Ancestral features that have lost their utility can be preserved in development.*

- Development depends on embryonic rather than adult environment.
  - *Embryos may be similar because they have similar ways of life.*
  - *Related embryos may be different when they have different ways of life.*
  - *Function and ancestry both contribute to the development of form.*

- The control of development is similar in most animals.
  - *Development is controlled by genes.*
  - *Animals have a common set of development genes.*

Sex is a uniquely biological process.

- Most organisms can recombine genetic information.
  - *Sex is not required for reproduction.*
  - *Sex and reproduction are different kinds of process.*
  - *Males are necessary for genetic recombination.*
  - *Sex and genetic recombination facilitate adaptation.*

- Sexual genomes are infested by genetic parasites.
  - *The genome is much larger than it needs to be.*
  - *Much of our genome has no clear function.*

- Sexual organisms often evolve bizarre structures and behaviors.
  - *Males and females often look different.*
  - *Males are sometimes bizarre.*
  - *Males are endangered by their sexual ornaments.*

There is variation at all levels of organization.

- Most populations contain a great deal of genetic variation.
  - *All individuals are different.*
  - *Most genetic variation has little effect.*

- Most communities contain a very large number of species.
  - *Ecological communities are food webs.*
  - *Many species have very similar ways of life.*
  - *Closely related species are often ecologically similar.*

● FURTHER READING

Here are some pertinent further readings at the time of going to press. For relevant readings that have been released since publication, visit the book's Online Resource Centre at **www.oxfordtextbooks.co.uk/orc/bell_evolution/**

Section 1.1   Dawkins, R. 2009. *The Greatest Show on Earth: the Evidence for Evolution*. Simon & Schuster, New York.

Section 1.2   Stearns, S.C. and Hoekstra, R. 2005. *Evolution*. Oxford University Press, Oxford.

● QUESTIONS

Evaluate the following theories of Earth history in the light of the material presented in this chapter.

1. All living and extinct species came into existence recently by means of a single act of creation.

2. Species have been created on several occasions during a very long span of past time.

3. All living species are the modified descendants of a single ancestral cell.

4. Living species descend from several unrelated ancestors.

5. Organisms from distant worlds are continually arriving on Earth from space.

6. We do not have enough information to know which of these theories is correct.

You can find a fuller set of questions, which will be refreshed during the life of this edition, in the book's Online Resource Centre at **www.oxfordtextbooks.co.uk/orc/bell_evolution/**

The Engine of Evolution

2

In everyday life we have an intuitive grasp of how many physical processes operate. We can throw a stone to hit a distant object, get out of the way of a moving vehicle, or pour water into a glass. Even so, the laws that govern the motion and interaction of objects took many hundreds of years to formulate correctly. Evolution is much further removed from common experience, because evolutionary change often occurs over time periods much longer than human lives, and because we seldom use evolutionary processes in our daily activities. It is only very recently in human history that the laws governing evolution have been accurately worked out. This chapter introduces you to the two basic processes that underlie evolutionary change: variation and selection. Because they are less familiar than physical processes, we shall explain them in a simple abstract way, before proceeding to a more detailed and concrete account in subsequent chapters.

2.1 Mutation. *Mutation supplies the raw material for evolution but does not determine its direction.*

In a perfect world, offspring would be perfect copies of their parents, alike in all respects. If this were so, evolution could not happen, because there would be no basis for change. In fact, it is never true. Offspring inherit the genome of their parents, but they always inherit a slightly changed version. The changes that have occurred are called mutations. They are of many different kinds. Some involve a single nucleotide, substituting one for another; others are more extensive and involve the deletion, or the duplication, or the rearrangement of whole genes or regions of chromosomes. Mutations introduce the variation that fuels evolution. They occur in every organism, without exception. Mutations occur in bacteria, and they occur in humans—at about the same rate. Most mutations occur during the copying of DNA, but they can also occur in cells between divisions, because of exposure to mutagenic agents such as sunlight or toxic chemicals. If they occur in somatic cells they can have serious consequences for individuals—such as cancer—but they do not contribute to evolutionary change because they are not transmitted to offspring. Mutations that occur in germ line cells (eggs and sperm) are transmitted to offspring and so provide a reservoir of variation for future change. This section describes some of the basic properties of mutations.

2.1.1 Mutation is not appropriately directed.

Mutation is inevitable. Each gene in the genome of an individual is a copy of the corresponding gene carried by one of its parents. It is not necessarily a perfect copy, however, because any copying procedure will introduce errors at a certain rate. If you typed out this page, for example, it is quite likely that your version would contain some errors. You could reduce the number of errors by proofreading the page and correcting them, but even so you will miss some of them. Even this book, which has been proofread by several people, is likely to contain a few typographical errors. The error rate can be reduced to zero only by putting an infinite amount of effort into identifying and correcting errors, which is of course impracticable. The same principle applies to copying the sequence of nucleotides from the template DNA strand of the germ line of the parent to the new strand of the single cell, the zygote, from which the offspring develops. Hence, mutation is not a process requiring some special explanation: it will inevitably occur at some rate that depends on the energy and time devoted to avoiding or correcting errors.

A mutation is a unit genetic change. There are four kinds of mutation: genes can be altered, duplicated, deleted, or rearranged. The simplest kind of mutation is the insertion of a non-complementary nucleotide in the new strand when DNA is replicated. Another possibility is that a nucleotide is skipped or an extra nucleotide added, which will alter the position of the subsequent nucleotides in the reading frame of the gene. More radical changes involve the deletion of whole genes, or the duplication of a gene, so that the new genome lacks a given gene entirely, or has two copies of it. In rare cases, indeed, the entire genome might be duplicated, so that the offspring has two complete copies of the parental genome. Large sections of DNA may occasionally suffer other kinds of accident: a block of genes might be transferred from one chromosome to another, for example, or turned around so that the order of genes is inverted. All these kinds of mutation are caused by discrete alterations of the genetic material and are usually transmitted in their altered form to future generations.

Mutation is not usually adaptive. Mutation is sometimes said to be random. This is not quite correct. Some genes, and some locations within genes, are more likely to mutate than others. For example, mitochondrial genes have a higher mutation rate than nuclear genes. Moreover, exposure to certain environmental factors, such as ultraviolet radiation or chemicals such as benzene, increases rates of mutation dramatically, and in general mutations often tend to be more frequent in stressful conditions. The important point, however,

is that a particular stress does not induce mutations that specifically alleviate the effects of the stress. For example, ultraviolet radiation (a stress) increases the rate of a particular kind of mutation, the insertion of an incorrect nucleotide. If this occurs in a germ cell it will lead to a genetically altered offspring that may differ from its parent as a result. It is most unlikely, however, that the offspring will respond differently to ultraviolet radiation; and even if it did, it is just as likely, or more likely, to be more sensitive as to be more resistant. All adaptation is based ultimately on mutation, but mutation alone cannot explain adaptation.

2.1.2 Most mutations have little effect.

Many mutations are neutral. Later in this chapter, we shall use some simple word games to illustrate some of the basic principles of evolution. One of these involves changing one word into another by altering a single letter at a time, with each alteration resulting in an actual word. For example, "boat" can be changed to "mast" like this: boat → moat → most → mast. This is an interesting game because it is quite difficult, as the English language (and any other language) is brittle, in the sense that a random change will almost certainly produce a meaningless sequence of letters. By contrast, the genetic language is quite tough, because a random change in a gene is likely to have little effect on the protein it encodes, still less on the individual in which this protein is expressed. There are two reasons for this. The first is that the genetic code is redundant, since there are $4^3 = 64$ triplet codons but only 20 amino acids: therefore, most mutations involving the substitution of one nucleotide for another are unlikely to have any effect on the protein expressed by a gene. The second is that many changes in the amino acid sequence of a protein will not greatly affect its function. This is because the biological activity of a protein depends mostly on the few amino acids at its active site, and may not be greatly affected by alterations in other regions.

Mutations with very severe effects are rare. This is not always true: there are many examples of changes that seem irrelevant to the activity of a protein and yet have a profound effect on the wellbeing of the individual. By and large, however, most mutations have little if any consequence. This has been proven experimentally by removing whole genes, the most severe possible kind of mutation. In well-studied model organisms, such as yeast, it is possible to delete every gene in the entire genome, which means that a complete collection of different strains can be created in which each strain lacks exactly one gene. About 20% of these strains cannot grow—the missing genes are clearly essential. Most of these genes encode proteins crucial to some fundamental and irreplaceable cellular function, such as the manufacture of ribosomes. The other 80%, however, grow quite normally. Hence, most mutations have only slight effects on fitness. If this were not the case, then evolution would be impossible, or very difficult, because almost any change would lead to a serious loss of function.

Beneficial mutations occur occasionally. Most mutations may have little effect, but what effect they have is usually deleterious. It is easy to appreciate that any complex instrument, such as a piano or a computer, is unlikely to be improved by a random alteration of one of its components. This should apply with redoubled force to a much more complex entity, such as a frog. We know it does, because when we compare the genes of related species, we find that the nucleotide sequences of corresponding genes are usually strikingly similar. It follows that most of the innumerable variants that have arisen by mutation have been eliminated by selection. Nevertheless, species do gradually diverge, so as to become specialized for somewhat different ways of life. This involves mutations that lead to altered proteins that are better than the original versions in some circumstances. This process can be observed in the laboratory by exposing populations to a stress

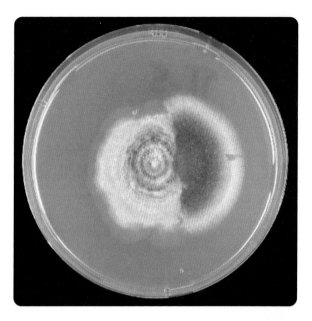

Figure 2.1 Seeing a beneficial mutation. A colony of *Aspergillus*, a filamentous fungus, was started with a single propagule put on an agar plate. It grew into the colony with concentric growth on the left. Early in its growth a mutation occurred in a cell at the three o'clock position. This mutation was beneficial, increasing the radial rate of expansion and overgrowing the filaments on either side.
Image kindly supplied by Dr Sijmen Schoustra.

(such as high temperature or a new kind of nutrient) and then observing their evolutionary response over hundreds or thousands of generations. In many cases they succeed in adapting to their new environment through rare beneficial mutations. In the laboratory we can clock the occurrence of these mutations, measure their frequency, and estimate their effects. An example of a beneficial mutation observed in laboratory culture is illustrated in Figure 2.1. The fundamental processes of evolution can be studied just like the dynamics of any other physical process. We shall describe how to isolate and characterize beneficial mutations in Chapter 13.

2.2 **Selection.** *Selection screens the variation made available by mutation and thereby changes the genetic composition of the population.*

Mutations do not systematically lead to improvements (in fact, usually the reverse), and therefore mutation alone cannot cause adaptation. Instead, mutation leads to variation, so that a population contains many different kinds of individual. A few of these may be better than the rest, at least in some circumstances, and are able to grow and reproduce more rapidly. This creates a process of

Figure 2.2 The power of selection. Gray wolf (left) and Yorkshire terrier (right).

selection that is responsible for driving evolution in a particular direction.

2.2.1 Selection is caused by variation in fitness.

Selecting exceptional individuals alters populations. If your cat or dog has a litter you might not be able to place all the kittens or puppies in new homes. If you could choose to keep only a few of them, perhaps you would prefer the bigger ones, or the more obedient, or those with brown coats. When they grow up they are likely to retain the features you chose them for, so they will be systematically different from their parents. Moreover, the differences that you noticed among the members of the litter, and used to make your choice, might have a genetic basis. This would mean that litter-mates carried different versions of the same gene, originally arising by mutation, and as a result differ in appearance. In this case, the individuals you chose will transmit these differences to their offspring. By choosing a few exceptional individuals you will have altered the characteristics of the animals in your care.

Now, suppose the same thing happens again, and you make the same choice. It will naturally have the same effect, leading to offspring that are on average different from their parents in a particular direction. They will necessarily be even more different from their grandparents. Continuing to choose exceptional individuals as parents, generation after generation, will result in a greater and greater divergence from their ancestor, until the appearance or behavior of the line may be changed beyond recognition.

This process is called *artificial selection*. When it is applied to hundreds or thousands of animals over many generations, it is capable of altering whole populations in remarkable ways. The variety of dogs shows that specialized types of animals are readily produced by artificial selection. The largest dogs (such as Newfoundlands and Great Danes) weigh fifty times as much as the smallest (such as Yorkshire terriers and Chihuahuas), and many breeds have been selected to carry out quite specific tasks such as guarding (mastiffs), hunting (foxhounds), herding (sheepdogs), carrying (retrievers), pulling (huskies), and so forth. Figure 2.2 shows one dramatic example (I have to declare an interest: the modern Yorkshire terrier was perfected by Mary Anne Foster, my great-great-grandmother). Figure 12.10 gives more details about the human-directed evolution of modern dogs.

The most extensive and important experiments in artificial selection are carried out by farmers and agronomists, who modify livestock or crop plants by choosing individuals with desirable characteristics, such as a high yield of milk or grain, and allowing only these individuals to breed. This has had two important effects. The first is to create domesticated animals and plants that are distinctively different from their wild ancestors: for example, the evolution of maize from teosinte has involved not only modifications of the grain and the ear but also major changes in plant architecture. The second effect is a large increase in yield: modern maize varieties yield about four times as much grain as those of sixty or seventy years ago. Artificial selection shows how easily and how extensively organisms can be modified over quite short periods of time.

Scarcity of resources leads to reproductive competition. In natural populations the situation is quite different, of course, because nobody is deliberately choosing individuals to survive and reproduce. Evolution will often occur quite automatically, however, through a process of natural selection. This is the consequence of two general principles: scarcity and variation.

Not all individuals who are born can live and reproduce, because the resources available to a population (such as food or enemy-free space) are nearly always in short supply. This is because populations have an inherent tendency to increase in numbers. If resources are abundant then individuals will grow quickly and produce many offspring. As the population becomes larger each individual's share of resources must become smaller, so it will grow more slowly and produce fewer offspring than before. Eventually the population will become so large that each individual has barely enough resources to replace itself in the next generation. At this point the population is in balance with its environment. Most natural populations, most of the time, must therefore live at the edge of subsistence. Because resources are in short supply, individuals must compete intensely with one another to secure enough to grow and reproduce. Most will fail, because the resources that one succeeds in obtaining are no longer available to the others.

Variation leads to natural selection. Individuals vary in all sorts of ways that affect their appearance and behavior. We can choose any feature we wish to modify through artificial selection, but in natural populations a feature will be modified only if it affects the ability of individuals to obtain resources in competition with others. What kind of feature is involved will depend on circumstances: in a population of plants, for example, it might be the length of a taproot (if water is scarce) or the concentration of tannin in the leaves (if herbivores are abundant), and so forth. In any particular conditions, however, there will be some feature or combination of features that determines whether or not an individual is likely to survive. Those individuals that vary in the appropriate direction (by having a longer taproot than average, for example) are more likely to survive. In any given generation, therefore, those which survive and reproduce will be systematically different from those which were born, because they will include an enhanced proportion of individuals that differ in some particular way from the average. There is therefore a process of natural

selection, akin to artificial selection, that will operate quite automatically, provided that resources are scarce and that individuals vary in their ability to acquire them.

Mutation and selection cause evolution. Individuals vary for all sorts of reasons, such as whether they received enough nourishment when they were young, whether they were injured when growing up, and so forth. All of this variation will be screened by natural selection. Some part of it may be due to mutations that affect the ability of individuals to garner resources and thereby to survive, to grow or to reproduce. This constitutes genetic variation, and means that the features that contributed to the success of individuals can be transmitted to their offspring. The modification that has been caused by natural selection is then permanently incorporated in the population.

For example, mutations in the *PAH* gene, which encodes an enzyme active in the liver, may make it more difficult for affected people to metabolize the amino acid phenylalanine. This results in the accumulation of phenylalanine and its products, often leading to intellectual impairment, seizures and psychiatric disorders. Affected individuals are less likely to survive than individuals who have an intact copy of the gene. The frequency of the mutation will then be less among adults than among newborn individuals of the same generation. Perhaps those which do survive also produce fewer offspring than normal because they are sick or malnourished. The frequency of the mutation will then be lower among newborn individuals of the next generation than among adults of the current generation. In either case, the frequency of the mutation will tend to fall from one generation to the next.

Conversely, a mutation that increased survival, growth, or reproduction would tend to increase in frequency. Lactase persistence is an example. Lactase is an enzyme that hydrolyzes the milk sugar lactose to yield the simple sugars glucose and galactose. It is encoded by the *LCT* gene, which is normally down-regulated after weaning, so that adults are unable to metabolize lactose. Mutations in the upstream promoter region of *LCT* prevent effective down-regulation and thereby make it possible for adults to digest milk. These mutations seem to have spread recently in European populations, at about the time of cattle domestication 5000–10,000 years ago. They also spread in other pastoralist populations, such as the Dinka of southern Sudan. Such mutations would clearly provide an advantage when coupled with cattle rearing, by providing a new source of nutrition. Figure 2.3 shows the great variation in frequency of one of the mutations associated with lactase persistence among human populations, reflecting recent natural selection caused by cultural differences.

Hence, when there are heritable differences among individuals in their ability to survive or reproduce the population will evolve.

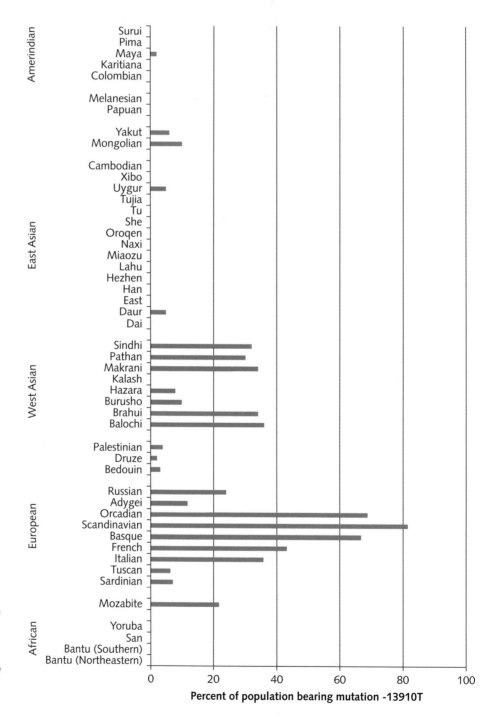

Figure 2.3 The frequency of a mutation in the *LCT* gene associated with lactase persistence varies widely among human populations as the result of recent natural selection based on the adoption of cattle-rearing.

Data based on Bersaglieri et al. 2004. *American Journal of Human Genetics* 74: 1111–1120, Table 1.

Evolution is caused by heritable variation in fitness. We can summarize this concept in the form of three linked propositions. First, the tendency of populations to increase in numbers creates scarcity and hence competition for resources. Second, variation in competitive ability leads to selection. Third, heritable variation and selection cause evolution. This sequence is so simple and elegant that it is sometimes mistaken for a purely logical construct that cannot be tested and is therefore unscientific. This is incorrect, because whether or not there is heritable variation for the ability to acquire the resources necessary for growth is a question of fact, not of logic. It is not always the case that heritable variation exists. However, it will very often be the case, and therefore most populations, most of the time, will be modified to a greater or lesser extent through the operation of natural selection.

This conclusion can be stated even more briefly by using the concept of *fitness*, which plays an important role in evolutionary biology. Fitness is a rate of proliferation, or reproduction. It is a property of any defined category of individuals, for example all those individuals bearing a defective copy of the lactase gene. The *relative fitness* of this type is its rate of reproduction relative to the average of the population. If its fitness is greater than average it will tend to increase in frequency, and if less than average it will tend to decrease. Changes in the frequency of different types will occur only if they differ in fitness, and these changes will be transmitted to the next generation only if they are heritable. Hence, we can restate our fundamental principle in this way: evolution is the consequence of heritable variation in fitness.

There are three aspects of fitness that are crucial in understanding how evolution works.

- The first is that fitness is a purely relative concept. It does not matter how successfully any given type of individual survives, or how fast it grows; all that matters is its growth and survival relative to the average of the population, since this alone determines how it will change in frequency. There is a well-known anecdote that illustrates the crucial importance of relative fitness. A wise man lived with his disciple in a remote forest. One day when they were out gathering berries, the two were suddenly confronted by a ferocious bear. As they fled through the trees the bear was clearly gaining, and the disciple panted despairingly to the sage, "I am afraid, sire, that it is no use: we cannot outrun the bear." But the sage only replied, "It is not necessary that I run faster than the bear. It will be enough if only I can run faster than *you*."

- Secondly, fitness is the only character that is selected. Any type with greater fitness than average will necessarily tend to increase in frequency. This is not true of any other character. This is an important point because we tend to think of some kinds of feature as being inherently superior. Selection does not necessarily favor larger, stronger, faster, brighter or smarter types, however, because they are not necessarily more fit. They may be in some circumstances, but not in others.

- Thirdly, other characters are selected only if they contribute to fitness. An active diurnal predator such as a swordfish, a hawk, or a leopard needs speed, strength, visual acuity, and good camouflage. A predator that lies in ambush (like a moray eel) or poisons its prey (like a sea anemone), or constructs a trap (like a spider) does not necessarily need any of these features. Internal parasites like flukes and tapeworms have evolved to eat their victims from the inside rather than the outside. Sight, coloration, and motility no longer contribute to fitness and are not maintained by selection, so flukes and tapeworms evolve to become blind, colorless, and almost immotile as the result of mutations that disrupt the development of structures such as eyes and limbs. These mutations would have been deleterious in the free-living ancestor of a parasitic organism, and so would have been eliminated by selection. Once a parasite no longer needs eyes or limbs, however, such mutations are no longer deleterious and so will tend to accumulate, especially if developing and maintaining these structures would require resources that could more profitably be utilized in upgrading adaptations to a parasitic way of life, such as evading host immune responses. In short, the direction in which a character evolves depends entirely on how it contributes to fitness, which in turn depends on the way of life of the species.

Natural selection acts through differences in fitness. Some simple numbers help us to understand how selection acts. Look at the Goldfish Pond Model in your Supplementary Material. This is a simple spreadsheet-based simulation of a population of goldfish living in a pond. The ecology of the pond is quite straightforward. Every spring the adults spawn and then die. The young goldfish grow during the summer, eating small animals like worms and snails. Some of them are eaten by the perch that also live in the pond. The rest become adults provided there is enough food for them; if too many young goldfish survive the perch, however, the excess die of starvation. The surviving adults then reproduce and the cycle is repeated. At some point, a mutation that produces a silvery body color occurs in a single egg. This makes the young fish more difficult for the perch to see, so it is less likely to be eaten and has a better chance of surviving to reproduce. The gene will therefore tend to spread in the population as the silver type replaces the normal gold individuals. The advantage might be quite small; let us suppose that the silver type is just 5% more likely to escape the perch. If the pond supports 100 adult goldfish and each adult produces 1000 offspring, then the initial frequency of the unique silver mutation is only 1 in 100,000. After 50 generations have passed, the silver fish will constitute about 1% of the population. They are still rare, but their frequency has risen by a thousandfold. After 100 generations they have passed the 10% mark, and after 200 generations about 95% of the population is silver. This shows that even a rather modest advantage can drive rapid change. More intense selection will naturally cause even faster change: if the silver type is 50% more likely to escape being eaten, it will exceed 95% of the population in just 25 generations. You can use the simulation to show how altering the properties of the population, such as how many offspring are produced, will affect the dynamics of natural selection.

2.2.2 Selection acting on heritable variation causes permanent change.

In the goldfish pond there is just one character ("color") and two states of this character ("gold" and "silver") that determine survival. If each gene encoded a single character and each character were encoded by a single gene, then we could simply replace "character" by "gene," and "character state" by "gene variant." In situations where this is possible, the process of evolution is relatively straightforward and easy to understand. In most cases, however, there are two major complications that make it necessary to reformulate the basic argument.

- The first is that not all differences in character state among individuals are caused by differences in the genes which those individuals bear. If one plant is short and another tall, the two may bear different variants of a gene or genes influencing height; but equally, the short plant may have grown under crowded conditions with insufficient nutrients. Differences in phenotype do not necessarily imply differences in genotype.

- The second consideration is that the relationship between genotype and phenotype may be very complicated. As a general rule, each gene encodes a single protein, and each protein is encoded by a single gene. However, any given protein may have manifold effects on the phenotype; and moreover, any given aspect of the phenotype may be affected by many different proteins. Thus, different variants of a gene are likely to cause variation in several characters; and the variation in any given character may be caused by variants of several genes.

In very simple systems, it is enough to recognize that evolution will occur as the consequence of differences in the rate of reproduction among variant gene sequences. In the more complex organisms with which we are usually concerned, we must retain this basic principle while reformulating it in terms of the evolution of characters and the selection of individuals. Here is a framework for thinking about evolution that sets out the main ideas (Figure 2.4 gives a simplified cartoon version).

a) Populations of individuals have a tendency to increase exponentially in numbers, whereas there is a finite supply of the resources necessary for reproduction.

b) As these resources become depleted, there will be a struggle for existence among individuals competing for opportunities to reproduce.

c) Individuals will vary to a greater or lesser extent with respect to all characters, because the process of hereditary transmission is not perfectly precise and so mutations are inevitable.

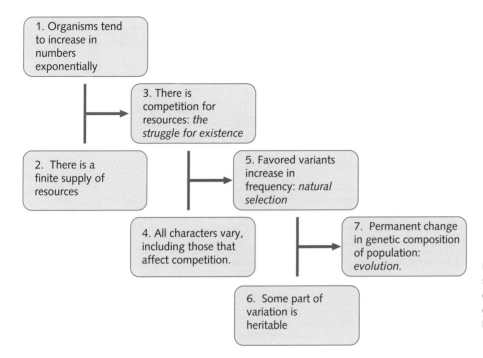

Figure 2.4 A simplified scheme for inferring evolution from the facts of competition, variation, and heritability.

d) They will therefore vary with respect to those characters that affect reproduction.

e) Some individuals will possess character states that enable them to reproduce more successfully than average, in the environment they occupy.

f) The offspring of these individuals will be disproportionately represented in the next generation: this is the process of selection.

g) If the successful individuals differed from the average of the population because they bore different variants of certain genes, these variants will have been transmitted to their offspring, who will then express the same aberrant character state as their parents.

h) Because this character state is more frequent in the population than it was in the previous generation, the average of the population will have changed, as a consequence of the increase in frequency of the variants responsible for encoding the altered character state.

i) This permanent genetic change in the composition of the population is the process of evolution.

This is a rather long-winded account of what happened in the goldfish pond. It is set out like this so that you can see how every step in the argument follows logically from the previous step, provided that we can measure variation and heritability.

Populations of individuals have a tendency to increase exponentially in numbers, whereas there is a finite supply of the resources necessary for reproduction.	Each goldfish can produce 1000 offspring, but the amount of food available in the pond is fixed.
As these resources become depleted, there will be a struggle for existence among individuals competing for opportunities to reproduce.	If the population increases beyond the capacity of the pond to support it, the goldfish will compete among one another for food.
Individuals will vary to a greater or lesser extent with respect to all characters, because the process of hereditary transmission is not perfectly precise and so mutations are inevitable.	Young goldfish will bear a range of mutations.
They will therefore vary with respect to those characters that affect reproduction.	Mutations causing changes in coloration affect the liability to be captured and eaten by perch.

Some individuals will possess character states that enable them to reproduce more successfully than average, in the environment they occupy.	Individuals with a silver body color are more likely to survive, and thus more likely to reproduce, when perch are present.
The offspring of these individuals will be disproportionately represented in the next generation: this is the process of selection.	Individuals with a silver body color produce on average more offspring than those with a normal gold body color.
If the successful individuals differed from the average of the population because they bore different variants of certain genes, these variants will have been transmitted to their offspring, who will then express the same aberrant character state as their parents.	The offspring of silver individuals will themselves be silver, provided that body color is heritable.
Because this character state is more frequent in the population than it was in the previous generation, the average of the population will have changed, as a consequence of the increase in frequency of the variants responsible for encoding the altered character state.	The frequency of silver individuals will be greater in the offspring generation than in the parental generation.
This permanent genetic change in the composition of the population is the process of evolution.	The population will evolve to become predominantly silver in color.

2.2.3 Selection can produce rapid changes in populations.

You might have come into this course thinking that evolution is a very slow process that happens over long periods of time. It is certainly true that radical changes in body plans and lifestyles require very long periods of time indeed, at least in terms of human lifespan. A great deal of evolutionary change, however, can occur very rapidly—rapidly enough that we can see it within our own lives, or even within the brief span of an experiment in the laboratory. The goldfish pond is just an imaginary example, of course. But there are other much more important evolutionary changes that are taking place all around us and that have a direct impact on our lives.

Harvesting causes evolution in fish populations. Let's stay with fish for the time being. Most heavily fished stocks—of salmon on the west coast of British Columbia, say, or cod on the eastern coast of North America—have declined in abundance because of fishing pressure. It is less well known that the fish themselves have changed. If they are caught in drift nets (like salmon) they are likely to have become slimmer. This is because slimmer fish of a given age are more likely to be able to pass through the meshes of the net. If they are caught by trawls or baited lines (like cod) they are likely to reproduce at younger ages. This is because individuals that delay reproduction are likely to be caught before they

can reproduce. These are genetic changes that are caused by the selection that we have imposed on the population. By harvesting a particular segment of the population, we have unintentionally set in motion a process of evolutionary change.

Antibiotics cause evolution in bacterial populations. A more somber example is provided by bacterial diseases in hospitals. Administering antibiotics to infected patients creates very strong selection for resistance among pathogenic bacteria. Only the resistant bacteria survive. The consequence has been that the effectiveness of most antibiotic therapy has been seriously compromised by the emergence of resistant types. This is not because bacteria have somehow become acclimatized or accustomed to antibiotics. It is because bacterial populations have evolved genetic sources of resistance, often within a few months or years following the introduction of a new therapy.

Fish stocks can change appreciably over a single human lifetime; bacterial populations can evolve in a few months. Selection is capable of driving very rapid evolutionary change.

2.2.4 Selection has no foresight.

In retrospect, it is tempting to think that every adaptation is an evolutionary "solution" to an environmental "problem." It is as though the final state were in some way prefigured, so that the population advances towards it as we would drive to a destination using a roadmap. But nature has no roadmap and has no destination in mind. Instead, any random change that produces an improvement will tend to spread, through natural selection, and the eventual outcome of this process may be a well-adapted population that gives the appearance of having moved deliberately towards a particular combination of characters. Selection is like an inefficient engineer, however, capable of seeing no more than one step ahead, and consequently it often produces imperfect or incomplete adaptations. We will explain this by a series of simple models which start out with a roadmap and then progressively abandon it.

2.3 Cumulation. *Selection acting on heritable variation necessarily incorporates a ratchet-like advance in adaptation.*

It is easy to understand how a single mutation (like that causing silver body color) can spread through a population. It is more difficult to understand how characters that depend on changes in many genes can evolve. The key is to realize that, at any given time, selection always modifies the changed state that has been produced by the previous episode of selection. This is the principle of *cumulation*.

2.3.1 Random mutation alone cannot cause adaptation.

You might think that a purely random process is unlikely to lead to the appearance of complex, highly integrated organisms, and you would be perfectly correct. It is easy to show that random changes can never lead to complex structures within reasonable periods of time. Word games are an entertaining way of showing the correct way of thinking about how evolution is caused by selection. They are not meant to be taken literally, of course—they are only intended to illustrate one particular point, that randomness by itself is not an effective agent of evolution, whereas randomness combined with selection provides solutions easily and rapidly. This is the great central idea of evolution, so understanding how it happens is crucial. Let us start with the simplest kind of game: given a word with a given number of letters, how can we get some other word of the same length by changing one letter at a time? Turn to the program WordChange in your Supplementary Material to play the game. It is a little like biological evolution, which involves changing one nucleotide at a time in the string of instructions provided by a gene.

For example, if we start with a very short word like "an," how long will it take to find another word of the same length, say "it?" We start copying the word "an" but every time we may substitute a random letter for "a" or "n" (or both) with some given probability. This is like a mutation rate, which may be quite small. Suppose it is 1 in 1000. Invoke the Random option and run the program. You have specified the starting sequence (the word "an"), and the program then works like this:

1. Copy the sequence, with a certain probability for each symbol of changing it to a different random symbol.
2. See whether the new sequence is the sequence you want; if so, stop.
3. If not, try (1) again using the new sequence.
4. Remember the best sequence (that has most letters in common with the target sequence) you have got so far. This is the improvement that has been accomplished by random change when you decide to stop.

These instructions can be implemented by a completely blind agent, like a computer, that has no means of foretelling what the right answer will be. It will infallibly generate the correct sequence—but it may take a long time. Since there are 26 possible letters there are $26^2 = 676$ combinations of two letters. If each is changed at random every 1000 tries, it will take about 676,000 attempts to produce "it." Sometimes it will take longer, sometimes not so long, but on average this will be about right. This is obviously quite a long time! Now, give the program a harder task. We start with the word "word" and want to arrive at the word "gene." How long will this take? You can work this out quite easily, because there are 26 different possibilities for each letter, so $26^4 = 456,976$ different sequences of four letters. With a mutation rate of 0.001, this will take about 456,976,000 iterations to produce "gene." This is now beginning to take time—and to generate more complex sequences is clearly out of the question. You will conclude that evolution cannot possibly happen this way, and you will be correct. What are we doing wrong?

2.3.2 Selection establishes a new point of departure for change.

To understand how evolution really happens, we introduce a new rule: if a random change has made a meaningful English word that more nearly resembles "gene" this sequence replaces the previous one, and is used as the starting point for the next round of randomization. In the same way, an improved protein will replace an inferior version, and will then itself serve as the basis for further improvement. This is the Darwinian option of modification through descent, which again begins with the input sequence and then works like this:

1. Copy the sequence, with a certain probability for each symbol of changing it to a different random symbol.

2. See whether the new sequence is the sequence you want; if so, stop.

3. See whether the new sequence is closer to the sequence you want.

4. If it is, look it up in the dictionary to see whether it is a meaningful word.

5. If not, try (1) again using the old sequence.

6. If so, adopt the new sequence as the standard and then try (1) again.

7. Remember the best sequence (that has most letters in common with the target sequence) you have got so far. This is the improvement that has been accomplished by Darwinian change when you decide to stop.

This looks rather similar to the previous method, and as before it can be implemented by a completely blind agent. You will find that one of two things will

happen. The first is that "word" changes into "gene" with blinding speed, through the intermediate words "wore," "gore," and "gone." Instead of about half a million steps, it will take just four steps. How can this be?

Figure 2.5 illustrates the difference between random and Darwinian evolution. You will see that step 3 in the random procedure has been replaced by steps 3–6 in the Darwinian procedure. This ensures that only an *improved* sequence is taken as the next step in the search for a still better sequence. This is analogous to the replacement of a less fit (gold-colored goldfish) by a more fit (silver-colored goldfish) type in the population, which can occur very rapidly, relative to the length of time required for a new beneficial variant to appear. By adopting the improved sequence as the new standard, we no longer need to screen all the possible permutations of letters, but only those including the improvement. If the change that caused the improvement is changed back, producing the previous inferior sequence, this does not matter, because it will not be adopted. It is like a harmful mutation in a population of animals or plants, which will occur from time to time but will never spread. Hence, the majority of changes need not be scrutinized, because they are not improvements, and therefore would not spread in the population. Consequently, every improvement eliminates 25/26 of all possible combinations, so that the number of possibilities that remain is only 1/26 as large after each improvement as before it. Adopting each improvement as the new standard means that change will be cumulative, moving towards the optimal sequence in a short series of steps, each somewhat better than the one before.

There is a second possibility, however: "word" changes to "wore" and then to "were," from which no further improvement can be made. Selection will usually lead to an improved sequence through a short series of steps, but it will not necessarily lead to the best sequence. This is because we are allowed to change only one letter at a time. In real organisms, the equivalent constraint is being able to change only one nucleotide (or at most a few nucleotides) of DNA from parent to offspring. Hence, potential improvements that require many independent changes to be made at the same time will not be realized, because the necessary variant sequences will never occur in the population. Selection acting on blind variation is the only process that can systematically improve

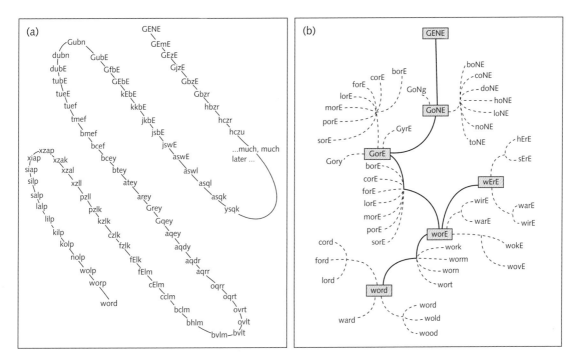

Figure 2.5 The evolution of GENE from WORD by a random process (left) and by a Darwinian process (right).

the attributes of natural populations, but it is not a very effective way of engineering complex structures, and this is why adaptation is often far from perfect (look back to Section 1.3.4).

2.3.3 Cumulative change can be astonishingly rapid.

The number of steps needed to change one word into another of the same length by selection is equal to their length—the number of letters in each, or a few more. For words of increasing length, therefore, the number of steps will be 1, 2, 3, …. If we just change letters at random, the number of combinations we shall have to screen increases as a power of the length of the word: 26, 26^2, 26^3, …. The power of selection to speed up adaptation, relative to random change, increases with the length of the sequence. Hence, complex structures can be generated by selection even though they would never appear through random change alone. Moreover, selection can give rise to complex structures surprisingly quickly.

You can use WordChange to show that selection becomes more and more effective for longer words, but it becomes difficult to find appropriate pairs of words with more than five or six letters.

To demonstrate how rapidly complex structures can evolve through selection, turn to the program PhraseMaker in your Supplementary Material. The object is now to generate a phrase or sentence that may consist of any number of words and symbols from a random (and usually meaningless) sequence of the same length. Any change that puts the correct letter or symbol at a given position is reckoned as an improvement. For example, if you select the Random option with a mutation rate of 0.001 and you type the 30-character phrase "the true north strong and free," the program will begin with a random 30-character sequence such as "eoc$?nvydqp/ hx,!qs,mj?'mg::-muk". When you run the program, the first improvement will occur quite quickly (after about 100 generations), but the second only much later (after about 10,000 generations), the third later still (after about a million generations), and so forth. The complete message will be discovered in about 10^{78} generations. My laptop iterates about a million generations per minute, so this will take about 10^{66} years, which is to say that it would never be observed even if you employed a billion laptops for a billion years. Now invoke the Darwinian option. You will find that the correct phrase is generated in less than a minute. Darwinism works.

2.3.4 Selection finds much better solutions than random change.

The notion of "improvement" in these word games is rather artificial, of course. We define the end-point of evolution and then attempt to find sequences that most resemble it—we have a roadmap, in effect. A more realistic model for how evolution actually happens would be to use a sequence that has a definite function that depends on some overall property of the sequence. A clock face, for example, is a sequence of this sort: it is useful to the extent that each of the 12 positions is occupied by the corresponding number. A slightly more complicated example is provided by a sequence that will be familiar to many of you: the arrangement of 20 numbered sectors around a dartboard. The number that labels each sector is the number of points scored when a dart thrown at the board lands in that sector, and in simple versions of the game the object is to score as many points as possible with a given number of throws. The sequence on the conventional board that is normally used looks rather odd: starting from the top, it reads 20, 1, 18, 4, 13, 6, 10, 15, 2, 17, 3, 19, 7, 16, 8, 11, 14, 9, 12, 5, and so back to 20. A little practice will show you that this unusual arrangement is designed to reward accurate play and to punish unskillful play, by putting high and low numbers next to one another, while the highest numbers are scattered in different parts of the board. An unskillful player aiming for the 20, for example, would be likely to get 1 or 5 instead, so their average score is only about 26/3 = 8.67. This cannot be improved much by aiming for a different part of the board: the 19, for example, at the bottom of the board, is flanked by 3 and 7, so the expected score of an unskillful player is 29/3 = 9.67. One property of this sequence is therefore that any group of three adjacent sectors has about the same total score as any other such group, as illustrated in Figure 2.6. On a perfect board, each group would have exactly the same average, and an unskillful player would have the same average score no matter where they aimed. To put this in another way, a perfect board has the property that the variation of average scores among groups of three adjacent sectors is zero. We can easily calculate the difference between any given board and the perfect board, with better boards having smaller values. A randomly arranged board has an average value of 10.5, whereas the conventional board is much better, with a value of 3.7. Can this be bettered by either random or Darwinian evolution?

Turn to the program DartBoard in your Supplementary Material to design a board. The initial board is arranged randomly, but note that unlike the word

The numbers are arranged so that the expected value of a poorly aimed throw is about 10 in any region of the board. This is equivalent to the *similarity of overlapping triplets*

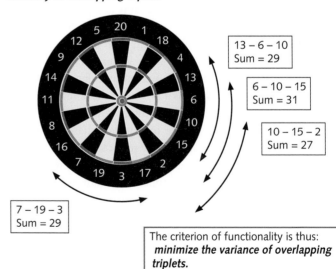

13 – 6 – 10
Sum = 29

6 – 10 – 15
Sum = 31

10 – 15 – 2
Sum = 27

7 – 19 – 3
Sum = 29

The criterion of functionality is thus: *minimize the variance of overlapping triplets.*

Figure 2.6 The dartboard: a functional circular sequence of integers.

games, no final sequence is defined. There is no road-map. Hence, we are not asking how long it takes to arrive at an end-point that has been decided in advance, but rather how much improvement will be made by a given process of change. If you invoke the Random option, a pair of numbers will be swapped at random in each generation to create a new sequence. In the next generation this new sequence is again changed at random, and this process is repeated until 1000 consecutive generations have passed with no further improvement being made. The total number of different swaps for any given board is only 190, so a sequence of 1000 consecutive failures almost certainly means that no further improvement can be made by a single swap. The Darwinian option retains any improved board as the basis for further change, as in the word games, and again gives up after 1000 unsuccessful attempts.

It is interesting to compare the output of random and Darwinian change. They both take about the same length of time (a few hundred cycles) to obtain a sequence that cannot be improved. Surprisingly, the outcome of random change is usually a sequence that is somewhat better than the conventional board, with a score of between 2 and 3. This is probably because our simple criterion does not capture all the desirable attributes of a dartboard (putting very low scores next to the highest scores, for example, so as to penalize players who are only moderately skillful). The outcome of Darwinian change is almost always a much better arrangement, however, with a score of less than 1. The very best sequences, with scores of about 0.2, are produced only by Darwinian evolution, never by random evolution. The best that has yet been found is the almost perfect sequence 20, 9, 4, 18, 8, 7, 17, 6, 10, 15, 5, 13, 14, 3, 16, 12, 2, 19, 11, 1. Thus, when changes to a structure have functional consequences, selection can readily generate highly improbable structures that perform a given task very well.

2.4 Incorporation. *The cumulative nature of evolutionary change means that current adaptation often bears the signature of the past.*

If organisms were perfectly engineered, their present state would tell us nothing about their history. But selection is far from being a perfect engineer, because it can only modify the previous state of the population. This has an important implication: we can use the current state of the population to retrace its evolutionary history.

2.4.1 New adaptations involve modifying old adaptations.

Evolution is a process of cumulative change because each new improvement, once it has been fixed in the population by natural selection, serves as the starting point for the next episode of variation and selection. To understand how this leads to a characteristic feature of complex adaptations, consider two ways of playing WordChange.

- First, change every letter at random in every cycle. (You can do this by specifying a mutation rate of 1.) If the target word is ever generated, it will appear abruptly and will, of course, be completely different from the original word. Hence, the improved version of a sequence contains no information about its ancestor, when evolution takes place by random change.

- Next, change a single letter at random occasionally. (You can do this by specifying a low mutation rate, say 0.01 or 0.001.) When the target word is reached it will again be unlike its ancestor (unless you deliberately made it similar). In this case, however, the program records the entire series of steps taken to derive the target word, each representing a slight improvement on the one before. Hence, the cumulation of altered letters leads to an historical process of change. The "fossil record" of words preserved by the program reveals the mechanism responsible for the appearance of the target word. This is why the fossil record of real organisms is convincing evidence of Darwinian evolution, as we described in Section 1.2.

Finally, there is one further lesson that can be extracted from the word game. If you type in the same word for the target as for the initial word, the program always operates in a Darwinian fashion, but will accept *any* meaningful word as an improvement on its predecessor. If you type in a short word (such as "word") then you are likely to trigger a long succession of dozens or hundreds of different words. The final word is unlikely to resemble the initial word, but, as before, the fossil record of the process will show clearly how each improved version was derived by a single change from its immediate ancestor. At the other extreme, if you type in a long word (like "appropriate" or "benevolent") there will never be any improvement—the final version is identical to the ancestor. The most interesting cases are in between. Try "computer," for example. You will find that this leads to "communes," "commutes," or sometimes "composes" (or another aspect of the same word, such as "commuted"), after which no further change is possible. The eight letters of "computer" thus fall into two categories. One includes the four letters that can be changed (p, u, t, and r) to give rise to new meaningful sequences. These correspond to functional changes during adaptation. The second includes the four letters that cannot be changed (c, o, m, and e). These correspond to ancestral characters that are retained even in remote descendants, which we described from real organisms in Section 1.3.3.

2.4.2 Similar agents of selection may lead to different evolutionary outcomes.

In the Dartboard model, we specified a criterion for the effectiveness of a sequence, without specifying any particular target sequence. The Darwinian process of random variation and selection then has three main features. The first is that the evolved sequences often perform a given task very well, just as real organisms are often highly adapted (Section 1.3.2). Secondly, some evolved sequences perform much less well than others, showing that adaptation is far from perfect, as discussed in Section 1.3.4. The third feature is simply that the outcome of selection varies from case to case: if the model is run twice, some degree of improvement will evolve in both cases, but the actual sequences that evolve will almost certainly be quite different. This is another consequence of the cumulative nature of evolutionary change. There are two reasons for this.

The first is that the outcome of selection depends on the state of the ancestor, because every improvement builds on the previous state. Hence, adaptation will be constrained by ancestry, as described in Section 1.3.3. The ancestral state may or may not affect the level of adaptation that can be attained, but it will always influence the way in which adaptation occurs.

Secondly, different outcomes may evolve even when similar ancestors experience similar forces of selection. The reason for this is that the sequence of beneficial mutations is unpredictable. Thus, "word" may evolve to "gene," but may get stuck at "were." Here there are just two endpoints; in more complex situations, such as the Dartboard model, there may be dozens or hundreds of endpoints. The course of a river provides a useful analogy. Its initial course, close to its source, is strongly influenced by very slight variations in relief; a minor bump or slope predisposes its course in one direction or another, and each slight shift of course shuts off a host of future possibilities. A difference of a few feet one way or another near the source may mean that hundreds of miles downstream it flows on one side of a mountain range rather than the other. The flow of a river is therefore an historical process, in which events at any point influence its whole future course. Two rivers that arise in the same meadow may yet follow very different courses, diverging further and further, until they eventually debouch into different oceans. The patterns of relief that cause their divergence may be slight, but their effects are cumulative and irreversible.

2.5 **Evolution.** *Evolution in real time confirms the Darwinian mechanism.*

The games and toy models that we have used to show how evolution works have severe limitations because none of them faithfully reflects all the features of real evolving populations. Each of them is intended to illustrate one particular aspect of evolutionary change, even though in other respects it may be unrealistic. We do not have to rely on them as evidence for the mechanism of evolution, however, because we can arrive at the same conclusions (although with much more work) by studying real populations evolving in real time in the laboratory. This is not usually feasible for large organisms that live for months or years, but it can be done routinely by using microbes that pass a whole generation in an hour or less. This is the field of experimental evolution.

2.5.1 A virus can adapt to novel conditions.

The simplest natural self-replicators are viruses. Qβ is a virus that infects bacteria. It uses RNA, rather than DNA, as its genetic material. In cellular organisms, RNA is used only as messages to decipher the DNA code, and there is no machinery for replicating RNA. The Qβ RNA encodes a number of proteins, including one that specifically catalyzes the replication of Qβ RNA—the Qβ replicase. The host cell provides the rest of the apparatus for producing the protein, and a supply of raw material from which new Qβ genomes can be constructed. The virus can thus be thought of as a small wormlike creature with an unusually simple morphology that is completely specified by the sequence of nucleotides in a single RNA molecule. Simple though it is, it can be simplified further. The viral genome encodes several kinds of protein used to transmit itself from one host cell to another, but not while replicating within the host cell. Once inside the cell, all that is needed is the replicase, together with a supply of nucleotides got from the host. Because the replicase can be isolated and purified from infected bacteria, these simple requirements can be provided in a culture tube. A solution of replicase and nucleotides provides a chemically defined environment in which Qβ RNA will replicate itself as though it were inside a bacterial cell.

The growth of self-replicating RNA molecules was the basis of one of the most remarkable experiments in modern biology. It was performed by a group of scientists in Sol Spiegelman's laboratory at the University of Illinois in 1967. They designed their experiment around the following question:

"What will happen to the RNA molecules if the only demand made on them is the Biblical injunction, *Multiply*, with the biological proviso that they do so as rapidly as possible?"

The elementary considerations that I have outlined are enough to provide an answer. The initial inoculum of viral RNA encounters a new and strange world, an unusually benign world in which it did not have to deal with the usual problems of parasitizing complex and hostile bacterial cells. Being provided with replicase and nucleotides, it begins to increase exponentially in numbers. This increase is soon checked by the finite supply of these resources. The tendency to increase in numbers while resources are in short supply creates competition, because not all can prosper: there will arise, as one might put it, a *struggle for existence*. But the growing population is necessarily diverse, and not all variants will be equally able to replicate themselves. Those which replicate more rapidly will increase in frequency, replacing their competitors: this process can be referred to as the *selection* of the more rapidly-replicating types. Because each type tends to reproduce itself, selection will involve the replacement of some lineages by others, or in other words will cause a permanent change in the genetic composition of the population, which constitutes *modification through descent*, or evolution. The experiment will result in the evolution of RNA molecules that are better *adapted* to the novel conditions of growth furnished by culture tubes: better able, that is to say, to replicate themselves at high rates in this novel environment.

Experimental cultures of virus rapidly adapt to novel conditions. One change that is easy to demonstrate, and whose effect on rates of replication is easy to appreciate, is that Qβ gets smaller. The intact virus with which the experiment is originally inoculated is a chain of 3300 nucleotides, encoding the proteins necessary for functioning as a transmissible intracellular parasite. After 70 transfers, the evolved variant, with its much greater rate of replication, is much smaller, usually about 550 nucleotides in length. The reason is very simple: other things being equal, a smaller molecule will be replicated more rapidly than a larger one. In the benign environment of the culture tube, more than 80% of the viral genome is unnecessary, and variants which lack the unnecessary sequences are favored by selection by virtue of their greater rates of replication. What eventually remains is not a random fragment—random pieces of Qβ RNA are unable to replicate—but rather a minimal sequence that supports efficient replication. The experimenters, however, made no attempt deliberately to select small molecules. Rather, they contrived a situation in which novel types with higher rates of replication would evolve, and small size evolved as a side-effect of this procedure because of its correlation, in these circumstances, with greater rates of replication.

Adaptation to stressful conditions involves a step-like series of genetic changes. After 70 or so transfers, Qβ has become fairly well-adapted to the comfortable environment of the culture tube. What if we now make it uncomfortable? RNA molecules can be made uncomfortable in a number of ways: one is to add to the culture medium a small amount of ethidium bromide, a substance that binds to RNA molecules and makes it difficult for them to replicate themselves. A very low concentration of ethidium bromide is at first sufficient to inhibit RNA replication almost completely. Within a dozen transfers—fewer than a hundred rounds of replication—however, the molecules have evolved a resistance to the drug, and by raising the dosage every dozen or so transfers, and thus maintaining a continual process of selection for resistance, it is possible to produce types able to grow at concentrations twenty times as high as that which was originally sufficient to suppress growth entirely (Figure 2.7(a)).

The simplicity of the system makes it possible to analyze the genetic basis of this physiological change. In one experiment, for example, resistance to ethidium bromide was caused by altering three nucleotides at different positions along the RNA molecule (Figure 2.7(b)). From a physiological or genetic point of view, this is as deep as any analysis can be pursued: nucleotide sequence is the ultimate character, from which all other attributes are eventually derived. From an evolutionary point of view, however, the fundamental issue is the manner in which these alterations occur in time. There are two possibilities.

- The first is that the original population was so diverse that it contains all possible sequences, including the sequence that happens to be successful when ethidium bromide is present in the growth medium. Evolution through selection is then simply a matter of *sorting* this initial variation, until the best-adapted sequence has replaced all others.

- The second possibility is that the original population includes only a small fraction, perhaps an extremely small fraction, of all possible sequences, so that the well-adapted sequence that eventually predominates evolves through the *sequential* replacement of sequences by superior variants that have arisen during the course of the experiment.

The evolution of resistance to ethidium bromide occurs sequentially, with the three altered nucleotides being substituted one at a time, so that the well-adapted sequence that eventually evolves is built up in a step-like manner (Figure 2.7(c)). It is easy to understand why this will almost always be the case. The number of possible combinations of nucleotides is so large that it is inconceivable that any but the tiniest fraction of them will be present in the original population. The evolution of any but the simplest modifications will generally involve the *sequential substitution* of several or many slight alterations, leading eventually to a state that was not originally present.

Adaptation involves a particular sequence of unit genetic changes. The three alterations conferring resistance not only occur one after another, they also occur in the same sequence whenever the experiment is attempted. This is because only one of the alterations confers resistance when it occurs by itself; the other two increase resistance once the first has spread through the population. Their spread is therefore contingent on the prior establishment of the first alteration; until this has occurred, they do not cause increased adaptation by themselves. They do not, in other words, have independent effects on resistance to ethidium bromide; rather, their effect depends on the presence of another alteration. Hence, the evolution of adaptation does not only require that the appropriate alterations should be substituted sequentially, but also that they should be substituted in a particular sequence, because only this sequence—or several sequences, out of the very large number of possibilities—involves a continuous increase in adaptedness.

A simple process of evolution resembles a word game. Qβ is such a simple creature that it looks rather like a long word or phrase made up of only four letters, and its evolution in culture tubes plainly resembles

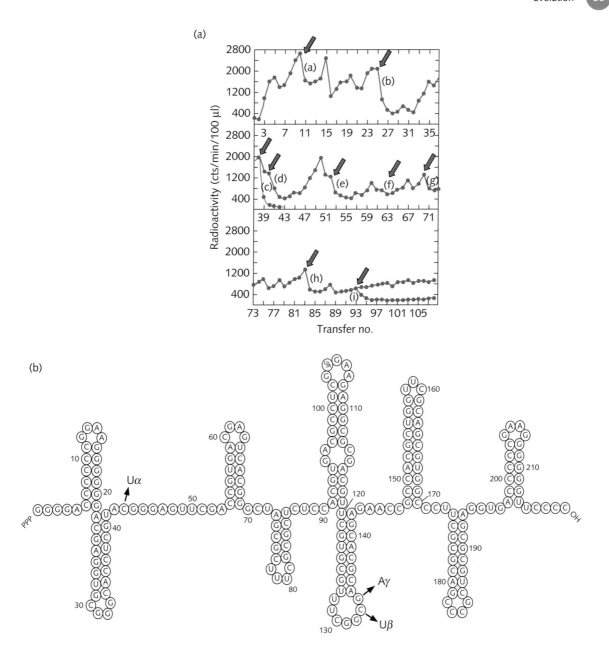

Figure 2.7 Experimental evolution of a self-replicating nucleic acid. (a) Collapse and recovery, through natural selection, of a population exposed to increasing doses of ethidium bromide. Dosage was increased at the times indicated by the red arrows. (b) The structure of the evolved resistant molecule. This is the resistant variant of one of the small molecules evolving from the intact Qβ virus in permissive conditions through serial transfer. The three nucleotide substitutions responsible for resistance are indicated. *(Continued)*

the games we have used to exemplify evolutionary principles. A highly improbable sequence, greatly superior to the ancestor, evolves very rapidly in every experiment, although the particular sequence differs from experiment to experiment. This is like the Dartboard game. In stressful conditions, adaptation involves a changing one letter at a time in a particular order, like the WordChange game. The stepwise approach to the final sequence quickly generates a novel message, as in the PhraseMaker game.

(c)

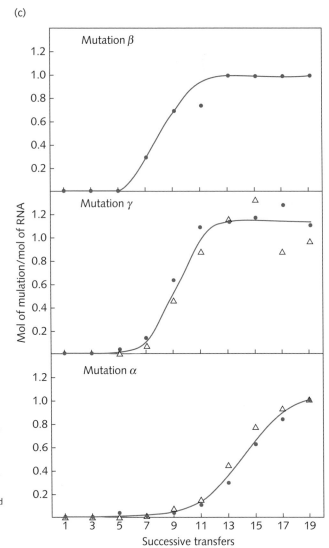

Figure 2.7 (*Continued*)   (c) The succession of beneficial mutations leading to resistance. This is the sequence in which the three mutations indicated in the molecular structure diagram are substituted through natural selection.

Part (a) reprinted from Saffhill et al. In vitro selection of bacteriophage Qβ ribonucleic acid variants resistant to ethidium bromide. *Journal of Molecular Biology* 51 (3): 531–539 © 1970. Parts (b) and (c) reprinted from Kramer et al. Evolution in vitro: sequence and phenotype of a mutant RNA resistant to ethidium bromide. *Journal of Molecular Biology* 89 (4): 719–736 © 1974, with permission from Elsevier.

2.5.2 **Bacteria can evolve new metabolic processes.**

In more complex, cellular organisms it is more difficult to demonstrate the fine details of the genetic changes underlying adaptation. Very careful experiments, however, have shown that the same principles continue to govern evolutionary change.

One of the classical studies in experimental evolution is the selection of new amidases in the soil bacterium *Pseudomonas* by Patricia Clarke, an English evolutionary geneticist. An amidase is a kind of enzyme, a protein whose shape allows it to bind amides, small organic molecules that bacteria can use as a source of carbon and energy. An amide has the chemical formula $RCONH_2$, where R stands for a chain or ring of carbon atoms. The amidase catalyzes the hydrolysis of the amide to produce ammonia and a carboxylic acid, which enters central metabolism.

An efficient amidase has two properties. The first is obvious: it should have the right shape to hydrolyze the amide efficiently. This is determined by the structural gene that encodes the amidase protein. The second, which is just as important, is that the amidase should be produced when the amide is present, but not otherwise. This is because it would be wasteful to produce something for which there is no call. An enzyme that is produced only when it is required is

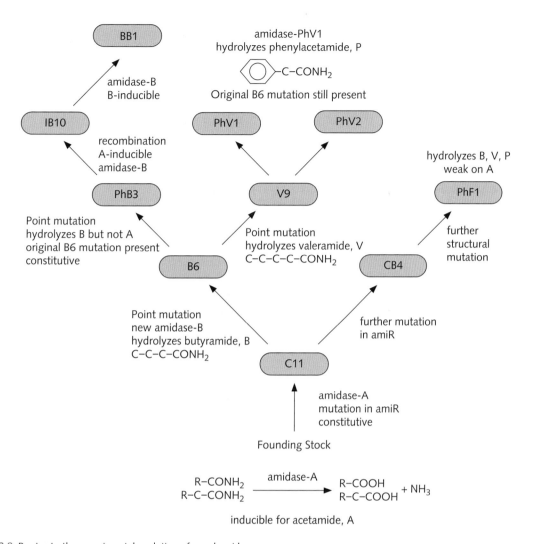

Figure 2.8 Routes to the experimental evolution of novel amidases.

Data from Clarke, P.H. 1984. Amidases of *Pseudomonas aeruginosa*. In R.P. Mortlock (ed.), *Microorganisms as Model Systems for Studying Evolution*, pp. 187–232. Plenum Press, New York.

said to be inducible. This requires regulatory genes whose products (other kinds of protein) switch the amidase gene on or off according to whether its substrate is present or absent. A structural gene that is always switched on is said to be constitutive. This condition is caused by mutations in the regulatory genes that stop them working properly.

The simplest amides are acetamide and propionamide, which have two and three carbon atoms, respectively. (Acetamide is responsible for the characteristic odor of mouse cages.) These are both substrates and inducers for the amidase, because they are common substances and bacteria have evolved to be well adapted to using them.

Experimental populations evolve new enzymes by unit genetic changes. The simplest amide that is not normally used for growth is the four-carbon butyramide, which is hydrolyzed only about 2% as fast as acetamide. The experimental evolution of new amidases is shown in Figure 2.8. It is easy to isolate mutants that overproduce the amidase and might therefore be able to grow on butyramide despite their inefficient utilization of the substrate. Some are regulatory mutants that express the amidase constitutively. Others are mutations in the promoter region of the amidase structural gene that cause increased rates of transcription. However, neither necessarily permits rapid growth, because butyramide, far from being an inducer, actually represses amidase synthesis. Effective adaptation requires one of two further changes. The first is another change in gene regulation, giving rise to the so-called CB strains, which causes a much higher rate of production of amidase. The second is the production of an altered amidase with higher activity towards butyramide. These B mutants, because they are able to hydrolyze butyramide efficiently, remove it from the medium and thus prevent it from repressing amidase synthesis. One such mutant, B6, was studied in detail. It produced an amidase B that differed from the original amidase A in a single amino acid residue, the replacement of serine by phenylalanine in the seventh position from the N-terminus of the protein. This was in turn caused by a single nucleotide change (UCU to UUU), in the appropriate codon of the amidase gene. This provides an example of how the simplest unit genetic change can make the difference between life and death.

New metabolic capacities evolve by a step-like series of genetic changes. The B6 strain was then selected on growth media containing more complex amides. A second mutation in the structural gene permitted growth on the five-carbon amide valeramide. These V mutants could in turn be used to select PhV mutants able to grow on phenylacetamide, which contains an aromatic ring. These mutants have three alterations in the amidase. Phenylacetamide is neither a substrate nor an inducer for the original amidase system, so that by this point a genuinely new metabolic capacity had evolved. It had evolved through a cumulative process of successive substitution: the original B6 mutation is still present in the V strains, and both B and V mutations are present in the PhV strains, showing unequivocally how new capacities evolve through the stepwise modification of prior states.

There may be several alternative routes to adaptation. This is not the only way in which the ability to metabolize phenylacetamide can evolve. Another class of PhB mutants arose directly from the B6 strain, by a second change in the amidase structural gene. These are able to metabolize either phenylacetamide or butyramide. However, they have acquired one capacity at the expense of losing another: they are now unable to grow on the original

substrate, acetamide. They are still constitutive, but it is possible to select a strain that is induced by acetamide (although it cannot utilize it) by recombination with the basal stocks. A further regulatory mutation then produces a strain that is induced by butyramide. The result is a strain that both hydrolyzes butyramide efficiently and is appropriately induced by it.

These beautiful experiments show how the principles that are illustrated by games and toy models actually operate in real organisms. They demonstrate rapid adaptation through natural selection, the stepwise nature of evolution, cumulative change, and a diversity of outcomes, just as our simple models predict. There are many more examples of how we can study the mechanism of evolution in real time, which are described in more detail in Chapter 13.

2.5.3 Protozoans rapidly adapt to hostile conditions.

Outside the laboratory it is much more difficult to trace the course of adaptation in detail, although this is now becoming possible with the improvement of genomic technology. The most detailed studies have investigated organisms that cause human disease, and one of the best worked-out cases involves *Plasmodium*, the parasite responsible for malaria. This is a single-celled eukaryotic organism ("protozoan" is the old-fashioned term) that lives in the bloodstream, where it enters red blood cells and consumes their hemoglobin. Malaria was effectively treated, until recently, with chloroquine, a drug developed in the 1940s which kills the parasite by interfering with its ability to process hemoglobin.

Adaptation can occur rapidly in natural populations. Resistance to chloroquine was first observed in south-east Asia, about a decade after its introduction. The resistant strains spread slowly, dispersed by the mosquitos that transmit the disease, reaching Africa in the mid-1970s. Meanwhile, resistance evolved independently in South America, Papua New Guinea, and the Philippines. By the 1990s chloroquine had become ineffective, and was replaced by sulfadoxine/pyrimethamine. This new drug in turn failed within about five years as new resistant strains evolved. Chloroquine was a relatively long-lived drug that prolonged hundreds of millions of lives in its 20-year career. Nevertheless, *Plasmodium* was eventually able to evolve resistance to it. By analyzing archived body fluids, it has been shown that resistant genotypes appear and become abundant very quickly, within two or three years. Hence, there is a relatively long period of time during which fully resistant genotypes are being assembled and dispersed, followed by rapid spread.

Chloroquine resistance requires several mutational changes. The gene primarily responsible for chloroquine resistance is *pfcrt*, which encodes a long transporter protein located in the digestive vacuole of the parasite. As shown in Figure 2.9, the amino acid positions crucial for resistance are 72, 74, 75, 76, and 220. (Other positions, and at least two other genes, make minor contributions.) Using the standard code in which each amino acid is represented by a single letter, we can write the normal type as CMNKA, which you can think of a short word that means "susceptible" in *Plasmodium* language. The crucial mutation is K (lysine) to T (threonine) at position 76, which alters the net charge on the protein. Other mutations are almost always associated with resistance, however. Two highly resistant genotypes are SMNTS (three mutations) and CIETS (four mutations), so the relatively long period of time taken to assemble resistant genotypes is presumably because a rather complex series of changes is required. In fact, eight or nine mutations are probably involved in all.

Resistance evolves through successive substitution of unit genetic changes. Chloroquine was usually administered as three successive daily doses, which pumped up serum concentrations to a high level. In the following days the concentration slowly declined. Hence, a strain of the parasite resistant to low levels of chloroquine would out-compete the wild type a certain time, say five days, after the treatment has stopped and drug had dissipated but not completely disappeared. Once this low level of resistance had evolved, a second mutation conferring a higher level of resistance could spread because it out-competed the first mutant on the fourth day after treatment; and so forth. In this way a complex genotype with a

fairly high level of resistance could be built up, step by step, until a final mutation made the parasite resistant to the therapeutic dose. This would then spread very rapidly through the population because of its ability to grow in all conditions. We can infer the sequence of mutational changes by following changes in the frequency of genotypes, or by scoring their levels of resistance. A likely sequence of changes leading to a fully resistant genotype is then: CMNKA→ CMNTA → CMNTS → (CMDTS) → CIDTS → CIETS, where all of these genotypes have been identified in natural isolates except the one in brackets.

There is more than one outcome of evolution. However, SMNTS is also a highly resistant genotype. This suggests a more complex evolutionary scheme, as can be seen in Figure 2.10.

This looks just like the WordChange game—WORD can evolve to GENE but may get stuck at WERE—because both are evolutionary processes that create complex adaptation through cumulative change guided by selection acting on random variation.

Adaptation is constrained by functional interference. All three cases that we have described illustrate another general principle of adaptation. In every case, selection in certain conditions causes adaptation to those conditions, but at the expense of adaptation in other conditions.

- The bacteriophage Qβ rapidly adapts to life in culture tubes, but at the expense of being able to infect bacteria. This is because selection for rapid replication has led to the shedding of genes essential for infection, such as the gene encoding the envelope protein. This illustrates the principle of *functional interference*: the characters (such as shorter genomes) that increase fitness in some conditions (culture tubes) necessarily reduce it in others (bacteria).

- Selection of *Pseudomonas* for the ability to metabolize novel amides is effective, but the adapted strains now grow poorly on the original substrate, acetamide.

- The chloroquine-resistant strains of *Plasmodium* are rapidly out-competed by susceptible strains in the absence of the drug.

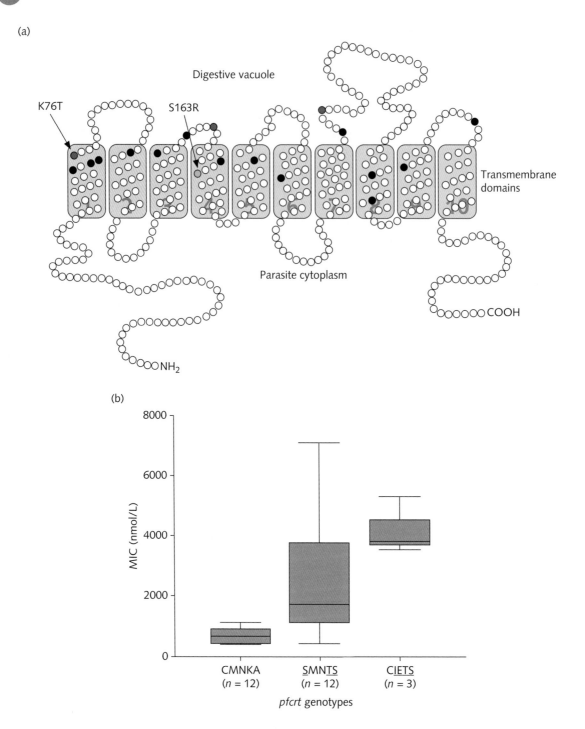

Figure 2.9 The gene primarily responsible for chloroquine resistance is *pfcrt*, which encodes a long transporter protein located in the digestive vacuole of the parasite, illustrated in (a). Filled circles are known mutations associated with resistance. The growth of clones bearing versions of *pfcrt* is shown in (b); the MIC is the minimum inhibitory concentration of chloroquine for each mutant.

(a) from Bray et al. Defining the role of PfCRT in *Plasmodium falciparum* chloroquine resistance. *Molecular Microbiology* 56 (2): 323–333 © 2005, John Wiley and Sons. (b) from Mittra et al. 2006. Progressive increase in point mutations associated with chloroquine resistance in *Plasmodium falciparum* isolates from India. *Journal of Infectious Diseases* 193 (9): 1304–1312 by permission of Oxford University Press.

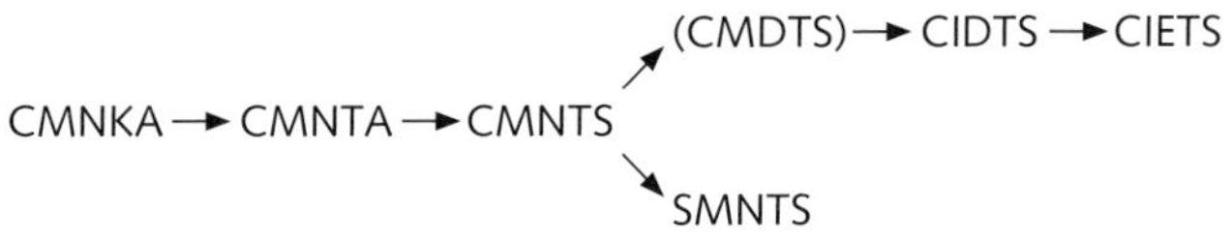

$$CMNKA \rightarrow CMNTA \rightarrow CMNTS \nearrow (CMDTS) \rightarrow CIDTS \rightarrow CIETS$$
$$\searrow SMNTS$$

Figure 2.10 Sequence of changes leading to the SMNTS resistant genotype.

In all these cases, the evolved strains are inferior to the ancestor, in the ancestral conditions of growth. Selection does not produce strains that are in all respects superior to their ancestor, but rather to strains that are better adapted to particular conditions, at the expense of being worse in the conditions to which their ancestor was well adapted. Adaptation does not produce perfection, but only enhanced fitness in particular conditions.

Self-replicating systems have a common set of properties. These three cases—phage in culture tubes, bacteria growing on new substrates, malarial parasites challenged by a drug—exemplify a series of generalizations about the behavior of systems of self-replicators that was anticipated by simple word games.

- Heritable variation in the rate of replication causes evolution through selection.
- Heritable variation arises as random, or undirected, alterations of nucleotide sequence; it does not in itself direct the course of evolution.
- The rate of replication is the only attribute that is selected directly.
- Characters that cause changes in the rate of replication will be selected indirectly, and may evolve as a consequence.

- Evolution proceeds through the sequential substitution of superior variants, not exclusively by sorting pre-existing variation.
- A given state can evolve from a prior state only if the two states are connected by a continuous series of modifications, each of which is individually advantageous.
- Selection causes the modification of prior states of organization, but cannot abruptly give rise to wholly novel states; the course of evolution is an historically unique process conditioned by the fortuitous occurrence of particular variants.
- Adaptation caused by selection in given conditions is likely to be associated with loss of adaptation in other conditions.

The way that these general laws, and others yet to be discussed, mold the evolution of living organisms will be modulated by the developmental, physiological, genetic, and ecological circumstances in which they operate. We close this chapter by emphasizing that natural selection acting on blind variation is not only capable of driving the rapid evolution of complex adaptations, but also leaves unmistakable traces of its action that can be checked by studying experimental populations in the laboratory or natural populations in the field.

● CHAPTER SUMMARY

Mutation supplies the raw material for evolution but does not determine its direction.

- Mutation is not appropriately directed.
  - *Mutation is inevitable.*
  - *A mutation is a unit genetic change.*
  - *Mutation is not usually adaptive.*
- Most mutations have little effect.
  - *Many mutations are neutral.*

  – Mutations with very severe effects are rare.
  – Beneficial mutations occur occasionally.

Selection screens the variation made available by mutation and thereby changes the genetic composition of the population.

- Selection is caused by variation in fitness.
  - *Selecting exceptional individuals alters populations.*
  - *Scarcity of resources leads to reproductive competition.*
  - *Variation leads to natural selection.*
  - *Mutation and selection cause evolution.*
  - *Evolution is caused by heritable variation in fitness.*
  - *Natural selection acts through differences in fitness.*
- Selection acting on heritable variation causes permanent change.
- Selection can produce rapid changes in populations.
  - *Harvesting causes evolution in fish populations.*
  - *Antibiotics cause evolution in bacterial populations.*
- Selection has no foresight.

Selection acting on heritable variation necessarily incorporates a ratchet-like advance in adaptation.

- Random mutation alone cannot cause adaptation.
- Selection establishes a new point of departure for change.
- Cumulative change can be astonishingly rapid.
- Selection finds much better solutions than random change.

The cumulative nature of evolutionary change means that current adaptation often bears the signature of the past.

- New adaptations involve modifying old adaptations.
- Similar agents of selection may lead to different evolutionary outcomes.

Evolution in real time confirms the Darwinian mechanism.

- A virus can adapt to novel conditions.
  - *Experimental cultures of virus rapidly adapt to novel conditions.*
  - *Adaptation to stressful conditions involves a step-like series of genetic changes.*
  - *Adaptation involves a particular sequence of unit genetic changes.*
  - *A simple process of evolution resembles a word game.*
- Bacteria can evolve new metabolic processes.
  - *Experimental populations evolve new enzymes by unit genetic changes.*
  - *New metabolic capacities evolve by a step-like series of genetic changes.*
  - *There may be several alternative routes to adaptation.*
- Protozoans rapidly adapt to hostile conditions.
  - *Adaptation can occur rapidly in natural populations.*
  - *Chloroquine resistance requires several mutational changes.*
  - *Resistance evolves through successive substitution of unit genetic changes.*

- *There is more than one outcome of evolution.*
- *Adaptation is constrained by functional interference.*
- *Self-replicating systems have a common set of properties.*

● FURTHER READING

Here are some pertinent further readings at the time of going to press. For relevant readings that have been released since publication, visit the book's Online Resource Centre at **www.oxfordtextbooks.co.uk/orc/bell_evolution/**

Section 2.1   Bell, G. 2008. *Selection: The Mechanism of Evolution*. 2nd ed. Oxford University Press, Oxford.

● QUESTIONS

Write a commentary on the following criticisms of evolution through natural selection.

1. Complex integrated structures cannot arise through chance mutation.

2. There has not been enough time for animals and plants to have evolved through natural selection.

3. If natural selection were effective, living organisms would be perfectly adapted.

4. Natural selection cannot work because any successful variant would be diluted out by crossing with normal members of the population.

5. In order for complex adaptations to evolve, several mutations would have to occur at the same time in the same individual, which is very unlikely.

6. Almost all mutations are deleterious because a random change will almost certainly damage a complex, integrated organism, so selection may preserve old types but cannot promote new types.

7. Evolution by natural selection means the survival of the fittest, which is a tautology.

8. The speed of adaptation shows that appropriate mutations often arise to meet the demands of a changing environment.

9. Evolution through natural selection is only one theory, and we should also teach other theories, such as special creation.

You can find a fuller set of questions, which will be refreshed during the life of this edition, in the book's Online Resource Centre at **www.oxfordtextbooks.co.uk/orc/bell_evolution/**

PART 2
History

The outcome of evolution on Earth is the multitude of animals, plants, and other kinds of organism that inhabit sea and land, together with the even greater number that lived in the past and are now extinct. The core of evolutionary biology is the discovery that all these organisms, living or extinct, are related to one another by an unbroken chain of ancestry stretching far back into the history of the Earth. The first items in the syllabus of evolution are thus the diversity and the descent of living organisms. Both are summarized by the phylogenetic tree, which depicts the sequential branching of lineages over time. The "tree of life" is the most familiar kind of evolutionary diagram, and appeared very early in the literature of evolution. It traces the course of evolution, while enabling us to construct a natural classification of organisms in terms of their degree of relatedness to one another. This part of the book begins with a general account of phylogenetic trees, describing how they arise from lineage dynamics and how they can be drawn using information about the features of organisms. It continues with a phylogenetic account of modern diversity that identifies the main innovations that have evolved in the lineage leading to our own species. The final chapter in this part of the book deals with the history of our lineage and the features of our ancestors.

The Tree of Life

3

The purpose of this chapter is to show that all living organisms are related to one another and to describe how these relationships can be worked out.

3.1 Evolution is the flux of lineages through time.

Evolution is the process by which organisms become modified over time. New organisms do not appear spontaneously, of course: they are produced by old organisms. The basic unit that undergoes modification is thus the lineage—the long succession of parents and offspring, extended over time.

3.1.1 The lineage is the chain of descendants through time.

Individuals, populations, and lineages are three fundamental aspects of organisms.

The individual is the unit of physiology and genetics. The most familiar biological category is the individual. This is because we are strongly individualized organisms: each one of us has separate bodies, identities and fates. Physiology, genetics, and developmental biology are built on organisms like us: mice, fish, flies, and so forth. Individuals do not evolve, but they are important in evolution because they are the units on which the agents of selection actually operate. The land snail *Cepaea nemoralis* is an example of a strongly individualized organism. It is a European species that has been introduced to eastern North America and has been intensively studied by evolutionary biologists. Its shell is highly variable. The ground color spans a range from yellow, through pink, to brown, and may be overlain by one or more dark bands. Both color and banding are inherited in a simple Mendelian fashion. Snails are eaten by visual predators such as thrushes and blackbirds, which tend to choose the most conspicuous individuals. Against the background of dark wet leaves of the forest floor in late fall, for example, a yellow unbanded shell is easy to see, whereas a brown banded shell will be inconspicuous. When a bird detects a yellow shell and consumes the snail this removes a yellow-shelled individual from the population and so alters the genetic composition of the population. All evolutionary change ultimately stems from events that happen to individuals.

Although this may sound obvious, there are two important reservations to be made. The first is that not all organisms develop as completely discrete individuals like snails. We have already mentioned beech trees, where new stems readily arise as suckers from roots, so that separate stems are physically linked underground and nutrients can be transferred between them. Each stem is not a physiologically separate individual, but the stems are not as fully integrated as the different tissues in the body of a snail. There are similar examples from animals. Corals grow as colonies of minute animals, called zoids, each of which occupies a small calcareous chamber with cytoplasmic connections to its neighbors. The colony expands through the asexual proliferation of the zoids. Neither the individual coral animal nor the colony exactly corresponds with our notion of an individual. In organisms like these, a mutation arising in a single zooid may cause it to reproduce faster, at the expense of other zooids in the same colony. Hence, the details of how selection occurs depend on the way in which an organism develops.

Secondly, even highly individualized organisms consist of colonies of cells that are often capable of dividing. Other kinds of tree, such as maple trees, reproduce entirely by seed, and each stem is separate. The growth of the tree, however, is based on cell division in apical meristems (the growing points at the tips of the branches) which are to some degree isolated from one another. A mutation in one branch that makes the leaves it bears less edible might favor that meristem over others borne by the same tree. Our own bodies are much more highly integrated than trees, requiring tight control over cell division in different tissues. When a mutation causes a cell to escape from this control it forms a tumor, within which selection will favor cells that divide more rapidly. This drives an evolutionary process that has no permanent result (because the tumor cells are not part of the germ line) but which explains some features of how cancer develops and spreads within the body.

The population reflects the past operation of selection. The population is a set of individuals of the same species, usually thought of as occupying the same place at the same time. "The population of goldfish in Walden Pond," for example, refers to all of the fish of a particular species living together at a particular place at the time when they are being described. Populations are important in evolutionary biology because they reflect the outcome of events that happen to individuals. When yellow snails are eaten disproportionately by birds, the genetic composition of the population is altered, and when this is repeated generation after generation there will be an evolutionary shift from one variety of snail to another. Although the individuals have long died, the population continues to exist, and its properties will reflect the outcome of selection in the past.

The concept of a population suffers from the same drawback as the concept of an individual. Populations are seldom completely isolated, because most receive a continual stream of immigrants and send out an equivalent number of emigrants. Most organisms live in "metapopulations" consisting of more

or less densely inhabited sites separated by uninhabited areas across which they exchange migrants. The changing composition of the population will therefore reflect events in other parts of the species' range as well as changes in local conditions of growth. The dispersive larvae of marine invertebrates that were discussed in Section 1.4.2, for example, ensure that most offspring will emigrate to distant sites. Poplar trees release clouds of fluffy fruits that are likely to be blown many miles before settling on the ground. In cases like this, a local population may have little genetic continuity from one generation to the next.

Lineage dynamics drive evolutionary change. The lineage is the chain of ancestors and descendants that propagates itself over time. It is a more subtle concept than the individual or the population, because one cannot usually view, at the same time, all the individuals making up a lineage. Moreover, it can be viewed in two aspects, either forwards or backwards in time.

In the first place, consider some focal individual who has lived in the past. It belongs to an asexual species; sex introduces complications that will be discussed later. The *lineage forwards-in-time* from this individual comprises this individual and all its

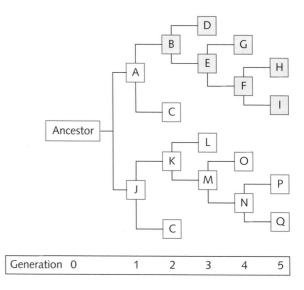

Figure 3.1 Lineage dynamics drive evolutionary change. The shaded boxes represent the lineage forward-in-time stemming from individual B.

descendants, as shown in Figure 3.1. There may be no descendants: this individual died without issue. Or there might have been some, all of which failed, so that they left no living descendants. In either case, this lineage is now extinct. Alternatively, some or many of the current population may have descended from this individual. The number of distinct lineages that currently exist depends on how far we travel into the past. If we go back a single generation, each family is a distinct lineage and there may be thousands of them. If we go back further, many lineages will have become extinct and relatively few will have survived. The effect of selection is that some lineages expand, whereas others dwindle and eventually disappear. It is the waxing and waning of lineages that underlies the change in the genetic composition of populations over time. Evolution is lineage dynamics.

Secondly, consider a focal individual who is living at present. The *lineage backwards-in-time* of an individual comprises itself and all its ancestors, as shown in Figure 3.2. For any individual living at the present time (or at any other time) we could make a list of all their ancestors, generation by generation. The lineage backwards-in-time does not tell us anything directly about how evolution happens, but it provides a useful device for cataloguing the ancestry of a population because it is not necessary to take into account

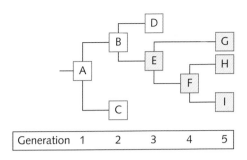

Figure 3.3 The coalescent. All the individuals currently alive (G, H, I, in generation 5) descend from a single individual that lived in the past (E, in generation 3).

any of the many lineages that have become extinct within the span of time being surveyed.

There is an interesting and useful extension of this concept. Suppose we take all of the individuals now living, and project their ancestry back into the past, as in Figure 3.3. The number of ancestors of the current population must continually decrease, as some individuals in each generation die without leaving descendants. Hence, the current population must all descend from a single individual that lived at some time in the past. This individual is called the *coalescent*: the individual from which all individuals now living descend. It follows that all other individuals at this period in the past contribute nothing to the ancestry of the current population. The coalescent summarizes the evolutionary dynamics of any non-recombining lineage: genes, asexual organisms, and sexual species.

3.1.2 Descendants come to differ from their ancestors.

Offspring differ from their parents because genetic alterations are transmitted, and their own offspring will bear further alterations. Hence, the differences between members of the same lineage will tend to increase over time. How rapidly this occurs will depend on the rate of mutation and the strength of selection. Obviously, a high rate of mutation will lead to greater differences between parents and offspring, but most mutations are deleterious, so offspring that bear many mutations are unlikely to survive. In this way, selection will limit the effect of mutation and moderate the rate of change in the lineage.

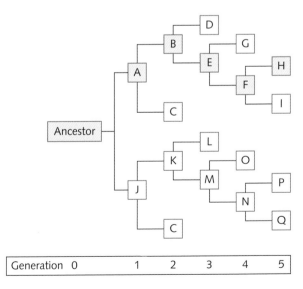

Figure 3.2 Lineage dynamics drive evolutionary change. The shaded boxes represent the lineage backwards-in-time stemming from individual H.

Mutations are individually rare but collectively frequent. The most straightforward way of estimating the rate of mutation is from the frequency of children afflicted with a genetic disorder caused by a dominant allele who are born to unaffected parents. This must be the result of a new mutation, and typically leads to estimates for the mutation rate of about 10^{-5} per gene per generation. The average gene consists of roughly 2000 base pairs, so the fundamental rate of mutation is about 5×10^{-9} per nucleotide per generation. This must be divided by the number of cell divisions in the germ line to give the mutation rate per cell division. This is bound to be an underestimate, because only a small fraction of mutations will cause disease. More representative mutations, causing a range of morphological or biochemical phenotypes in genetically well-studied organisms such as yeast, fruit flies, and mice, arise at a rate of about 2×10^{-10} per nucleotide per cell division. Again, only mutations that have readily detectable effects will be counted.

Another way of estimating mutation rates is to sequence a gene in two closely related species whose time of divergence is known (because their fossil record is well known, or because they have evolved on two recently formed islands, for example). This is most informative when a pseudogene is used—a gene that no longer encodes a functional protein, so that mutations are not eliminated by selection. This procedure gives estimates of about 2×10^{-8} per nucleotide per generation for the human genome. Since there are about 30 cell divisions in the human germ line, this figure implies a mutation rate of about 10^{-10}–10^{-11} per nucleotide per cell division. Thus, different methods have different limitations and give somewhat different estimates, but these are generally in the region of 10^{-9}–10^{-10} per nucleotide per cell division. This is a very small number, but to appreciate its significance it must be scaled by the size of the genome and by the number of individuals in the population.

The human genome contains 3.3×10^{9} nucleotides, but not all of these encode proteins. We have only about 25,000 protein-coding genes, so a more realistic estimate of the functional genome size is about $25,000 \times 2000 = 5 \times 10^{7}$ nucleotides. If we take the fundamental mutation rate to be 10^{-9} per nucleotide per cell division the overall rate of mutation per genome per generation is about $(5 \times 10^{7}) \times 30 \times 10^{-9} = 1.5$. This is not a very reliable figure because of the approximations used to derive it, but it is unlikely to be far out. Hence, offspring will often bear a novel mutation in a functional gene. Similar calculations for other animals and plants give a similar result. Microbes have smaller genomes, so their genomic mutation rate is less (about 0.01 per generation for *E. coli*), but their shorter generation time makes up for this. Thus, despite the high fidelity of DNA replication, enough mistakes are made to generate considerable variation among the members of a lineage. Most of these mutations are likely to be slightly deleterious, but the minority of beneficial mutations provides the crucial raw material for evolution.

The human population comprises about 5×10^{9} individuals, so a rate of mutation of 2×10^{-8} per nucleotide per generation implies that every single-nucleotide mutation is already present in this population. Microbial populations are still larger. If you take 10 mL of nutrient broth and add a few yeast cells, there will be about 6×10^{8} cells the following day, so that most single-nucleotide mutations will already be present in that small culture tube. In such large populations the supply of mutations is so great that simple kinds of evolutionary change (such as antibiotic resistance) can occur very quickly indeed. Conversely, many large organisms have total populations of only a few tens of thousands of individuals. Evolution will be much slower because there is likely to be a long wait before any given mutation occurs.

Selection restricts the accumulation of mutations. The accumulation of variation in a lineage depends on how effectively selection removes mutations. A mutation that completely prevents reproduction is removed in a single generation, of course, although the same mutation might occur in another individual in the next generation. A slightly deleterious mutation, on the other hand, is quite likely to be transmitted and may persist for many generations, although it is not likely to become very frequent. Even if we suppose that a population were perfectly adapted at some time, deleterious mutations are certain to occur (Section 2.1.1) and will therefore tend to accumulate in the population. Individuals bearing more mutations are less likely to survive and reproduce, however, which provides a brake

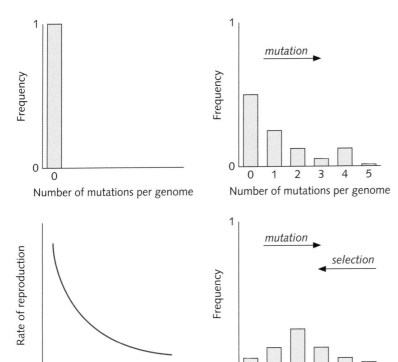

Figure 3.4 Selection restricts the accumulation of mutations. The number of mutations per genome reflects a balance between mutation and selection.

Reproduced from Bell, G. 2008. *Selection: The Mechanism of Evolution* (OUP).

to the increase in the load of mutations. Figure 3.4 illustrates how the opposed processes of mutation and natural selection lead to a dynamic equilibrium in which most individuals bear a modest number of deleterious mutations. None of these mutations are likely to be frequent in the population; rather, there are many rare mutations, of which most individuals carry a few.

Whether or not a mutation is severely deleterious depends to a large extent on which gene it affects. The genes responsible for ribosome assembly and the basic features of transcription and translation, for example, are highly intolerant of mutation; most mutations that cause changes in the structure of the proteins encoded by these genes are lethal. This indicates that beneficial mutations are exceedingly rare, so that evolution will be very slow. Such genes are highly conserved in evolution, having similar sequences even in distantly related organisms. Because they evolve very slowly they are useful in mapping lineages that diverged from one another in the distant past. Most genes are less sensitive, with many variants that are more or less equivalent in

fitness, or vary in fitness according to the conditions of growth. Genes responsible for activities such as carbon metabolism, signal transduction or transport fall into this category. Beneficial mutations are likely to be more frequent, so they usually evolve more rapidly and can be used to map lineages over moderate timescales of tens or a few hundreds of millions of years. The most rapidly evolving genes are concerned with disease resistance and mating behavior, which may be quite different even in closely related species and can be used to map lineages only on very short timescales.

Neutral mutations convey a purely historical signal. The effect of a mutation also depends on where it occurs within a gene. If it modifies a nucleotide without affecting protein structure it is likely be neutral or only very weakly selected. Neutral mutations are particularly useful in mapping lineages because their variation is the outcome of a random sequence of changes that is unlikely to occur twice in the same way. Consequently, they convey a purely historical signal that reflects the true pattern of descent.

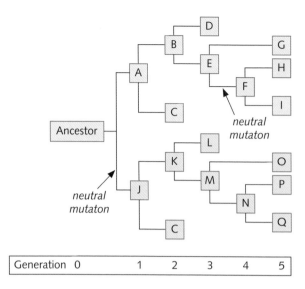

Generation	0		1	2	3	4	5

Figure 3.5 Neutral mutations convey a purely historical signal.

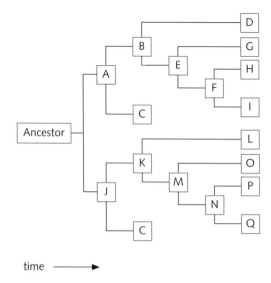

time ———▶

Figure 3.6 Individuals and species both have lineages. Here, the boxes can be taken to represent the radiation of species from a single ancestral species over time.

Figure 3.5 shows how they can be used to identify individuals that share a common ancestor. Mutations that affect protein structure, on the other hand, may be beneficial, resulting in an adaptation that might occur independently in several lineages. Consequently, they are less reliable as indicators of descent.

3.1.3 Ancestral and descendant forms are connected by an unbroken series of intermediates.

Individuals and species both have lineages. After a sufficient lapse of time, descendants may be so greatly modified from their ancestors that they are recognized as belonging to a different species. Consequently, we can in principle draw the lineage of a focal species either forwards or backwards in time. A species lineage of this sort can be thought of as a shorthand way of representing the combined lineages of the much larger number of individuals that lived during some long period of time. Provided that there is no hybridization, the lineages of individuals in different species are as separate as the lineages of individuals descending from asexual ancestors. Consequently, a diagram connecting ancestors with descendants, such as Figure 3.6, can be used to represent the radiation of species from an ancestral species just as it can be used to represent the expansion of a clonal population from an ancestral individual.

The flow of genetic information from ancestor to descendant must necessarily be unbroken from the founder of a lineage to its current members. Moreover, evolution is very generally cumulative, with each new adaptation building upon and modifying the previous one. Hence, a continuous series will have existed of intermediate forms linking any group of organisms with their ancestors. In experimental populations this can be demonstrated directly, with the "fossil record" of an evolving lineage stored in the freezer and full genetic information accessible. In the field, however, it is usually difficult to identify even a few of these intermediates, because only a small fraction of individuals become fossilized, and of these only a small fraction are ever found.

Almost all lineages are incompletely known. Even for individuals, very few lineages are known completely for more than two or three generations; examples might include closed flocks of chickens on experimental farms, and some of the European royal houses. The pedigrees of more obscure families are usually impossible to work out in detail, because the necessary documents never existed or have been irretrievably lost. An immense effort has been made to investigate the life of William Shakespeare, for

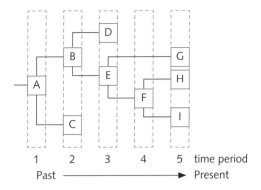

Figure 3.7 Almost all lineages are incompletely known.

impossible to know whether any species is directly ancestral to another. In this example, species B is in fact directly ancestral to species H whereas species C and D are not, but in any real situation we would have no way of knowing this. On the other hand, all of these species belong to the lineage forwards-in-time of their common ancestor A. The fossil record should always be interpreted in this way. Given a geological sequence of fossils that documents the transition from one kind of animal to another, we may be confident that all of them are members of the same lineage, in the sense of all belonging to the lineage forwards-in-time of a recent common ancestor, but we never know that one species is directly ancestral to another, and in fact this will seldom be the case. This limits our knowledge of the precise sequence of events involved in evolutionary transitions. Nevertheless, in organisms that fossilize readily and have been intensively studied it is possible to work out in some detail how major groups evolved.

example, but all we know of his ancestry is a few facts about his father and mother and his paternal grandfather. Almost all lineages are very incompletely known, with more blanks than names.

When we are attempting to work out an evolutionary lineage from the fossil record the names will refer to species rather than to individuals, but the rule remains the same—more blanks than names. To understand the implications of this, look at the species lineage illustrated in Figure 3.7. All of the species are imagined to have existed at different times in the past, although only a few have been preserved as fossils. These are recognized as being related to one another by their similarity, but the fossil record is normally so imperfect that it is impossible to trace any lineage backwards-in-time. Consequently, it is equally

Mosasaurs illustrate medium-scale evolutionary change. A good example of a well-documented evolutionary transition involving major changes in structure is illustrated in Figure 3.8. It is provided by the mosasaurs, marine reptiles that flourished towards the end of the Cretaceous period, 90–65 Mya. They were very large, active carnivores that grew up to 15 m long. When their fossils were first found, over 200 years ago, they were thought to be crocodiles

Varanid lizard	Aigialosaur	Mosasaur

Hinge in lower jaw	Hinge in lower jaw	Hinge in lower jaw
Vertebrae with oblique articulation	Trunk vertebrae with oblique, tail vertebrae vertical articulation	Vertebrae with vertical articulation
Feet digitate	Feet digitate, smaller	Feet paddle-like

Figure 3.8 Mosasaurs illustrate medium-scale evolutionary change.

or whales, but it was soon realized that they were most similar to lizards, and especially to the monitor lizards that still live in tropical countries.

Monitor lizards belong to the family Varanidae, and are related to alligator lizards and the Gila Monster of Arizona. These are collectively known as anguimorph lizards, and the earliest known members of the group lived in the Upper Jurassic period about 155 Mya, although there is then a long gap in the fossil record before the appearance of varanids in the Cretaceous. Varanids are large lizards with well-developed limbs and a long muscular tail. They can run fast in short bursts, and often attack relatively large prey such as rodents, birds, frogs and other lizards. To accommodate a large prey item, the bones in the middle of each half of the lower jaw are only loosely articulated, allowing the jaw to bow out on either side.

Aigialosaurs were amphibious lizards that evolved from an anguimorph lineage at the base of the Upper Cretaceous period, about 95 Mya. They are known to be related to varanids because both have trunk vertebrae with an oblique articulation, unlike other lizards. The tail vertebrae of aigialosaurs have a vertical articulation, however, and the zygapophyses (the pegs that hold the vertebrae together) are reduced in size; these modifications would allow the tail to be flexed more powerfully when used in swimming. The limbs have become smaller, although there is little change in their structure. The skull is longer, and the loose articulation in the middle of the lower jaw has become a true hinge that would have allowed much larger prey to be manipulated. These animals mark the beginning of the transition to an aquatic way of life.

The first mosasaurs appeared about 88 Mya. They show numerous adaptations to life at sea and almost all are obligately aquatic animals. The pectoral girdle is less strongly ossified, and the digits become flattened to form a paddle-like limb. This is used for maneuvering rather than for locomotion, and the orientation of the limb has changed to allow it to be held close to the body, while the structure of the elbow joint has changed so that the paddle can be used as a control surface. The articulation of the trunk vertebrae has become vertical (like a fish), while the tail is broad and flattened, with enlarged hemal spines supporting a caudal fin that is chiefly responsible for locomotion. The lower jaw hinge is retained, and the articulation of the tips of the two

halves of the lower jaw becomes looser, allowing yet larger prey to be swallowed. Moreover, there are now curved teeth on the roof of the mouth, as well as along the edges of the jaws, so that large prey can be inched back in the mouth to the point where they can be swallowed. The very earliest mosasaurs do not exhibit all of these changes: *Dallasaurus*, for example, which lived about 92 Mya, has a generally mosasaur-like skeleton but retains aigialosaur-like limbs that would permit movement on land.

Later in the Cretaceous, mosasaurs evolved into highly adapted deep-sea predators that lived far from land and gave birth to live young at sea. They must have dived to considerable depths, as some fossil skeletons show the characteristic necrosis caused by nitrogen bubbles, the cause of the diver's "bends." The paddle evolved by a shortening of the humerus, changes in musculature, and the addition of extra phalanges (finger joints). Body size increased, and some forms became very large: *Mosasaurus* is the largest known marine reptile, with an overall length of 15–17 m, including a skull over a meter long. This presumably allowed it to swallow almost anything it met, and the complex lower jaw hinge of its ancestors has become non-functional. It lived towards the end of the Cretaceous period, 65 Mya. At this point a diversity of mosasaurs had evolved, but all were victims of the great mass extinction event at the Cretaceous–Tertiary boundary. There are no mosasaurs alive today, but about 35 species of monitor lizards are still found in tropical Africa, southern Asia, and Australia.

Whales illustrate large-scale evolutionary change. The mosasaur lineage shows how a terrestrial lizard can evolve into a fully marine form in a fairly short period of time. The varanid ancestry of mosasaurs is clearly apparent, however, and except for the limbs the structural changes that took place are not very profound. Much more sweeping changes were involved in the evolution of another secondarily aquatic group of tetrapods, the whales, which we have described briefly before (Section 1.3.2). Representatives of the lineages leading from ancient terrestrial mammals to modern whales, with some of the structural modifications that evolved in these lineages, are illustrated in Figure 3.9.

Modern whales, dolphins, and porpoises are a clearly demarcated group of mammals that are highly

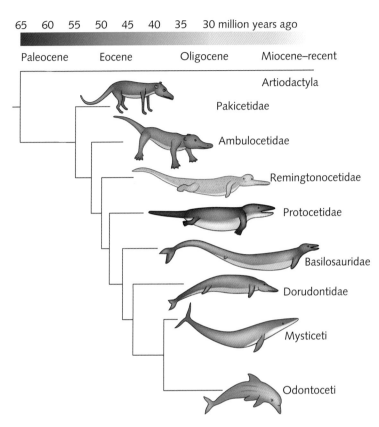

	Double-pulley ankle	Paddle-like forelimbs	Hind limbs absent	Tall fluke	Blow hole at rear	Lateral eyes	Mandibular hearing	Undifferentiated teeth
Pakicetidae	+	–	–	–	–	–	–	–
Ambulocetidae	+	+/–	–	–	–	–	+/–	–
Remingtonocetidae	+	+/–	–	–	–	+/–	+	–
Protocetidae	+	+/–	–	–	+/–	+/–	+	–
Basilosauridae	0	+	+/–	+/–	+/–	+	+	–
Dorudontidae	0	+	+/–	+	+/–	+	+	–
Mysticeti	0	+	+	+	+	+	+	+
Odontoceti	0	+	+	+	+	+	+	+

Figure 3.9 Whales illustrate large-scale evolutionary change.

Tree by Felix G. Marx, University of Bristol. Images of cetaceans adapted from National Geographic's *The evolution of whales* by Douglas H. Chadwick, Shawn Gould, and Robert Clark. Re-illustrated for public access distribution by Sharon Mooney ©2006. Open source.

adapted for life at sea. They do not much resemble any other group of mammals, and there are no living intermediate forms, so their origin and evolution have long been mysterious. Recent fossil discoveries have made it possible to reconstruct whale evolution in some detail. We will follow it backward in time, beginning with a familiar modern species, the sperm whale. It is a gigantic deep-diving squid predator whose trade-mark cranial bulge houses an apparatus that produces high-intensity sound that is used to locate and perhaps to stun its prey. Whales do not hear sounds in the same way as terrestrial mammals. We use an external ear bearing a tympanum that vibrates in response to sound waves in air, setting in motion a series of small bones that transmit the vibration to the middle ear. This does

not work very well under water because the external ear fills with water, creating a pressure difference across the tympanum. Sound waves under water are readily transmitted by bone, however, whereas bone reflects sound waves in air. Sperm whales hear through their lower jaw. In all mammals, the nerves and blood-vessels of the teeth on the lower jaw pass through a canal called the mandibular foramen. In sperm whales this foramen is very large and houses a fat pad that extends to the middle ear. Sound waves are conducted by the bone of the lower jaw to this fat pad, which passes the vibrations through the mandibular foramen to the middle ear, where they are received by the tympanic bone and passed on to a lever-arm system similar in principle to that of terrestrial mammals. The same kind of hearing is possessed by other kinds of toothed whale, such as dolphins and beaked whales. Sperm whales have several other distinctive morphological features; for example, they have no teeth on their upper jaw, with simple peg-like teeth for catching squid on the lower jaw only.

The earliest fossil sperm whales lived in the early Miocene period, with similar forms extending back to the middle Oligocene, about 30 Mya. They are numerous and diverse in the Miocene and Pliocene, together with dolphins, beaked whales, and other toothed whales. Another kind of tooth-bearing whale first appeared in the late Eocene, about 33 Mya. These animals have an expanded maxilla that forms the origin of the baleen plates in their descendants, modern forms such as right whales and rorquals. They have a completely different mode of feeding, in which small crustaceans are trapped by the baleen plates as the whale swims through the surface waters of the sea, scraped off by the tongue, and passed down the narrow gullet. The adults are completely toothless, but teeth appear during development, as we would expect (Section 1.4.1), and the fossil evidence shows how the lineages of toothed whales and baleen whales separated in the late Eocene.

Shortly before this, dolphin-like animals called dorudontids appeared in the middle Eocene, about 30 Mya. The reconstruction in Figure 3.9 shows that they were unmistakably whales, with paddle-like forelimbs, minute hindlimbs, a tail fluke, a nasal opening ("blowhole") towards the rear of the skull, laterally placed eyes, and a mandibular foramen specialized for sound conduction. They are not quite like modern whales, however. The nasal opening is not as far posterior. The hindlimbs are strongly reduced and the pelvic girdle is absent, but most of the bones of the limb are still present, whereas in modern whales they are reduced to a few simple rods embedded in the body wall. The teeth are differentiated in the normal mammalian fashion, rather than being simple uniform tusks or pegs. Basilosaurs are similar except for their extremely elongate bodies. These animals are clearly highly specialized for an obligately marine life and include or are closely related to the ancestors of modern whales.

Many of the characteristics of these archaic whales are found in protocetids, marine mammals which lived about 37 Mya. They have elongate skulls with lateral eyes roofed by a large flat projection of the frontal bone, a blowhole, and a large mandibular foramen. In other respects, however, they are less highly specialized for open-water swimming. In particular, they have relatively large hindlimbs and a fully ossified pelvic girdle, and could almost certainly have supported their weight on land. The lateral digits of forelimbs and hindlimbs were not weight-bearing, and presumably served as the framework for webbed hands and feet. They had no tail fluke, and instead swam by paddling with the hindlimbs. The ankle has one very surprising feature. It consists of a series of small bones that are grooved so as to form a double pulley for the muscles to move the foot. This is a unique feature of artiodactyls, the even-toed ungulates such as horses and deer, where it facilitates rapid locomotion. Since it is very unlikely that such a distinctive structure would have arisen independently, this shows that whales share a common ancestor with artiodactyls (see Section 3.3.1).

Ambulocetids are similar animals that lived about 38 Mya. They had large powerful hindlimbs that could be flexed by the spinal muscles to provide a power stroke when swimming. They could also walk on land, however, and were amphibious rather than primarily aquatic. Their remains are found in shallow marine deposits with fossils of marine plants and littoral mollusks, showing that they lived close to the shoreline. They have many whale-like features, including an expanded (but not greatly expanded) mandibular foramen, together with an artiodactyl-like ankle structure. These animals represent one of the first stages in the transition from terrestrial animals to fully aquatic forms.

The pakicetids are found in riverine deposits from the same area, dated to about 52 Mya, in association with freshwater fish and terrestrial mammals such as opossums. The structure of the ear bones and teeth are similar to those of later whales, but they had long thin limbs and short hands and feet, with no indication of adaptation for swimming. They represent the earliest known phase in the evolution of whales, an almost completely terrestrial animal about the size of a large dog. Their antecedents cannot be traced; it is possible that they descended from some extinct group of artiodactyls such as the mesonychids, hoofed carnivores common in the Eocene, or from a yet undiscovered common ancestor of whales and some other artiodactyl group.

3.2 Lineages are tree-like or web-like.

The series of parents and offspring that constitutes a lineage may be generated in two different ways. In an asexual population every lineage is isolated from other lineages, and the relationships between ancestors and descendants form a branching tree. The same applies to sexually isolated species. Sexual parents, on the other hand, mingle their genomes in producing offspring, giving rise to a web-like pattern of ancestry. Hybridization between species would produce a similar pattern in a phylogenetic tree. Hence, lineages vary between strictly tree-like and entirely web-like, with any intermediate state being possible according to the degree of mingling of lineages during descent.

3.2.1 Any two lineages have a most recent common ancestor.

Sister species have an exclusive recent common ancestor. Figure 3.10 shows both the common ancestry of whales with artiodactyls and the relationships among different species of whale. Evidently the blue whale and the fin whale are

very similar, judging by the genes that were used to estimate the tree, and are likely to be closely related. In fact they are more closely related to one another than to any of the other species in the diagram. (It is always possible, of course, that there is another species, not included in the survey, that is even more closely related to the blue whale than the fin whale is. Any statement about relationships refers to the particular group of species that has been studied.) What we mean by being more closely related is that the blue whale and the fin whale have a recent common ancestor that is not the ancestor of any of the other species in the diagram. This is represented in the diagram by the convergence of their lines of descent to form a lineage forwards-in-time that does not include any other species. We summarize this by saying that the blue whale and the fin whale are sister species.

What is then the sister species of the North Atlantic minke whale? It is not the blue whale or the fin whale, since these are sister species to one another. Nor is it the gray whale, since the most recent common ancestor of the gray whale and the minke whale is also the

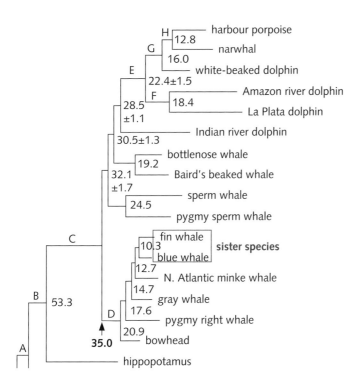

Figure 3.10 Sister species have an exclusive recent common ancestor. Numbers are molecular estimates of divergence times in My.

Reprinted from Arnason et al. Mitogenomic analyses provide new insights into cetacean origin and evolution. *Gene* 33: 27–34 © 2004, with permission from Elsevier.

ancestor of the blue whale and the fin whale. Instead, the sister taxon of the minke whale is a group of two species, {blue whale + fin whale}. Similarly, the sister taxon of the gray whale is a group of three species, {blue whale + fin whale + minke whale}.

Sister taxa can be defined at any level. For example, Figure 3.11 shows how the same phylogenetic tree identifies the sister taxon of the baleen whales (blue, fin, minke, gray, right, and bowhead whales) as the toothed whales (sperm, beaked, and bottlenose whales, plus dolphins and porpoises). In short, any two lineages have a most recent common ancestor. If this is not the ancestor of any other extant lineage, the two lineages are sister taxa. This gives us a simple and unambiguous way of describing how any two lineages are related to one another.

3.2.2 Evolutionary diversification is a branching process.

Nesting sister taxa produces a tree of descent. The gene sequences show that about 10 Mya a population of an unknown species of whale became divided into two groups which evolved into two modern species, the blue whale and fin whale. Perhaps more than one

division took place, but if so all the other species have become extinct, leaving the blue and fin whales as the only surviving lineages. The diversity of whales, and all other organisms, is ultimately generated by this process of lineage-splitting, or speciation. The process of speciation is described in Chapter 7.

More ancient speciation events gave rise to lineages which themselves went through further rounds of speciation to give rise to entire groups of species. Thus, about 35 Mya the population that was the most recent common ancestor of all modern whales divided into two lineages, one of which gave rise through successive episodes of speciation to the baleen whales and the other to the toothed whales. The origin of major divisions in this way is called cladogenesis.

Cladogenesis gives rise to a tree of descent, also called a phylogenetic tree. Figure 3.12 is an estimate of this tree based on gene sequences for selected whales and ungulates. The tree shows the lineage forwards-in-time of the most recent common ancestor of {whales + ungulates}. The phylogenetic tree is a set of nested sister taxa that describes the relationship between any two species or any two lineages. This allows us to arrange species in a natural way, in terms of their descent, rather than using some artificial

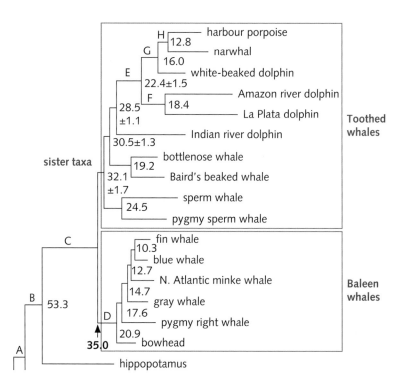

Figure 3.11 Sister taxa can be defined at any level.

Reprinted from Arnason et al. Mitogenomic analyses provide new insights into cetacean origin and evolution. *Gene* 33: 27–34 © 2004, with permission from Elsevier.

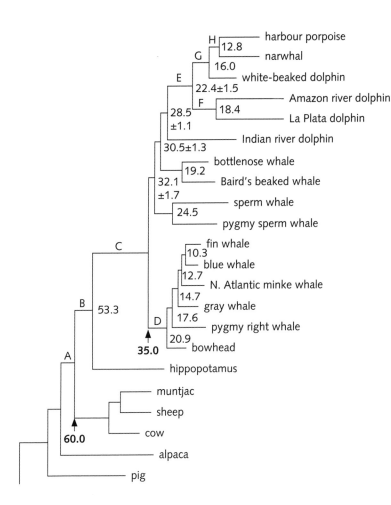

Figure 3.12 Nesting sister taxa produces a tree of descent.

Reprinted from Arnason et al. Mitogenomic analyses provide new insights into cetacean origin and evolution. *Gene* 33: 27–34 © 2004, with permission from Elsevier.

criterion of similarity. The use of phylogenetic trees in classification is discussed in more detail in Section 3.3. Thus, "baleen whales" and "toothed whales" are more informative categories than "big whales" and "small whales" because they reflect the history of the animals themselves rather than an arbitrary distinction. The phylogenetic tree gives a clear depiction of evolutionary diversification as a branching process giving rise to lineages consisting of species.

3.2.3 Sexual populations form a web: sexual species form a tree.

Representing evolutionary history as a phylogenetic tree makes one strong assumption: lineages can divide but cannot merge. This will only be true for entities that neither fuse nor recombine. The most important of these are:

- lineages of individuals within strictly asexual populations

- lineages of species which are sexually isolated from all others

- lineages of genetic elements in sexual species that do not recombine.

These all represent special cases, and in practice some degree of fusion and recombination will often occur. The relationships between entities will then resemble a web to some extent rather than a strictly branching tree.

Sexual lineages involve complex patterns of descent within species. The main difficulty of interpreting lineages in sexual populations will be evident immediately you try to draw your own lineage. The simple lineage in Figure 3.13 extends over five generations, with the author of this book as the focal individual. It is complete, in the sense that everyone in the direct line of descent is represented. However, it does not include other kin, such as aunts and nephews,

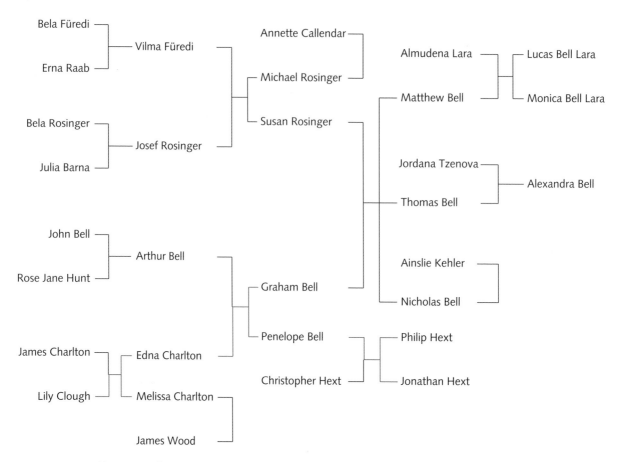

Figure 3.13 Sexual lineages involve complex patterns of descent within species.

and if it did it would quickly become very complicated. This is because everyone has two parents in a sexual population. An asexual lineage is much simpler. For example, an asexual lineage backwards-in-time has a single ancestor per generation, whereas everyone in a sexual population had, in principle, 2^G ancestors G generations ago (2 parents, 4 grandparents, 8 great-grandparents, and so forth). In practice this cannot be strictly true; otherwise the number of your ancestors 20 generations ago would be greater than the total human population at that time. This is the consequence of assuming that each of your ancestors contributed to your ancestry through a single chain of descendants (as they would in an asexual population). Since this cannot be the case, some of your ancestors must be related to you along several lines of descent. This is the same as saying that if your lineage were extended backwards in time far enough, we would find at some point that your mother and father shared a common ancestor.

This principle can be illustrated by an even more distinguished evolutionary biologist: Charles Darwin married Emma Wedgwood, who was his first cousin, and the couple produced ten children. Since they had both grandparents in common, each of their grandparents was related to each of their children by two routes—through Charles' mother, and through Emma's father, as traced in Figure 3.14. This complex interweaving of lines of descent has important implications for evolution in sexual populations, which we will explore in Chapter 15.

Sex carves out trees because it suppresses gene exchange. Species remain distinct because their members do not interbreed. Hence, although the lineages of individuals within sexual populations are web-like, the lineages of the sexual species themselves are tree-like when estimated from gene sequences. In fact, the lineages of sexual species behave like the

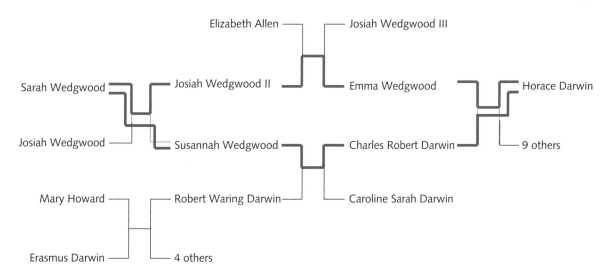

Figure 3.14 Charles Darwin married his first cousin Emma Wedgwood; consequently, each of their children (such as Horace) is related to a maternal grandparent (such as Sarah Wedgwood) through two paths of descent, labelled blue and red. Remarkably, Josiah Wedgwood III and Caroline Sarah Darwin also married.

lineages of individuals within an asexual population. In practice this is often not completely true, because of hybridization. The species lineage is then no longer strictly branching. Moreover, using different genes may yield somewhat different estimates of the tree if some are recombined more frequently than others. Low levels of hybridization, however, will not greatly affect the tree-like character of species lineages.

It is important to distinguish the phylogenetic tree from other kinds of tree diagrams, which may be used to express the overall similarity among objects, the administrative structure of an organization, or the hierarchy of moves possible in a game. The phylogenetic tree is not used for any of these purposes. It is used solely to use common ancestry as the criterion for expressing relatedness. The key concept is that of sister taxa: two taxa are more closely related to one another than either is to a third taxon if they have a recent common ancestor that is not shared by that third taxon.

The ancestry of bacteria is a web because they exchange genes. Bacteria reproduce asexually and hence a bacterial lineage should strictly be branching and treelike. This is often the case. It came as a surprise, however, to learn that different strains of the same kind of bacterium may have radically different genomes. Different strains of the common gut bacterium *E. coli*, for example, may share fewer than 30% of their genes.

There are two main reasons for such variation in bacterial genomes. The first is that cells can often take up DNA from the environment, where it has leaked out from dead bacteria, and incorporate it into their own genome. Some DNA elements, in fact, have evolved to be used in this way: they contain sequences that facilitate uptake and integration, plus a structural gene and a promoter, so they work right away. They spread because they enhance the growth rate of the cells they enter. The second reason is that other DNA elements are able to manipulate the behavior of the cell they are in causing it to fuse with another cell and transfer a copy of the element to it. These may bear genes that are useful to the cell (such as genes conferring resistance to antibiotics) or they may just be successful genetic parasites. They are described further in Chapter 11. In either case, a relatively high rate of horizontal transmission of genetic material between cells creates mosaic genomes in which genes have been acquired from a range of sources. Moreover, these routes of horizontal transfer may move genes between very distantly related bacteria, unlike sexual exchange between individuals of the same species. There is a core genome consisting of genes that are seldom if ever transmitted horizontally and whose sequences will yield consistent estimates of a branching, tree-like phylogeny. Other genes, however, are frequently shuffled among lineages, yielding inconsistent phylogenies and a

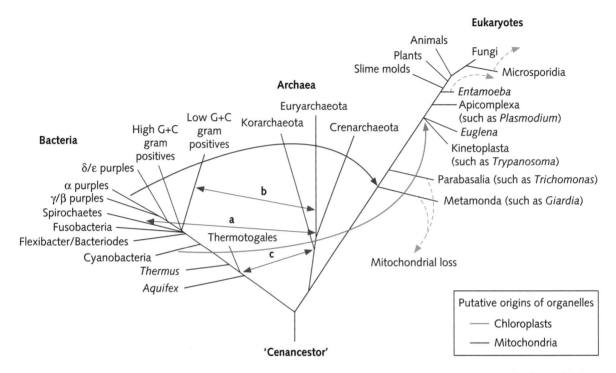

Figure 3.15 The ancestry of bacteria is a web because they exchange genes. Green and purple lines represent transfer of plastid and mitochondrial lineages to eukaryote cells. Broken lines indicate loss of mitochondria from some eukaryote lineages. Double-headed arrows show transfer of genes between Bacteria and Archaea.

Reprinted by permission from Macmillan Publishers Ltd: James R. Brown, Ancient horizontal gene transfer. *Nature Reviews Genetics* 4 (2) © 2003.

web-like pattern of relatedness. These processes are discussed in more detail in Section 6.2. Figure 3.15 shows how whole genomes can be transferred to very distantly related taxa, sometimes resulting in the evolution of quite different kinds of organism.

The origin of eukaryotes in this fashion from bacterial ancestors is discussed in Section 8.2. Clearly, evolutionary processes in bacteria may be different to those in animals and plants.

3.3 Organisms are classified by relatedness, not by similarity.

Naturalists have been classifying animals since the very beginnings of science. The Greek philosopher Aristotle, who lived over two thousand years ago, divided all animals into two large groups, depending on whether or not they had red blood. After the rebirth of scientific biology many centuries later, more and more complex schemes of classification were invented. All of them relied on some set of features (such as red blood) to group together similar species. This has a grave weakness: the choice of features to assess similarity is always subjective. With the advent of evolutionary biology in the mid-nineteenth century, classification was revolutionized by the use

of relatedness, rather than similarity, as the basis for grouping species together. The fact of evolution makes it possible to construct a natural classification that expresses the genealogical relationships among all kinds of organisms.

3.3.1 The criterion of classification is common ancestry.

Suppose that you were given a range of tools, and asked to put them into categories. You would probably choose categories that were convenient in some way. For example, you might put machine tools (like chainsaws and lathes) into one category, hand tools (like hammers and screwdrivers) into another, and consumables (like nails and screws) into a third. Alternatively, you might decide to separate gardening tools, construction tools, household tools, and so forth. It would depend on the motive you had for creating the categories, as the owner of a store, for example, or simply as someone arranging a garage. No single scheme has clear priority over another.

Now, here is a list of animals: how should we classify them?

Mosasaur

Whale

Chameleon (a lizard)

Frog

Deer

There are many ways of grouping them, and each will have some merit. Mosasaurs and whales are aquatic animals, for example, whereas lizards, frogs, and deer live (mostly) on land. Alternatively, frogs have no tail as adults, whereas the others do. Alternatively, mosasaurs are extinct whereas the others are not. Any such classification may have some utility—whether you are likely to meet one of these animals when diving, for example. But there is no single classification based on similarity which is better than all others.

A natural classification expresses the evolutionary relationships of groups. It is based on the fact that organisms are unlike tools in one crucial respect: any two have descended from some common ancestor, more or less remote in time. There is therefore a single universal scheme for classifying all organisms: two organisms belong in the same category, excluding a third, if they are more closely related to one another than either is to the third. This is a natural system of classification because it reflects descent rather than similarity. It places me in the same category as my sister, who is shorter and of different gender, to the exclusion of my friend, who is of about the same height and of the same gender.

Natural classifications can be expressed by phylogenetic trees, as explained in Section 3.2. Here is the correct phylogenetic tree for the five vertebrates on the list:

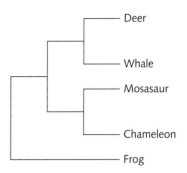

For example, we have seen that mosasaurs evolved from terrestrial lizards (Chapter 1 and Section 3.1.3). There are three meaningful groupings in the phylogeny:

(a) deer with whale (both are mammals)

(b) mosasaur with chameleon (both are lepidosaurs)

(c) deer, whale, mosasaur, and lizard together (all are amniotes).

Provided these groupings are retained, it does not matter how we draw the diagram. For example, this diagram is exactly equivalent to the first diagram:

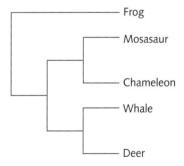

Take a moment to check that all three groupings have been retained and no new ones added. The second diagram therefore conveys the same phylogenetic information as the first. Note that this is true even though the order of the names from top to bottom has changed completely.

3.3.2 The basis of evolutionary classification is the monophyletic group.

Using the concepts that we developed in Section 3.2, we can state the general principle of natural classification in more formal terms: organisms are classified in the same group only if they are members of the same lineage forwards-in-time. By itself, however, this principle does not lead to an unambiguous classification. To make this clear, here is the same phylogeny with labels for the three most recent common ancestors of the three natural groups:

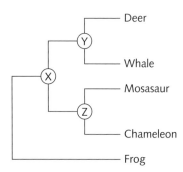

First, suppose that we include chameleon, whale and deer in one group (all are extant) and mosasaur in another (extinct). Chameleon, whale, and deer certainly evolved from the same common ancestor X. Mosasaurs also evolved from X, however, and there is no phylogenetic reason to exclude them from the group of {chameleon + whale + deer}.

Secondly, suppose that we include mosasaur and whale in one group (marine), and chameleon and deer in another (terrestrial). Mosasaurs and whales again share the common ancestor X. Mosasaurs also have an ancestor Z, however, which is also the ancestor of chameleons. Likewise, whales have an ancestor Y which is also the ancestor of deer. Hence, there is no phylogenetic reason to exclude either chameleons or deer from the group of {mosasaur + whale}.

To avoid these inconsistencies, we need a stricter criterion for membership of a natural group. A *monophyletic group* comprises all the descendants of a common ancestor. This simple and unambiguous definition is the basis of the natural classification of organisms of all kinds.

Fortunately, we do not need to list all the descendants of every common ancestor in order to arrive at the correct natural classification. Rather, all the species that we are interested in should be placed into monophyletic groups. Adding more species will create more groups, but those already identified will continue to be valid.

Many familiar groups are not monophyletic and are therefore not valid natural groups.

- Fish: tetrapods evolved from lobe-finned fish but are not included in the group.
- Reptiles: birds share a common ancestor with dinosaurs (and more remotely with crocodiles) but are not included in the group.
- Homeotherms (warm-blooded animals): the common ancestor of mammals and birds was also the ancestor of lizards, snakes, and crocodiles, which are not included in the group.

We can continue to refer to "fish", "reptiles," and "homeotherms" when it is convenient to do so, but they are not monophyletic groups and are no longer used in classification.

3.3.3 Related taxa share derived characters.

It is clear why we should use monophyletic groups as the basis for an evolutionary classification. But how can we recognize them? To explain how this can be done, we return to the WordChange game. The initial word is still "word," but there are now several outcomes: "barn," "cork," "dare," "darn," and "bare." How should we classify them on evolutionary principles?

Here is one solution:

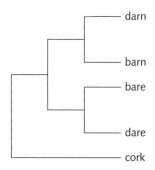

This shows how the four letters of each word are used quite differently in working out the phylogenetic tree. Bear in mind that the goal of the classification is to identify the relationships among groups of organisms—or, in this case, words.

- The letter "r" occurs in the same position in all the words. This is because it has been retained from the ancestral word, "word." It is a primitive character shared by all descendants ("primus" means "first" in Latin). A shared primitive character cannot tell us anything about relationships.

- The letters "b" and "d" distinguish "barn" from "darn" and also distinguish "bare" from "dare." They serve only to identify differences between words at the highest level of classification. They are unique derived characters that tell us nothing about the relationships among groups ("derived" means that they have changed from the primitive, or ancestral, state).

- The letters "o" and "a" distinguish one group from another: cork from {bare + dare + barn + darn}. Likewise, "e" and "n" distinguish groups at a higher level: {bare + dare} from {barn + darn}. These are shared derived characters: shared, because members of the same group have them in common, and derived, because they differ from the ancestral state.

The conclusion is that *only shared derived characters are useful in building a phylogenetic tree.* To confirm this, run the problem through WordChange and confirm that you almost always find the pathway illustrated in Figure 3.16. This shows how shared derived characters signpost the route of evolution and therefore provide the basis for working out the phylogenetic tree.

After a little thought, however, you might also notice that "dare" and "darn" are very similar, as are "bare" and "barn." This suggests an alternative classification:

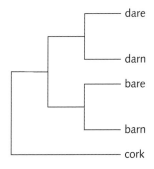

This also uses shared derived characters as the basis for classification, although this time they are different: "d" and "b" are phylogenetically informative, whereas "e" and "n" are not. Why is this not just as good as the first attempt? Figure 3.17 shows the evolutionary reconstruction of the phylogeny.

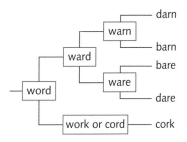

Figure 3.16 The pathways leading from the ancestral state "word" to the five derived states.

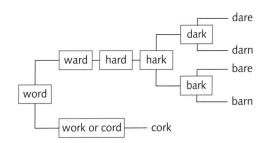

Figure 3.17 An alternative evolutionary pathway.

This classification is equally justified, on the basis of shared derived characters. It differs in only one respect. Count up the number of changes in the two schemes that are needed to produce the end result. Nine changes are needed in the first case; eleven in the second. The first phylogeny is therefore simpler, in the sense that we need to invoke fewer evolutionary changes in order to produce the observed result. In general, we prefer the simpler hypothesis over the more complex. This is the principle of least evolutionary change, or, for short, parsimony. It is justified in real phylogenies when change is rare, so that two or more changes are much less likely to occur than a single change.

3.3.4 Relationship can be inferred from character state.

We have identified three core principles for classifying organisms according to their evolutionary relationships:

- Only monophyletic groups are recognized.
- Only shared derived characters are phylogenetically informative.

- The number of evolutionary steps should be minimized.

How can we use these to classify our set of five real organisms? We first need to know the state of a series of characters in each of the organisms concerned. We can use any characters we wish, but to begin with we shall use morphological characters which can take only one of two possible states, either present or absent. Here is a list of eight characters that are easily scored for the five organisms we have chosen.

A. Limb with digits

B. Egg with amnion

C. Upper temporal fenestra (an opening in the temporal bone of the skull)

D. Single bone in lower jaw

E. Hinge in lower jaw

F. Extensible tongue

G. Ruminant gut

H. Vestigial hindlimbs

Each of these characters is either "present" or "absent" in each taxon, as shown in Figure 3.18. We can now use this information in one of two ways.

- First, we might classify the organisms by using some measure of overall similarity. For example, we could count up all the characters that are "present" and use this total to find groups whose members are more similar to one another than they are to members of other groups. This is called a *phenetic* approach, because it is based solely on phenotypic resemblance, and the tree diagram it leads to is called a phenogram. This does not consistently reflect the real course of evolutionary descent, because a given character state may evolve independently in two or more different lines of descent. For example, if we used "large/small" and "terrestrial/aquatic" as two-state characters, we would classify chameleon and frog in one group, mosasaur and whale in a second group, and deer in a third group.

- Secondly, we could classify the organisms into monophyletic groups by using shared derived character states as evidence of ancestry. This is called the *cladistic* approach, because "clade" is another word for a monophyletic group. A phylogenetic tree inferred by cladistic analysis is often called a *cladogram*. The problem is to work out how to use character state as an indication of descent.

To begin with, what is the ancestral state of each of the characters? In the WordChange game, we were given "word" as the ancestral state, but for real organisms we do not know what it is, and must find a way of inferring it. Take an animal that is known to be related to the group being classified but is not a member of it. Since all five are tetrapods, an appropriate

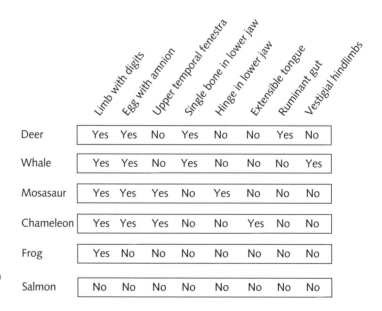

Figure 3.18 Character state represents modification through descent.

	Limb with digits	Egg with amnion	Upper temporal fenestra	Single bone in lower jaw	Hinge in lower jaw	Extensible tongue	Ruminant gut	Vestigial hindlimbs
Deer	Yes	Yes	No	Yes	No	No	Yes	No
Whale	Yes	Yes	No	Yes	No	No	No	Yes
Mosasaur	Yes	Yes	Yes	No	Yes	No	No	No
Chameleon	Yes	Yes	Yes	No	No	Yes	No	No
Frog	Yes	No	No	No	No	No	No	No
Salmon	No	No	No	No	No	No	No	No

choice would be a bony fish, such as a salmon. The salmon here serves as an "outgroup" that is not itself classified but is used to assist in the classification of the five tetrapods. All eight characters we are using have the state "absent" in the salmon (for example, it does not have limbs with digits). We infer that the most recent common ancestor of tetrapods and bony fishes had "absent" as the state of all eight characters. This enables us to reconstruct the sequence of evolutionary events, the order in which the groups we shall identify diverged from one another.

We can now proceed to interpret the characters.

- Character A (Limb with digits) has the same state in all five animals, and we infer that it was present in their most recent common ancestor. It is a primitive character state that does not help us to classify the animals.

- Each of the characters E–H takes the same state in all animals except one. They are unique derived characters that define a particular kind of animal but do not give us any information about its relationship to the others.

- The remaining characters B–D are shared derived characters that can be used to define monophyletic groups.

- Character B (Egg with amnion) is absent in the frog and present in all four others. It is a unique feature of frogs that distinguishes them from all the other animals in the group. Since we know that "absent" is the ancestral character state, however, it does not merely identify frogs as a particular kind of animal, but also shows that the first divergence in our phylogeny separates frogs from the others.

- Character C (Upper temporal fenestra) is a shared derived feature of mosasaur and chameleon that unites them in the same group. Likewise, character D (Single bone in lower jaw) is a shared derived character of whale and deer. This completes the phylogenetic classification by identifying {mosasaur + chameleon} and {whale + deer} as monophyletic groups, diverging from one another after their common lineage had diverged from the frog lineage.

The phylogenetic tree that we have inferred is shown in Figure 3.19. The sequence of changes in character

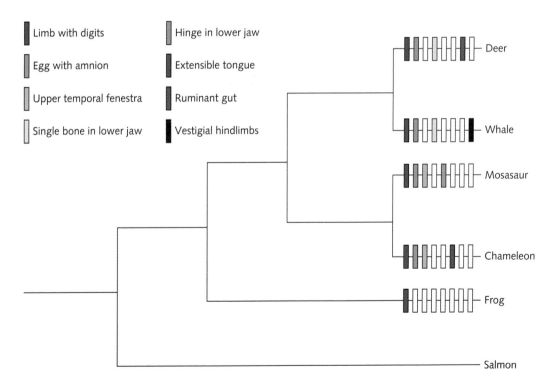

Figure 3.19 **Relationship can be inferred from character state.**

state during the evolution of the group is shown in Figure 3.20. You can confirm that this is a very parsimonious interpretation because it involves only seven changes in character state, the fewest possible.

Although cladistic analysis is simple in principle, it becomes very tedious (or entirely impracticable) to do by hand when more than about 10 species are involved. For larger data sets, computers are always used to derive phylogenetic trees. In the Supplementary Material for this chapter, you will find instructions for downloading and implementing free software for phylogenetic analysis. These include a worked example using the powerful software package PHYLIP to obtain a cladogram for the five tetrapods in our example. You can check that this gives the same result that we obtained by intuitive arguments. We urge you to go further and try more complicated examples, in order to improve your grasp of how evolution produces diversification.

If you could survey the whole process of evolutionary change, by directly sampling organisms over millions of years, then you could write out the phylogeny of a group without error—just as you can for artificial examples like WordChange. In practice you cannot, and must use cladistic methods to infer the real pattern of change. The cladogram that you infer from the character states of existing organisms is an hypothesis, not a fact. There are several reasons that a cladogram may not be an accurate portrayal of the real course of evolutionary change.

The first is that the method depends on the fundamental distinction between primitive and derived characters. We define these by reference to an outgroup known to be closely related to, but not a member of, the group we are analyzing. But how do we obtain this knowledge, since it implies a previous classification involving a more distant outgroup, and so forth? We said that the choice of bony fish as the outgroup for tetrapods was uncontroversial, but why are we so sure? One answer is that it is consistent with a large suite of other characters defining gnathostomes (vertebrates with jaws) which are shared by bony fishes and tetrapods. A better answer is that bony fish appeared in the fossil record before tetrapods, and a series of fossil forms connecting the ancestors of the two modern groups has been found.

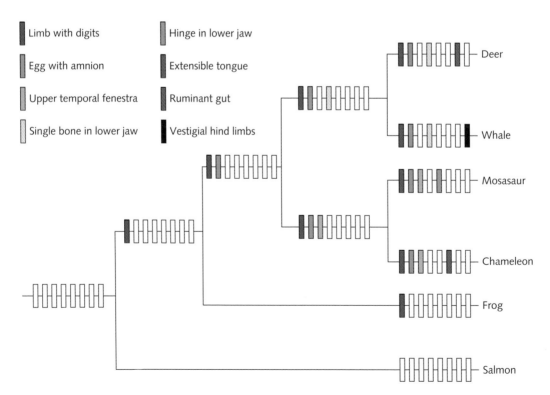

Figure 3.20 The sequence of change can also be inferred from character state.

It is a better answer because it supplies completely independent evidence of evolutionary descent. In this case it is indeed uncontroversial; but where this quality of evidence is not available the choice of an outgroup will be more difficult.

A second objection is that a particular character state may be derived independently in two or more lineages, notwithstanding our assumption that evolutionary change is rare. For example, suppose that one of the characters that we chose to classify our five animals was "flipper-like limbs." This would group whales with mosasaurs. We know that this is incorrect, both from a host of other characters and from the fossil record. We also know why it is wrong: flipper-like limbs evolve repeatedly as a device to scull or steer massive bodies through water. This is an example of parallel evolution: similar-looking structures evolve independently in different lineages in response to similar selective agents (Section 1.3.1). In this case, we can easily detect parallel evolution because we understand the dynamic principles involved in the locomotion of large animals in water. In other cases, the adaptive significance of a structure may not be as obvious, and parallel evolution will distort the phylogenetic analysis.

Thirdly, note that one character, "Vestigial hindlimbs," is really the loss of a character, "Functional pelvic girdle and hindlimbs." It occurs not only in whales, but also in snakes, legless lizards, and amphisbaenid amphibians, none of which are closely related to whales. This is another example of parallel evolution, but involving losses rather than gains: a complex evolved structure is lost when the way of life of a lineage no longer provides effective selection for its maintenance. It gives rise to similar difficulties for phylogenetic analysis.

Nevertheless, the independent gain of similar complex structures and the independent loss of such structures are surely not equivalent. Loss must evolve more readily than gain, since mutations that degrade a function no longer required are much more frequent than mutations that create a new function in response to changed conditions. Our notion of parsimony should be sharpened, so that gain of function is less likely than loss of function. In practice, this means that we should weight a shift from ancestral to derived state (for example, gain of limbs) much more heavily than the reversion of derived to ancestral state (for example, loss of limbs).

Finally, if you experiment with character sets you will very soon find that PHYLIP returns several or many equally parsimonious trees, not just one. You can choose between them only by adding more information (more characters), or, better, by seeking independent evidence (preferably, fossil sequences). If this is not practicable, then it is simply not possible to settle on any particular tree as the basis for an unequivocal classification.

These reservations lead to a simple conclusion: the phylogenetic tree for a particular group that we have derived from observations of character states, manually or by computer, is an hypothesis about the process of evolution that is continually under revision, by adding further characters, or by adding further species, or by refining analytical techniques. Some relationships are already so strongly supported that they are unlikely to change much; others shift a little, while leaving the basic structure of the tree little changed; others are much more volatile and will be finally resolved only when more data or better techniques become available. Mapping the tree of life is a progressive and absorbing project.

3.3.5 The ultimate character is gene sequence.

The continuity of the germ line enables us to map lineages. The somatic cells of an individual differentiate into muscle, nerve, or organs, function for a while, and then die when the individual dies, leaving no descendants. The sole exception is the germ line, which forms gametes and gives rise to the next generation. Most gametes are unsuccessful, of course, and die like somatic cells. All of the individuals in the next generation, however, descend from the successful minority of germ cells. In this sense, the germ line is potentially immortal, since the germ cells are the physical basis of the lineage. When we recognize a lineage, indeed, we are recognizing some kind of continuity between generations. This is not a continuity of material, because the germ cells develop into a body that subsequently makes a new crop of germ cells. Rather, it is a continuity of information: what is transmitted through the germ cells from one generation to the next is the information for making new individuals, primarily through the chemical instructions encoded in the genes.

This information is necessarily passed on every time individuals reproduce. It is passed on from parents to children, and from great-great-grandparents through four generations to the great-great-grandchildren they will never see. Because this flow of genetic information can never be interrupted, it can be traced back beyond our own family, and even beyond our own species, into the distant past. Hence, all living organisms belong to the same single universal lineage, because all descend through an unbroken series of intermediates from the same common ancestor.

Although genetic information is always transmitted from one generation to the next, however, it is usually transmitted imperfectly. The genomes of offspring are not perfect copies of the genomes of their parents, and the alterations that have occurred are themselves transmitted to future generations. The history of a lineage is thus recorded in the genes of its members. Now that we can sequence genes and genomes, we can calibrate the flow of genetic information, work out the course of evolutionary change and demonstrate how different kinds of organism are related to one another, even when we have little or no fossil evidence. There are many advantages of using genetic data.

- The same gene can be found in many (or all) organisms, so that we are not limited by characters that are found only in a restricted number of organisms, such as ankle bones or digits.
- Each nucleotide in a gene contributes a separate character, so phylogenies can be estimated from hundreds or thousands of characters.
- Each nucleotide takes only one of four discrete states, simplifying the analysis of the data.
- The nucleotide sequence is transmitted directly from ancestor to descendant, rather than being expressed through development.

In this section, we shall describe in some detail the estimation of a phylogeny from genetic data, using the artiodactyl/cetacean lineage as an example. You can implement all the analyses we describe by following the instructions in the Supplementary Material (Artiodactyl Phylogeny using PHYLIP).

Protein-coding genes can be used to infer phylogeny. The most straightforward information is the nucleotide sequence of a gene, or of part of

a gene, or of several genes. Figure 3.21(a) shows a sequence of 30 nucleotides in a gene encoding casein, the milk protein found only in mammals, for 13 artiodactyls, plus a perissodactyl, the horse, to serve as an outgroup.

The gaps every 10 nucleotides are put in to make the sequences easier to read and compare. It is still quite difficult to make sense of it. To make it clearer, we can write out the sequence for the outgroup and then show only changed nucleotides for the other species. This data matrix is demonstrated in Figure 3.21(b).

Most of the positions are not phylogenetically informative.

- Some (such as 151) share the outgroup state and are therefore shared primitive characters in all species.
- One (165) differs from the outgroup, but has the same state in all members of the ingroup, so it serves only to characterize artiodactyls without distinguishing different artiodactyl groups.
- Others (such as 150) are unique derived characters that serve only to identify a species without indicating its relationships.

There remain 11 nucleotides that are potentially informative, and lead to inferences about relationships.

- Positions 171, 172, 186, and 189 are shared derived characters that diagnose (pig + peccary) as a monophyletic group.
- Position 185 diagnoses (toothed whale + baleen whale) as a monophyletic group.
- Position 187 diagnoses (cow + sheep + goat) as a monophyletic group.
- Position 188 diagnoses (goat + sheep + cow + pronghorn + deer + giraffe) as a monophyletic group.
- Positions 159, 162, and 173 diagnose (goat + sheep + cow + pronghorn + deer + giraffe + chevrotain) as a monophyletic group.

At this point we encounter difficulties.

- Position 186 diagnoses (camel + hippo) as a monophyletic group, whereas position 166

(a)

```
                 150        160        170        180
HORSE          CAGTTAGGCC ACATGTCCAA ATTCCTCAAT GGCAAGTCCT
GOAT           CAGTTAGGTC ACCTGCCCAA ACTCTTCAAT GGCAAGTTTT
SHEEP          CAGTTAGGTC ACCTGCCCAA ACTCTTCAAT GGCAAGTTTT
COW            CAGTTAGGTC ACCTGCCCAA ATTCTTCAAT GGCAAGTTTT
PRONGHORN      CAGTTAGGTC ACCTGCCCAA ATTCTTCAAT GGCAAGTCTT
DEER           CAGTTAGGTC ACCTGCCCAA ATTCTTCAAT GGCAAGTCTT
GIRAFFE        CAGTTAGGTC ACCGGCCCAA ATTCTTCAAT GGCAAGTCTT
CHEVROTAIN     CAGTTAGGTC ACCTGCCCAA ATTCTTCAAT GGCAAGTCCC
TOOTHED WHALE  AAGTTAGGCC ACATGCCCAA ATTCCTCAAT GGCAATTCCT
BALEEN WHALE   CAGTTAGGCC ACATGCCCAA ATTCCTCAAT GGCAATTCCT
HIPPO          CAGTTAAGCC ACATGCCCAA ATCCCTCAAT GGCAAGCCCT
CAMEL          CAATTAGACT ACATGCTCAA ATTCCTCAGT GGCAAGCCCT
PIG            TAGCTGGGCC ACATGCTCAA AAACCTCAAT GGCAAGACCA
PECCARY        CAGTTAGGCC ACATGCCCAA AAACCTCAAT GGCAAGACCA
```

(b)

```
                 150        160        170        180
HORSE          CAGTTAGGCC ACATGTCCAA ATTCCTCAAT GGCAAGTCCT
GOAT           ........T. ..C..C.... .C..T..... ......TT.
SHEEP          ........T. ..C..C.... .C..T..... ......TT.
COW            ........T. ..C..C.... ....T..... .....TTT.
PRONGHORN      ........T. ..C..C.... ....T..... ......T.
DEER           ........T. ..C..C.... ....TC.... ......T.
GIRAFFE        ........T. .CG.C..... ....T..... ......T.
CHEVROTAIN     ........T. ..C..C.... ....T..... ......C
TOOTHED WHALE  A......... ..C....... .......... ....T.
BALEEN WHALE   .......... ..C....... .......... ....T.
HIPPO          .....A.... ....C..... ..C....... ....C.
CAMEL          .A...A.T.. ....C..... ......G... ....C.
PIG            T..C.G.... ...CT..... ....AA.... .....A..A
PECCARY        .......... ......C... ....AA.... .....A..A
```

Figure 3.21 (a) Partial sequence of the casein gene in artiodactyls. (b) Partial sequence of the casein gene in artiodactyls, showing only those sites that differ from the outgroup (horse).

diagnoses (pig + camel) as a monophyletic group. But we have previously concluded from four positions that (pig + peccary) is a monophyletic group. These diagnoses are incompatible. This problem arises because one or more of the assumptions of the analysis have not been met; for example, position 166 may have changed state twice, once in the camel lineage and once in the pig lineage. This leads to ambiguity: there will be two or more equally parsimonious trees, and we have no means of deciding between them. *Ambiguity* is best reduced or removed by increasing the number of species in the sample.

• No further monophyletic groups can be diagnosed from the available data. Hence, we can say nothing about the relationships between the groups we have already identified. Nor can we completely determine the structure within clades such as (goat + sheep + cow + pronghorn + deer + giraffe + chevrotain). This is because we have too few phylogenetically informative characters, which leads to lack of resolution. *Resolution* is best enhanced by increasing the number of informative characters in the sample.

We can check these inferences by running them through PHYLIP. As we expected, it cannot identify a single most parsimonious tree, but instead gives three equally parsimonious trees. We cannot decide between them, so we instead identify the features that all these trees have in common. This is called the consensus tree, which is shown in Figure 3.22. Doing this introduces another subjective procedure, which is deciding what we mean by "consensus." Here, we have adopted the strictest view, that all monophyletic groups diagnosed in the consensus tree are diagnosed by all the most parsimonious trees. More lenient alternatives are often used, for example the tree supported by a majority of the most parsimonious trees, which may be necessary when a great many are available. Nevertheless, even the strictest criterion by no means guarantees that we have identified a correct classification. To be more confident, we need to use more data.

Using the same methods for the full sequences of two casein-encoding genes generates the fully resolved phylogeny shown in Figure 3.23. This has three particularly interesting features.

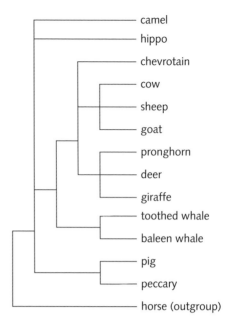

Figure 3.22 The consensus phylogenetic tree inferred from the casein gene sequences.

From Gatesy et al. 1996. Evidence from milk casein genes that cetaceans are close relatives of hippopotamid artiodactyls. *Molecular Biology and Evolution* 13 (7): 954–963, by permission of Oxford University Press.

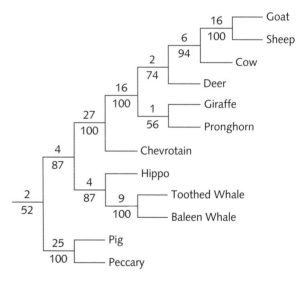

Figure 3.23 The phylogenetic tree inferred from complete casein gene sequences. The number above each dichotomy is the number of nucleotide changes in the descendent lineages. The number below is a statistical measure of the confidence that can be placed on the correct identification of sister groups at this point, with a maximum of 100%.

From Price et al. A complete phylogeny of the whales, dolphins and even-toed hoofed mammals (Cetartiodactyla). *Biological Reviews* 80 (3): 445–473 © 2007, John Wiley and Sons.

- The monophyletic groups that we identified in the much smaller data set are all confirmed by the more extensive analysis. (This was lucky. It is not unusual for phylogenetic hypotheses to be revised in the light of fuller evidence.)

- Cetacea is confirmed as a clade within Artiodactyla.

- The sister group of Cetacea is identified as the hippos.

Other kinds of genetic element give a similar or identical result. We can test the robustness of these hypotheses by using different kinds of genetic data. In order to compare gene sequences from different species, it is first necessary to align them—to make sure that we are comparing nucleotides related by descent. This is often difficult, especially for distantly-related species or rapidly-evolving genes. It also creates an opportunity, however, because the process of aligning gene sequences reveals regions where a few nucleotides have been inserted or deleted. These are called "indels" for short. Because indels are unlikely to happen at the same position in different lines of descent they provide an additional and independent source of phylogenetic data. Because each indel is either present or absent, they can be treated as presence/absence characters just like morphological characters such as digits or an

amnion. Figure 3.24 shows the data matrix for 34 indels in eight nuclear genes.

By now, you will begin to recognize shared derived characters such as those defining a clade of pigs (position 5) or a clade of artiodactyls excluding pigs (position 4). Running the data through PHYLIP (see Supplementary Material) produces the phylogeny that resembles the casein-based phylogeny, especially in supporting a (whale + hippo) clade within Artiodactyla.

Retroposons are parasitic genetic elements that replicate through their ability to insert copies into new locations in the genome. Their behavior is described in Chapter 11. Every new insertion generates a novel, stable phylogenetic signal that is unlikely ever to occur twice at the same position in the genome. LINEs (long interspersed nuclear elements) and SINEs (short interspersed nuclear elements) are types of retroposon with a characteristic genetic signature that can be used in phylogenetic analysis. Figure 3.25 illustrates the data for presence/absence of insertions at 20 sites in the artiodactyl genome (see Supplementary Material, Artiodactyl Phylogeny).

The data are less extensive (fewer species, and no outgroup) and less complete ("?" means uncertain) than for nucleotides or indels because they are harder to collect. Consequently, there are many equally parsimonious trees, and the consensus tree

	5	10	15	20	25	30	
HORSE	00000	00000	00101	10100	00000	00000	0000
GOAT	01010	00000	11100	01101	00100	01010	0000
SHEEP	01010	00000	11100	01101	00100	01010	0000
COW	00110	00000	01110	00101	00100	01010	0000
PRONGHORN	00010	00000	01100	10101	00100	01010	0000
DEER	00010	10100	01100	10101	00110	01010	1001
GIRAFFE	00010	00000	01100	10101	00100	01010	0000
CHEVROTAIN	10010	00000	00100	10101	00000	00000	0000
TWHALE	00010	00000	00001	10110	10000	00010	0000
BWHALE	00010	00000	00001	10110	10000	00010	0000
HIPPO	00010	00000	00001	10100	00000	00010	0000
CAMEL	10010	01000	00101	10100	01001	00011	0100
PIG	00001	00000	00101	10000	00000	10000	0010
PECCARY	00001	00000	00101	10000	00000	10000	0010

Figure 3.24 Occurrence of a sample of 34 short insertions or deletions ("indels") in artiodactyl genomes. Matte, C.A. et al. 2001

	5	10	15	20
SHEEP	00000	?0111	11111	11100
COW	00000	00111	11111	11100
DEER	00000	001?1	11111	?1100
GIRAFFE	?0000	00111	11111	11100
CHEVROTAIN	?0???	?????1	0???1	10?00
TWHALE	11111	110?1	01100	0?100
BWHALE	11111	11011	01100	0??00
HIPPO	0?011	11011	01100	0?100
CAMEL	00000	00000	00000	0?000
PIG	000?0	000?0	00??0	0?111
PECCARY	?????	?????	?????	???11

Figure 3.25 Occurrence of a sample of 20 retroposon insertions in artiodactyl genomes.

Nikaido, H., Rooney, A.P. and Okada, N. 1999.

is poorly resolved. Nevertheless, it supports several of the groupings found in the other analyses, notably (whales + hippos) as a monophyletic group within Artiodactyla.

Robust phylogenies can be estimated by combining different kinds of information. Obtaining molecular data is often laborious, so most analyses, like those we have described, involve rather few species or characters or both. By combining trees from several studies, however, it is possible to estimate a "supertree" that best represents the relationships among a large group of species by using the relationships among more restricted samples from individual studies using different species and characters. Figure 3.26 shows a supertree for the Artiodactyla, based on nearly 200 studies using both morphological and molecular characters.

This supports many of the relationships we worked out from smaller data sets, and clarifies others. The main features of the phylogeny with respect to the groups we have used before are as follows.

- Camels and pigs are basal clades.
- Cattle are the sister group of {sheep + goat}.
- Deer are the sister group of {cattle + sheep + goat}.
- Chevrotain, pronghorn, and giraffe are closely related and difficult to resolve.

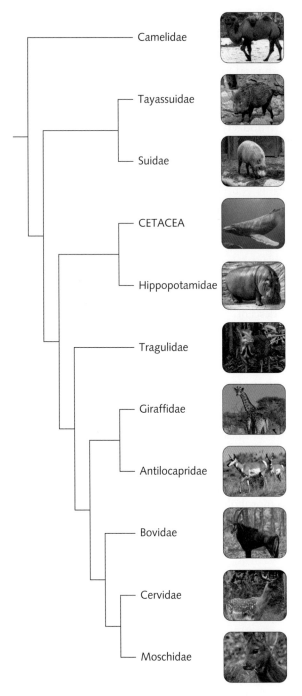

Figure 3.26 Robust phylogenies can be estimated by combining different kinds of information. This is a "supertree" for families of artiodactyls, estimated by combining 200 studies using morphological and molecular characters.

- Whales form a monophyletic clade within Artiodactyla.
- Hippos are the sister group of whales.

The supertree is an hypothesis that will continue to be revised as new data become available, and the relationships of some species (especially the difficult ones, such as the pronghorn) will undoubtedly be re-evaluated. As knowledge advances, however, our confidence in the main features of the tree increases. This is especially true when we have completely independent sources of evidence that point to the same conclusion. The fossil sequence shows clearly the transition from terrestrial artiodactyls to marine whales (Section 3.1.3), so the relationship between whales and artiodactyls should still be discernible in modern forms. The agreement between the fossil evidence, based on the shape of ankle bones in groups that became extinct long ago, and the molecular evidence, based on genetic data such as nucleotide sequence, indels or LINE insertions, shows how independent lines of evidence can be used to work out major evolutionary transitions and to map the relationships among living organisms.

● CHAPTER SUMMARY

Evolution is the flux of lineages through time.

- The lineage is the chain of descendants through time.
  - *The individual is the unit of physiology and genetics.*
  - *The population reflects the past operation of selection.*
  - *Lineage dynamics drive evolutionary change.*
- Descendants come to differ from their ancestors.
  - *Mutations are individually rare but collectively frequent.*
  - *Selection restricts the accumulation of mutations.*
  - *Neutral mutations convey a purely historical signal.*
- Ancestral and descendant forms are connected by an unbroken series of intermediates.
  - *Individuals and species both have lineages.*
  - *Almost all lineages are incompletely known.*
  - *Mosasaurs illustrate medium-scale evolutionary change.*
  - *Whales illustrate large-scale evolutionary change.*

Lineages are tree-like or web-like.

- Any two lineages have a most recent common ancestor.
  - *Sister species have an exclusive recent common ancestor.*
  - *Sister taxa can be defined at any level.*
- Evolutionary diversification is a branching process.
  - *Nesting sister taxa produces a tree of descent.*
- Sexual populations form a web: sexual species form a tree.
  - *Sexual lineages involve complex patterns of descent within species.*
  - *Sex carves out trees because it suppresses gene exchange.*
  - *The ancestry of bacteria is a web because they exchange genes.*

Organisms are classified by relatedness, not by similarity.

- The criterion of classification is common ancestry.
- The basis of evolutionary classification is the monophyletic group.
- Related taxa share derived characters.
- Relationship can be inferred from character state.
- The ultimate character is gene sequence.
  - *The continuity of the germ line enables us to map lineages.*
  - *Protein-coding genes can be used to infer phylogeny.*
  - *Other kinds of genetic element give a similar or identical result.*
  - *Robust phylogenies can be estimated by combining different kinds of information.*

● FURTHER READING

Here are some pertinent further readings at the time of going to press. For relevant readings that have been released since publication, visit the book's Online Resource Centre at **www.oxfordtextbooks.co.uk/orc/bell_evolution/**

Section 3.1 Gatesy, J. and O'Leary, M. 2001. Deciphering whale origins with molecules and fossils. *Trends in Ecology and Evolution* 16: 562–570.

Baum, D.A., DeWitt Smith, S. and Donovan, S.S.S. 2005. The tree-thinking challenge. *Science* 310: 979–980.

Thewissen, J.G.M., Cooper, L.N., George, J.C. and Bajpai, S. 2009. From land to water: the origin of whales, dolphins, and porpoises. *Evolution Education Outreach* 2: 272–288.

Section 3.2 Doolittle, W.F. and Bapteste, E. 2007. Pattern pluralism and the Tree of Life hypothesis. *Proceedings of the National Academy of Sciences of the USA* 104: 2043–2049.

Section 3.3 A useful short introduction to cladistics analysis: <http://www.ucmp.berkeley.edu/clad/clad3.html>

● QUESTIONS

1. Define a lineage and distinguish a lineage forwards-in-time from a lineage backwards-in-time. How do sexual lineages differ from asexual lineages?

2. Explain how deleterious mutations do not continue to accumulate indefinitely in a population.

3. Suppose that you come across a statement such as this: "The extinct species *Somethingus primus* is the ancestor of the living species *Otherus secundus*." Comment on the meaning and limitations of this statement.

4. "Marine tetrapods have evolved from terrestrial ancestors through a series of intermediate forms." Describe how this statement is supported by the comparative anatomy of either (a) mosasaurs or (b) whales.

5. Can a phylogenetic tree be completely defined as a nested series of sister taxa? Provide a justification for or against this proposition.

6. Define a monophyletic group and explain why monophyletic groups are the basis of a natural classification of organisms.

7. Are these two phylogenetic trees completely equivalent? Justify your answer.

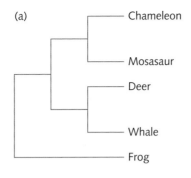

(a)
- Chameleon
- Mosasaur
- Deer
- Whale
- Frog

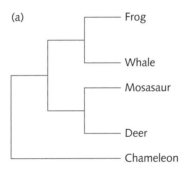

(b)
- Whale
- Deer
- Mosasaur
- Chameleon
- Frog

8. Are these two phylogenetic trees completely equivalent? Justify your answer.

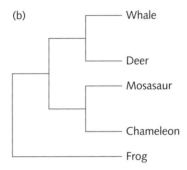

(a)
- Frog
- Whale
- Mosasaur
- Deer
- Chameleon

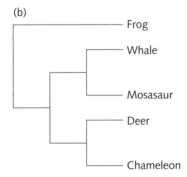

(b)
- Frog
- Whale
- Mosasaur
- Deer
- Chameleon

9. Define a "shared derived character," distinguishing it from unique characters and primitive characters. Explain why shared derived characters alone are useful in identifying ancestry.

10. Here is a character matrix for the six species A–F. Each of the eight characters 1–8 is coded "0" for absent or "1" for present. The ancestral state is "0.". Identify the phylogenetically informative characters, infer the sequence of change, and draw the cladogram.

Character		1	2	3	4	5	6	7	8
Species:	A	1	0	0	1	1	0	1	0
	B	0	0	0	0	1	0	0	0
	C	0	1	0	0	1	1	1	0
	D	1	0	1	0	1	0	1	0
	E	0	1	0	0	1	0	1	1
	F	0	0	0	0	0	0	0	0

11. Here is a character matrix for the six species A–F. Each of the eight characters 1–8 is coded "0" for absent or "1" for present. The ancestral state is "0." Identify the phylogenetically informative characters, infer the sequence of change, and draw the cladogram.

Character		1	2	3	4	5	6	7	8
Species:	A	1	1	0	1	0	1	0	0
	B	1	0	0	1	0	0	1	0
	C	0	0	0	0	0	0	0	0
	D	0	0	1	1	0	0	0	0
	E	1	0	0	1	0	1	0	1
	F	0	0	1	1	1	0	0	0

12. Here is a character matrix for the six species A–F. Each of the nine characters 1–9 is coded "0" for absent or "1" for present. The ancestral state is "0.". Identify the phylogenetically informative characters, infer the sequence of change, and draw the cladogram.

Character		1	2	3	4	5	6	7	8	9
Species:	A	1	1	1	0	0	0	0	0	1
	B	0	0	0	1	0	1	0	0	1
	C	0	1	0	0	0	0	1	0	1
	D	0	1	1	0	1	0	0	0	1
	E	0	0	0	1	0	0	0	1	1
	F	0	0	0	0	0	0	0	0	0

13. Discuss the advantages and drawbacks of using information about morphology or gene sequence in estimating phylogenetic trees. Is there any drawback in using both to construct a supertree?

14. What is a natural classification? Describe the phenetic and cladistic methods of classification and how these are related to the concept of a natural classification.

15. Summarize the evidence that Cetacea (whales) is a monophyletic group within the more inclusive monophyletic group of Artiodactyla (even-toed ungulates).

16. A given scheme, based on the principles outlined in this chapter, provides a natural classification of a group of organisms. What evidence might arise that would falsify this classification?

You can find a fuller set of questions, which will be refreshed during the life of this edition, in the book's Online Resource Centre at **www.oxfordtextbooks.co.uk/orc/bell_evolution/**

The Diversity of Life

4

The branching of lineages over time has given rise to all the different kinds of living things that we see around us. In this section we shall review biodiversity by tracing our own lineage backwards in time. This approach was suggested by Richard Dawkins' book, *The Ancestor's Tale*, which can be consulted for a much fuller treatment. It is an approach that does not give a very balanced view of diversity, because it emphasizes animals like us at the expense of larger groups such as insects or fungi. On the other hand, it underlines two important facts. The first is that phylogenetic trees are "branchy" at all scales, because lineages have continually diversified at all periods in the history of life. The second is merely that our own species in common with all others has a long evolutionary history that can be traced back to our most distant relatives.

As we descend the tree of life, branch by branch, we shall be describing more and more inclusive clades. If we were to pause at each branch point, this would be an impossibly long chapter. Our choice of branch points must therefore be to some degree arbitrary, and in general we have chosen points that mark the divergence of familiar groups, such as mammals or animals.

Each clade that we describe comprises a group of lineages that have diverged from their common ancestor by adapting to different ways of life. This process of adaptive radiation may occur at any scale, from a small group such as "apes" to a much larger one such as "vertebrates." Wherever possible, we have emphasized the evolutionary principles underlying adaptive radiations by identifying the innovations associated with diversification.

An *innovation* is a shared ancestral character that facilitates the adaptive radiation of a clade. Innovations mark the developmental and structural shifts responsible for major evolutionary transformations.

We can make a useful distinction between *potentiating innovations* and *implementing innovations*.

- A potentiating innovation is a character that, once established in an ancestral lineage, enables other characters to diverge in descendant lineages. The example illustrated in Figure 4.1 is the vertebrate amnion. Tetrapods with an amniotic egg can follow a completely terrestrial existence, and consequently are free to evolve into the thousands of species of reptiles, birds, and mammals with every way of life from burrowing to flapping flight.

- An implementing innovation is a character that, once established in an ancestral lineage, may itself become divergently specialized in descendant

Figure 4.1 The amniote egg is a *potentiating* innovation.

Lizard eggs © Biosphoto/ Thierry Van Baelinghem. Echidna egg © Peggy Rismiller/Mike McKelvey. Lizard courtesy of Aconcagua. Echidna courtesy of KeresH. Bird courtesy of JJ Harrison. Bird eggs courtesy of Bjørn Christian Tørrissen. Turtle courtesy of Bernard Gagnon. These files are licensed under the Creative Commons Attribution-Share Alike 3.0 Unported license. Turtle eggs courtesy of Yosri. This file is licensed under the terms of the GNU Free Documentation License.

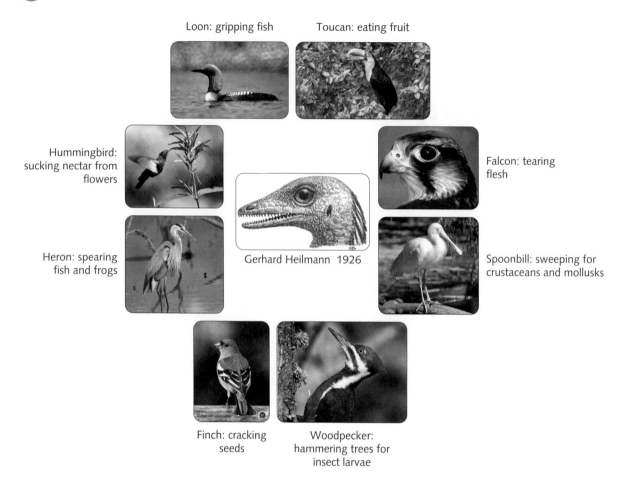

Loon: gripping fish

Toucan: eating fruit

Hummingbird: sucking nectar from flowers

Falcon: tearing flesh

Gerhard Heilmann 1926

Heron: spearing fish and frogs

Spoonbill: sweeping for crustaceans and mollusks

Finch: cracking seeds

Woodpecker: hammering trees for insect larvae

Figure 4.2 The bird beak is an *implementing* innovation. The beak of modern birds was derived from the narrow toothed snout of early forms such as Archaeopteryx.

lineages. The example illustrated in Figure 4.2 is the beak of birds. This is a structure that can itself readily be modified by natural selection for gripping, tearing, hammering, stabbing, sucking, filtering, and so forth, underlying the evolution of the many kinds of modern bird.

The distinction between potentiating and implementing innovations is useful but not immutable, and an innovation may be both potentiating and implementing. For example, the tetrapod limb powers terrestrial locomotion and thus enables tetrapod lineages to adapt to many ways of life on land, which involves extensive modification of all parts of the body. At the same time, the limb itself is often modified, for example as a wing for flying, or a spring for jumping, or a shovel for burrowing, or a pillar for supporting a massive body.

The separate phylogenetic trees of the clades we describe can be fitted together so as to make a tree for the whole of life—although some branches will be much more detailed than others. This can be compared with the universal tree documented by the Tree of Life project (<www.tolweb.org>), which gives a conservative account of current phylogenetic knowledge. The diagrams in this book show fully resolved trees that are supported by the best available data. We emphasize, however, that each tree constitutes an hypothesis that is liable to be modified as knowledge increases. Phylogenetics is a live science where hypotheses are generated, modified, or discarded as the evidence leads us towards more complete understanding.

4.1 Ourselves: *Homo sapiens*

Innovation: language and culture

All people are the same, yet no two persons are the same. We are the same because we belong to a single species, which means that any two mature people of different gender can mate and produce fully viable and fertile offspring, whereas we cannot do so with individuals of any other species. We are different because every person is unique, with attributes that are not found in any other person, and a genome that has never previously occurred and will never again be precisely repeated. Even identical twins differ in their attributes from accidents of development or upbringing, and differ genetically from mutations. A glance around any classroom will provide a small sample of human variation. There are three main causes of the differences we see.

- Differences attributable to *gender*, such as pelvic structure. These are caused by genetic switches encoded on the sex chromosomes, in some cases overlain by complex cultural factors.
- Differences attributable to *upbringing*, such as language. These are environmental differences, expressed in both genders. They may be modified by genetic differences.
- Differences attributable to *inheritance*, such as the color of the iris. These are genetic differences, likewise expressed in both genders. They may be modified by environmental differences.

Here we are concerned solely with genetic differences, because we want to know how the current diversity of people evolved during the history of our species. We can apply the phylogenetic methods explained in Chapter 3 to our rapidly growing knowledge of the human genome to interpret human diversity.

People are often classified by their appearance into races, which means groups whose recent ancestors lived in widely distant parts of the world. Because we are a single species, there are no fixed genetic differences between any large groups of people, whether races or not; if there were any, they would be broken down immediately by interbreeding. However, there is considerable variation in allele frequency from place to place at many loci. Consequently, it is usually possible to identify where an individual's ancestors came from, provided that several variable loci have been scored.

The totality of genetic variation within any given species, including our own, can always be broken down into two components:

- variation among individuals within populations
- variation among the average values of populations.

The balance of human variation depends on whether protein or DNA is taken as the measure, and on which protein or gene sequence is used. As a rough rule of thumb, about 85% of variation in protein-coding genes is expressed among individuals within a population, and about 15% among populations. The among-population variation is about equally divided between regional variation (such as the peoples of North Africa or Eastern Europe) and continental variation (including races).

Some genes, such as those associated with lactose tolerance or skin pigmentation, are strongly differentiated among populations, and the patterns of allele frequency that we measure have probably been created largely by natural selection. In other cases, allele frequencies differ because they have drifted apart in populations that have been genetically isolated from one another. Drift provides a purely historical signal that enables us to recognize groups that have been separate in the past.

Drift is reinforced when a small group of people move into an area and subsequently expand. Some alleles that are generally rare will be over-represented in this group by chance, and will give it a characteristic genetic signature. On the other hand, marriage between people from different groups will tend to homogenize allele frequencies over wide regions, and will thereby reduce genetic differentiation. Current patterns of human diversity have been created by the combined effects of drift, selection, migration and intermarriage. Large-scale (continental) structure is usually attributable to barriers such as oceans and mountain ranges that are crossed only infrequently by adventurous bands and that prevent regular intermarriage with other populations.

Gene frequencies usually change gradually across large areas because of the smoothing effect of intermarriage. Two populations living nearby are likely to have very similar constellations of genes, because of the regular and frequent gene flow between them. Two populations that live far apart, on the other hand, are likely to be almost isolated from one another and may have quite different balances of alleles. Hence, the genetic difference between any two populations will tend to increase with their distance apart. The amount of difference that we find among populations will thus depend on how we sample them: for samples taken less than 100 km apart less than 1% of genetic differences among individuals will be attributable to them coming from different populations, whereas for samples taken thousands of kilometers apart on different continents this fraction rises to about 15%.

The sequences of mitochondrial, Y-chromosome, X-chromosome, or autosomal genes can all be used to construct phylogenies by cladistic methods. The results, as sketched in Figure 4.3, are broadly congruent when large samples of genes are used. The deepest branch separates African from non-African populations, as expected from the "out of Africa" model of human evolution (Section 5.1). The non-African populations fall into two major groups, one including Pacific, Australian, and south-east Asian peoples, the other including north-east Asian, west Asian, European, and Amerindian peoples. This pattern reflects the pathways of migration after the irruption of human bands into Asia about 0.1 Mya. Nevertheless, it must be borne in mind that diagrams such as Figure 4.3 are not like cladograms representing the divergence of separate species. Genes have been continually exchanged among all human groups through more or less frequent intermarriage, and a more realistic version of human descent would be an intricate web of lineages.

The spread of *Homo sapiens* is associated with a complex material culture involving carved tools, personal ornaments, and both abstract and figurative art. This level of symbolic expression almost certainly required the parallel appearance of language with a complex grammar and an indefinitely large range of meaning. Language is the characteristic innovation of *Homo sapiens*, the crucial advance that made modern human success possible. The most likely candidate for the potentiating innovation that gave the human

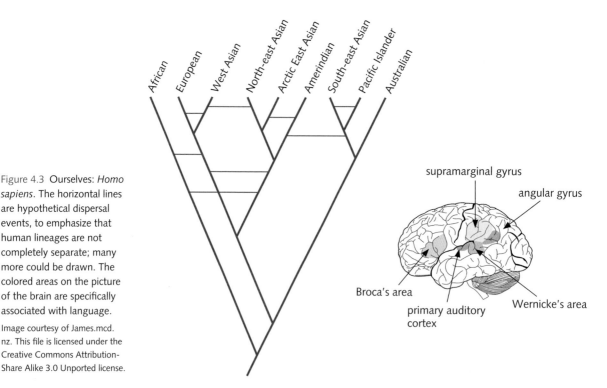

Figure 4.3 Ourselves: *Homo sapiens*. The horizontal lines are hypothetical dispersal events, to emphasize that human lineages are not completely separate; many more could be drawn. The colored areas on the picture of the brain are specifically associated with language.

lineage the capacity to develop language and culture is imitation, or observational learning. This leads to a process of cultural evolution: individuals imitate a skillful act (that is, learn to do it from seeing it done), occasionally modify it, and are in turn imitated by individuals in the following generation. The ability to copy the behavior of other, successful individuals can drive a cumulative increase in skill throughout a society. It is found to a lesser degree in birds (learning song dialects) and apes, but it is only in humans that it has led to complex societies based on a learned division of labor.

Interestingly, language itself evolves, with modern languages descending from ancestral tongues that are no longer spoken. The mechanism of language evolution, however, is quite different from biological evolution. Although the capacity to speak a language is inherited, the particular language that a person speaks is learnt, and changes and modifications are transmitted by learning rather than by genetic inheritance. A useful way of thinking about language evolution that emphasizes its underlying similarity with biological evolution is to imagine how words are copied from one brain to another, always with some rate of error, occasionally creating variants that spread through a population and replace the previous version. An ongoing example in the English language is the regularization of irregular verbs. Rare irregular verbs such as writhe (past participle wrothe) and chide (chid) have become regularized (writhed, chided) in the last few hundred years. The past participle of "help" changed from "holpen" to "helped" between Chaucer (about 1400) and Shakespeare (about 1600). You can actually witness the language evolving in your own lifetime. The next irregular past participle to vanish, for example, may be wed/wed, which is rapidly being replaced by wed/wedded.

4.2 *Homo + Pan*

Sister clade added: *Pan* (chimpanzees)
Innovation: no distinctive innovation

Our nearest relatives are the chimpanzees. There are two species: the common chimpanzee, *Pan troglodytes*, which occurs broadly in central and western Africa, and the pigmy chimpanzee or bonobo, *Pan paniscus*, which lives in central Africa south of the Congo River. These two closely related species are the sister group of *Homo sapiens*, as shown in Figure 4.4. Curiously, there is no agreed formal name for the clade that includes only ourselves and our nearest living relatives; nor is there any morphological character that defines this clade.

Comparison of the human and chimp genomes has shown that they differ by a little more than 1% of nucleotides in single-copy DNA. This figure has led to the observation that we are "98.4% chimp," with the implication that the genetic differences between humans and chimps are very small. So they are, relative to other animals. However, two reservations need to be made. The first is that there are genetic differences other than single-nucleotide substitutions, especially the insertion or deletion of short sequences of DNA. When these are taken into account, the divergence between chimp and human genomes rises to about 4%. Secondly, even 1% divergence implies that tens of millions of nucleotides are different. It is certainly not true that 98.4% of genes are the same

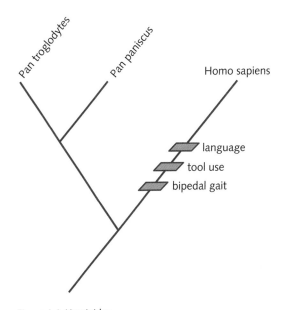

Figure 4.4 Hominidae.

in humans and chimps—on the contrary, most genes are different.

The anatomical difference between humans and chimps are quite modest: an upright posture and the loss of most body hair are among the most obvious, and may have evolved to enhance endurance running to hunt down large mammals such as antelopes. There are less obvious differences, too, such as the degeneration of many of the genes responsible for olfaction, as our sensory world shifted from smelling to seeing. The most characteristic attribute of humans, however, is the possession of language, which has enabled us to construct symbolic representations of the world, and, after many thousands of years, to develop the technologies by which we measure and control our physical environment. This has been associated, in ways that are not entirely clear, with the expansion of brain size from about 400–440 cm^3 in chimpanzees to about 1340–1440 cm^3 in humans.

The sequencing of human and chimp genomes creates the opportunity to identify the genes that make us human. This is by no means an easy task, despite their similarity, because of the very large number of genes whose sequence or structure has changed substantially. Moreover, rather slight alterations of regulatory genes can produce downstream effects that are difficult to predict. There will not be a single "human gene," of course. There may be a relatively small number of genes, however, involved in the crucial transition to human anatomy, language, and intelligence.

A striking case of language impairment has been described from an extended family where the distribution of affected individuals suggests the segregation of a single dominant gene. The disorder affects the understanding of complex sentences and the generation of inflections to modify verb tense, without being associated with any substantial reduction of nonverbal intelligence. It has been mapped to a single locus *FOXP2* that encodes a transcription factor (a protein that regulates gene expression by binding to DNA), where the genetic lesion is a point mutation causing the alteration of a single amino acid in the DNA-binding domain. FOXP2 is a highly conserved protein with only three amino acid differences between human and mouse. Two of these three substitutions, however, occur in humans but not in chimpanzees (or other apes), suggesting rapid evolution of the gene in the human lineage. It should not be imagined that *FOXP2* is the "language gene" that is alone responsible for human language, as there may be many others that remain to be discovered. It is rather the best available example of a gene associated with language skills that has evolved rapidly in the human lineage and whose disruption causes a specific language impairment.

Several loci are known to be involved in the development of a brain of normal size, with mutations causing a condition called microcephaly. For example, nonsense mutations in *ASPM* are associated with severe reduction in brain size to about 430 cm^3, accompanied by mild to moderate mental retardation. This gene has undergone several amino-acid substitutions in the lineage leading to humans, again suggesting rapid evolution of a crucial human attribute.

4.3 Hominoidea

Sister clades added: gorillas; the other apes
Innovation: the brachiation syndrome

Apes are large-brained tailless primates. They (and we) have a broad chest with the shoulder-blade on the back, a shortened lumbar region, a fully rotatable shoulder, and a highly flexible wrist. These are specializations that facilitate brachiation, a mode of moving through the forest by swinging from one bough to another suspended from the hands. We do not brachiate, but the separation of the functions of hands and feet in our ancestors was the potentiating innovation that facilitated the evolution of bipedal locomotion and manual dexterity in the human lineage. The phylogeny of apes is shown in Figure 4.5.

• Gorillas are the animal most closely related to humans, except for chimpanzees. They are very large apes, with males weighing 200–240 kg. The two species are both restricted to equatorial Africa. The Mountain Gorilla (*Gorilla beringei*) lives in cloud forest and is mostly terrestrial, moving around by knuckle-walking, and eating leaves

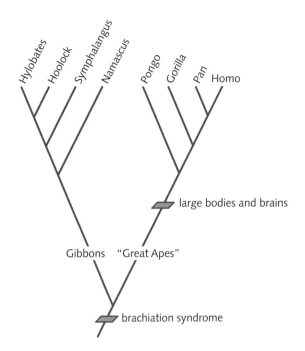

large bodies and brains

Gibbons   "Great Apes"

brachiation syndrome

Figure 4.5 Hominoidea: the apes.
Image courtesy of Tu7uh. This file is licensed under the Creative Commons Attribution 3.0 Unported license.

and shoots. The Western Gorilla (*Gorilla gorilla*) is a more slender animal that is more arboreal, climbing trees to obtain fruit. Both species live in roving family groups with a single adult male, several females, and their offspring.

- Orangutans are specialized arboreal apes found in Borneo (*Pongo pymaeus*) and Sumatra (*Pongo abelii*). Their diet is based on fruit, but they will also eat bark, shoots, and insects. They are for the most part solitary, coming together only to mate.

- Gibbons are small arboreal apes with long arms and hook-like hands specialized for brachiation. They are monogamous and live in family groups each defending a small territory that includes the fruiting trees that supply them with food. There are a dozen living species, all restricted to tropical south-east Asia.

Thus, apes are characteristically animals of the forest, eating mainly fruit. Humans are exceptional in living in open country and eating large quantities of meat.

Populations of African apes are threatened in most areas by human activities and pressure for land. The forests of western equatorial Africa were until recently a secure refuge, but even these are now affected by hunting, war and mechanized logging. Moreover, both gorillas and chimpanzees have been killed in large numbers by outbreaks of Ebola fever, which may almost completely destroy populations over a large area. In 2002–2003, an Ebola outbreak in Republic of Congo killed about 4000 gorillas. The situation is no better in Indonesia. In Sumatra, logging in the principal reserve for orangutans reduced their numbers by about 40% in less than ten years. About a third of the Borneo population was killed by drought and fire in 1996–1997. The large apes that are our nearest relatives may become very rare or even extinct in the wild within the next century.

4.4 Primates

Sister clades added: monkeys, tarsiers, and lemurs
Innovation: flexible limbs with grasping hands

Primates are arboreal mammals that are adapted to clambering about in trees. The limbs are flexible and free to move in almost any plane, rather than being restricted to the fore-and-aft movement characteristic of terrestrial mammals. They have a prominent shoulder-blade (clavicle) that connects shoulder with trunk to allow the animal to hang suspended from a bough. The big toe and thumb diverge from the other digits so that foot and hand are capable of grasping boughs, and the tips of the digits are protected by flat nails rather than claws, which would get in the way. Acute sight is necessary for rapid locomotion in the canopy, and the eyes are usually large and directed forward to give overlapping visual fields and stereoscopic vision. The olfactory organs so important to terrestrial mammals are reduced, to the point where the sense of smell becomes almost rudimentary in apes and humans. The relationships of the main groups of primates are shown in Figure 4.6.

- Lemurs are "wet-nosed" primates (Strepsirrhini) with a relatively acute sense of smell. Most are nocturnal and all have a reflecting layer, the tapetum lucidum, overlying the retina. It reflects light back onto the photoreceptors, thereby increasing the acuity of night vision. Cats also possess a tapetum, which is why their eyes shine in the dark. The tapetum is a good example of the parallel evolution of similar structures in different groups.

- The tarsier is another nocturnal primate, but lacks a tapetum; this reminds us that useful structures often fail to evolve. It has very long hindlimbs and digits, using them to spring out and grasp insects and even small birds in flight.

- New World monkeys (Platyrrhini) live in South America and have flat noses with sideways-facing nostrils. They are specialized for arboreal life and rarely descend to the ground. Many have a highly prehensile tail which is used for grasping boughs, and the thumb is correspondingly reduced or even lost.

- Old World monkeys (Catarrhini) live in Africa and Asia and have narrow noses with nostrils that

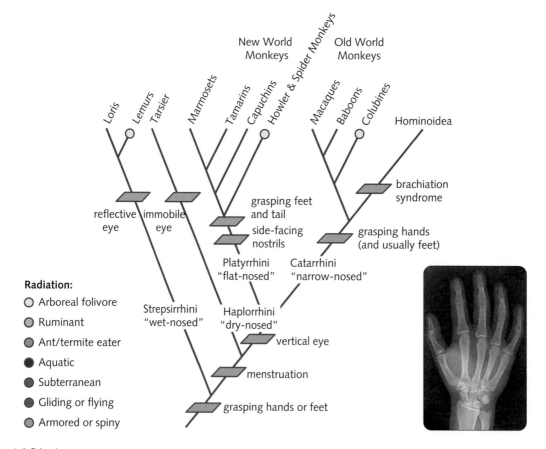

Figure 4.6 Primates.

face forward and downward. They have a broader range of lifestyles than the New World monkeys, and some (baboons) live in the open far from trees. The tail is often small or even absent, and both big toe and thumb are generally well-developed and opposable, except in some species where this ability has been lost, as it has been in human feet.

Leaves are an obvious source of food for arboreal animals, but they contain a high proportion of structural carbohydrate such as cellulose that is difficult to digest. Consequently, folivores often evolve a large hindgut chamber, the caecum, to house symbiotic bacteria capable of breaking down cellulose to sugars that can be used in central metabolism. The phylogeny of the Primates illustrated in Figure 4.6 shows that leaf-eating has evolved independently in at least three groups of primates: lemurs, howler monkeys, and columbine monkeys.

4.5 Eutheria

Clades added: rodents and relatives; Laurasiatheria; Afrotheria + Xenarthra

Innovation: placentation

The placenta is the deciduous organ made of embryonic tissue that surrounds the growing embryo and is shed at birth. It is the characteristic developmental innovation of Eutheria that makes it possible for one individual to develop inside another until it is well advanced towards independence.

The conceptus of mammals differentiates very early in development into two kinds of tissue, the spherical trophoblast surrounding the inner cell mass that will form the embryo. The trophoblast is responsible for implanting the conceptus into the wall of the uterus and mediating the initial contact between fetal and maternal tissue. The embryo comes to be invested by four membranes, mostly derived from the inner cell mass: yolk sac, amnion, chorion, and allantois. The yolk sac endoderm becomes associated with the trophoblast, with a layer of mesoderm containing branches of the vitelline blood vessels lying between them, forming a yolk-sac placenta. The allantois subsequently expands into the fluid-filled cavity in which the embryo is growing until it comes into contact with the trophoblast, forming a chorioallantoic placenta that is supplied by the umbilical blood vessels. The replacement of the yolk-sac placenta by a chorioallantoic placenta is the general path of development in eutherian mammals, but there are many variants: in humans, for example, only a chorioallantoic placenta is formed.

The zone of contact between fetal and maternal tissue is the most sensitive region of the placenta, because it must simultaneously serve two opposed functions: to increase contact between fetus and uterus, in order to maximize nutrient flow, and to reduce contact, in order to protect the fetus against attack by the maternal immune system. The interhemal membrane, consisting of apposed layers of fetal and maternal tissue, is the structure that regulates this interaction. There are three main versions of the membrane (with many minor variations).

- Epitheliochorial: the chorionic trophoblast lies against the uterine epithelium (for example, in Cetartiodactyla and Strepsirrhini).

- Endotheliochorial: the uterine epithelium has been lost in this region, and the trophoblast is in contact with the internal lining of the uterine capillaries (Carnivora and Proboscidea).

- Hemochorial: all maternal membranes are lost in this region, and the trophoblast is in direct contact with circulating maternal blood (Rodentia and Haplorhini).

This sequence involves increasingly intimate contact between maternal and fetal tissues, with a corresponding increase in the effectiveness of nutrient transfer to the fetus. However, it is evidently not an evolutionary sequence: the examples we have given show that the three modes of placentation are not phylogenetically successive innovations, each defining a clade. Rather, each must have evolved independently, in more than one clade. Although it is often tempting to imagine that a functional series reflects an evolutionary sequence, the phylogenetic distribution of modes of placentation shows that this is not necessarily true.

Eutheria have undergone an extensive adaptive radiation leading to a broad range of lifestyles. Recent molecular phylogenetic analysis suggests that all eutherians can be grouped into four main clades, as shown in Figure 4.7, that represent independent radiations on different landmasses early in eutherian history.

- The Afrotheria are typically African forms, including hyraxes, elephants, sirenians, aardvarks, tenrecs, and golden moles.
- The Xenarthra are typically South American forms, including armadillos, anteaters, and sloths.
- The Laurasiatheria are typically Eurasian and North American forms, including insectivores, ungulates, whales, carnivores, and bats.

- The fourth group includes rodents and primates, with their close relatives. It is not yet firmly established that this group ("Euarchontoglires") is monophyletic, and unlike the others it is not associated with a particular geographical region.

The independent adaptive radiations of Eutheria display a remarkable degree of parallel evolution, dominated by a restricted set of evolutionary themes, especially among Afrotheria and Laurasiatheria, that are illustrated in Figure 4.7.

- Shrew-like insectivores. Minute, long-snouted mammals actively foraging for insects by day. This theme is represented by shrews (Soricinae) in the Laurasiatheria and by shrew tenrecs (Oryzorictinae) in the Afrotheria.
- Hedgehog-like insectivores. Bulky insectivores with a covering of sharp spines: hedgehogs

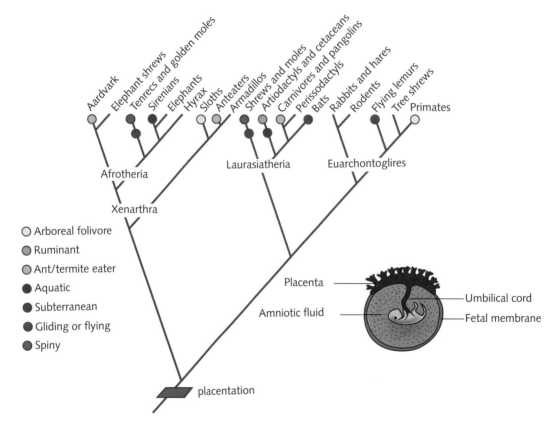

Figure 4.7 Eutheria.

(Erinaceinae, Laurasiatheria) and tenrecs (Tenre-cinae, Afrotheria).

- Mole-like insectivores. Subterranean forms with spade-like forelimbs, dense short fur, and reduced eyes and ears: moles (Talpinae, Laurasiatheria) and golden moles (Chrysochlorinae, Afrotheria).

- Ant- and termite-eaters. These mammals have strong claws to rip anthills apart, long snouts, and extremely long tongues, often detached from the hyoid and extending back into the thorax, and coated with sticky saliva. Teeth are reduced, lack enamel, or are completely absent: pangolins (Pholidota, Laurasiatheria), aardvark (Orycteropodidae, Afrotheria), and anteaters (Myrmecophagidae, Xenarthra).

- Aquatic mammals. Large fusiform mammals with flipper-like limbs and tail flukes, rearward nostrils, and other adaptations for an obligately aquatic way of life: whales and dolphins (Laurasiatheria), and manatees and dugongs (Sirenia, Afrotheria).

4.6 Mammalia

Sister clades added: Metatheria (marsupials); Monotremata
Innovations: milk, differentiated teeth

Mammals are warm-blooded tetrapods that suckle their young, usually with teats that express nutritive milk synthesized in the mammary glands. These attributes make them exceptionally independent of environmental conditions, able to flourish in all parts of the world except the interior of Antarctica. Their teeth are complex and differentiated, rather than being simple cones. The molar teeth bear several cusps or ridges, arranged in a pattern adapted to a particular diet. The molar, premolar, canine, and incisor teeth perform different chewing, grinding, slicing, piercing, and cutting actions and can be modified for a vast range of food items to graze, browse, crush seeds, grip prey, shear bones, and so forth. The limbs are rotated to lie beneath the trunk, parallel with the body, and have become modified in different groups for running, jumping, burrowing, climbing, and swimming. This flexibility underlies the wide adaptive radiation of the mammals.

Figure 4.8 shows the relationships and radiation of the three major clades of mammals.

- Monotremata. The monotremes are egg-laying mammals that lactate but have no defined nipples. The reproductive ducts, urinary duct, and gut all empty into a common chamber, the cloaca. The only living monotremes are the duck-billed platypuses and the echidnas, both restricted to Australia and New Guinea.

- Marsupialia (Metatheria). The marsupials are live-bearing mammals that suckle their young for most of its development in a pouch. They have undergone an extensive adaptive radiation in Australia, with some New World forms such as opossums (Didelphidae).

- Eutheria. The placental mammals.

The pattern of parallel evolution found in placental mammals is echoed in the radiation of monotremes and marsupials, which has given rise to many forms with similar adaptations to placental mammals following the same way of life.

- Shrew-like insectivores. The shrew opossums (Caenolestidae) of the Andes are larger than placental shrews, but likewise pursue insects and other small prey.

- Mole-like insectivores. The marsupial mole (Notoryctidae) is a subterranean insectivore found in the deserts of Western Australia. It is very similar in appearance to the African golden moles; in addition, its pouch opens to the rear, so as not to fill with sand when burrowing.

- Ant- and termite-eaters. The numbat or banded anteater is an Australian marsupial with the long snout, sticky tongue, degenerate dentition, and powerful limbs of comparable ant-eating placental mammals. Among monotremes, the echidnas (Tachyglossidae) have a similar combination of characters, and are also covered in hedgehog-like spines.

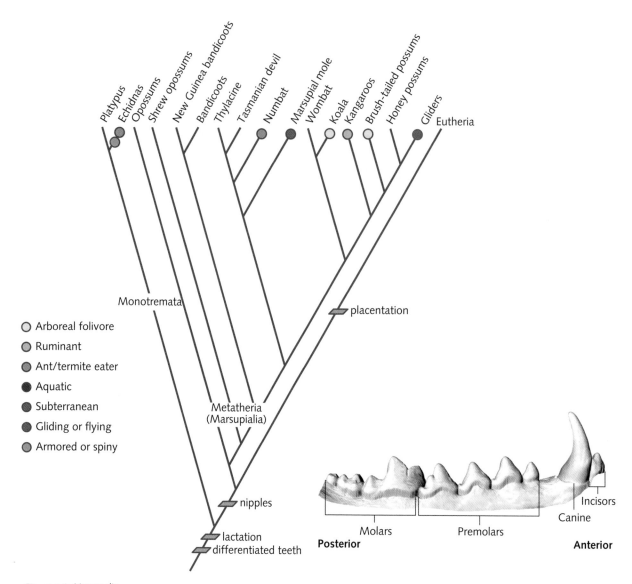

Figure 4.8 Mammalia.

Image reprinted by permission from Macmillan Publishers Ltd: P. David Polly, Evolutionary biology: development with a bite. *Nature* 449 (7161) © 2007.

- Large browsers and grazers: herbivorous mammals with dentition adapted to cropping and grinding leaves and herbs, and a ruminant-like intestine supporting a community of symbiotic bacteria capable of digesting plant structural carbohydrates. Placental browsers and grazers include deer and cattle (Artiodactyla, Eutheria), while their marsupial counterparts are the kangaroos (Macropodidae, Metatheria).

- Arboreal folivores. The diet of arboreal folivores is likewise associated with a complex gut including a caecum. This is characteristic of both placental types (sloths, Folivora; also certain primates, see above) and their marsupial counterparts (brushtail possums, Phalangeridae; koala, Phascolarctidae).

- Carnivores. The thylacine (marsupial wolf, or Tasmanian tiger) (Thylacinidae) was a strictly carnivorous marsupial up to about 2 m in length that preyed on kangaroos, possums and wombats. Its skull is strikingly wolf-like, albeit with a narrower muzzle. It is almost certainly extinct, although possible sightings are still occasionally reported. The Tasmanian Devil (Dasyuridae) is the largest extant marsupial carnivore, about the size of a corgi.

- Gnawing and burrowing animals. Wombats (Vombatidae) are stout-bodied marsupials with incisor teeth capable of gnawing tough vegetation. They are similar to large placental rodents such as groundhogs and cavies.

- Gliders. The flying lemurs (Dermoptera) and flying squirrels (Petauristinae) both have a flap of skin extending from wrist to ankle that can be extended to provide a gliding plane. The same arrangement is found among marsupials in the flying phalangers (Petauridae) and the Greater Glider (*Petauroides*).

The remarkable parallels between Australian, African, and Laurasian radiations of mammals shows that there is a limited number of "professions" open to mammals. Some are only rarely filled: for example, the only mammals that feed primarily on nectar are the minute honey possums of Western Australia and a few bats in the south-western USA. Others are often but not always filled: there are no obligately aquatic marsupials, for example. Some professions, like eating ants and termites, or burrowing underground, have evolved in all three faunas, but seem to have a limited capacity, as only a few species take them up. Finally, there are large and populous "professions," such as browsers and grazers with ruminant-like stomachs, which are always well-stocked.

4.7 Amniota

Sister clade added: {Archosauria + Lepidosauria}; Chelonia
Innovation: the amnion

The amnion is a closed fluid-filled sac enclosing the embryo that provides a replacement for the ocean or the pond where fish and amphibians' eggs develop. Protected by a tough, porous shell, it is the potentiating innovation that provides the basis for a completely terrestrial way of life. The three main lineages of amniotes are shown in Figure 4.9.

- The Archosauria. These are reptiles with socketed teeth, although these are often lost, to be replaced by a horny bill. The two surviving archosaur lineages are very different in appearance.

  – Crocodiles (Crocodylia) are large predatory reptiles with elongated and flattened skulls. There are only about a dozen living species, although they are found on all continents. They live in or near water and have many adaptations to a semi-aquatic way of life, such as a secondary palate that prevents water from entering the airstream.

  – Birds (Aves) are bipedal archosaurs with warm blood, feathers, and powered flight. There are thousands of species that follow many ways of life.

Crocodiles and birds illustrate the same evolutionary principle as ungulates and whales. If you like, it is the converse of parallel evolution. They share a common ancestor, to the exclusion of all other living groups, but their appearance and ways of life are completely different. Their pattern of diversification has also been different. The basic crocodilian way of life—seizing large vertebrates in or near water—is associated with a conservative morphology including a rather lizard-like shape, elongate skull, dermal armor plates, and long, flattened tail that varies only in detail among members of the group. Birds have been a much more "evolvable" group that includes forms as disparate as ostriches and humming birds, or eagles and penguins.

- The Lepidosauria. These are reptiles with teeth fused to the jaw, and a body covering of overlapping scales. There are two modern lineages.

  – The tuatara (Rhynchocephalia) is a rather stoutly-built, lizard-like reptile found only on small islands off the coast of New Zealand. They are mainly nocturnal animals that live in burrows and emerge to feed on invertebrates and small vertebrates.

  – The group of squamate reptiles (Squamata) includes lizards, snakes, and amphisbaenids. They are almost all carnivores, although there is a great diversity of shapes and sizes.

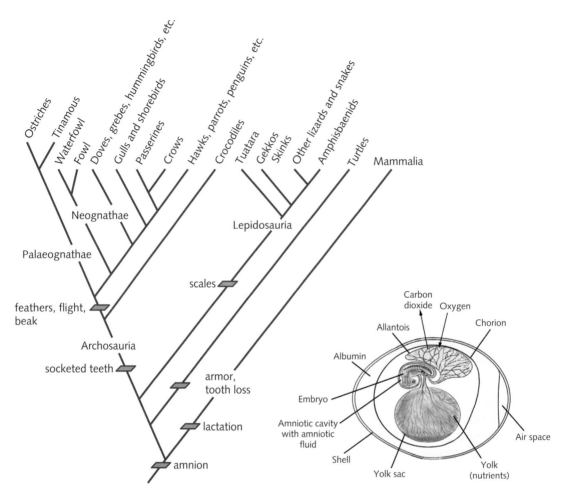

Figure 4.9 Amniota.

Image reproduced with permission from Dept. Biol., Penn State.

They are basically quadrupedal, with sprawling limbs, but limbless forms have evolved independently in several groups. The most familiar are the snakes, but there are several families of lizard in which the limbs are strongly reduced or lost. Amphisbaenids are peculiar legless squamates specialized for a completely subterranean existence.

These two groups of squamates underline the point that sister lineages may have utterly different evolutionary fates, in this case the Rhynchocephalia being restricted to one or two species living on a few small islands whereas the Squamata has diversified into thousands of species living all over the world.

• The Mammalia, which we have already described.

The threefold division into Archosauria, Lepidosauria, and Mammalia does not coincide with the traditional division into reptiles, birds, and mammals. This is because it is based on relatedness rather than on similarity. "Reptiles" is not a monophyletic group, because it excludes birds, which are more closely related to crocodiles than crocodiles are to lepidosaurs. We can continue to use "reptile" in common speech, but it is incorrect in technical language because it gives a misleading impression of phylogenetic relationships.

There is one remaining group of amniotes, the turtles (Chelonia), which is difficult to place. They are very distinctive animals, with their characteristic complete armor and toothless beak, and they occupy terrestrial, freshwater, and marine habitats. They lack the openings in the skull characteristic of all other reptiles, mammals, and birds, which suggests they may be an early-branching lineage. We have shown them branching from the main amniote stem immediately after mammals, but other possibilities have not yet been excluded.

The phylogeny of birds is still being actively debated. When an ancestral group radiated very rapidly at some time in the past, descendant groups are all about the same age, and the features that distinguish descendant groups all arose at about the same time. Consequently, the branching order of these groups is difficult to discern, either with molecular or with morphological characters. There are two basic clades: the Palaeognathae, which includes the large flightless ratites (ostrich, emu, cassowary, and rhea) and the tinamous, and the Neognathae, which includes all other birds. The most deeply branching lineage of Neognathae are the fowl and waterfowl, but above this level relationships in the remaining clade, Neoaves, are not firmly established. We have provisionally adopted the suggestion that the Neoaves has two sister clades, Metaves and Coronaves. The Metaves includes doves, nighthawks, flamingos, tropicbirds, swifts, hummingbirds, grebes, and some less familiar groups; the Coronaves comprises all others, including songbirds. This provides another example of independent adaptive radiations involving extensive parallel evolution.

- Soaring plunge divers: tropicbirds (Metaves) and gannets (Coronaves).
- Swimming divers: grebes and loons.
- Littoral filter-feeders: flamingos and spoonbills.
- Stealth waders: sunbitterns and bitterns.
- Nectarivores: hummingbirds and sunbirds.
- Aerial insectivores: swifts and swallows.
- Large-eyed nocturnal predators: frogmouths and owls.

4.8 Tetrapoda

Clade added: Amphibia
Innovation: the pentadactyl limb

A bony limb made up of three sections is the characteristic tetrapod innovation. The upper section is a single long bone which is hinged to the second section, a pair of parallel long bones, by a joint that allows movement in a plane. The third section is the pentadactyl hand, or foot, consisting of five digits each made up of a series of joints allowing it to bend. The hand articulates with the second set of long bones by a complex joint made up of many small bones that can permit rotation in any plane. Finally, the upper part of the limb articulates with a girdle by a ball-and-socket joint that may also permit considerable freedom of movement. The pelvic girdle becomes fused to the vertebral column by specialized ribs because the hindlimbs provide most of the power to push the body forward. The pectoral girdle becomes detached from the skull because a land animal needs to move its head independently of the forepart of the body. The limb as a whole is a morphological innovation making it possible to support the weight of the body and to move freely on land.

- It is a potentiating innovation because it makes the whole range of terrestrial ways of life available to tetrapods. Most of the adaptations that are associated with consuming grasses, leaves, seeds, nectar, insects, or other vertebrates do not directly involve modifications of the limb, but the possession of limbs is an essential precondition for them to evolve.

- It is also an implementing innovation, because it is a remarkably evolvable structure that can be modified to serve a wide variety of functions. The sprawling limbs of salamanders that allow them to waddle slowly on land have evolved into more specialized structures for jumping in frogs, running in ungulates, digging in moles, and flying in birds and bats.

The sister group of the amniotes is the amphibians. Amphibians are terrestrial animals that are nevertheless tied to water by the permeability of their eggs and their skin.

- Amphibians reproduce by eggs that lack extra-embryonic membranes to protect the embryo and must therefore be laid in water. Some species are viviparous, the embryos developing inside their mother's body with the aid of a yolk sac, and others are able to reproduce on land by incubating the eggs in damp soil. The familiar red-backed salamander (*Plethodon*) of eastern North America, for example, lives and breeds in the forest without ever visiting ponds or streams. Conversely, others such as the clawed toad (*Xenopus*) or the giant salamander (*Cryptobranchus*) live in water all their lives. Amphibian eggs are very sensitive to salt and cannot develop in seawater.

- Amphibians do not drink, but instead take up water through their skin. This enables them to live in damp soil or sand without a need for standing water, but the permeability of their skin also makes them very sensitive to evaporative water loss. For this reason, most species can live only in moist places. Some amphibians, such as spadefoot toads, are able to live in arid habitats, and even in deserts, by burrowing into damp layers of sand, by evolving a tolerance for desiccation (desert toads can survive losing 40% of body weight), or by using the bladder (another tetrapod innovation) to hold a store of water. The permeability of the skin also makes adult amphibians sensitive to salt. A few species can forage along the beach, but all are killed by immersion in seawater.

Because of these constraints, there are no marine amphibians, and many islands remain unpopulated by amphibians because they are unable to cross the sea. We have previously emphasized the great variety of animals that have evolved to follow different ways of life. However, this variety is far from being unconstrained. The marine amphibian is an example of a kind of animal that has never evolved, perhaps because of the fundamental constraints imposed by amphibian development and physiology.

There are three main kinds of living amphibian, whose relationships are illustrated in Figure 4.10.

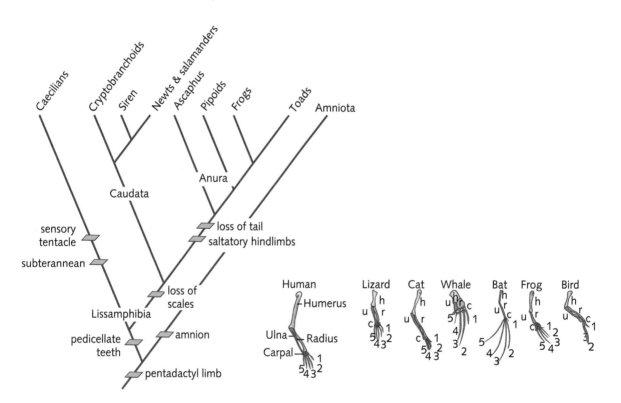

Figure 4.10 **Tetrapoda.**

Image redrawn from Hall and Hallgrimsson (2014), *Strickberger's Evolution* Fifth Edition, with permission from Jones & Bartlett Learning.

- The frogs and toads (Anura) are so familiar that they scarcely need a general description. Their life cycle involves one of the most dramatic events in tetrapod development, the transition from one body plan to a completely different one. Anurans develop from a characteristic limbless aquatic larva, the tadpole, with internal gills and a horny beak for scraping algae from plants and rocks. This metamorphoses into the adult form by remodeling the body under the influence of thyroid hormone. Limbs appear, the tail regresses, the head and mouth alter shape, and internal organs such as heart and gut are modified. One deeply-branching anuran, the tailed frog *Ascaphus*, retains a short tail throughout adult life.

- The newts and salamanders (Caudata) have a long trunk and tail, small limbs, and quadrupedal locomotion. The larva has external gills, but otherwise resembles the adult form. Neoteny has independently evolved in several salamander lineages, through mutations in thyroid hormone receptor genes: individuals fail to metamorphose, but instead become sexually mature as obligately aquatic animals that resemble giant larvae. Examples in different families of amphibians include mudpuppies, axolotls, and the cave-dwelling olm.

- The caecilians (Caecilia or Gymnophiona) are limbless amphibians that live underground in tropical rain forests. They are superficially segmented and resemble large earthworms in appearance. The skull is small and solid, the eyes covered with skin and the tail very short as adaptations to a subterranean way of life. In lieu of eyes, they bear unique chemosensory tentacles in pits on the skull (see Section 1.4.1).

The highly specialized morphology of Anura and Caecilia obscures the relationship among the major amphibian groups. We have retained the traditional arrangement, which also has molecular support, with Caecilia as the most deeply branching clade, but it is also possible that Caecilia and Caudata are sister groups.

4.9 Sarcopterygii

Clades added: coelacanths and lungfishes
Innovation: the monaxial fin; the lung

The potentiating innovation behind the evolution of the tetrapod limb is the monaxial fin of the lobe-finned fishes, the coelacanths and lungfish. In ray-finned fishes the fin is a web of skin extending between parallel bars of cartilage or bone that extend from a broad, low base. In coelacanths the fin is a fleshy lobe consisting of a short series of stout bones terminating in a fan of smaller elements with a fringe of fin rays. In lungfish the leaf-like fin is supported by a long, jointed bony axis with numerous small side-branches. In neither case has the fin itself been extensively modified for different ways of life, although lungfish can use their fins to "walk" underwater, which may be useful in shallow pools.

Air-breathing has evolved in many groups of fish as an adaptation to low oxygen levels. In ray-finned fishes the accessory respiratory structure is usually a modified swim bladder, which develops as an unpaired sac from the dorsal surface of the pharynx. In lungfish the lungs develop as paired pouches from the ventral surface of the pharynx, posterior to the gills. They are used for breathing when the water level falls or becomes deoxygenated. As a similar organ is found in a group of basal ray-finned fish, the African bichirs (*Polypterus*), it is not clear whether lungs evolved independently in the Sarcopterygii or whether they were present in the common ancestor of ray-finned and lobe-finned fish. In either case, the retention of lungs in the lungfish-tetrapod lineage was an essential pre-condition for terrestrial life.

There are two main groups of sarcopterygians besides the tetrapods (Figure 4.11).

- A single species of coelacanth (*Latimeria chalumnae*) has been found in several locations around the coasts of East Africa and Madagascar, where small groups live in marine caves at depths of

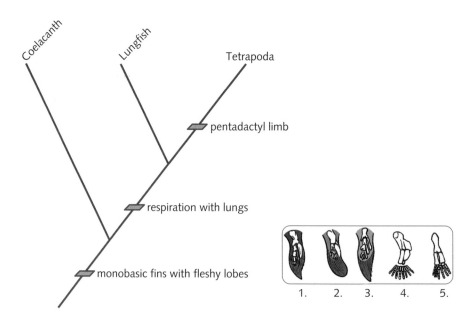

Figure 4.11 Sarcopterygii.

100–200 m. A second population, named as a different species, was found off Sulawesi, Indonesia, in 1997. The coelacanth is a stoutly built fish up to about 2 m in length that eats squid and other fish. It is well known as the last surviving descendant of a group that was much more abundant and diverse in the past, and which was thought to be long extinct until the discovery of the African coelacanths in 1937. It appears to be a rare animal, with a population size of a few hundred individuals. It is very unlikely, however, that so small a population could persist for tens of millions of years. Either we are just in time to witness the final demise of an ancient lineage, or there remain many other populations yet to be discovered.

- There are three genera of lungfish (Dipnoi) in tropical fresh waters. The Australian lungfish (*Neoceratodus*) lives in permanent rivers where oxygen levels are always fairly high. It has fully functional gills and does not seem to rely much on breathing air. The South American (*Lepidosiren*) and African (*Protopterus*) lungfish live in swamps where oxygen levels may become very low. Their gills are reduced, they breathe often, and if prevented from breathing they may asphyxiate. Both can survive in burrows when the pools and streams they live in dry up.

4.10 Osteichthyes (bony fishes)

Clade added: Actinopterygii (ray-finned fishes)
Innovation: the bony endoskeleton

Calcium salts are used by a wide variety of animals to form hard mineralized skeletons for defense (like shells) or offense (like teeth). Invertebrates use calcium carbonate, as crystalline calcite or aragonite. Only vertebrates use calcium phosphate, as crystalline hydroxyapatite. When deposited in an organic matrix of collagen, it forms bone. As dermal bone, it is laid down in dense connective tissue to form flat plates as the basis for skulls and shoulder blades. This is a very ancient vertebrate feature. It can also be formed by the ossification of an originally cartilaginous element, which is how backbones and long bones are formed. These form most of the axial skeleton of bony fishes and tetrapods, a complex system of levers activated by muscles capable of providing both powerful and delicate movement. Endoskeletal bone

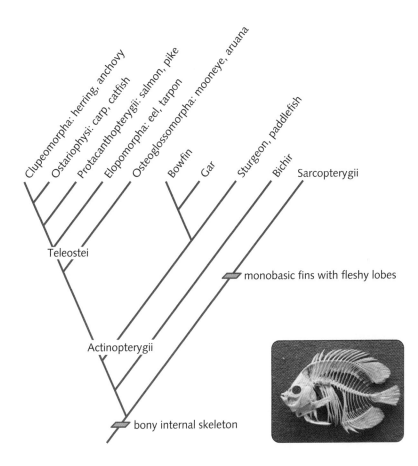

Figure 4.12 **Osteichthyes.**

Image courtesy of Dave Russ. This file is licensed under the Creative Commons Attribution-Share Alike 2.0 UK: England & Wales license.

is the material base for the body plans of bony fishes and their tetrapod descendants.

The adaptive radiation of bony fishes, shown in Figure 4.12, provides many examples of convergent evolution to similar ways of life, as we saw in mammals. The redhorses and suckers (Catastomidae) of North America, for example, have fusiform bodies with large, toughened ventral mouths specialized for foraging in gravel and coarse sediments; the same general arrangement is found in the barbels (Cyprinidae) of Eurasia and the catfish (Plecostomidae) of Central and South America. Another example is the eel-like body form, with reduced fins and serpentine locomotion, that has evolved independently in many groups of fish. Here are two less familiar examples of convergent evolution.

- Antifreeze (see also Section 6.4.1). Fishes that live in very cold seawater can survive only if they are able to protect their body fluids from freezing. Freezing kills cells because ice crystals puncture the cell membrane. This can be prevented by

wrapping the growing crystal with a glycoprotein; this requires a very particular amino acid sequence, because the spacing of the hydrogen-bonding units of the protein must match exactly the spacing of water molecules in the ice crystal lattice. A suitable protein has evolved both in the Arctic cod of the northern hemisphere and the Antarctic notothenoid fish of the southern hemisphere, both of which can function normally in water at –2 °C, with no means of regulating their body temperature. The antifreeze proteins that allow them to do this are almost identical. However, they have evolved quite independently. The notothenoid antifreeze is a modified trypsinogen, whereas the Arctic cod antifreeze is encoded by a locus with no relationship to trypsinogen genes. Moreover, the processing of the gene product is also different in the two groups. Thus, the same adaptive solution has emerged from two different evolutionary routes.

- Electrolocation (see also Section 4.11). Weakly electric fish can generate an electric current of about 1 volt using specialized cells called electrocytes.

These are arranged in an electric organ in the tail like miniature batteries connected in series. When they fire, they produce an electric field in the surrounding water. This field is detected by electroreceptors embedded in the skin, which are able to perceive small alterations in the field caused by nearby objects which conduct electricity differently from water. The fish use this ability to navigate and to detect prey in dark or turbid water where sight is ineffective. It has evolved independently in two distantly related groups of fish: the elephantfish (Mormyridae) in Africa and the knifefish (Gymnotidae) of South America. These families belong to different clades of bony fish, the Mormyridae to Osteoglossomorpha (a deeply branching clade that includes mooneyes and the gigantic Amazonian aruana), and the Gymnotidae to Ostariophysi (which includes more familiar fish such as carp, minnows and catfish). The organs that generate and perceive electric current are quite different in the two families, owing to their different ancestry, but they share a number of striking morphological similarities, in particular the evolution of structures that make it possible to swim without bending the body, so as to keep the body axis of the fish parallel to the axis of the electric field during active movement.

There are three main groups of ray-finned fishes, whose radiation and relationships are illustrated in Figure 4.12. These groups differ greatly in their diversity: two have only a few living species, whereas the third contains more species than any other comparable group of vertebrates.

- The bichirs (Polypteriformes) have fleshy pectoral fins and a respiratory swim bladder. A few species live in African swamps. They resemble lobe-finned fish, and indeed when viewed from above while propelled by their fins in shallow water, or on wet mud, they even resemble salamanders.

- Sturgeons, gars, and bowfins seem to form a distinct clade, although the relationships among these fish are not conclusively resolved. All are large fish with poorly-ossified skeletons whose bodies are armored with bony plates or scales. Gars (Lepisosteiformes) are large predatory fish with elongate jaws found in fresh and brackish waters in North America. Bowfins (Amiiformes) are stoutly-built freshwater fish, now represented by a single North American species. Both gars and bowfins can breathe air, the gars with a vascularized swim bladder and the bowfins with a paired lung arising from the ventral surface of the pharynx. Sturgeons (Acipenseridae) typically feed in coastal waters of Eurasia and North America, entering rivers to spawn, although some are found entirely in freshwater. They have large ventral mouths equipped with barbels, which they use to locate prey in soft sediments. Paddlefish (Polyodontidae) are related to sturgeons; they are large freshwater fish found in North America and China that are distinguished by a long flattened snout used to filter zooplankton.

- All the remaining ray-finned fish are teleosts (Teleostei), a very diverse and heterogeneous group found in marine and freshwater habitats throughout the world. They have homocercal tails in which the spine ends at the caudal peduncle; the other groups described above have heterocercal tails in which the spine extends into the upper lobe of the tail.

Teleosts have highly kinetic skulls with many moveable elements. In the simplest terms, the jaw skeleton of most teleosts is modified so that the maxilla acts as a lever to protrude the premaxilla, giving delicate control over mouth movements that enables them to capture individual prey. In any particular case the anatomical arrangement of bones and muscles is more complicated, with a dozen or more elements involved in the control of feeding movements. The teleost jaw is an implementing innovation that has allowed the group to diversify into a very wide range of specialized ways of life. These can be grouped into several main themes. For example, muscles operating close to the fulcrum of the jaw generate rapid movement with a small mechanical advantage. This is adapted for the capture of relatively small active prey by fish with long narrow jaws, like gars or needlefishes. The converse design gives a powerful crushing bite in fish like bowfins, carp and piranha. This neatly illustrates two evolutionary principles. The first is that adaptation cannot be perfect because features that are useful in different ways of life are often in conflict: a jaw can be fast or powerful, but it cannot be both at the same time. Secondly, functionally similar jaws have evolved independently in several lineages of bony fishes.

4.11 Gnathostomata

Clades added: Chondrichthyes (sharks and rays)
Innovation: jaws

The jaw is a device for opening and closing the mouth forcefully, using a pair of hinged elements operated by muscles inserted on the side of the skull. The margin of the jaw is a ridge of hard tissue often bearing hard bony teeth. Jawless animals can feed on fluids and small soft particles only. Animals with jaws are able to tackle large and tough prey, using their jaw and teeth to pierce, cut, slice, crush, and grind solid tissue. Hence, jawed vertebrates are able to capture and process food items as diverse as seeds, leaves, fish, snails, shrimps, insects, and ungulates. The jaw is the potentiating innovation that made possible the radiation of aquatic and terrestrial vertebrates, including subsequent radiations such as the teleosts, based on particular versions of jaw architecture.

Sharks and their relatives (Chondrichthyes) are mainly large marine fish with entirely cartilaginous skeletons. They are very rarely found in freshwater, and never develop lungs. They have ventral mouths usually armed with sharp teeth formed from dermal denticles, the classic "Jaws" appearance. These are used to cut and tear lumps of flesh from large prey such as marine mammals. Other sharks feed on small fish and shrimps, using their teeth to cut up these smaller prey items. Several clades have evolved flattened tooth plates used to crush mollusk shells. A few species, including the largest pelagic sharks, simply swim with the mouth open and strain out and swallow small crustaceans and zooplankton. The radiation of sharks has been modest, however, relative to that of bony fishes, perhaps reflecting the lesser potential of their jaw to be modified for different ways of life.

Typical pelagic sharks share many features in common with other large marine vertebrates such as tuna, whales, and ichthyosaurs, including a streamlined body and lunate tail fin (Section 1.3.1). Two less familiar examples of convergent evolution are placentation and strong electrical discharges.

- We may think of the placenta as being a uniquely mammalian innovation, but in fact similar structures have repeatedly evolved in other groups, such as lizards (skinks, Scincidae) and bony fishes (guppies, Poeciliidae). Perhaps surprisingly, advanced modes of placentation have also evolved in some live-bearing sharks, such as hammerhead sharks. The embryo in these species is initially dependent on yolk and a nutritive "milk" secreted into the uterus. The yolk sac then differentiates into a true placenta, a vascular organ consisting of a mixture of maternal and embryonic tissue that mediates the flow of nutrition and waste products between mother and offspring and is capable of sustaining the young during a prolonged gestation. The young sharks are born with placenta and umbilical cord still attached. The placenta is a complex organ, with over 40 genes known to be involved in regulating its development. It must overcome formidable and fundamental difficulties, such as the immunological screening of the embryo. Nevertheless, a placenta has evolved independently in several viviparous groups of gnathostomes. There are, of course, many other viviparous groups in which a placenta has failed to evolve, but the close similarity of placentation in sharks and mammals is striking evidence that complex adaptations evolve more readily than might have been imagined.

- Another remarkable example of convergent evolution is provided by strong electric currents produced by bony and cartilaginous fishes (see Section 9.3.3). The basic anatomy is a hypertrophied version of the structures responsible for the weak electric discharges produced by gymnotids and mormyrids for electrolocation (see Section 4.10). The electric organ of strongly electric fish contains electroplaques, which are stacks of coin-shaped cells that develop from neuromuscular tissue.

Sodium ions are pumped out of these cells, creating an electrical potential across the cell membrane. When the permeability of the membrane is altered (by a nervous impulse), sodium ions flow into the negatively charged cytoplasm, resulting in an electric current. Arranging more electroplaques in series gives a higher voltage; arranging more in parallel gives greater amperage. The electric "eel" (*Electrophorus*, Gymnotidae) is a South American species living in shallow fresh waters that has massive electric organs in the tail. It can deliver about 400 volts at 1 amp (0.4 kW), sufficient to stun large prey that it can then eat. Ironically, it uses its perception of the weak electric impulses of other gymnotids to stalk its prey. The electric catfish are tropical African forms in a distantly related family (Malapteruridae) that have independently evolved electric organs capable of developing a potential of about 340 volts. More surprisingly, the same capability has evolved in the electric rays (Torpediniformes), a family of cartilaginous fishes related to the familiar rays and skates, found worldwide in warm and temperate oceans. They have large electric organs on either side of the head, capable of generating up to about 200 volts; the Pacific ray is said to generate 40 volts at 20 amps (1 kW). These rays fold their pectoral fins around their prey, then kill it with an electric discharge. It seems that there is an almost universal capacity among fish for detecting weak electrical impulses in water that has convergently evolved in some groups as a means of navigating and detecting prey, and in yet others as a device for stunning or killing prey.

Chondrichthyes are much less diverse than Osteichthyes. Figure 4.13 shows the two main clades, one of which contains all the familiar sharks and rays.

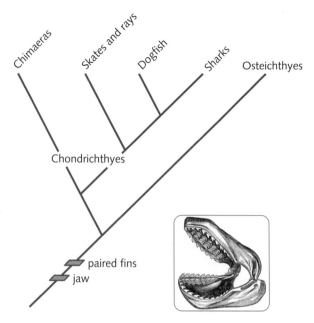

Figure 4.13 Gnathostomata.

- The chimaeras (Holocephali) are rare deep-sea forms that are the sister group of all other cartilaginous fishes. The upper jaw is firmly fused to the brain case and, like the lower jaw, bears a large tooth plate. This provides a powerful grinding apparatus for processing benthic invertebrates such as gastropods, echinoderms, and crabs.

- The rays (Batoidea) are also benthic feeders but have enormously expanded pectoral fins. These enable a bottom-dwelling fish to resist currents, and they eliminate the give-away shadow cast by a round-bodied fish resting on the bottom. The sharks and dogfish (Squalimorpha) include all other cartilaginous fish and are typically pelagic predators with fusiform bodies and a large heterocercal tail fin, although there are also many benthic forms.

4.12 Chordata

Clades added: Agnatha, Cephalochordata, Urochordata
Innovation: the notochord

The notochord forms the axial skeleton of the embryo of chordates. It provides a firm but flexible structure permitting undulatory swimming by the alternate

contraction of muscles on opposite sides of the body. In one deeply branching clade (Cephalochordata) it is retained in the adult; in other chordates it is lost

(Urochordata) or replaced by the vertebral column (Vertebrata). It acts during development to regulate development in both in an anterior-posterior axis and in a dorsal-ventral axis, by specifying cell fate through secreted growth factors. It is the potentiating innovation that permits the evolution of specialized structures such as fins and limbs in different regions of the body of an elongate bilaterally symmetrical animal.

- The amphioxus (*Branchiostoma*, Cephalochordata) is a small (ca. 4 cm) transparent marine animal shaped like a scalpel blade. It has a notochord, dorsal nerve cord, and pharyngeal gill slits. Water is pulled into the mouth by ciliary action and passes through the pharyngeal slits, where food particles are trapped by a mucus sheet, and into a spacious chamber called the atrium, from which it is expelled through a posterior opening. Amphioxus has chevron-shaped myomeres, like vertebrates, and can burrow into coarse sand by eel-like flexion of its body. It looks like a very simple version of a fish that lacks brain, skull, jaws, ears, heart, and kidneys. There are only about 30 species of amphioxus: evidently, the notochord in itself does not open up many new ways of life.

- Tunicates (Urochordata) have a similar mode of feeding, by drawing in a stream of water through the inhalant siphon, sieving it through pharyngeal perforations into an atrium, and finally expelling it through an exhalent siphon. The adults appear quite different in form, however: they have a sack-like body enclosed in a tough protein–polysaccharide tunic, and are firmly and immovably attached to the substrate. Many tunicates are colonial, with new adults being formed by budding to create a group of individuals that remain connected to one another with living tissue. The chordate affinity of tunicates is shown most clearly by their larva, which is a very small (1 mm) swimming animal with a clearly formed notochord and dorsal nerve cord. They look like miniature tadpoles and can swim quite fast (ca. 1 cm s^{-1}) by flexing the tail. They do not feed, but seek suitable settlement sites where they metamorphose into the adult form. Tunicates have undergone a modest radiation, with some pelagic forms (salps, Thaliacea) that use muscular force to drive a water current

powerful enough to serve for locomotion as well as feeding. There are about 3000 species of tunicate in oceans throughout the world.

The tunicates have long been considered to be the sister group of a clade (Pharyngotremata) that comprises amphioxus and the craniates, as shown in Figure 4.14. Recent molecular analyses suggest that amphioxus may be the basal clade, with tunicates and craniates as sister groups. In either case, there is a remarkable divergence between the adult body plans of tunicates and craniates. We have emphasized convergent evolution as demonstrating the power of natural selection to drive the evolution of superficially similar adaptations to a common way of life in unrelated lineages. The converse phenomenon is divergent evolution, when very different kinds of organism evolve from a common ancestor. Few organisms could be more different in form or in lifestyle than a tunicate anchored to a rock and the fish swimming by it, but their development unambiguously reveals their evolutionary relationship.

The craniates (Craniata) are defined by their head, which bears a skull containing the brain and other organs such as ears and eyes. The brain develops from

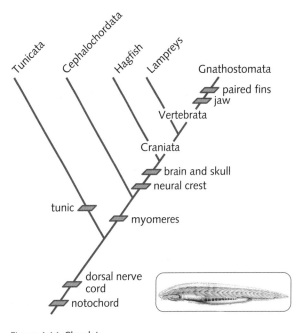

Figure 4.14 Chordata.

the anterior region of the dorsal nerve cord, which itself develops from a strip of ectoderm running between the notochord and the overlying epidermis. Transcription factors secreted by the notochord and epidermis interact to induce the neural ectoderm to thicken and curl up to form a tube. The cells along the rims of this strip continue to divide rapidly, and as the rims fuse to close the tube these cells delaminate by moving out of the neural ectoderm to form a loose tissue that migrates down the sides of the embryo, under the epidermis. This migration is highly directional: cells from a particular region of the neural crest migrate to a specific locality, where they differentiate into a particular cell type. It involves very large numbers of cells, especially in the head region, and is involved in the formation of a wide range of tissues, especially ganglia, but also including the pharyngeal arches and their derivatives, thymus and thyroid glands, the eye and inner ear, the teeth, and the connective tissue of the heart and great blood vessels. The remodeling of the head as a battery of sensory organs and feeding structures, derived in large part from neural crest cells, is the crucial potentiating innovation in the development of craniates.

● There are two groups of craniates without jaws, the hagfish (Myxinoidea) and the lampreys (Petromyzontida). The hagfish have sharp horny teeth borne on a plate of cartilage within the mouth cavity. This can be protracted, causing the teeth to open, and retracted, causing the teeth to close, to give a rasping action that strips the flesh from dead fish and whales. There are about 70 species, found mainly in cold oceans in both hemispheres. The lampreys have a specialized mouth that is capable of holding on to a living fish while rasping through its skin to obtain tissue and body fluids. Lampreys

reproduce in freshwater, migrating to the sea to feed; there are about 40 species worldwide. Hagfish and lampreys are traditionally grouped together as Agnatha, as they lack jaws and other attributes of gnathostomes, such as paired fins and scales, and share other attributes, such as horny teeth. This interpretation is supported by the most extensive molecular analyses. The alternative view is that lampreys are vertebrates. Although the notochord persists in the adult, it bears vertebral elements along its dorsal surface, and lampreys share other features with gnathostomes, such as radial fin muscles and specialized heart innervation. This is currently the consensus view (Figure 4.14), but may be revised as more molecular and developmental evidence accumulates.

It was observed some time ago that vertebrates have four times as many copies of some genes, such as the *Hox* cluster, as invertebrates. This suggested that the entire genome was duplicated twice, early in vertebrate evolution. It soon became apparent that most genes did not follow the 4:1 rule, but this does not refute the double-duplication theory, because the great majority of duplicated genes might be subsequently lost or modified. Complete genome sequences allow the theory to be tested, because the remaining duplicated genes should continue to occupy the same physical locations. As yet, there are too few sequences available for the critical basal groups, but the most careful analyses seem to favor the idea that early vertebrate history involved two rounds of gene duplication. This could have been important in the subsequent radiation of the vertebrates, as the spare copies of duplicated genes can be selected for new functions, without impairing their original function (Section 6.3.1).

4.13 Deuterostomia

Clades added: Echinodermata, Hemichordata

Innovation: pharyngeal gill slits

Animal eggs develop into a hollow ball of cells, the blastula; at some point on its surface a dimple called the blastopore forms that pushes inside to create a two-layered embryo, the gastrula. The developmental fate of the blastopore distinguishes the two

great clades of multicellular animals: in protostomes it is a slit that rolls into a tube whose open ends become mouth and anus, whereas in deuterostomes it becomes the anus alone, with the mouth formed as a new opening where the ingrowing blastopore meets the gastrula wall opposite. Chordates are deuterostomes. The other characteristic features of deuterostomes are radial cleavage, the formation of the coelom by outpocketing from the gut, and indeterminate development. These are only generalizations, to which there are many exceptions, but the list emphasizes the primacy of development in signposting the major branching points of evolution.

The pharyngeal gill slits are an important innovation of deuterostomes. The notochord of chordates and the hemichordates gill slits are replaced by other structures in most adult forms, but can be recognized at some stage in development. The sister group of chordates is the Ambulacraria, which comprises hemichordates and echinoderms, as shown in Figure 4.15. The hemichordates share some chordate features, such as the gill slits, although these have been completely lost in echinoderms. The endoderm of the pharyngeal slits expresses the same regulatory genes in hemichordates and chordates, leading to the conclusion that these structures were present in the common ancestor of deuterostomes and have been lost in echinoderms. Hence, chordates, hemichordates, and echinoderms comprise a distinct clade, the Deuterostomata.

• The two groups of hemichordates (Hemichordata) are the acorn worms (Enteropneusta) and the pterobranchs (Pterobranchia). Acorn worms are medium-size (10–200 cm) solitary worm-like animals that use a distinctive muscular proboscis to burrow in the intertidal zone. They use cilia to sweep food particles into the mouth, or swallow sand and sediment, sieving out food particles with a long series of pharyngeal slits. There are about 70 species. Pterobranchs are minute (1–4 mm) stalked sessile animals that bud to form colonies. They use ciliated tentacles to pass food particles to the mouth, and sieve them with a single pair of pharyngeal slits. About 30 species have been named, mostly from deep water.

Food particles are trapped and transported to the gut by a ciliated groove in the ventral wall of the pharynx, the endostyle. In lampreys, a similar groove is converted into the thyroid gland at metamorphosis, so the hemichordate endostyle may be homologous with the vertebrate thyroid. Hemichordates also possess a short rod of cells extending into the proboscis that has been likened to a notochord, but it develops differently and probably evolved independently. They have a diffuse nervous system with no brain or dorsal nerve cord.

• The echinoderms (Echinodermata) are a much more diverse and abundant group, with about 7000 species of starfish (Asteroidea), brittle stars (Ophiuroidea), sea urchins (Echinoidea), sea

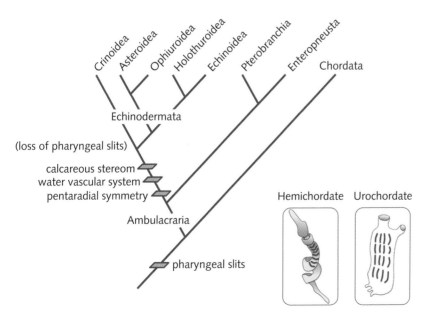

Figure 4.15 Deuterostomata.

cucumbers (Holothuroidea), and crinoids (Crinoidea). The adults have a distinctive fivefold symmetry, although they develop from bilaterally symmetrical larvae. Their body plan is based on a unique water vascular system that provides a hydrostatic skeleton to power the rows of tube feet used in locomotion, protected by a skeleton of calcareous plates. They have neither brain nor central nervous system, nor any other morphological structure linking them with chordates. The vast difference in adult body plan between the two groups is no doubt associated with differences in how development is regulated: echinoderms have only a single *Hox* cluster, and the genes do not seem to be expressed in the collinear fashion of vertebrates and arthropods. Working out the evolutionary basis of the wide divergence in deuterostome body plans is currently an active field of research.

4.14 Bilateria

Clade added: Protostomia

Innovation: directed locomotion; cephalization

Sessile organisms conserve energy, take full advantage of a favorable location, and can usually prevent competitors from establishing themselves nearby. There are disadvantages to being fixed in one place, however. If it depletes the local food supply, a sessile organism cannot forage for food; it cannot seek another site if conditions deteriorate; it cannot flee from predators; nor can it search for mates. There are many ways of life that only motile organisms can follow, and consequently there will be selection to harness the potential of muscle and nerve for active locomotion. Motile organisms normally have a preferred direction of movement; this favors an elongation of the body in the plane of movement, with a leading and a trailing end. Movement over a surface in turn creates a distinction between the side of the body in contact with the surface and the side in contact with the surrounding medium. Finally, an elongate organism that is always the same way up necessarily has right and left sides of the body. These simple considerations set up three axes of symmetry: anterior–posterior, dorsal–ventral, and left–right. Animals with this bilaterally symmetrical body plan are the Bilateria. This is a general description rather than a definition: some animals have bilateral symmetry but are not Bilateria (such as ctenophores) while some Bilateria are not bilaterally symmetrical, at least as adults (such as echinoderms).

The evolution of bilateral symmetry creates the opportunity for regional differentiation of the body along the axes of symmetry, and hence the evolution of more complex body plans. The primary developmental innovation that takes advantage of this opportunity is mesoderm, the layer of cells appearing after gastrulation between ectoderm and endoderm. The mesoderm is responsible for the formation of muscle and most of the internal organs. Animals with mesoderm are said to be triploblastic, that is, developing from three "germ layers" of tissue; those with ectoderm and endoderm alone are diploblastic (as described in Section 4.15). Mesoderm typically arises from cells migrating from the margins of the blastopore to form cavities in the interior of the embryo (schizocoely), or from cells lining cavities formed as outpocketings of the gut (enterocoely). In either case, one consequence of mesoderm formation is the appearance of a body cavity or cavities, the coelom. The differentiation of mesoderm and the arrangement of body cavities are the main features of the range of body plans found in Bilateria.

The most strongly differentiated axis in bilateral animals is anterior–posterior, because sensory and feeding structures are best placed at the front, where they will encounter external objects first, while the gut and gonoducts discharge at the rear. Hence, bilateral animals are usually differentiated from front to rear into regions such as head, trunk, abdomen, and tail, each with its characteristic complement of structures. This differentiation is mediated by transcription factors encoded by regulatory genes. The best-known are the *Hox* genes, which occur in clusters at specific sites in the genome (Section 1.4.3). The expression of a particular *Hox* gene in a particular part of the body during development governs what structures will be formed there. Orthologous *Hox* genes are expressed in comparable regions of the body, even in distantly related animals. Genes at one end of the cluster are expressed first, with expression passing down the cluster until those at the opposite end are expressed last. In animals that develop from front to rear, this means that the genes are arranged in the same order as the body regions they are expressed in (3′ being anterior and 5′ posterior). The collinearity of the *Hox* cluster and the conservation of the *Hox* code are remarkable observations that stimulated the intense research effort in evolutionary developmental biology over the last decade. This is described in more detail in Chapter 10. Here, we introduce the *Hox* genes and related elements as the underlying genetic machinery responsible for the radiation of Bilateria.

The phylogeny of the Bilateria has been controversial for over a century. Most organisms can be immediately placed into a sharply defined phylum, leaving only a handful of species of uncertain affinities. The relationships among phyla are much more difficult to establish with confidence. The older approach used comparative morphology and embryology to delineate a succession from simpler to more complex body plans. The underlying concept is that the earliest animals must have been simpler than modern forms, so that in general evolution will proceed in the direction of more complex and highly integrated organisms. Some of the major trends used to work out the relationships among phyla were:

- Radial → bilateral → cephalized
- Acoelomate → pseudocoelomate → coelomate
- Gastric cavity → through-gut
- Diploblastic → triploblastic
- Cell groups → tissues → organs
- Open body cavity → segmentation

On this basis, the most deeply-branching bilaterians are the flatworms (Platyhelminthes), which have neither a body cavity nor a through-gut. Their descendants are the pseudocoelomates, or Aschelminthes, whose body cavity is not a true coelom because it is not lined with mesoderm. This group includes nematodes, rotifers, and several minor groups. The most recently derived group comprises the true coelomates. These include schizocoelous phyla such as mollusks, annelids, and arthropods, which form the Protostomia, and enterocoelous phyla such as chordates and echinoderms, which form the Deuterostomia.

The application of DNA sequencing to phylogenetic analysis has confirmed many of the key features of the classical interpretation, for example the basal position of Porifera (Section 4.15), the diploblasts as sister group to the triploblasts (Bilateria), and the monophyletic Deuterostomia. However, it has led to an extensive reinterpretation of the Protostomia. The modern view is shown in Figure 4.16. In particular, the Platyhelminthes are secondarily simplified

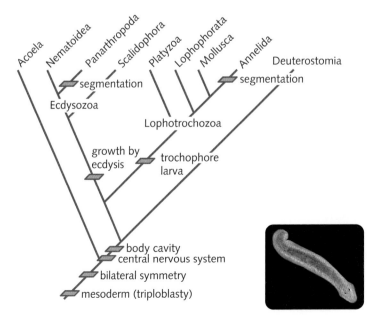

Figure 4.16 Bilateria.

Image courtesy of Eduard Solà. This file is licensed under the Creative Commons Attribution-Share Alike 3.0 Unported license.

protostomes, Aschelminthes is suppressed, and the Protostomia is arranged in two large clades, Ecdysozoa and Lophotrochozoa.

- Ecdysozoa. These are protostomes with a cuticle that must be shed periodically in order to permit growth.

    – Scalidophora. A small clade of marine protostomes characterized by an eversible proboscis armed with spines that they use for capturing prey. Priapulida is a small phylum (10 species) of unsegmented marine worms with a spacious coelom used as a hydrostatic skeleton in burrowing. Kinorhyncha (140 species) are minute annulated animals living in coarse marine sands. Loricifera (20 species) is a recently discovered phylum that likewise lives in marine sediments.

    – Nematoidea. The nematodes (Nematoda) are a highly diverse (24,000 species) phylum including both free-living and parasitic species. All have a narrow, tapered, unsegmented body protected by a thick collagenous cuticle, and move by serpentine undulation. They are extremely abundant and ecologically important, especially in soil communities. Most free-living forms graze on bacteria and fungi, moving food through the gut by rhythmic contractions of the pharynx. The Nematomorpha (320 species) is a related group of very long, thin worms lacking a gut that are internal parasites of arthropods.

    – Panarthropoda. The arthropods are an enormously diverse group (about 1 million species) of segmented protostomes. All possess a thick chitinous cuticle and jointed limbs. Their diversification is based on a flexible developmental program governing the regional specialization of segments. The main clades within Panarthropoda are:

        o Lobopods: two basal groups with fleshy unjointed limbs. The Onychophora are worm-like animals living in the litter of rainforests. The Tardigrada are minute (<1 mm) animals found in ephemeral water such as wet moss.

        o Chelicerata (spiders and scorpions) have mouthparts on two head segments and walking legs on thoracic segments, with no appendages on the abdomen.

        o Myriapoda (millipedes and centipedes) have antennae on the second segment and mouthparts on segments 4–7, with a proverbially long series of walking legs on the posterior segments.

        o Crustacea (crabs, shrimps, copepods, etc.) have a complicated structure based on two pairs of antennae, 3–4 mouthparts, and a series of posterior segments that bear walking or swimming appendages.

        o Hexapoda (insects) have one pair of antennae, three pairs of mouthparts on segments 4–6, three pairs of walking legs, and 1–2 pairs of wings on thoracic segments, with no appendages on the abdomen.

- Lophotrochozoa. Protostomes with a lophophore (tuft of ciliated tentacles used for feeding) or a trochophore larva.

    – Annelida (earthworms, leeches, ragworms, tubeworms, etc.). A diverse phylum (more than 10,000 species) of segmented worms. Most excretory and reproductive organs are repeated in each segment, which also bears hollow unjointed chitinous tubes (chaetae) used for locomotion.

    – Mollusca (snails, slugs, clams, chitons, squid, etc.). A very diverse (more than 100,000 species) phylum of unsegmented protostomes with a body divided into a head, bearing tentacles, a muscular foot used in locomotion, and the visceral mass housing the organs.

    – Lophophorata. The phoronids (Phoronida) and brachiopods (Brachiopoda) are rarely seen marine protostomes bearing a lophophore. Brachiopods are enclosed in a bivalve shell and look superficially like a clam.

    – Platyzoa. A group of different kinds of small organisms that may or may not be a clade. It includes the rotifers (Rotifera) and most flatworms (Platyhelminthes), as well as less familiar phyla such as Gastrotricha and Gnathostomulida.

There are a few groups whose position remains uncertain. Chaetognaths (Chaetognatha) are a small phylum of worm-like marine organisms with sharp grasping teeth that are important predators on fish larvae. At present they cannot be satisfactorily related to any other group in Bilateria. The bryozoans (Ectoprocta or Bryozoa) are minute colonial animals bearing a lophophore that are currently assigned a basal position within Lophotrochozoa, which on morphological grounds seems unlikely. The phylogeny of protostomes remains a topic of much debate.

The most debatable issue of all is the identity of the most basal group of Bilateria. With the flatworms removed from this position, there has been a renewed search for their successor. The difficulty in identifying suitable candidates is that a simple body plan may well evolve secondarily. The Mesozoa, for example, are minute worm-like organisms of fewer than 100 cells, but they are internal parasites of marine invertebrates and are probably secondarily simplified as an adaptation to living in body fluids. *Xenoturbella* is a marine worm that is little more than a ciliated fluid-filled bag without a through gut, coelom, excretory structures, gonads, brain, or central nervous system, but from molecular data it seems to be an aberrant deuterostome. The most convincing candidates are the acoels (Acoela and Nematodermatida, perhaps not a clade), which are small solid-bodied marine worms formerly associated with Platyhelminthes. Morphological and molecular analyses support a basal position, making them the sister group of all other Bilateria. Moreover, they have fewer *Hox* genes than other Bilateria.

However, there may well be no extant clade that preserves the ancestral condition of Bilateria. The most plausible ancestral state of Bilateria may be a small acoelomate marine worm resembling the modern acoels, but there is at present no decisive evidence that identifies acoels as the least-modified descendants of the most recent common ancestor of Bilateria. It is by no means inevitable that the ancestral condition will persist in the modern biota.

4.15 Metazoa

Clades added: Cnidaria; Ctenophora; Porifera
Innovation: extracellular matrix; somatic division of labor

The Metazoa are the multicellular animals. They are one of the five groups (Metazoa, Chlorophyta, Rhodophyta, Phaeophyta, Fungi) that have independently evolved regular multicellular form with somatic division of labor among cells or tissues. Becoming multicellular requires two fundamental innovations.

First, a multicellular body requires that cells should remain associated rather than separating after division. Cells may be directly connected to one another, or they may be connected indirectly through an extracellular matrix (ECM). Both systems already exist in Porifera, and they are retained, and elaborated, in all other Metazoa. The ECM is a universal feature of metazoans that is not found in their unicellular relatives. It is a compound material, somewhat akin in this respect to fiberglass, in which long fibrils are embedded in an amorphous ground substance. The gel-like ground substance is formed from unbranched polysaccharide chains of many different kinds, each made up of a unique repetitive disaccharide. The typical fibrous substance of metazoan ECM is collagen, a protein based on an alpha-chain in which every third residue is glycine. This repetitive structure causes it to fold into a triple helix of three intertwined alpha-chains. Other types of collagen have non-repetitive sequences interrupting the series of triplet repeats and form non-fibrillar structures. The collagen of the ECM forms a continuous intermolecular network consisting of both fibrillar and non-fibrillar components stabilized by cross-linking covalent bonds that provides the mechanical stability of tissues.

Only two kinds of collagen are known from Porifera: free fibrils in the mesohyl and the non-fibrillar spongin that forms the skeleton and attaches the animal to the substrate. In other metazoans there has been an extensive radiation of different kinds of collagen based on the modular structure of the collagen molecule.

The second basic feature of a multicellular body is that it should have a definite form arising from the differentiation of cell types, their arrangement into tissues, and their strictly limited proliferation. This is ultimately based on the primary distinction between:

- the soma, consisting of sterile vegetative cells that are responsible for vegetative functions such as digestion, secretion, protection, and transport;

- the germ line, consisting of reproductive cells that continue to proliferate and which alone are capable of developing into new individuals.

This primary division is always accompanied in metazoans by the secondary differentiation of somatic cells into a range of types, each specialized for a particular function. The human body includes somewhat more than 100 morphologically distinct kinds of cell; sponges have only four or five.

Each of these cell types must develop in the appropriate region of the body so as to form functional tissues and organs. The development of a metazoan involves the formation and folding of sheets of cells, beginning with the characteristic metazoan event of gastrulation. All of this requires a high degree of regulation, which is supplied by transcription factors encoded by regulatory genes such as those of the Hox and paraHox families. It is mediated through the ECM, which is far from being an inert filler, and which provides the equipment for cell–cell signaling such as the animal-specific tyrosine kinase gene family.

The expression of multicellular animal body plans thus requires the deployment of a complex battery of cellular, genetic, developmental, and physiological systems. The most deeply branching Metazoa are the sponges, which have few cell types, a simple ECM, no regular tissues, develop without gastrulation and have no *Hox* genes. Cnidaria are more complex, develop by gastrulation, form true tissues, and possess a restricted range of *Hox* genes. It is in the Cnidaria that muscle and nerve cells that are characteristic of all other metazoans appear for the first time. They form the basis of the subsequent elaboration of metazoan body plans in the Bilateria.

The relationships of the basal clades of Metazoa are illustrated in Figure 4.17. Some aspects of this phylogeny (such as the basal position of Porifera) are well established, whereas others (such as the sister groups of Placozoa and Ctenophora) remain debatable.

- Porifera. The body plan of sponges is based on chambers containing flagellated cells that draw a current of water through the animal from which edible particles can be trapped by the collar cells, which have a mucilaginous collar that traps bacteria. Sponges lack epithelium or tissues and always appear as basal metazoans in phylogenetic analyses. It is not clear whether they constitute a monophyletic group, as we have shown them. An alternative arrangement supported by recent molecular analyses suggests that Demospongiae (freshwater and marine sponges with a spongin skeleton stiffened by silicaceous spicules) and Hexactinellida (deepwater marine sponges with syncitial somatic cells and silicaceous spicules) may constitute a clade to the exclusion of later-branching Calcarea (marine sponges with calcareous spicules).

- Placozoa. There is a clutter of dubious organisms close to the divergence of diploblastic metazoans from the sponges. The best-known are the Placozoa, flattened discoid animals capable of slowly creeping over hard substrates. It is possible that they are the surviving little-modified descendants of a benthic cnidarian larva ancestral to Bilateria, but this remains controversial.

- Cnidaria. The Cnidaria are radially symmetrical, diploblastic animals with muscle and nerve, but without a central nervous system or any degree of cephalization: they are "headless animals." Their basic body plan is a column of cells surrounding a blind gut and crowned with tentacles equipped with characteristic stinging cells, the nematocysts, which paralyze their prey. As a group, they have an extraordinary life history in which a benthic phase alternates with a pelagic phase. The benthic phase is the polyp, like the freshwater *Hydra*, which is attached to the substrate and extends its tentacles into the water column. The pelagic phase is the medusa, a jellyfish that is capable of swimming slowly by contractions of a muscular umbrella of tissue. The polyp reproduces to give rise to medusae, while the medusa reproduces to give rise to polyps. In some groups (Anthozoa, the sea anemones) the polyp is the dominant phase, whereas in others (Scyphozoa, the jellyfish) it is the medusa. In Hydrozoa

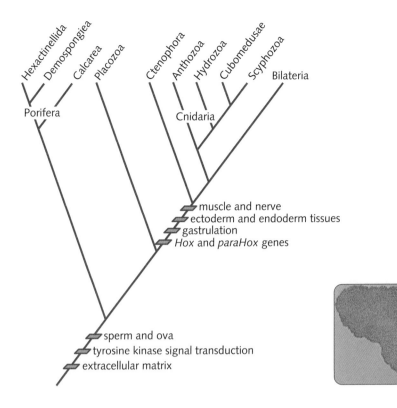

muscle and nerve
ectoderm and endoderm tissues
gastrulation
Hox and *paraHox* genes

sperm and ova
tyrosine kinase signal transduction
extracellular matrix

Figure 4.17 Metazoa.

Image courtesy of Oliver Voigt.
This file is licensed under the
Creative Commons Attribution-
Share Alike 3.0 Unported license.

(corals, and many less familiar forms) both phases are usually prominent, so that the same life cycle includes two different body plans. This radical dichotomy has no close parallel among free-living Bilateria.

- Ctenophora. The Ctenophora are likewise radially symmetrical predators, but are often elongate and swim by means of fused groups of cilia that propel the animal like oars. Their relationship with Cnidaria has long been debated. Both groups are diploblastic and were traditionally included in a single group, the Coelenterata. The DNA evidence suggests that they are a separate and probably earlier-branching clade, but their precise relationships have yet to be determined beyond doubt.

4.16 Unikonta

Clades added: Choanoflagellata; Fungi; Amoebozoa

Innovation: chitin

The attributes required for the evolution of multicellularity did not all appear abruptly in the common ancestor of Metazoa, but rather evolved from simpler precursors. It had long been guessed that the unicellular sister group of animals is the Choanoflagellata. These have a single flagellum beating within a mucilaginous collar, enabling them to trap and consume bacteria, and are similar in appearance to the collar cells of sponges. This bold speculation has been strongly confirmed by molecular analyses, which consistently identify choanoflagellates as the sister group of metazoans. The first step in the evolution of multicellularity is represented by colonial forms possessing an ECM that allows similar cells to remain associated with one another. Many choanoflagellates are colonial, and some appear to have a rudimentary soma–germ distinction between cells in a common matrix. The most likely route from a unicellular organism towards a multicellular animal thus involves a large benthic colonial choanoflagellate as the immediate ancestor of sponges.

As we explore still deeper branches in the tree of life, it becomes more and more difficult to discern monophyletic groups that are clearly defined by shared derived characters. The membership of the major groups of eukaryotes and the relationships between them are impossible to discern from morphological characters, and it is only recently that detailed molecular analyses have begun to reveal them. These analyses have consistently supported a clade consisting of animals, fungi and a few other organisms, which has been called "Unikonta," referring to the possession of a single posterior flagellum. Hence, the sister group of animals (plus a few unicellular relatives, such as choanoflagellates) is the fungi, despite their utterly different body plan, as shown in Figure 4.18. Both groups are multicellular, but whereas animals have evolved massive three-dimensional bodies specialized for ingestion, fungi have evolved extensive one-dimensional bodies (mycelia) specialized for absorption.

- Chytridiomycota. The chytrids are typically unicellular parasites with a chitinous cell wall. They attack algae, plants, animals and other fungi and are suspected of contributing to recent declines in amphibian abundance and diversity.

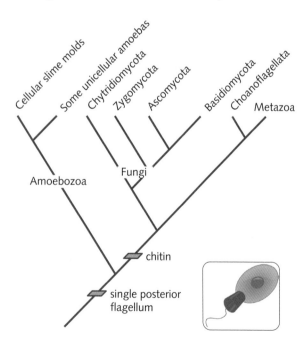

Figure 4.18 Unikonta.

- Zygomycota. The molds are typically saprophytic mycelial fungi growing on decaying plant and animal material. The most familiar member is the black mold that forms on stale bread.

- Ascomycota. A diverse group of fungi that includes both unicellular and multicellular forms. The unicellular ascomycetes are collectively called yeasts, and include the familiar baker's and brewer's yeast. Multicellular forms include truffles and morels, as well as important plant pathogens such as the powdery mildews.

- Basidiomycetes. Mushrooms and similar fungi, such as puffballs and bracket fungi, are basidiomycetes. The group also includes some unicellular forms, as well as plant pathogens such as rusts and smuts.

The classification of fungi is based on their sexual phases, with Ascomycota and Basidiomycota forming well-supported clades with distinctive patterns of spore formation. On the other hand, it seems likely that neither molds nor chytrids are monophyletic, but rather constitute, together with some other groups, a diverse assemblage of fungi whose relationships are still unclear. Chytrids have the single posterior flagellum that is the inferred ancestral state for Unikonta, but other fungi have lost the flagellum and lack any motile stage.

Animals and fungi use chitin, not cellulose, as a structural fiber in exoskeletons. Chitin is a polymer of glucose, like cellulose, but with a side branch containing nitrogen substituted for a hydroxyl group. It is the main structural fiber in the cell walls of fungi, the cuticle of nematodes and the exoskeleton of arthropods. It is also found in many other protostomes, for example in the radula of mollusks. Chitin is a light, pliant, tough material that provides good protection as an exoskeleton, and can be hardened by incorporating calcium carbonate, as in a crab carapace. It is only about half as stiff as cellulose, however: tall trees could only evolve in a lineage with the ability to synthesize cellulose.

- Amoebozoa. The sister group of the clade that includes animals and fungi appears to be a group of amoebas, naked cells able to move by internal cytoplasmic flow. They feed by phagocytosis, engulfing food particles and digesting them in the cytoplasm. The cellular slime molds can aggregate when food runs out to form a multicellular "slug" with a simple soma–germ differentiation, illustrating a mode of multicellular development utterly different from that of metazoans. There are superficially similar amoeboid organisms in several other groups whose classification is often uncertain.

4.17 Eukaryota

Clade added: Bikonta

Innovations: compartmentalized cell; cytoskeleton; organelles

The potentiating innovation of eukaryotes is a large cell with a flexible wall, a complex system of folded internal membranes and a cytoskeleton based on actin microtubules. The replacement of the rigid murein cell wall of bacteria with glycoprotein was particularly important because it enabled the eukaryote lineage to adopt phagotrophy, the ingestion of solid food particles. This not only opened up new ways of life, such as eating bacteria, but also led to the evolution of symbiotic relationships with bacteria that had been eaten but not digested, the ancestors of the mitochondria and chloroplasts of modern eukaryote cells.

Large cells are normally at a severe competitive disadvantage because small cells with short diffusion distances work faster and hence replicate more rapidly. The eukaryotic membrane system enables enzymes to be produced and deployed close to their work sites in the cytoplasm, reducing diffusion times and increasing the rate of growth and replication. The membranes also enclose compartments within the cell, notably the nuclear compartment, where genes borne on linear chromosomes are sequestered. Other compartments enclose the mitochondria and plastids, which originated as endosymbiotic bacteria and still possess the remnants of their ancestral genomes, usually borne on circular chromosomes. The evolution of eukaryote types by serial endosymbiosis is described in Chapter 8.

A second important potentiating innovation is the sexual cycle. This is a dual process involving syngamy, the complete nuclear fusion of two sexually compatible gametes to form a zygote, followed by meiosis, the reductional nuclear division of the germ cells derived from the zygote that restores ploidy. As a result, the eukaryote life cycle involves an alternation between haploid and diploid nuclear states. Syngamy and meiosis usually result in extensive genetic recombination through crossing over between homologous chromosomes. Sex and recombination have profound evolutionary consequences that are discussed in Chapter 15.

Before extensive genetic information became available it was usual to recognize five "kingdoms" of life: the bacteria, plus the four eukaryote groups of animals, plants, fungi, and protists. It is now clear that "protists" do not constitute a monophyletic group, and the five-kingdom interpretation has been abandoned. The divergence of the major groups of eukaryotes took place so long ago, however, that their identity and relationships are still the subject of vigorous debate. The interpretation we give in Figure 4.19, which is supported by current evidence but by no means finally decided, is that there is a primary branching that separated the Unikonta (animals, fungi, and amoebas), with a single posterior flagellum, from the Bikonta (all other eukaryotes), with a pair of flagella, one anterior and the other posterior.

One major innovation of bikonts is the use of cellulose as the major structural fiber of the cell. Five major bikont clades are recognized in recent molecular analyses.

- Alveolata. The alveolates include three important groups of single-celled eukaryotes.

  - Ciliata. Ciliates are very abundant predators of bacteria in marine and freshwater habitats. They are often very large cells, up to 1 mm in length, propelled by sheets of cilia. To operate so large a cell they have two kinds of nuclei: the micronucleus, which is copied from parent to offspring, and a macronucleus, which contains several copies of each gene and controls metabolism.

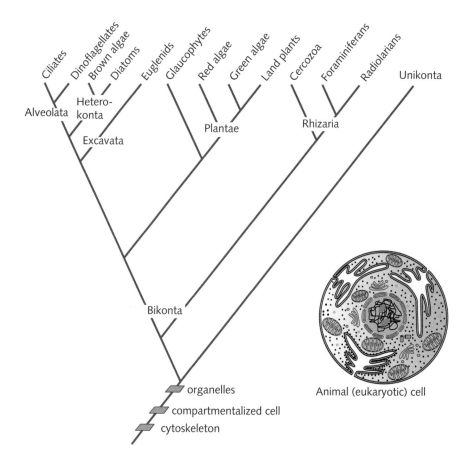

Figure 4.19 Eukaryota.

Illustration credit Lynn Gancher,
College of DuPage.

– Dinoflagellata. Dinoflagellates may be either predatory or photosynthetic organisms, and may be both at the same time. Many dinoflagellates produce extremely potent neurotoxins, perhaps as a defense against grazing zooplankton, and when they bloom they can poison fish, birds, shellfish, and, indirectly, people.

– Apicomplexa. This is a large group of unicellular parasites of animals that are responsible for several human diseases, including malaria.

• Heterokonta. Most heterokonts (or stramenopiles) are algae whose photosynthetic organelle consists of a unicellular red alga (now greatly reduced) and its chloroplast. Photosynthetic heterokonts use chlorophyll c, as well as accessory carotenoid pigments that give them a characteristic brownish color.

– Chrysophyta. These are unicellular phytoplankton that make a major contribution to global productivity. The diatoms (Bacillariophyceae) are photosynthetic organisms that live within a siliceous box, and are one of the most abundant and diverse groups of protists in marine and freshwater habitats.

– Phaeophyta. The brown seaweeds are common multicellular algae on most rocky shores, especially in cold oceans. Kelps can grow to large size—up to 40 m or more in length—and form dense forests that provide a habitat for hundreds of species of invertebrate.

– Oomycetes. The oomycetes are plant pathogens that were formerly classified as fungi. Some attack freshwater invertebrates and even fishes and frogs. The causative agent of the potato famine in Ireland was an oomycete.

• Excavata. The most familiar free-living excavates are the euglenid algae, green unicells that readily lose their chloroplasts and become heterotrophic.

Others are parasites, including the trypanosomes, which cause sleeping sickness.

• Plantae. This is a large group of photosynthetic organisms that includes green algae and land plants. All contain chlorophyll a and store starch.

– Rhodophyta. The rhodophytes are common seaweeds containing accessory pigments that give them a red color. They differ from other multicellular seaweeds in having no flagella or centrioles.

– Chlorophyta and Charophyta. These groups have both unicellular and multicellular members, and are especially common in freshwater. Together with all plants except rhodophytes, they utilize both chlorophylls a and b.

– Bryophyta. The most familiar non-vascular plants are the mosses, usually small soft plants that form mats in damp terrestrial habitats. The sphagnum mosses of the north temperate zone are one of the most extensive plant communities in the world. Unlike all other land plants, mosses are haploid during most of their life cycle.

– Pteridophyta. The ferns are vascular plants with true leaves that reproduce by means of spores. They are often found in damp places, but occupy a much broader range of habitats than mosses. Club mosses (Lycopodiophyta), horsetails (Equisetophyta), and whisk ferns (Psilophyta) are other groups of spore-bearing vascular plants related to ferns.

– Spermatophyta. The seed plants include all the familiar flowering plants (Angiospermae) and conifers (Gymnospermae), as well as less familiar groups such as cycads, gnetophytes, and gingkos. The success of vascular plants on land is partly attributable to their use of lignin as a structural fiber in association with cellulose, allowing a very tall, tree-like habit to evolve.

• Rhizaria. The rhizarians are amoeboid protists that capture bacteria and diatoms with long thin pseudopodia, but also often contain a variety of endosymbiotic algae. The most important groups are the Foraminifera and Radiolaria, both of which are very abundant in the surface waters of the ocean. Other rhizarians are amoeboid or flagellated unicells living in soil and sediments.

The diversity of eukaryotes other than animals, plants, and fungi was long obscured by the traditional label of "protists." It is only with the advent of molecular phylogenetics that we have been able to begin to understand the full range of body plans and ways of life exhibited by microscopic eukaryotes. Only the larger groups have been mentioned here; there are many smaller groups with distinctive ways of life that contribute to eukaryote diversity.

4.18 Life

Sister clades added: Bacteria, Archaea
Innovation: cell, metabolic systems

There are two large groups of prokaryotic organisms, the Bacteria and Archaea. These are the ancestors of Eukaryota, and Figure 4.20 gives one view of the relationships among the three groups, but the route by which eukaryotes evolved from non-eukaryote forebears has not yet been clearly demonstrated.

• Archaea. The archaea have only recently been recognized as a distinct group, differing from bacteria in their cell membrane, cell wall, ribosomes, and flagella. They are abundant in extreme environments such as hot springs and highly saline or acid water. Some are capable of growing at 122 °C and at a pH of 0. They also occur in more normal environments, but their importance has probably been underestimated because they are difficult to culture.

• Bacteria. Bacteria are the more familiar prokaryotes, present in almost all habitats, including our own bodies. They have no nucleus or organelles, so processes such as electron transport occur across the cell membrane. The cells are so small (typically only about 1% as large as a eukaryotic

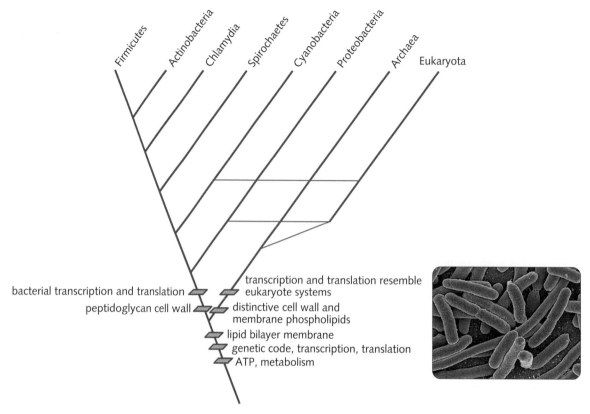

Figure 4.20 Life.

cell) that this is sufficient to maintain high rates of metabolism. Outside the cell membrane is a cell wall made of peptidoglycan that protects the cell. Many bacteria are motile, using a rotary flagellum quite unlike the flagellum of eukaryotes.

The diversity of Bacteria and Archaea is so great that we cannot even begin to summarize it here. Instead, we make two general points about the groups.

The first is that their abundance and diversity is not just greater than that of eukaryotes, it is enormously greater. A handful of soil may contain 10^{11} bacteria, more than ten times greater than the human population of the whole world. The total number of bacteria on the surface of the Earth is about 4×10^{30}. Of all individual organisms (unicellular or multicellular), 99% are bacteria. Moreover, the extent of genetic diversity is far greater among bacteria than among eukaryotes. Bacteria are asexual and do not have clearly defined species, but if we consider strains that differ by more than 10% of their genome to be

different kinds, then it is likely that there are more than 10^{10} different kinds of bacteria on Earth, thousands of times more than the number of species of eukaryotes. Clearly, bacteria and eukaryotes may often evolve in quite different ways.

Secondly, the scope of bacterial diversity is reflected by their metabolic versatility. The basis of metabolism is the free energy released by oxidation-reduction reactions. These involve the transfer of electrons from the electron donor (or energy source) to the electron acceptor. Hence, metabolic diversity is reflected by the range of substances that can be used as electron donors, and the range of reactions that are involved. Eukaryotes have only a narrow range of capabilities: they can ferment or respire some organic compounds, or they can use chloroplasts (which were originally cyanobacterial endosymbionts) for photosynthesis. Bacteria exploit a very broad range of energy sources: hydrogen, sulfur, hydrogen sulfide, thiosulfate, ferrous iron, ammonia, nitrite, carbon dioxide, and many more exotic substances. Besides that, heterotrophic bacteria can metabolize almost any carbon compound, including crude oil and pesticides. The global biogeochemical cycles that govern the availability of carbon, nitrogen, sulfur, phosphorus, and other essential substances depend on bacterial activity. Hence, the evolution of bacterial metabolism has had profound consequences for the biosphere as a whole.

We have now descended the whole tree of life, to the deepest divisions of life itself. These divisions, and the others that we have met on our way down, are not merely human constructs that assist in classification. They are the outcome of the incessant process of separation of lineages during the history of the Earth. The next step toward understanding modern diversity is thus to follow these lineages back through time, tracing their ancestry and discovering new groups that have flourished long ago but are now extinct. This is the subject of the next chapter.

● CHAPTER SUMMARY

	Clade	Sister clades added	Innovation
4.1	Ourselves: *Homo sapiens*		Language and culture
4.2	*Homo + Pan*	*Pan*	No distinctive innovation
4.3	Hominoidea	Gorillas; the other apes	The brachiation syndrome
4.4	Primates	Monkeys, tarsiers, and lemurs	Flexible limbs with grasping hands
4.5	Eutheria	Rodents and relatives; Laurasiatheria; Afrotheria + Xenarthra	Placentation
4.6	Mammalia	Metatheria; Monotremata	Milk, differentiated teeth
4.7	Amniota	Archosauria + Lepidosauria; Chelonia	The amnion
4.8	Tetrapoda	Amphibia	The pentadactyl limb
4.9	Sarcopterygii	Coelacanths and lungfishes	The monaxial fin; the lung
4.10	Osteichthyes	Actinopterygii	The bony endoskeleton
4.11	Gnathostomata	Chondrichthyes	Jaws
4.12	Chordata	Agnatha, Cephalochordata, Urochordata	The notochord
4.13	Deuterostomia	Echinodermata, Hemichordata	Pharyngeal gill slits
4.14	Bilateria	Protostomia	Directed locomotion; cephalization
4.15	Metazoa	Cnidaria; Ctenophora; Porifera	Extracellular matrix; somatic division of labor
4.16	Unikonta	Choanoflagellata; Fungi; Amoebozoa	Chitin
4.17	Eukaryota	Bikonta	Compartmentalized cell; cytoskeleton; organelles
4.18	Life	Bacteria, Archaea	Cell, metabolic systems

● FURTHER READING

Here are some pertinent further readings at the time of going to press. For relevant readings that have been released since publication, visit the book's Online Resource Centre at **www.oxfordtextbooks.co.uk/orc/bell_evolution/**

Section 4.1   Berwick, R.C., Friederici, A.D., Saint, R. and Miller, D.J. 2013. Evolution, brain and the nature of language. *Trends in Cognitive Sciences* 17: 89–98.

Fisher, S.E., Lai, C.S.L. and Monaco, A.P. 2003. Deciphering the genetic basis of speech and language disorders. *American Review of Neuroscience* 26: 57–80.

Cavalli-Sforza, L.L. and Feldman, M.W. 2005. The application of molecular genetic approaches to the study of human evolution. *Nature Genetics Supplement* 33: 266–275.

Section 4.2   Li, W-H. and Saunders, M.A. 2005. The chimpanzee and us. *Nature* 437: 50–51.

Section 4.3   Andrews, P. 1992. Evolution and environment in the Hominoidea. *Nature* 360: 641–646.

Section 4.4   Perelman, P., Johnson, W.E., Roos, C. et al. 2011. A molecular phylogeny of living primates. *PLoS Genetics* 7: e1001342.

Section 4.5   Springer, M.S., Stanhope, M.J., Madsen, O. and de Jong, W.W. 2004. Molecules consolidate the placental mammal tree. *Trends in Ecology and Evolution* 19: 430–438.

Section 4.6   Warren, W.C., Hillier, L.D.W. and Graves, J.A.M. 2008. Genome analysis of the platypus reveals unique signatures of evolution. *Nature* 453: 175–185.

Section 4.7   Fain, M.G. and Houde, P. 2004. Parallel radiations in the primary clades of birds. *Evolution* 58: 2558–2573.

Section 4.8   Roelants, K., Gower, D.J., Wilkinson, M. et al. 2007. Global patterns of diversification in the history of modern amphibians. *Proceedings of the National Academy of Sciences of the USA* 104: 887–892.

Section 4.9   Fricke, H. and Hissmann, K. 2000. Feeding ecology and evolutionary survival of the living coelacanth *Latimeria chalumnae*. *Marine Biology* 136: 379–386.

Section 4.10   Westneat, M. 2004. Evolution of levers and linkages in the feeding mechanisms of fishes. *Integrative and Comparative Biology* 44: 378–389.

Section 4.11   Velez-Zuazo, X. and Agnarsson, I. 2011. Shark tales: a molecular species-level phylogeny of sharks (Selachimorpha, Chondrichthyes). *Molecular Phylogenetics and Evolution* 58: 207–217.

Section 4.12   Cameron, C.B., Garey, J.R. and Swalla, B.J. 2000. Evolution of the chordate body plan: new insights from phylogenetic analyses of deuterostome phyla. *Proceedings of the National Academy of Sciences of the USA* 97: 4469–4474.

Section 4.13   Smith, A.B. 2004. Echinoderm roots. *Nature* 430: 411–412.

Section 4.14   Edgecombe, G.D., Giribet, G., Dunn, C.W. et al. 2011. Higher-level metazoan relationships: recent progress and remaining questions. *Organisms, Diversity and Evolution* 11: 151–172.

Section 4.15   Ball, E.E., Hayward, D.C., Saint, R. and Miller, D.J. 2004. A simple plan – cnidarians and the origins of developmental mechanisms. *Nature Reviews Genetics* 5: 567–577.

Section 4.16   Ruiz-Trillo, I., Burger, G., Holland, P.W.H. et al. 2007. The origins of multicellularity: a multi-taxon genome initiative. *Trends in Genetics* 23: 113–118.

Section 4.17   Keeling, P.J., Burger, G., Durnford, D.G. et al. 2005. The tree of eukaryotes. *Trends in Ecology and Evolution* 20: 670–676.

Section 4.18   Alers, T. and Mevarech, M. 2005. Archaeal genetics – the third way. *Nature Reviews Genetics* 6: 58–73.

● QUESTIONS

1. Distinguish between a potentiating innovation and an implementing innovation. Assign each of the following structures to one category or the other, giving reasons: the lung of sarcopterygians; the teeth of mammals; the pentadactyl limb of tetrapods.

2. Describe the descent of the major groups of Prostomia, indicating the sister-group relations between Annelida, Arthropoda, Mollusca, and Nematoda. To what extent does the phylogeny correspond with major innovations such as segmentation?

3. Describe the descent of the major groups of Metazoa, indicating the sister-group relations between Arthropoda, Cnidaria, Mollusca, and Porifera. To what extent does the phylogeny correspond with major innovations such as body cavities?

4. Draw a reasonable phylogeny for *either* Amniota *or* Bilateria. Indicate the shared derived characters of major clades for the group you have chosen, and identify an innovation characteristic of the group.

5. Compare and contrast the evolution of organisms and the evolution of languages.

6. Identify the sister group of humans and describe the attributes that distinguish it from humans.

7. Identify the main innovations that have evolved in the lineage descending from the most recent common ancestor of Primates to modern humans.

8. Distinguish between "parallel evolution" and "adaptive radiation" using examples from birds.

9. Describe the independent evolution of either arboreal folivores or ant eaters in different groups of Mammalia. What general conclusions can you draw about the process of evolution?

10. Describe the independent evolution of either placentation or electroreception in different groups of Vertebrata. What general conclusions can you draw about the process of evolution?

11. Describe the adaptive radiation of Actinopterygii or Mammalia in terms of the implementing innovations that characterize either group.

12. Provide examples of the constraints that have prevented a group from adapting to a particular way of life.

13. Describe how pharyngeal slits have evolved in different groups of Deuterostomia and what functions they serve.

14. Comment on the idea that progressive increase in complexity can be used to infer ancestry, with reference to the attributes of Metazoa.

15. Would you expect any extant lineage within a clade to preserve the attributes of the most recent common ancestor of the clade?

16. "Because monotremes are an early-diverging group of mammals they preserve ancestral features such as egg-laying." Discuss this proposition.

17. Identify the main characteristics of Eukaryota that distinguishes this group from prokaryotes.

18. Given a microscope and a sample from soil, lake or sea, how many groups of eukaryotes would you be likely to find and what would they be?

19. If you were classifying the biota of a recently discovered Earth-like planet, would you expect to find the same branching pattern of relationships between groups that we find on Earth?

20. If you were classifying the biota of a recently discovered Earth-like planet, would you expect to find similar groups, in particular bacteria, eukaryotes, and echinoderms?

You can find a fuller set of questions, which will be refreshed during the life of this edition, in the book's Online Resource Centre at **www.oxfordtextbooks.co.uk/orc/bell_evolution/**

5 The Ancestry of Life

In the previous chapter, we organized living diversity in terms of clades, showing their diversification in terms of branching diagrams that express how one group is related to another through their common ancestor. Each branching represents a real event in the history of life, and each common ancestor, now extinct, really existed in the past. A thoroughly evolutionary treatment, therefore, needs to supplement our catalogue of diversity in two important ways.

- First, we must attach a time line to divergence. How long ago did one lineage separate from another?

- Secondly, in the long perspective of geological time, our account is very incomplete, because a review of existing groups ignores the many other kinds of organisms that flourished in the past but have not survived until the present day. What did these extinct lineages look like?

In this chapter, we shall retain the previous arrangement of clades, but shall step further and further backwards in time, to uncover the evolutionary history of the modern groups and describe their ancient ancestors.

Why do you need to know about organisms that are long extinct and that no longer play any part in our lives? The answer is that you cannot understand the present without knowing about the past. Modern society would make little sense, if you did not know how it came about. The present situation cannot be understood in isolation from the past. In the same way, modern biological diversity is only the latest act in a long drama in which lineages rise and fall. In this chapter, we shall trace our own lineage backwards in time, past the same landmarks, to uncover the roots of the modern groups we have described.

Geological dates can be estimated from sequence and isotopic composition. The same layer of rock can often be found in many places throughout the world, where it is recognized by the minerals or fossils it contains. The order in which different layers, called strata, were formed is shown by their position relative to one another, the uppermost layer being younger. Only a few of the hundreds of strata are present at any given site, but by correlating strata at many sites the complete sequence of their formation can be deduced. There are occasional inconsistencies, for example caused by the folding or shearing of rocks that forces older strata on top of younger strata, but these are thoroughly understood, and the order in which strata were formed is securely established. Strata are grouped into geological periods, during which conditions remained more or less similar. The geological table is the list of these periods, in the order of their occurrence. Knowing the period in which a rock was formed immediately gives its relative age, so the table provides the indispensable framework for understanding Earth history.

The absolute age of rocks is estimated from the quantities of certain isotopes produced by radioactive decay. For example, strontium-87 (^{87}Sr) is produced by the radioactive decay of rubidium-87 (^{87}Rb). The rate of this process can be estimated very precisely, and is unaffected by the range of temperature and pressure experienced on Earth. It does not change over time: no change has been found since estimates were first made nearly a century ago, and agree with rates of decay estimated from supernova light generated hundreds of thousands of years ago. Hence, the quantity of the parent element (such as rubidium) will fall exponentially through time, while the quantity of the daughter element produced by decay (in this case, strontium) increases. The quantity of the daughter element D in a mineral is therefore related to the quantity of the parent element P by the equation $D = P(e^{rt} - 1)$, where r is the constant rate of decay. Since we know D, P, and r, we can easily calculate the time t that has elapsed since the formation of the mineral.

This gives a very straightforward way of estimating geological ages, provided that we know that the daughter element was all produced radioactively. In many cases, however, it is likely that some was present when the rock was formed, so our simple method will overestimate the age of the rock. To correct for this, we use a slightly more complicated method called isochronic dating, which utilizes an isotope of a daughter element that neither is produced radioactively nor decays radioactively itself, in this case strontium-84 (^{84}Sr). It relies on the fact that most rocks are made up of crystals of several minerals. Minerals that contain more rubidium at the time of their formation, relative to the non-radiogenic isotope, ^{84}Sr, will subsequently contain more ^{87}Sr relative to ^{84}Sr. If the current ratio of ^{87}Sr to ^{84}Sr is S_t, and the current ratio of ^{87}Rb to ^{84}Sr is R_t, then $S_t = S_0 + R_t(e^{rt} - 1)$ for each mineral in the rock, where S_0 is the original ratio of ^{87}Sr to ^{84}Sr. Hence, the linear regression of S_t on R_t for a set of minerals from the same rock has intercept S_0 and slope $(e^{rt} - 1)$, giving us an estimate of the time t elapsed since the formation of the rock.

About a dozen decay pathways are in common use for dating rocks, some more useful for older and some for younger rocks, depending on the rate of decay. They give concordant results, and are in agreement with absolute ages known from other methods, such as growth rings in very old trees, or documented historical events. The absolute age of rocks from different periods, with the fossils they contain, can thus be estimated precisely, and the dates of most of the principal events in Earth history have been firmly established.

Biological dates can be estimated from genetic divergence. The age of a group of organisms can thus be estimated from the age of the rocks in which the earliest fossils belonging to this group have been found. This is the minimum age of the group, because still earlier representatives may not have fossilized, or their fossils may not have yet been found. Another way of estimating the age of the group is to sequence a gene, or genes, from species belonging to this group and from species belonging to its sister taxon. Neutral genetic differences will accumulate through time, and can be used to construct a cladogram that coalesces at the most recent common ancestor of the two groups. Hence, the branching order of lineages reflects their relative age. If the rate of neutral mutation were constant over genes and lineages, DNA sequences could also be used to estimate absolute ages, independently of fossils. This is unfortunately not the case, and in practice molecular ages are relative ages converted to years by calibrating them against well-established radiometric dates.

Climate and geography change radically over time. As we burrow backwards along our line of descent, it will be is useful to keep three things in mind.

- The first is simply that we shall be visiting scenes that happened very long ago. It is often thought that deep time is difficult for the human mind to grasp, but I am not sure that this is really true. It would be possible to fit a map of the world in half a page of this book, for example, and few would find this difficult to understand—to understand where they lived, that is, relative to other parts of the world. This would be a spatial scale of roughly 1:100,000,000. Now, the Earth is about 4500 My old, and human lifespan is about 40 years. This is a temporal scale of roughly 1:100,000,000, more or less the same as the map scale. If we understand the map, then, it should be as easy to understand the time chart—to understand *when* we live, relative to other parts of Earth history, in terms of human generations. We just lack practice.

- Secondly, the map of the world that we are familiar with is a temporary arrangement of land and sea, which have been arranged in very different ways in the past. The movements of land masses over the face of the Earth have had profound effects on evolution, allowing groups to evolve in isolation or bringing independently evolved groups into contact.

- Thirdly, conditions of life have also changed, in part because of continental drift. We live in exceptional times: a temporary period of relative warmth, during a period in which polar ice has advanced and retreated several times. I am writing these lines in a place (Montreal) that was under a kilometer of ice 12,000 years ago and may be again (global warming notwithstanding) 12,000 years hence. The fossil remains of tropical forests, and the animals and plants that lived in them, can be found in northern Canada and Greenland, where at present there is only barren rock and ice.

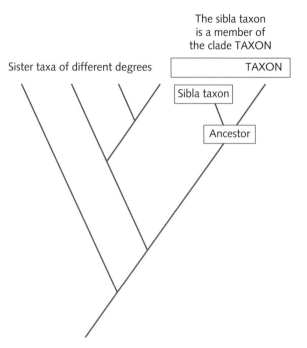

Figure 5.1 The sibla taxon.

The ancestral state is estimated from the sibla taxon. Any given clade descends (to the exclusion of all other clades) from a most recent common ancestor that lived at some time in the past. Unless this ancestor was an unusually abundant, widespread and easily fossilized species, it will probably never be found, because only a small fraction of extinct species are known as fossils (Section 3.1.3). Moreover, even if it were found it would be difficult, or impossible, to be certain that it was the ancestor. What may be found, however, are the fossil remains of a species that has ancestral character states, as inferred from the phylogeny, and that lived at about the same time as the ancestor, as inferred from the subsequent fossil record of the clade. This is the species that has the *sh*ortest *i*nferred *b*ranch *l*ength from the *a*ncestor, or sibla, to coin an acronym, as shown in Figure 5.1. The identity of the sibla species (or taxon) is likely to change, of course, as more fossils are found and we are able to home in closer to the unknown true ancestor. The current sibla species is interesting because it is the best available estimate of what the ancestor of the clade we are studying was like.

One last word about hard names: extinct organisms often have no familiar names, but instead receive very unfamiliar names usually made up from Latin and Greek roots by taxonomists. We shall use these formal names when referring to a

group, so you will be encountering Temnospondyli, Semionotoidea, Heterostraci, and many others. Do not be put off by these jawbreakers. We use them partly because there is little alternative, and partly because you will need them if you want to find out more, in the library or on the web. It is not necessary to memorize them all, however, unless you intend to become a professional paleontologist. They are there as markers that help us to describe how life has changed through time. The table of geological periods is essential, however, because it provides convenient names to anchor the narrative, with their dates. You should memorize it as a guide to Earth history, in the same spirit that you know that the printing press was invented about 500 years ago, rather than 5000 years ago or yesterday.

We shall now set out to trace the human lineage back through shifting continents and changing climates, along a line of ancestry from our closest relatives to the first living creatures to appear on the surface of the Earth.

Geologic Timescale

Era	Period	Epoch	Approximate duration (My)	Approximate number of years ago (Mya)
Cenozoic	Quaternary	Holocene	10,000 years ago to the present	
		Pleistocene	2	.01
		Pliocene	11	2
	Tertiary	Miocene	12	13
		Oligocene	11	25
		Eocene	22	36
		Paleocene	71	58
Mesozoic	Cretaceous		71	65
	Jurassic		54	136
	Triassic		35	190
Paleozoic	Permian		55	225
	Carboniferous		65	280
	Devonian		60	345
	Silurian		20	405
	Ordovician		75	425
	Cambrian		100	500
Precambrian			3,380	600

Image from *The Columbia Electronic Encyclopedia*, 6th ed. Copyright © 2007, Columbia University Press.

5.1 Ourselves: *Homo sapiens*

Coalescent: 0.2 Mya

Sibla taxon: *Homo erectus*

The oldest known fossils attributed to our species are two skulls embedded in volcanic pumice from the Omo valley in south-western Ethiopia. They are dated to 0.194 Mya. Over the next 80,000–100,000 years, *Homo sapiens* spread through Africa, but remained restricted to the continent. Starting about 0.1 Mya, however, *Homo sapiens* began to expand its range, with fossils being found in W Asia, then in SE Asia and Australia (c. 40,000 years ago), then in Europe (also about 40,000 years ago), and finally in the Americas (about 14,000 years ago).

These events have been interpreted in two ways. The first is that an ancestral species (*H. erectus*, see below) everywhere evolved in parallel towards a sapiens grade, so that modern human races have independent ancestry from different *H. erectus* populations. This is the "multiregional" model of human evolution. The second is that a small band of *H. sapiens* migrated north and passed out of Africa, subsequently migrating to all parts of the world and replacing other species of *Homo*. This is the "out of Africa" model.

It is very difficult to decide between these rival accounts of human evolution from fossils alone, as these record a transition from archaic to modern human types at many localities, which could be due either to parallel evolution or to gradual replacement, with or without interbreeding. The genetic data, however, seem unequivocally in favor of the "out of Africa" model. There are two main sources of information, both representing large genetic elements that do not recombine and that therefore have a strictly branching phylogeny like that of an asexual lineage. One is the mitochondrial genome, transmitted exclusively through the female line; the other is the non-recombining part of the Y chromosome, transmitted exclusively through the male line.

Given DNA sequences from individuals around the world, we can back-calculate the time when the lineage backwards-in-time coalesces to a single individual: the mitochondrial "Eve" and the Y-chromosome "Adam." These times are somewhat different: current estimates are about 0.23 Mya for mitochondrial DNA and about 0.10 Mya for Y-chromosome DNA. The discrepancy is probably attributable to differences in sexual biology, as a man can have more offspring than a woman and can therefore leave many more descendants. It should be emphasized that this does not mean that the entire human population consisted of a single woman 230,000 years ago (nor a single man 100,000 years ago). Rather, all people now alive bear mitochondria derived from a single woman who lived 230,000 years ago. If some other gene (non-recombining genetic element) were chosen for study, it would also descend from some single individual in the distant past, but from a different individual.

Mitochondrial and Y-chromosome sequences can also be used for tracing early human migrations, through the genetic structure of contemporary populations. Both lead to similar interpretations, although many details are still undecided. Figure 5.2 gives a sketch of migration routes consistent with these data. *Homo sapiens* initially expanded within Africa, occupying most of the continent by 100,000 years ago. There was some limited movement down the Nile valley into western Asia, attested by some very early fossils from Israel, but no wider spread. About 40,000–80,000 years ago there was a second expansion within Africa, when the original colonists were replaced everywhere except in the ancestors of the Biaka (West Pygmies) and Khoisan (Bushmen). The same expansion led to the decisive crossing into Asia about 40,000–70,000 years ago, probably from Ethiopia into the Yemen. This seems to have involved very small numbers of people, judging from the very restricted genetic diversity of non-African populations.

The routes followed by bands of *Homo sapiens* that subsequently migrated out of this beachhead were strongly influenced by climate change caused by glaciation. The more comfortable tropical route led around the coastline of southern Asia into Australia and New Guinea (reached about 44,000 years ago), which were more accessible then because the sea level was about 70 m lower than today, owing to the large amount of water locked up in ice sheets. The more northerly route trended north and west towards the Levant (about 44,000 years ago) and into central Asia (about 40,000 years ago), from where groups fanned out into India (about 34,000 years ago) and south-eastern Asia (also about 34,000 years ago).

Further expansion north and west became possible after the climate moderated about 30,000 years ago, when groups moved into Europe and north-eastern Asia. Arctic Asia was reached shortly afterwards, and it was probably from here that the Americas were populated across the Bering land bridge about 14,000–20,000 years ago, although both time and place have often been debated. At about the same time East Asia was populated and there were secondary migrations into Europe. The expansion of the ice sheets to their maximum extent 20,000–24,000 years ago forced human populations back southward, however, except for a few ice-free refugia such as the Iberian Peninsula in Europe. After deglaciation, which was complete by 7000 years ago, populations returned north, and the last remaining areas of the world were occupied. Vanuatu was reached about 4000 years ago from the northern coast of New Guinea, the more remote Pacific Islands by 3000 years ago, and finally New Zealand less than 1000 years ago. Actual population movements must have been much more complicated than this simple picture suggests, but our increasing knowledge of the genetic structure of human populations is allowing these ancient events to be reconstructed with much greater reliability than has hitherto been possible.

As our ancestors spread across the world, they found in many places abundant populations of large herbivores, such as the diprotodont marsupials of Australia, the giant ground sloths of South America, and the mastodons of North America. Many of them disappeared soon after the arrival of *Homo sapiens*, and there is a deepening suspicion (although the issue is still controversial) that they succumbed to improved hunting techniques and missile technology. Our ancestors also found that other humans had already occupied Eurasia—*Homo neanderthalis* in Europe and the Levant, and *Homo heidelbergensis* or *Homo erectus* in the rest of Asia. These, too, would be extinct before very long.

Our immediate ancestor was *Homo erectus*, which appeared in East Africa about 1.9 Mya. Early types are often named as a distinct species, *Homo ergaster*. In most respects all these types resembled *Homo sapiens*, with an upright stance and the manual dexterity to make sophisticated stone tools such as hand axes. Their skull was distinctively different, however, with a low forehead, prominent brow ridges and a receding chin. Having extended its range through

(a)

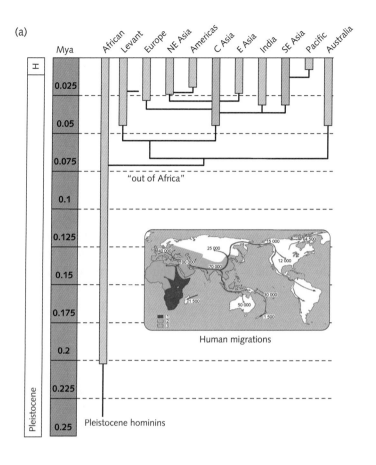

(b)

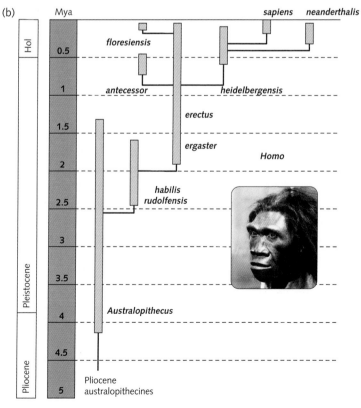

Figure 5.2 (a) *Homo sapiens*. (b) *Homo sapiens*. Sibla taxon: *Homo erectus*. In the diagrams in this section, filled bars represent range in time, and solid lines connecting bars represent inferred lines of descent.

most of Africa early in its history, *H. erectus* also migrated to tropical southern Asia. About 0.4 Mya it migrated north into temperate Asia and Europe, probably aided by the first use of domestic fire.

At about the same time, a larger-brained form, often recognized as the separate species *Homo heidelbergensis*, appeared in Europe and began to replace *H. erectus* over much of its range. Typical *H. heidelbergensis* was a strongly built hominid, again with a thick skull and prominent eyebrow ridges. Before it became extinct about 0.1 Mya, it had given rise to two other large-brained species, *Homo neanderthalis* and *Homo sapiens*. It is not yet clear whether these arose from different *H. erectus* populations, or whether a single lineage diverging from *H. erectus* later split into *neanderthalis* and *sapiens* lineages. In the first case, our sister species is *H. erectus*, whereas in the second case it is *H. neanderthalis*. At present, the evidence seems to favor *H. erectus* as our sister species, but future discoveries may eventually support either interpretation. *Homo neanderthalis* appeared about 0.24 Mya and became extinct about 30,000 years ago, after coexisting for a short time with *Homo sapiens* in Europe. *Homo erectus* also coexisted with *Homo sapiens* for a while, and lived on until about 40,000 years ago in Java. An extraordinary small-bodied island population of *Homo erectus*—"hobbits" in the popular press—survived on the Indonesian island of Flores until about 20,000 years ago.

The most ancient species in our genus was *Homo habilis*, which lived in East Africa from 2.4 Mya to about 1.4 Mya. It was a short, stocky form with relatively short legs and long arms that differed little from its ancestor *Australopithecus*; assigning it to one genus rather than the sister genus is, of course, a subjective decision with no phylogenetic implications. Some specimens are named as a separate species, *Homo rudolfensis*. Many characteristic features of modern human anatomy are poorly developed in *H. habilis*. It probably did not walk like subsequent species of *Homo*, and its hand was probably not capable of the precision and power grips that we use to manipulate objects. Nevertheless, it was responsible for the first major technological advance in human history, the manufacture of stone tools. These were sharp-edged flakes struck from a stone core that were used to butcher large animals. The earliest are from sites in Ethiopia and Kenya, and stone tools were made with increasing sophistication by species of humans until they were replaced by metal, only about 3400 years ago.

5.2 *Homo + Pan*

Coalescent: 4 Mya

Sibla species (for *Homo*): *Australopithecus africanus*

The immediate ancestor of *Homo* was *Australopithecus* (Figure 5.3), an ape-like hominid whose remains have been found in East Africa. It was fully bipedal, although its legs were relatively short, and it lived in savanna grassland and open woodland, where at least part of its diet was animal. The crucial transition from a climbing, forest-dwelling ape to a running, plains-dwelling human was made about 4 Mya. It led to a modest radiation of australopithecines in the late Pliocene. One set of lineages evolved a "robust" phenotype with a more heavily-built body and these species are often put into a separate genus, *Paranthropus*. They coexisted with more lightly-built "gracile" forms such as *Australopithecus africanus* and *A. garhi*. Most analyses derive the earliest *Homo* from a lineage close to *A. africanus*, but the fossil evidence is still too scanty for this to be firmly established.

The most recent common ancestor of humans and chimpanzees lived a little before this time, about 4–5 Mya. The fossil record of apes is very poor, because rainforest conditions are unfavorable for fossilization, although there are a few fragmentary remains of apes with a mixture of chimp-like and human features, such as *Sahelanthropus* from Chad (4–7 Mya), *Orrorin* from Kenya (ca. 4 Mya), and *Ardipithecus* from Ethiopia (4–5 Mya). The common ancestor of chimps and humans was probably a fruit-eating forest ape that resembling the two modern species of *Pan*, which diverged about 0.9 Mya. In the human line of descent, by contrast, there have been very striking changes associated with the new way of life originally adopted by *Australopithecus*. Obligate bipedal locomotion was made possible by extensive remodeling of the foot, lower limbs, and pelvis. It led

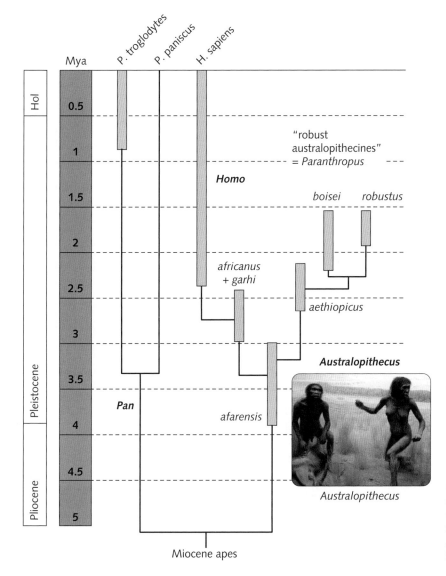

Figure 5.3 Hominidae. Sibla taxon: *Australopithecus*.

Image courtesy of Dschwen. This file is licensed under the Creative Commons Attribution-Share Alike 3.0 Unported license.

to the evolution of the hand as a precision gripping tool with complex articulations and an opposable thumb. Most of the body became naked through reduction in the size of hairs, although we do not know when this occurred, because hair does not fossilize well, nor why it was selected. The single most important trend was the remarkable increase in brain size, accompanied by changes in brain structure and the appearance of human intellect and culture. Since the mid-Pliocene, brain size in the human line of descent has increased exponentially at a rate of about 40% per million years.

5.3 Hominoidea

Coalescent: 21 Mya (Hominoidea); 14 Mya (living apes)

Sibla taxon: *Proconsul* (Hominoidea); *Kenyapithecus* (living apes)

Apes appeared in Africa during the early Miocene (22.4–17 Mya) as forest animals with mobile joints and grasping hands, capable of running along tree branches. The earliest ape is *Proconsul* (Figure 5.4), for which we have nearly complete fossils from Kenya. The ancestral lineage rapidly diversified into dozens of species that

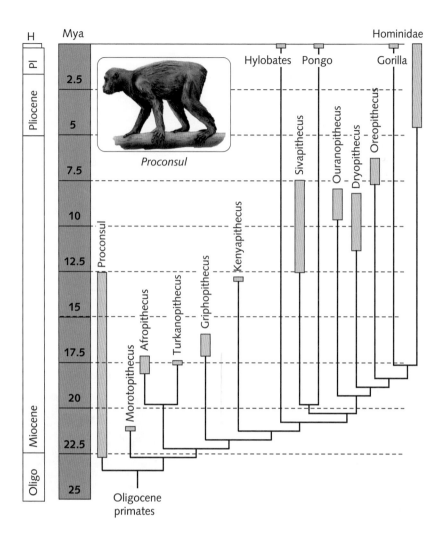

Figure 5.4 **Hominoidea.** Sibla taxon: *Proconsul*.

Image courtesy of Nobu Tamura. This file is licensed under the Creative Commons Attribution-Share Alike 3.0 Unported license.

varied widely in size (roughly from cat to gorilla) and diet (originally ripe fruit, subsequently nuts and leaves). Towards the end of this period sea levels fell, exposing a land bridge from Africa to Eurasia, and apes, in the company of many other African mammals, soon made the crossing and spread through Arabia into Europe and southern Asia. Here, they evolved the characteristic adaptations for brachiation, including a fully extendable elbow joint and powerful grasping hands and feet. The Asian lineage, *Sivapithecus* (from 13 Mya), diversified throughout India, China, and south-eastern Asia, giving rise to the modern orangutan (*Pongo*). The European lineage, *Dryopithecus*, was a large-brained ape resembling modern chimps and gorillas.

What happened next is less clear. Towards the end of the middle Miocene, about 12.4–8 Mya, apes were abundant and diverse in Eurasia, whereas fossils from this period are very scarce in Africa. In the late Miocene the climate became cooler and drier as the result of mountain-building in the Alps, the Himalaya, and East Africa. The tropical and subtropical forests of Europe, China, and East Africa were replaced by open woods and grassland, and apes retreated with their habitat into central Africa and south-eastern Asia, where they still survive. One possibility is that apes became extinct in Africa after the early Miocene radiation, and returned from Europe in the late Miocene. This is quite possible, with *Dryopithecus* supplying a plausible ancestral lineage. It relies on the "ape gap" reflecting a real absence of African apes in the later Miocene, however, and in recent years several ape fossils have been found in Kenya and Ethiopia from the crucial period of about 10 Mya. Hence, whether the lineage leading to modern African apes and humans evolved in situ or from a European ancestor is currently a matter for debate.

5.4 Primates

Coalescent: 44 Mya (fossils); 81 Mya (molecular)
Sibla taxon: *Plesiadapis*

The decline of ape diversity in the later Miocene was accompanied by an increase in the diversity of monkeys (Cercopithecoidea). One explanation of these events is that apes were severely affected by the retreat of the tropical forest and the decline in the supply of the ripe fruit that was the basis of their diet. Monkeys have broader diets, could exploit a wider range of plant parts, and so increased in abundance and diversity. This is consistent with the known facts, but it cannot be conclusively established because we cannot perform the replicated experiments that would be necessary to determine cause and effect. This is a flaw in all historical analyses: the facts may be clear, but the mechanism responsible for a given outcome is always disputable. We shall come across many such cases, as we cast back further in time and the evidence becomes fainter.

The common ancestor of monkeys and apes lived in the Oligocene, around 24–30 Mya. Figure 5.5

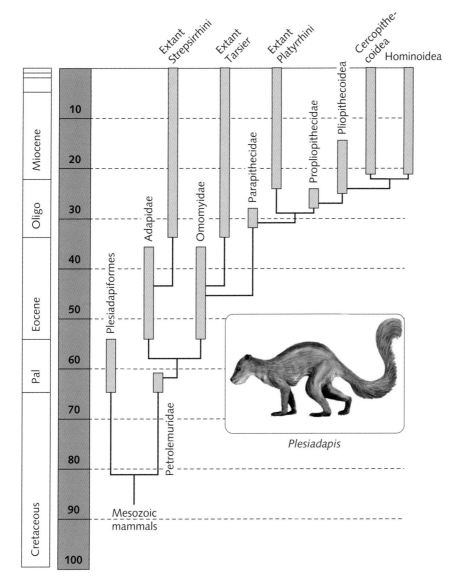

Plesiadapis

Figure 5.5 Primates. Sibla taxon: *Plesiadapis*.

Image courtesy of Nobu Tamura.
This file is licensed under the Creative Commons Attribution-Share Alike 3.0 Unported license.

shows a possible line of descent, although the fossils are too fragmentary for this to be traced conclusively. This ancestor seems to have descended from an Eocene clade, Omomyidae, resembling modern tarsiers. The earliest group of primates, and the presumptive ancestor of all later species, was the Petrolemuridae, an Asian group with grasping hands, nails rather than claws, and a bar of bone behind the orbit that enabled them to feed at the end of thin branches. They lived at the same time as the Plesiadapiformes, a group with claws rather than nails and rather rodent-like incisors that is sometimes included among the primates. It is possible that they eventually became extinct, leaving no descendants, through competition with similar but more efficient rodents (such as squirrels). Again, however, we emphasize that although the facts are clear (the fossils are complete and well-dated; rodents began to diversify at the same time) the interpretation is necessarily tentative.

5.5 Eutheria

Coalescent: 120 Mya (fossil), 134 Mya (DNA)
Sibla taxon: *Eomaia*

Primates and most other extant orders of placental mammals appeared soon after the beginning of the Cenozoic Era (65 Mya), in one of the most spectacular radiations in the whole fossil record. Forms that are clearly ancestral to modern carnivores, ungulates, bats, anteaters, rodents, and so forth are wholly absent 65 Mya and all present 50 Mya, together with several large groups that are now extinct.

The radiation occurred soon after the best-known mass extinction in Earth history at the K/T (for Cretaceous/Tertiary) boundary that separates the Mesozoic Era from the Cenozoic Era. This was almost certainly caused by the impact of an asteroid about 10 km in diameter which splashed down NW of the Yucatan peninsula about 65 Mya, causing immediate widespread devastation and releasing clouds of dust and debris that blocked the sunlight and brought about severe global cooling. Some kinds of organism were little affected: most genera of bony fishes, sharks, amphibians, and many marine invertebrates survived. About 40% of plant and coral genera disappeared, however, and some major groups were completely eradicated. Belemnoids (cephalopods resembling squid), ammonoids (shelled cephalopods), and rudists (reef-building clams) were abundant, diverse, and ecologically important groups that became extinct at the K/T boundary.

The most famous victims, however, were the dinosaurs. At least 100 genera of dinosaurs were alive at the end of the Cretaceous, constituting by far the most diverse and abundant group of large tetrapods. No tetrapod larger than about 20 kg survived the K/T event, however, and all dinosaurs abruptly disappeared. The subsequent radiation of mammals is a convincing example of ecological release: mammals could now proliferate because their main predator, large reptiles, had disappeared. They could diversify because the many ways of life that had previously been occupied by dinosaurs—large browsing herbivores, swift cursorial carnivores, and so forth—were now vacant, creating a selective advantage for lineages that were able to fill them, however ineffectual they may have been at first. Some of the mammal lineages living at this time evolved into the major groups of modern mammals; others flourished for a while before disappearing completely.

- Creodonts were the most abundant carnivorous mammals of the Eocene. They possessed the carnassial teeth necessary to shear flesh, although these are further back in the molar row than in modern Carnivora. Most were small animals about the size of a mink, but some were over a meter in length and were presumably capable of pulling down the large mammalian herbivores that were evolving at the same time.

- Condylarths were similar animals, appearing in the Paleocene, with large canine teeth but also with molar teeth capable of processing vegetation. Again, most were small but some were the size of bears and perhaps had a similar omnivorous diet. Several other ungulate-like groups evolved early in the placental radiation, in the Paleocene and Eocene (65–34 Mya), possessing low-crowned

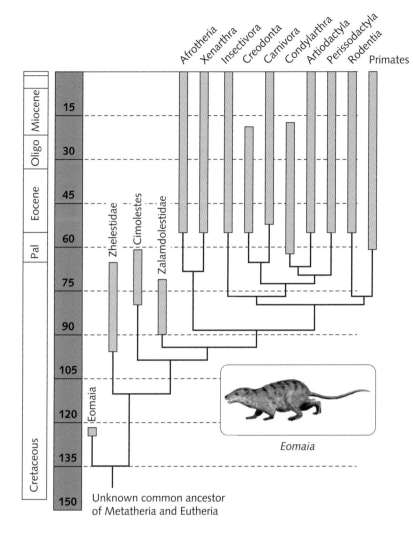

Figure 5.6 Eutheria. Sibla taxon: *Eomaia*.

Image courtesy of Nobu Tamura. This file is licensed under the Creative Commons Attribution-Share Alike 3.0 Unported license.

grinding cheek teeth and often reaching very large size. The North American uintatheres, for example, were ponderous horned animals about the size of a rhinoceros. It is difficult to escape the conclusion that they could never have evolved but for the previous extinction of dinosaurs.

The first eutherians appeared long before this radiation. *Eomaia* (Figure 5.6) was a small animal (about 10 cm in length and 25g in weight) that lived in the lower Cretaceous (125 Mya). It resembles marsupials in some respects and is likely to be close to the ancestry of modern placental mammals.

5.6 Mammalia

Coalescent: 210 Mya

Sibla taxon: *Hadrocodium*

The Cenozoic radiation followed a long Mesozoic history of mammals extending back to the late Triassic (210 Mya). These ancient forms are known mainly from their teeth, which are readily preserved and are diagnostic of mammals (from their differentiation and their cusped form), but which often tell little about the shape of the whole animal. More complete fossils found in the last 20 years, however, show that

Mesozoic mammals, though always small in stature, were more diverse than had previously been recognized. Hence, the Cenozoic radiation might have followed either of two pathways.

- The first is the "single-lineage" hypothesis: a single lineage of placentals surviving from the Mesozoic subsequently diversified to give rise to all modern placental groups. The most recent common ancestor of all extant placental mammals lived about 64 Mya.
- The second is the "many-lineages" hypothesis: the eutherians diversified during the Mesozoic, well before their Cenozoic radiation, giving rise to a number of lineages, each of which in turn gave rise to one of the characteristic groups of Cenozoic placentals. The most recent common ancestor of all placental mammals lived much more than 64 Mya.

The strongest evidence for the many-lineages interpretation is that fossils such as *Eomaia* date back over 100 My and the coalescent for extant placental mammals is estimated to be 100–110 Mya in most DNA analyses. This implies that there were many "ghost lineages" in the intervening period, which might be recovered as fossils but not recognized because they lack the shared derived characters of modern representatives of the clade.

A strong indirect argument for the many-lineages interpretation is the general constraint of functional interference. Consider the evolution of a group of graviportal mammals such as elephants, for example. This necessarily involves adaptations such as stout limbs, locking joints, and a strong vertebral column capable of supporting a massive body. A lineage that has acquired these adaptations is unlikely to evolve subsequently the slender limbs and specialized ankles and feet of a running mammal such as an antelope. The evolution of a toothless sloth into a raptorial carnivore is even more difficult to imagine.

On these grounds, the most likely interpretation of the Cenozoic radiation of placental mammals is that each of the modern orders descended from a different lineage already in existence 64 Mya. Each would then have become independently specialized for a different way of life, and the problem of functional interference would not arise. Before the great radiation, however, they would lack most or all of the shared derived character states that define the

modern groups, so the pattern of ancestry might be difficult or impossible to infer from morphology alone. Hence, the dichotomy between single-lineage and many-lineages interpretations cannot yet be conclusively resolved, and perhaps awaits the discovery of exceptionally well-preserved assemblages from the period immediately preceding the radiation.

During the Jurassic and Cretaceous most mammals were small rodent-like animals, some only a few grams in weight, the size of the smallest modern shrews. Many of them lived in trees, like small squirrels. The evolutionary changes were rather slow and undramatic. The cusps on the molar teeth shifted into a pattern where they ground against a flat area on their partner above or below. The animal can then chew its food with a transverse movement of the lower jaw relative to the upper jaw. This allows food, especially fibrous vegetation, to be processed before digestion much more effectively than the "slash-and-gulp" method that reptiles use.

The oldest known fossil metatherian (marsupial), *Sinodelphys*, comes from the early Cretaceous, and is of about the same age as the earliest eutherian (placental), *Eomaia* (125 Mya). The eutherian–metatherian coalescence is estimated to be 170–190 Mya in most molecular analyses. The living monotremes are the survivors of an earlier lineage known from middle Cretaceous fossils from Australia that branched off about 200 Mya in the late Triassic or early Jurassic.

Monotremes serve to illustrate an important point about the interpretation of phylogenetic trees. In popular accounts (and even in some textbooks) you will often come across some statement such as, "the monotremes are an early-diverging lineage which retain primitive characters such as laying eggs rather than giving birth to live young, a low body temperature, and diffuse milk production." Now, being early-branching, the monotremes are the sister taxon of all other extant mammals {Metatheria + Eutheria}. From our discussion of sister taxa (Section 3.2.1), it follows that monotreme lineage and the lineage of all other mammals have been evolving for exactly the same length of time. There is no reason, then, to infer that character states in monotremes are primitive, relative to all other mammals. Indeed, some characters, such as the bill or poison spur of platypuses, are notably derived. It is always an error to suppose that a species-poor clade necessarily retains primitive characters relative to its species-rich sister clade.

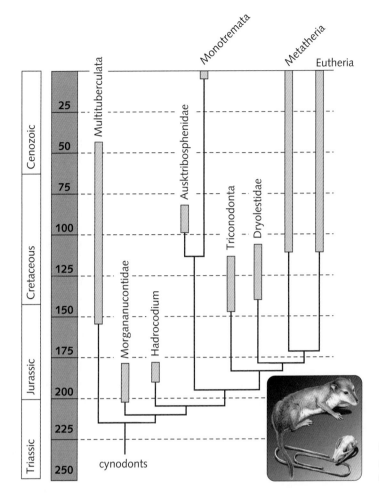

Figure 5.7 Mammalia. Sibla taxon: *Hadrocodium*.
Illustration courtesy of Mark A. Klingler/Carnegie Museum of Natural History.

The earliest traces of mammals come from the late Triassic (224–204 Mya), when a number of groups had evolved more or less complex dentition.

• Triconodonts, for example, had incisors, large canines, simple premolars, and cusped molar teeth. The molars had three sharp cusps arranged in a row, which made them better at shearing food. These cat-sized animals were probably omnivorous foragers rather like small raccoons.

The relationships among the various groups living at this time are still quite unclear, but one of them included the ancestor of all later mammals: the shrew-sized *Hadrocodium* (Figure 5.7) has been chosen to illustrate ancestral conditions.

5.7 Amniota

Coalescent: 330 Mya

Sibla taxa: *Protoclepsydrops* (Synapsida); *Hylonomus* (Diapsida)

The Cenozoic radiation that led to the evolution of most modern mammals naturally leads to the conclusion that mammals are a relatively recent offshoot from a basic reptilian stem. This is far from the truth. Mammals are distinguished from most reptiles (such as lizards and dinosaurs) by possessing a single opening in the skull (the synapsid state) rather than two (the diapsid state). This is how paleontologists can immediately tell what sort of animal a fossil skull belongs to. The synapsid ancestors of mammals, however, appear very early in the history of amniotes, and were a large and diverse group before the first dinosaur appeared. Mammals, therefore, descend from a very early branch on the amniote tree, rather than from a recently diverged clade.

- **Cynodonta.** The ancestors of mammals were the cynodonts, a group that radiated in the late Permian and Triassic (about 250 Mya) into a variety of large carnivores and herbivores. *Cynognathus*, for example, was a large predator, somewhat dog-like in appearance (its name means dog-jaw in Greek), with vertically-placed limbs that allowed it to run in a mammal-like fashion. It had a skull a foot in length, and its jaws bore large stabbing canines, and molar teeth capable of shearing flesh. *Cynognathus* and its relatives were abundant animals, but after they had died out in the middle Triassic (230 Mya) their descendants were much smaller creatures. Two families in particular, the tritylodonts and the trithelodonts, were very mammal-like in structure. The tritylodonts were weasel-shaped creatures that probably chewed plant material; the trithelodonts (or ictidosaurs) were tiny insectivorous forms. The lineage leading to mammals probably came from one of these two groups. Cynodonts died out in the Jurassic.

- **Therapsida.** The cynodonts were therapsids, a group that appears in the middle Permian (270 Mya) and is often referred to as "mammal-like reptiles." They are not reptiles at all, but rather synapsids whose teeth and gait are beginning to dimly resemble a primitive mammal. Some grew to large size: the dinocephalians, for example, reached the size of a rhinoceros. These very large animals were displaced from their way of life by diapsids such as dinosaurs in the Mesozoic, after being almost extinguished in the end-Permian mass extinction, which is described in Section 7.6.5.

- **Pelycosauria.** Therapsids evolved from pelycosaurs, a very early amniote group appearing in the late Carboniferous (about 310 Mya). The pelycosaurs were also large animals, with both carnivorous and herbivorous forms. One of the carnivores was *Dimetrodon*, whose bony dorsal sail has made it the most familiar fossil vertebrate that is not a dinosaur. As it happens, the skull and skeleton of *Dimetrodon* suggest that it was close to the ancestral lineage of therapsids. The genealogy of mammals

thus runs from an animal something like *Hadrocodium* in mid-Mesozoic back to tritylodonts (late Triassic–Jurassic), cynodonts (late Permian–Triassic), stem therapsids (middle Permian), and *Dimetrodon*-like pelycosaurs (early Permian), to the early pelycosaurs (late Carboniferous).

The other main branch of the amniotes is the Diapsida, which diverged from Synapsida in the middle Carboniferous, about 330 Mya. Modern diapsids are reptiles and birds. The oldest known diapsids were small lizard-like creatures that foraged in the undergrowth of the Carboniferous lycopod forests. When a tree fell, it left a stump whose interior would rot away to form a cavity that small animals could fall into but could not climb out of. These natural pitfall traps, preserved in the famous fossil cliffs at Joggins, Nova Scotia, have provided the remains of the earliest known truly terrestrial tetrapod, *Hylonomus* (Figure 5.8); only a terrestrial animal could be preserved in such a fashion. Throughout most of the Permian and Triassic the abundant large land vertebrates were pelycosaurs and therapsids, while diapsids remained small. Towards the end of the Triassic this situation reversed: the synapsids evolved into the small Mesozoic mammal lineages while the diapsids radiated into a range of large terrestrial and aquatic groups.

- **Ichthyosauria.** The ichthyosaurs were an early-diverging group of marine reptiles that were abundant throughout most of the Jurassic and Cretaceous. They had a dolphin-like appearance and were highly specialized for a fully aquatic lifestyle; we have met them already in Section 1.3.1.

- **Plesiosauria.** The plesiosaurs were another group of large aquatic reptiles that flourished in the middle and later Mesozoic. They are familiar from the long necks and paddle-like limbs of many forms. They are one of several groups with a euryapsid skull, having a single upper temporal opening, whose Permian ancestry is still obscure.

- **Archosauria.** The archosaurs evolved from the Triassic thecodonts, typically rather lightly built carnivores of moderate size. Many of them evolved more or less bipedal locomotion, with strong, vertically oriented hindlimbs and a large tail to balance the body.

  - **Crocodilia.** The crocodiles are closest to the ancestral archosaur condition, although they

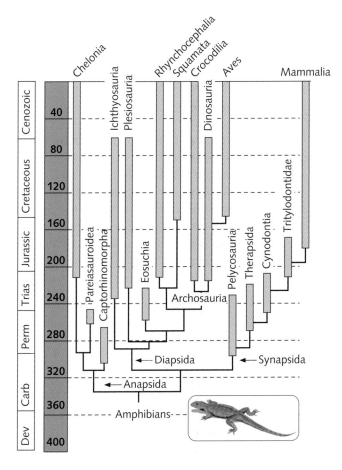

Figure 5.8 Amniota. Sibla taxon: *Hylonomus*.

Image courtesy of Nobu Tamura. This file is licensed under the Creative Commons Attribution-Share Alike 3.0 Unported license.

show many specializations to their semi-aquatic way of life. The earliest crocodilians date from the late Triassic, and the group became abundant and diverse in the Jurassic, with a range of specialized forms such as unarmored deep-sea crocodiles. The modern types (crocodiles, alligators, and gavials) emerged in the late Cretaceous and early Tertiary.

– **Dinosauria**. Almost every large terrestrial vertebrate of the Jurassic and Cretaceous, a period of more than 130 My, was a dinosaur. They have captured popular imagination since Victorian times, when it was first realized that their bones revealed the existence of huge reptiles in the remote past, and so changed forever our view of world history. There are two main groups, distinguished by the structure of the pelvic girdle.

o **Saurischia**. The classic bipedal carnivorous dinosaurs, including *Tyrannosaurus*, are all saurischians. Other lineages reverted to a quadrupedal pose and evolved into the gigantic herbivorous sauropods, including *Brontosaurus* and *Diplodocus*.

o **Ornithischia**. The ornithischians were all herbivores, rare in the early phase of the dinosaur radiation but becoming abundant and extremely diverse in the later Mesozoic. The ancestral forms were bipedal, and this pose was retained by the duck-billed dinosaurs that were abundant in the late Cretaceous. Other groups became quadrupedal, including the familiar armored dinosaurs, such as *Stegosaurus*, and the horned dinosaurs, such as *Triceratops*.

– **Aves**. The birds evolved from saurischian ancestors in the Jurassic. For many years the only known form close to the ancestor was *Archaeopteryx*, from the late Jurassic, which had fully feathered wings, but possessed teeth and a long tail, and lacked the pneumatic bones of modern birds. In the last decade, however, spectacular fossil finds in China have provided a series of transitional forms (see Section 8.1.3). "Dinosauria" is not a monophyletic group unless birds are included; hence, the cladistic interpretation of Aves is not that birds are descended from dinosaurs, but rather that they *are* dinosaurs.

- **Lepidosauria.** The ancestors of lizards and snakes were the eosuchians, small lizard-like diapsids of the upper Permian. They are the only comparable animals known from the early Permian, and may be ancestral to both lepidosaurs and archosaurs, which is how we have shown them in the tempogram. The surviving rhynchocephalian, *Sphenodon*, is in most respects similar to the most recent common ancestor of modern lepidosaurs. Members of the modern groups, including snakes, are known from the Cretaceous on.

- **Chelonia.** This leaves us with the turtles, which are notoriously difficult to place. Their solid anapsid skull is the ancestral condition for Amniota, and would justify placing them as a very early-diverging lineage, which is how we have shown them, but they may instead be the sister group of Diapsida. The earliest turtles are fully armored animals from the lower Triassic.

The anapsid reptiles of the late Carboniferous and Permian are collectively known as cotylosaurs. During the Permian they radiated into a modest range of small terrestrial types, with a few larger members such as the pareiasaurs, bulky terrestrial herbivores armored with bony plates that look like an early sketch for dinosaurs. The earliest known cotylosaur is *Hylonomus* of the Joggins lycopod forest, close to the ancestry of amniotes.

5.8 **Tetrapoda**

Coalescent: 374 Mya
Sibla taxon: *Acanthostega*

Our lineage can thus be traced back from Jurassic mammals through a more or less continuous series of intermediate forms to the Permian pelycosaurs. At this point, however, the narrative is interrupted. The main amniote clades are already distinguishable in the late Carboniferous: the lycopod stumps at Joggins have yielded both a synapsid (*Protoclepsydrops*) and a diapsid (*Hylonomus*). For a period of 20–30 My before this, however, the fossil record of terrestrial vertebrates is very poor. This is sometimes called "Romer's gap," after the unavailing efforts of the Harvard paleontologist A.S. Romer to find early Carboniferous tetrapods. The diversification that must have preceded the synapsid-diapsid split will remain obscure until better fossil localities are discovered.

- Seymouriamorpha. The seymouriamorphs were very reptile-like terrestrial animals related to cotylosaurs. *Seymouria* looked rather like a heavy-bodied, large-headed salamander with fangs whose remains are common in the Permian red beds of Texas. The ancestor of amniotes must have been similar in structure, although *Seymouria* itself appears too late in time to be a candidate. We know that seymouriamorphs were not themselves amniotes, however, because fossils of larvae with external gills have been found, indicating an aquatic egg and development resembling salamanders.

- Anthracosauria. The ancestors of cotylosaurs and seymouriamorphs were anthracosaurs, the "coal lizards" which lived in the Carboniferous forests and were preserved in the coal layers. Many of them were large aquatic animals that ate fish and were rather similar in appearance to crocodiles. Early Carboniferous anthracosaurs such as *Eoherpeton* were smaller animals resembling stocky, short-limbed salamanders, and these stand close to the line leading to amniotes.

- "Labyrinthodonts." Anthracosaurs had teeth which showed a complex labyrinth-like folding of dentine and enamel. This character state was inherited from their fish ancestors and is thus shared with many other groups of early amphibians, which are collectively called "labyrinthodonts."

  - Lepospondyli. The closest relatives of anthracosaurs were lepospondyls, generally small aquatic animals with a way of life similar to modern newts. Many of them had small limbs, or even lacked limbs altogether. Other evolved bizarre crescent-shaped skulls unlike those of any other vertebrate, whose function is still unknown.

– Temnospondyli. The other main group of labyrinthodonts was the temnospondyls; these included the stereospondyls and rhachitomes, typically large amphibians with massive heads, such as the well-known *Eryops*. This was a lower Permian amphibian about 2 m long with a broad, flat head bearing jaws studded with sharp teeth. It had sprawling limbs and could not have moved very fast on land; it probably had a semi-aquatic way of life and ate mostly fish.

• Lissamphibia. Modern amphibians most likely descend from rhachitome temnospondyls: in particular, larval labyrinthodonts, known from exceptionally well-preserved deposits, have external gills and are strikingly similar to larval salamanders and newts. The fossil record is very

incomplete, however, and the relationships of the major groups are at present unclear. Salamanders and frogs are sister taxa, but intermediate forms have not been found, and the fossil record of either group does not extend beyond the Jurassic. It is not even certain that Lissamphibia is monophyletic, as we have shown it: the Caecilia may be derived from the Carboniferous lepospondyls.

The early history of tetrapods is thus shrouded in uncertainty. There seems to have been an extensive radiation of amphibians once the crucial move onto land had occurred, but it has not yet been possible either to document this in detail or to trace the ancestry of modern amphibians with confidence. As we move further backwards to the time of the transition itself, however, the trail again becomes clearer, marked out by a series of forms spanning the transition from water to land. The earliest of these is our sibla taxon, *Acanthostega* (Figure 5.9).

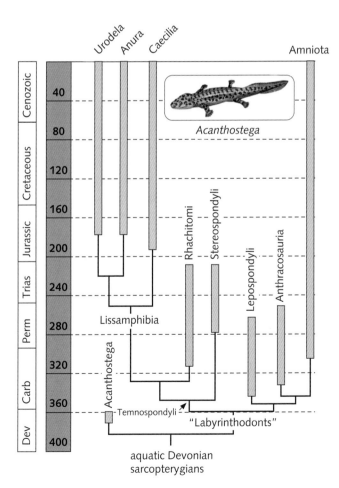

Figure 5.9 Tetrapoda. Sibla taxon *Acanthostega*.

5.9 Sarcopterygii

Coalescent: 415 Mya

Sibla taxon: *Osteolepis*

The evolution of tetrapods involved the transformation of a fish fin to a limb with digits. So long as the fossil record is seriously incomplete, we can make a clear distinction between fish and tetrapod. We now have so many fossils from the crucial transition period of the late Devonian, however, that it is impossible to draw a firm line between the two—as it should be, of course, since evolution involves a step-wise sequence of changes.

- *Acanthostega*. The remains of *Acanthostega* were discovered in east Greenland by a graduate student in 1971 and almost complete skeletons were subsequently collected in the 1980s. It was a rather bulky animal about a meter long, with a large head encased in a solid bony skull. Many of its attributes are clearly tetrapod: in particular, it had forelimbs with humerus, radius, and ulna. On the other hand, it had fish-like characters, too; for example, the notochord continues directly into the braincase, without the occipital condyle with which the spine articulates with the skull in all living tetrapods. The hindlimbs bear digits, but the foot is directed posteriorly, like the flipper of a fish. It has wrist and ankle joints, but neither have the characteristic articulatory facets of the load-bearing joints of terrestrial animals. In short, if you were to set out to draw an animal intermediate between a lobe-finned fish and a primitive amphibian, you would draw something very like *Acanthostega*. It lived in the latest upper Devonian (374–349 Mya).

- *Pederpes*. Moving one stage further than *Acanthostega* onto land requires a hindlimb with the foot directed forward, and wrist and ankle joints that bend so as to be able to propel the animal forward on land. This animal is *Pederpes*, an amphibian from the early Carboniferous (349–344 Mya).

- *Panderichthys*. Stepping backward into the water, we would expect to find an animal with a humerus that supported a fin rather than a limb, which had lost dorsal and anal fins, and which was able to breathe air (as shown by the presence of nostril bones). This animal is *Panderichthys*, which lived in the upper Devonian (384–374 Mya).

- *Eusthenopteron*. The fish whose descendants first moved onto land was close to *Eusthenopteron*, an abundant Devonian lobe-finned fish resembling a slim-bodied coelacanth, with a full complement of fins.

Sarcopterygians first appear in the upper Silurian and lower Devonian; *Osteolepis* (Figure 5.10) is a well-known middle Devonian genus.

The transition to terrestrial life seems to have involved paddling in shallow water, the forelimbs supporting the head while the hindlimbs were still fins or paddles capable of a power stroke, while a flat, tooth-studded skull served to catch fish or invertebrates. Most of the stages in this transition—many more than there is space to describe here—are now known from the fossil record.

The fossils document how this transition occurred, but they cannot explain why. Propping up the trunk to breathe air, or perhaps to bask on exposed rocks, might account for an oddly specialized kind of fish, but could not explain the subsequent radiation of amphibians. This was the consequence of the prior evolution of terrestrial plants. The first green plants to grow on land were algal crusts, and later mosses, but vertical stems based on vascular tissues and lignin were the potentiating innovation that led to permanent plant cover. Vascular plants first appeared in the middle Silurian as branched stems arising from a basal rhizome, with only rudimentary leaves. They are called rhyniophytes because exquisitely preserved specimens have been discovered in the chert deposits (think of a swamp transformed into a glassy matrix by silicate infiltration) of Rhynie, in Scotland. These diversified in the late Silurian and early Devonian, accompanied by terrestrial fungi and the first insects, whose remains are first found in the Rhynie cherts. Lycopods had evolved by the middle Devonian, as herbaceous or shrubby vegetation,

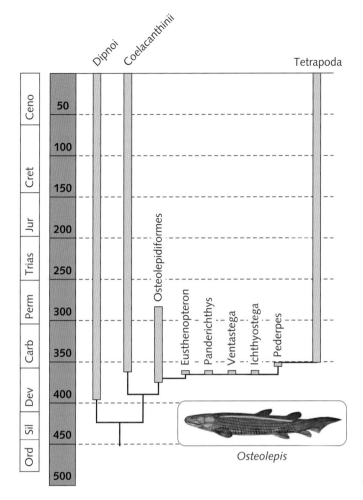

Figure 5.10 Sarcopterygia. Sibla taxon: *Osteolepis*.
Image courtesy of Nobu Tamura. This file is licensed under the
Creative Commons Attribution-Share Alike 3.0 Unported license.

together with ferns. By the late Devonian, lycopods had evolved into large trees, and for the first time forests appeared on land. By this time, insects were diversifying and the first productive terrestrial ecosystems had become established. It was only at this point that salamander-like or lizard-like tetrapods could make a living on land. The evolution of tetrapods from aquatic ancestors did not require only the transformation of fins to limbs, but also the prior evolution of plants, fungi, and insects to constitute the terrestrial ecosystem within which tetrapods could flourish.

5.10 Osteichthyes

Coalescent: 420 Mya
Sibla taxon: *Cheirolepis*

The first lobe-finned fishes and the first ray-finned fishes appeared in the lakes and swamps of the middle Devonian (400–384 Mya), with fragmentary material a little earlier. Their subsequent histories were very different. The dipnoans and crossopterygians were abundant during the later Paleozoic, but subsequently dwindled, until today only the lungfishes and the coelacanth survive. Their most abundant and diverse descendants are the tetrapods. The ray-finned fishes, on the other hand, became by far the most numerous aquatic vertebrates. Indeed, at present there are more species of ray-finned fishes than of all other vertebrates combined.

- Palaeoniscoidea. The earliest ray-finned fish were the palaeoniscoids, typically large-eyed, fusiform fishes with small, square, thick scales. *Cheirolepis* (Figure 5.11) is an early example. They became extinct in the early Cretaceous; their least-modified modern descendants are the bichirs. The main evolutionary trends during the early Mesozoic were the transformation of the originally heterocercal tail fin into the homocercal pattern, the thinning of the scales, the loss of a breathing lung, and above all the shortening and increased mobility of the jaw apparatus.

- Semionotoidea. These trends are apparent in the semionotids, rather heavy-bodied Mesozoic fish that may have included the ancestors of modern gars and bowfins. They were more pronounced in the pholodiphorans, small thin-scaled fish with a rather minnow-like appearance. The immediate ancestors of the teleosts were the leptolephomorphs, herring-like marine fish abundant in the Jurassic. The great radiation of the teleosts, that populated the oceans and rivers of the world with thousands of specialized forms, began in the Cretaceous, was little affected by the K/T mass extinction, and continued in the early Tertiary with the evolution of perches, sunfish, snappers, cichlids, blennies, and many more.

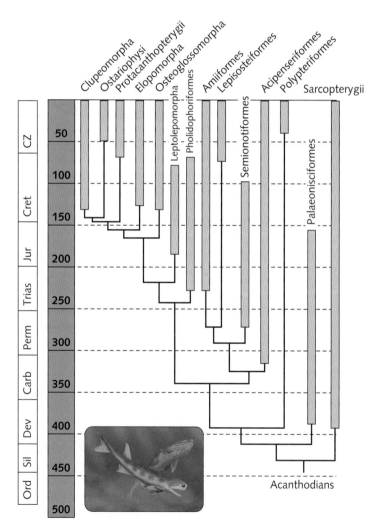

Figure 5.11 Osteichthyes. Sibla taxon: *Cheirolepis*.

5.11 Gnathostomata

Coalescent: 414 Mya

Sibla taxon: *Climatius*

Unlike the bony fishes, sharks and their relatives have been lungless marine vertebrates since their first appearance. Sharks are not a primitive group, as is sometimes supposed; in fact they are the last of the major groups of living fish to appear in the fossil record. Nor is the absence of bone a primitive character: the ancestors of cartilaginous fishes had plenty of bone, which was secondarily lost in the shark lineage.

- Hybodontoidea. Modern sharks and rays evolved from the hybodontoids of the Mesozoic, which were very shark-like in appearance, with powerful paired fins and a long heterocercal tail. They had both sharp-pointed teeth near the front of the mouth and blunt crushing teeth at the rear, so could eat either fish or mollusks. The Port Jackson Shark *Heterodontus* is a living form that resembles a hybodontoid.

- Xenacanthida. The xenacanthids were extraordinary freshwater elasmobranchs that became abundant in the Carboniferous but died out soon afterwards. They had paired fins rather like those of a lungfish, formed around a central axis, and perhaps able to scull the animal in shallow water. Their inability to breathe air, however, meant that no elasmobranch equivalent of the tetrapods could evolve.

- Cladoselachida. The basic split between holocephalans (chimaeras) and elasmobranchs (sharks and rays) had occurred by the lower Carboniferous, when chimaeras first appeared; they subsequently became moderately diverse in the Jurassic, before dwindling to their modern status. The common ancestor of chimaeras and elasmobranchs were Devonian forms such as *Cladoselache*. This was a fusiform fish about 1 m long with very broad-based paired fins capable of acting as little more than simple control surfaces.

- Acanthodii. The three great lineages of lobe-finned fish, ray-finned fish and cartilaginous fish were already distinct in the Devonian. Their common ancestry, however, cannot yet be traced confidently in the fossil record. The most likely suspects are the acanthodians, small freshwater fish that appeared in the Silurian about 420 Mya and were abundant in the lower Devonian. They were covered in small scales and had median fins supported by long spines—hence their common name of "spiny sharks." I have chosen a lower Devonian acanthodian, *Climatius* (Figure 5.12), as the sibla taxon, but the older, fragmentary fossils show that it lived some tens of millions of years after the common ancestor of gnathostomes. Acanthodians dwindled during the Carboniferous and died out in the lower Permian (about 300 Mya).

- Placodermi. The most abundant and diverse Devonian fish, however, were the placoderms. These very distinctive fish had both head and thorax encased in thick bony plates, leaving a long unarmored trunk and tail to provide propulsion. Many forms, such as the antiarchs, were rather flat-bodied fish that probably foraged for invertebrates on the bottom. *Bothriolepis* is a very common antiarch whose fossils can sometimes be found by the hundred. Other placoderms, however, were active pelagic predators. The best known are the arthrodires, some of which reached enormous size. *Dunkleosteus* (= *Dinichthys*) was a formidable marine predator that reached almost 10 m in length, with an armored head over a meter in length. The placoderms expanded into a wide range of marine and freshwater habitats during the Devonian, radiating into a great diversity of forms, but died out completely by the end of the period, about 340 Mya.

The history of jawed vertebrates, then, involves five major groups that became abundant in the Devonian period. The two earliest to appear, the placoderms and acanthodians, subsequently became extinct, while the cartilaginous fishes, the lobe-finned fishes (with their tetrapod descendants) and the ray-finned fishes diversified during the later Paleozoic and Mesozoic.

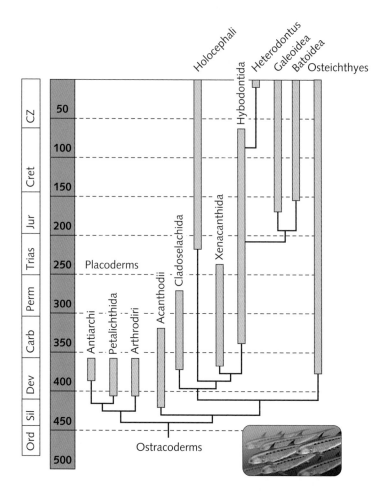

Figure 5.12 Gnathostomata. Sibla taxon: *Climatius*.

Image courtesy of Nobu Tamura. This file is licensed under the Creative Commons Attribution-Share Alike 3.0 Unported license.

5.12 **Chordata**

Coalescent: 430 Mya

Sibla taxon: *Myllokunmingia*

The pharyngeal gill slits of vertebrates are supported by jointed bars of cartilage or bone, attached to the skeleton of the head. They provide attachments for the muscles that open and close the gill slits. The most anterior bars are next to the mouth, and in one vertebrate lineage they and their musculature evolved the ability to open and close the mouth rather than the gill. This pair of jointed bones then constituted a jaw. This was a crucial event in vertebrate evolution: jawless vertebrates can only suck, whereas gnathostomes can bite. Consequently, a jaw enables animals whose ancestors could eat only sediment or small soft-bodied prey to exploit a much greater range of food.

- Osteostraci. In late Silurian times (about 420 Mya) a very distinctive group of fish evolved, having a

solid bony shell covering their head and shoulders completely. Osteostracans had rather flattened bodies, with eyes on the top of the head-shield, a small ventral mouth and a large gill chamber. They can be imagined as mobile filter-feeders, sieving water and sediment through their branchial apparatus in the classical chordate manner. They were jawless, but in other respects were quite fish-like in appearance. They swam by flexing the trunk and the long heterocercal tail, and could control their movements to some degree with paired pectoral appendages which are probably homologous with the pectoral fins of acanthodians.

- Heterostraci. To our eyes, heterostracans are even more bizarre. The front half of their body was again encased in bone, but they had neither median nor paired fins. Like the osteostracans, they disappeared at the end of the Devonian.

- Anaspida. These jawless fish of late Silurian and early Devonian times are collectively called ostracoderms ("shell-skinned"). They are clearly the outcome of a radiation that took place much earlier, as scattered fragments of bony plates have been found in Ordovician rocks (about 440 Mya) in Russia and the eastern Rockies. The earliest forms to have left a good fossil record are the anaspids, which were small freshwater fish with paired fins and a body covering of scales rather than bony plates. They appeared in the lower Silurian (about 430 Mya) and had become extinct by the end of the period.

- Conodonta. At the same time, a completely separate lineage of chordates was pursuing a very different way of life. Peculiar tooth-like structures ("conodonts") are extremely abundant in Paleozoic marine sediments from the middle Cambrian (400 Mya) onward. They have been intensively studied because they are useful in the oil industry for dating rocks, but until recently no body fossils had been found. A few well-preserved fossils with the teeth in place have now identified the conodonts as chordates: elongated animals with myomeres, a notochord, and a tail fin. Their body plan suggests that they were ecologically similar to chaetognaths: active epipelagic predators capturing small planktonic crustaceans and fish larvae. They disappeared completely in the end-Permian mass extinction.

- *Myllokunmingia*. The earliest chordates had cartilaginous skeletons and are preserved only in exceptional cases; modern lampreys and hagfish, despite being numerous, have almost no fossil record at all. Very early chordates have been found in the exceptionally well-preserved fauna of the Lower Cambrian (530 Mya) Chengjiang shales in Yunnan, China. *Myllokunmingia* (Figure 5.13) resembled a

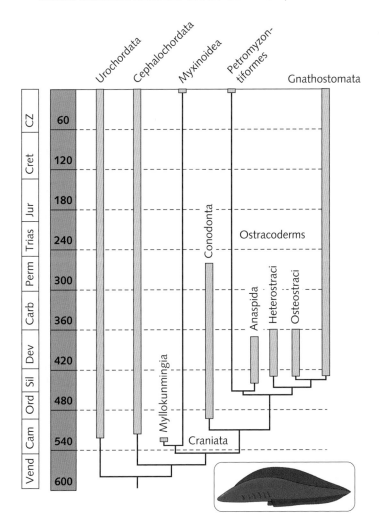

Figure 5.13 Chordata. Sibla taxon: *Myllokunmingia*.

small hagfish. It had a notochord, myomeres, fins, and a cartilaginous skull, and we have taken it to represent a form close to the ancestral state of craniates. It seems to have had an amphioxus-like way of life, sieving small food particles through its gill apparatus while lying half-buried in the sediment.

- Cephalochordata. *Pikaia* is a lancelet-shaped animal from the Middle Cambrian Burgess Shale fossil beds of British Columbia, which were made famous by Stephen Jay Gould in his book *Wonderful Life*. It had a notochord, V-shaped myomeres, and probably gill slits, but no skull.

It was probably a very early cephalochordate; no other fossil is known until the Permian.

Chordate evolution followed two radically different paths during the early Cambrian. One lineage became sessile and pumped seawater through its branchial basket, evolving into the urochordates (tunicates), which are known as fossils from the Lower Cambrian. The second took the alternative route of moving the basket through the sea and evolved the muscles, fins, sense organs, skeleton, and jaws to make this strategy more effective. It was this lineage that led from an animal like *Myllokunmingia* to fishes and tetrapods.

5.13 Deuterostomia

Coalescent: 530 Mya

Sibla taxon: *Haikouella*

The modern groups of echinoderms appeared during the Ordovician, and have an excellent fossil record because their covering of calcareous plate, the stereom, is so distinctive and easily preserved. Their ancestors lived in the Cambrian. Edrioasteroids were discoid animals looking rather like a starfish coiled on a plate; they are close to the ancestor of modern starfish, brittle stars and sea urchins. Crinoids probably descended from blastoids, stalked forms found in the Lower Cambrian (530 Mya). The oldest echinoderms are the helicoplacoids, Lower Cambrian animals shaped like a rugby football with spiral grooves carrying the water vascular system wrapped around the whole body.

- Homolozoa. At about the same time, however, there was a brief radiation of extraordinary animals that clearly have echinoderm affinities but cannot be easily related to any modern group. The homolozoans, or carpoids, had a rather flattened, disc-like body covered with calcite plates and penetrated by a series of pores. There is no trace of the usual pentaradial symmetry of echinoderms. Instead, some had a long feeding appendage, and seem to have crept along the sea bed holding this in front of them. Others had two appendages, one at the front and another at the rear. They seem to have been the equivalent in the echinoderm lineage of ostracoderms in the chordate lineage:

heavily armored animals trundling slowly over the sea floor, sieving small edible particles from the sediment. These peculiar animals died out in the Devonian, and nothing similar has been seen since.

- Graptolithina. The pterobranchs are an obscure group, but have a good fossil record because the test that houses each zooid is readily preserved. They seem to be closely related to the graptolites, which were extremely abundant in the Ordovician. They are usually found as thick, notched black lines in shales and slates. These are the tests of linear colonies of zooids that were suspended from a float; the zooids themselves are rarely preserved. A great diversity of growth forms evolved in the middle Paleozoic, but the entire group became extinct at the close of the Carboniferous.

- Yunnanozoa. The Chengjiang deposits where the earliest craniates have been found have also yielded a few animals that may be close to the ancestral state of chordates, and perhaps of all deuterostomes. They were soft-bodied, elongate organisms about an inch long that had gills and may have had segmented muscles and a dorsal fin. I have suggested one of them, *Haikouella* (Figure 5.14), as the sibla taxon for deuterostomes, but the interpretation of these recently discovered fossils is at present debatable.

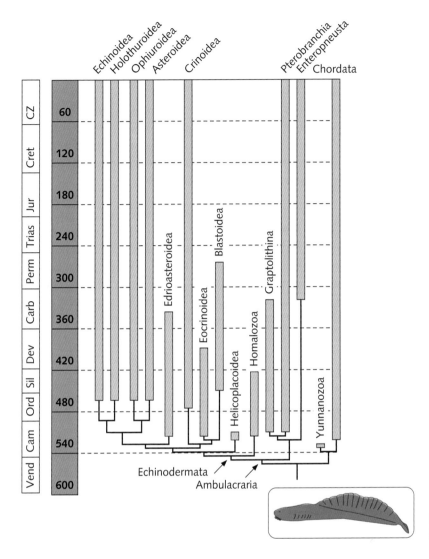

Figure 5.14 Deuterostomia. Sibla taxon: *Haikouella*.

- Vetulicolia. The even more enigmatic vetulicolians are another group of Chengjiang animals that may be close to the deuterostome ancestor. Their body had two parts, with one bearing structures resembling gill slits. Their interpretation is so uncertain at present that we have not shown them in the cladogram.

5.14 Bilateria

Coalescent: 544 Mya

Sibla taxon: *Kimberella*

The first half of the Cambrian was a period of very rapid diversification during which many of the major clades of marine animals appeared in the fossil record. Most of the elements of the Burgess Shale fauna (Middle Cambrian, 505 Mya) were already present in the Chengjiang fauna (Lower Cambrian, 530 Mya), but only a few occur in any earlier deposit. In particular, animals with mineralized exoskeletons that are readily fossilized, such as arthropods, mollusks, and echinoderms, are not found in late Proterozoic and very early Cambrian rocks where soft-bodied animals and delicate trace fossils have been preserved. Hence, there must have been an extensive radiation of animals with durable skeletons during the first 20 My of the Cambrian.

When the first well-preserved body fossils appear, they are either members of known groups, either living or extinct, or else they are members of a distinctive Cambrian fauna that died out before the end of the period. There are only a few plausible intermediates (such as Yunnanozoa and Vetulicolia) between major clades. The early Cambrian radiation of soft-bodied forms, which have not yet been discovered as fossils, was so rapid that the major groups were completely distinct by the time that readily fossilized skeletons had evolved. Figure 5.15 makes it clear that all the major modern clades with fossilizable hard parts had appeared by the middle Cambrian, including chelicerates, crustaceans, mollusks, brachiopods, hemichordates, and

echinoderms. The bryozoans are the only exception, not appearing until the early Ordovician. Some soft-bodied animals were also preserved, including priapulids, lobopods, annelids, and cephalochordates. Other soft-bodied groups, such as nematodes, appear only much later in the fossil record, probably because they are very unlikely to be fossilized. Small soft-bodied animals such as rotifers, gastrotrichs, acoels, and so forth are apparently unfossilizable, so that only the living representatives are known.

• Trilobita. There are too many extinct groups of Bilateria to describe them all, or even a substantial fraction of them. Perhaps the most prominent, however, are the trilobites. These were benthic marine arthropods with a flattened body consisting of a large head shield and a long thoracic lobe from which extended many lateral lobes bearing

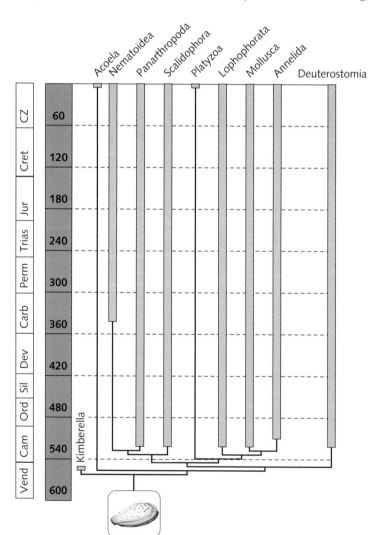

Figure 5.15 Bilateria. Sibla taxon: *Kimberella*.

gills and limbs. They foraged on the sea floor, eating small soft-bodied prey. These very abundant and distinctive animals flourished throughout most of the Paleozoic, but disappeared completely at the end of the Permian.

- "Trilobitomorpha." The early Cambrian arthropod faunas of Chengjiang and the Burgess Shale include several types that cannot be classified either with the trilobites or with the extant chelicerates, myriapods, and crustaceans. *Opabinia*, for example, was a segmented organism somewhat resembling a trilobite, but had a long proboscis ending in a beak, and five compound eyes on the top of its head. *Anomalocaris* was a large active predator with two spiny grasping appendages and a unique mouth that operated rather like an iris diaphragm. *Marrella* was an abundant small scavenger with a head shield bearing two pairs of long backward-directed horns. Other animals were probably not arthropods—but do not appear to belong to any other known group. For example, *Nectacaris* had a bipartite body consisting of an armored head set off from a long, muscular, segmented tail.

Hence, the early Cambrian radiation not only generated the bulk of modern animal diversity, but also produced many types that soon became extinct, leaving no descendants. The possible causes of this remarkable event have been much discussed.

- Elevated oxygen levels. Active locomotion by bilaterians requires high concentrations of dissolved oxygen. There is some evidence that atmospheric levels of oxygen became sufficient to power muscular locomotion in the late Proterozoic.
- Jaws and teeth. Predators armed with hard parts capable of capturing and processing prey first appeared in the early Cambrian. This would have led to the evolution of defensive measures, such

as shells, which would have led in turn to more massive armament.

- Hox genes. A limited set of Hox genes was already available. The modular or segmented construction of elongate bilaterians made it possible for an extended set of Hox genes, arising by duplication, to specify a very wide range of novel body plans.

It is quite likely that all of these are true. High oxygen levels were a precondition for deploying muscles effectively, the enlarged set of Hox genes provided the genetic basis for elaborate new body plans, and the co-evolutionary struggle of predators and prey provided the source of selection for ecological radiation.

There are abundant fossils in some deposits younger than Chengjiang, down to the base of the Cambrian. They are minute mineralized structures representing shells and perhaps tusks. Some can be assigned to familiar groups such as mollusks and brachiopods. Others belong to very early bilaterian groups that are now extinct, such as hyolithids, which were mollusk-like creatures quite common in the Cambrian. Many of them, however, are quite obscure. What they seem to be telling us is that the radiating bilaterian lineages that were evolving mineralized skeletons in the early Cambrian were very small, and presumably evolved from equally small soft-bodied creatures with little potential for fossilization.

Motile bilaterians leave trails and burrows in the sediment that can be preserved as trace fossils. The oldest worm tracks are found in the late Proterozoic, about 544 Mya. They are usually minute (less than 1 mm wide), which is consistent with ancestral bilaterians being very small. Body fossils from the Proterozoic are very rare and often difficult to interpret. The most bilaterian-like is *Kimberella*, which has been chosen as the sibla taxon for Bilateria. It has been interpreted as a small slug-like creature with a hard dorsal surface and a lobed ventral surface that crept over the sea floor.

5.15 Metazoa

Coalescent: 555 Mya

Sibla taxon: *Palaeophragmodictya*

Most cnidarians are flimsy creatures, and the fossil record of the group is correspondingly poor. The only exceptions are corals, anthozoans with

a durable mineralized skeleton, which appeared in the Cambrian and began to build coral reefs in the Ordovician. Modern corals date only from the

Triassic; during the Paleozoic, several distinct groups of corals appeared, radiated and eventually became extinct. As each appeared quite abruptly, it is likely that they represent the independent evolution of biomineralization in several soft-bodied anthozoan lineages.

The three main groups of modern sponges had all appeared by the Middle Cambrian. The oldest known body fossil of an identifiable metazoan is the hexactinellid *Palaeophragmodictya* (Figure 5.16), from the Ediacaran assemblage of South Australia (>540 Mya), so we have chosen this as the sibla taxon for Metazoa. It has also been interpreted as the attachment disc of an unknown animal. Scattered spicules have been found some 40 My further back, so the first stage in metazoan evolution was the radiation of sponges during this period.

- Archaeocyatha. Several other groups of sessile benthic metazoans appeared in the early Cambrian (about 424 Mya) and became abundant before dying out before the end of the period. Archaeocyathans were rather small cup-like organisms resembling two colanders, one placed inside the other, presumably pumping water from outside to inside like a sponge. They had a calcium carbonate skeleton, rather than spicules, and formed extensive reefs in the middle Cambrian. Chancelloriids

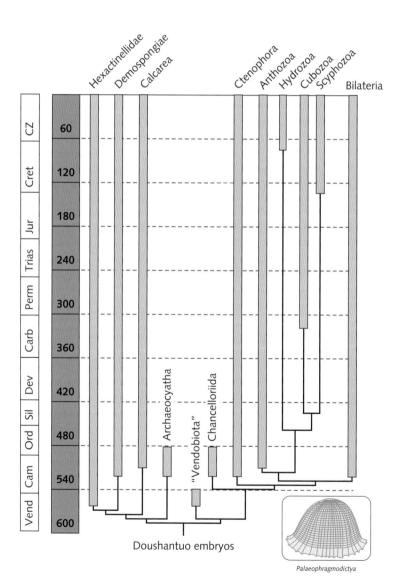

Figure 5.16 Metazoa. Sibla taxon: *Palaeophragmodictya*.

were another group of vase-shaped, sessile forms appearing in the lower Cambrian. There are several other early groups with a similar way of life, revealing a radiation of sessile metazoans before the radiation of the Bilateria. Their relationships are obscure: we have shown the archaeocyathans as the sister group of sponges and the chancelloriids as the sister group of cnidarians, but this is only one possible interpretation.

The oldest sponges occur in association with a distinctive community of organisms that flourished in the Vendian (or Ediacaran) period, during the last 24 My or so of the Proterozoic. The oldest example, from Mistaken Point in Newfoundland, dates from 565 Mya; there are other sites about 10 My younger in South Australia and the White Sea region of Russia. The most common organisms were frond-like forms with a long stalk attached to a basal holdfast. Others were large oval discs inscribed with radiating lines, elongate forms with a shield or plate at one end, and small circular discs enclosing a three-armed structure. None of them can be assigned with certainty to any modern group, and their relationships, though much debated, remain obscure.

The most ancient remains of metazoans are phosphatized embryos from Doushantuo, in China, which are dated at about 570 Mya. They show a pattern of cell division characteristic of early metazoan development, although the adults they developed into have not yet been found. Since metazoans appeared before the earliest known Vendian fauna, while this distinctive fauna persisted in widely separated regions of the world for 25 My, the simplest interpretation of the Vendian organisms is that they were indeed metazoans. I have placed them, purely for simplicity, as the sister group of all metazoans excluding sponges. There is no good reason to interpret them as constituting a monophyletic group, however: their diversity suggests that they may have included lineages ancestral to several later groups of metazoans, but the lines of descent cannot yet be traced in detail.

5.16 Unikonta

Coalescent: 740 Mya

Sister taxon to Metazoa: *Monosiga*

In this section I have departed from the usual practice of naming a sibla taxon for the clade, and instead named a choanoflagellate as a representative of the unicellular group closest to the ancestry of Metazoa (see Figure 5.17). This gives us a reliable marker of the unicellular–multicellular transition in our ancestry.

All three groups of multicellular fungi are known from the Devonian onwards, and may have played an important part in the development of terrestrial communities. They are already associated with early land plants in well-preserved material from the Rhynie chert. Some of the Rhynie fungi were mycorrhizae, the specialized fungi that form a web around the roots of plants. This is a mutualistic association in which the fungi supply nitrogen to the plant and receive sugars in return; mycorrhizae can have a dramatic effect on plant growth, and are essential for some plants to be able to grow at all. The oldest known terrestrial fungus is a mycorrhizal form from the Ordovician. Hence, it is likely that fungi were important in facilitating the evolution of land plants.

Fossil fungi are rare in pre-Devonian deposits, although zygomycetes appear in the Cambrian. Thus, the extensive radiation of ascomycetes and basidiomycetes probably took place in the context of the emerging terrestrial communities of late Silurian and Devonian. Apart from their symbioses with plants, fungi may well have been important elements of these communities in their own right. The largest land organism from the Late Silurian (420 Mya) until the appearance of lycopod trees 40 My later was *Prototaxites*, which consisted of an upright, unbranched column up to 8 m in height. It seems likely that it was a fungus, like a gigantic mushroom stalk. The vegetation it was growing amongst, before the appearance of vascular plants, seems to have been a mixture of mosses, algae and fungi, forming a low cover over the ground.

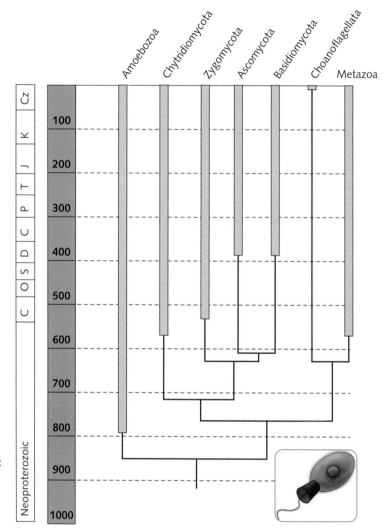

Figure 5.17 Unikonta. Sibla taxon (to Metazoa): *Monosiga*.

Image courtesy of Mateus Zica. This file is licensed under the Creative Commons Attribution-Share Alike 3.0 Unported license.

5.17 **Eukaryota**

Coalescent: 900 Mya

Sibla taxon: *Melanocyrillium*

Many unicellular eukaryotes have durable tests or shells with distinctive forms, such as the coiled calcareous shell of foraminiferans, the complex siliceous tests of radiolarians or the discoid scales of coccolithophores. These are preserved in enormous numbers and give a reliable fossil record for these groups. Most are quite recent: diatoms and dinoflagellates, for example, do not appear until the Mesozoic. The oldest bikonts are foraminiferans, chlorophytes, and

rhodophytes, which can be traced by unambiguous fossils back to the base of the Cambrian or a little beyond.

This suggests that the major eukaryote radiation occurred during the late Proterozoic, at about the same time as the Vendian animals. At this time, or shortly afterwards, most of the modern groups evolved and diversified, partly replacing what had been until then an almost exclusively bacterial

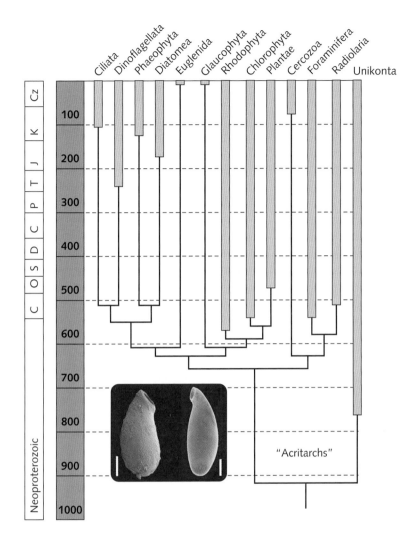

Figure 5.18 Eukaryota. Sibla taxon: *Melanocyrillium*.

Image reproduced from Porter, S., et al. 2003. Vase-shaped microfossils from the Neoproterozoic Chuar Group, Grand Canyon: a classification guided by modern testate amoebae. *Journal of Paleontology* 77 (3), with permission from Dr Susannah Porter.

biota. Whether by coincidence or not, this occurred shortly after a major glaciation that covered much of the Earth's surface with ice. The oldest known eukaryotes are then testate amoebas, dating back to 740 Mya, and I have chosen one of these (*Melanocyrillium*, from the Kwagunt Formation of Montana) as the sibla taxon (Figure 5.18).

There are remains of large cells with patterned walls and complex external processes, however, from much older rocks. These are collectively known as acritarchs, and interpreted as the cysts of eukaryotic algae. They appear as far back as 1800 Mya, but remained rather rare until a marked diversification that began a little before the Vendian. As an early example, *Tappania* was a large spherical cell, up to 140 μm in diameter, from which a number of slender processes extended a further 40 μm or so, found in shales dated to 1400 Mya. Moreover, much larger, multicellular fossils have been found in old rocks. The best known is *Grypania*, a ribbon-like structure found in coils up to an inch across, found in shales from Montana from 1440 Mya. *Bangiomorpha* is a well-preserved fossil with clearly visible cell structure that resembles modern red algae, dated to about 1200 Mya. Many other filamentous and frond-like fossils have been described from old rocks and identified as eukaryotes. It remains possible that they are bacteria with unusually complex cells or colonies. I am agnostic on the point, and have presented a conservative tree in which eukaryotes appear 900–1000 Mya and radiate into the modern biota about 540–700 Mya.

5.18 **Life**

Coalescent: 3240 Mya
Sibla taxon: *Archaeosphaeroides*

Bacterial communities were, of course, already well-established and diverse when the first eukaryotes appeared, and in exceptional circumstances, especially silicate infiltration of sediments to make cherts, they have been preserved so well that details of cell structure can be seen. The Bitter Springs chert of Australia (840 Mya) contains single coccoid cells, small colonies, and long multicellular filaments with tapering ends that resemble living cyanobacteria. Very similar communities go back a further billion years. The classic Gunflint chert of Ontario (2100 Mya) has the same range of growth forms, again comparable with cyanobacteria. It is now firmly established that bacteria

lived throughout the Proterozoic, without changing very much in external form, unless organisms like *Tappania* and *Grypania* are unusual kinds of bacteria.

Further back in time, the deposits become scarcer and the state of preservation deteriorates, but cells, colonies, and filaments have been described from many Archaean sites older than 2400 Mya. The very oldest cellular fossils have been described from the Apex chert from Western Australia, dated at 3444 Mya. The great age of these specimens, however, means that they are difficult to distinguish from bubbles and tracks made by purely physical processes. *Archaeosphaeroides* (Figure 5.19) is a spherical

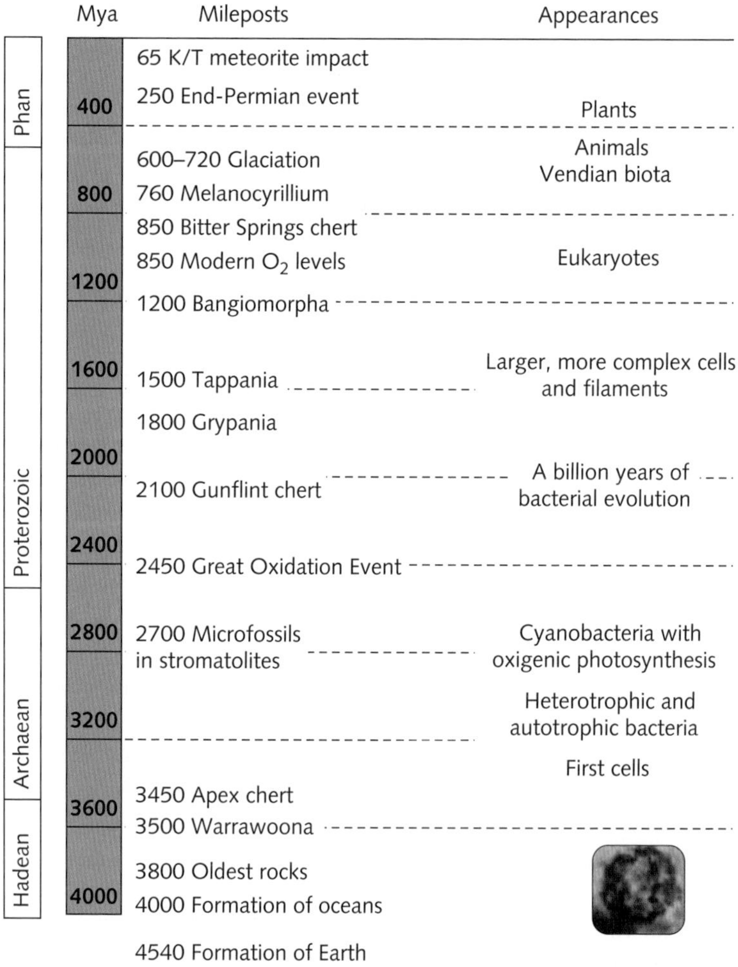

Mya	Mileposts	Appearances
Phan 400	65 K/T meteorite impact 250 End-Permian event	Plants
800	600–720 Glaciation 760 Melanocyrillium	Animals Vendian biota
1200	850 Bitter Springs chert 850 Modern O_2 levels	Eukaryotes
	1200 Bangiomorpha	
1600	1500 Tappania	Larger, more complex cells and filaments
2000	1800 Grypania	
	2100 Gunflint chert	A billion years of bacterial evolution
2400		
	2450 Great Oxidation Event	
2800	2700 Microfossils in stromatolites	Cyanobacteria with oxigenic photosynthesis
3200		Heterotrophic and autotrophic bacteria
		First cells
3600	3450 Apex chert 3500 Warrawoona	
4000	3800 Oldest rocks 4000 Formation of oceans	
	4540 Formation of Earth	

Figure 5.19 Life. Sibla taxon: *Archaeosphaeroides*.

Image reproduced from Schopf, J.W. Fossil evidence of Archaean life. *Philosophical Transactions B* 361 (1470) © 2006, The Royal Society.

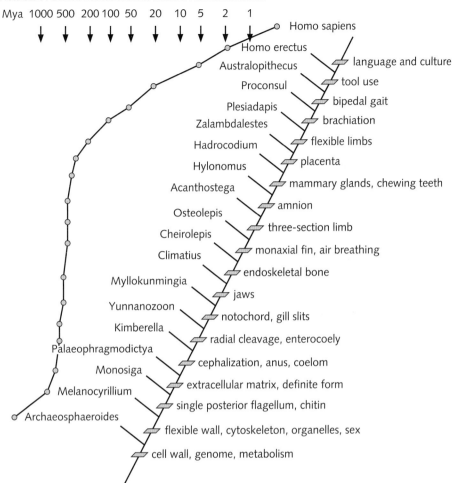

Proterozoic Paleozoic Mesozoic Eocene Miocene Pliocene Pleistocene

Mya 1000 500 200 100 50 20 10 5 2 1

Homo sapiens
Homo erectus
Australopithecus — language and culture
Proconsul — tool use
Plesiadapis — bipedal gait
Zalambdalestes — brachiation
Hadrocodium — flexible limbs
Hylonomus — placenta
Acanthostega — mammary glands, chewing teeth
Osteolepis — amnion
Cheirolepis — three-section limb
Climatius — monaxial fin, air breathing
Myllokunmingia — endoskeletal bone
Yunnanozoon — jaws
Kimberella — notochord, gill slits
Palaeophragmodictya — radial cleavage, enterocoely
Monosiga — cephalization, anus, coelom
Melanocyrillium — extracellular matrix, definite form
Archaeosphaeroides — single posterior flagellum, chitin
— flexible wall, cytoskeleton, organelles, sex
— cell wall, genome, metabolism

Figure 5.20 Summary of sibla taxa and major innovations in the lineage leading to humans. The line on the left connects dots representing the age of each named sibla taxon.

object about 10 μm in diameter found in the Sheba Formation of South Africa (3240 Mya). It resembles a coccoidal bacterial cell, and I have taken it as the sibla taxon for cellular life. The Proterozoic fossil record does not reveal the diversification of bacterial groups, however, so Figure 5.19 shows only some of the main events in their evolution. Figure 5.20 summarizes the ancestry of the human lineage that we have traced so far back in time.

One indication that these very ancient fossils are truly organic in origin is that they are associated with structures that resemble those produced by living bacteria. Stromatolites are macroscopic, mound-shaped structures formed on the sea floor by mucilage-producing cyanobacteria. They consist of many alternating layers of bacterial mat and sediment. They are quickly destroyed by burrowing invertebrates, and are found today mainly in hypersaline bays where few animals

can survive. They are common as fossils from the Proterozoic, before the evolution of bilaterians, and in cases where the stromatolite was mineralized by silica they contain recognizable cellular remains of the cyanobacteria that lived in them. Similar structures are found deep in the Archaean, among the oldest being the conical groups of stromatolites (think of an upturned egg carton) found at Warrawoona in Australia and dated to 3400–3500 Mya. These very ancient structures seldom contain recognizable microfossils, however, and similar structures can be produced by physical processes, so their biological origin, while rendered probable by their continuity with similar Proterozoic structures, is not established beyond all doubt.

The fossil record of bacteria extends far back in time, but does not tell us much about their diversification because their external appearance is rather uniform and uninformative. Bacteria are so active and abundant, however, that their metabolism has left deep traces on the face of the Earth, and some of these can still be read.

- Oxygenic photosynthesis. The most prominent geochemical record of early bacterial evolution is how the composition of the atmosphere has changed through time. The original atmosphere of the Earth was almost completely free of oxygen, and the first living cells must have had purely anaerobic metabolism. About 2440 Mya a number of geological indicators show that levels of oxygen began to rise. For example, the banded iron deposits that supply much of our iron ore were formed when ferrous iron dissolved in the oceans was oxidized to insoluble ferric iron, which then precipitated out as an iron-rich layer on the sea floor. The only process that could have generated the huge amounts of oxygen required is oxygenic photosynthesis. Hence, there must have been large populations of cyanobacteria living 2440 Mya. Moreover, ferrous iron would at first buffer the atmosphere against oxygen produced by photosynthesis, so cyanobacteria producing oxygen as a byproduct of carbon fixation must have evolved earlier, probably about 2800 Mya.

- Anoxygenic photosynthesis. These cyanobacteria had both Photosystem I and Photosystem II. They must have evolved from photosynthetic bacteria, which are still abundant today in anoxic conditions, that possessed only Photosystem I, using H_2 or H_2S as electron donors. Hence, bacteria using non-oxygenic photosynthesis must have lived earlier still, back to 3000 Mya or more.

Photosynthesis is a complex process that cannot have been the ancestral state for bacteria, and the earliest cells must have used some simpler system. There are two possibilities: using inorganic compounds as energy sources, or using organic compounds directly.

- Iron and sulfur bacteria. The earliest metabolic systems might have involved the use of inorganic chemical reactions. We can only guess at what they were. One very simple possibility, for example, is that ferrous iron reacts with hydrogen sulfide to form pyrite and molecular hydrogen: $FeS + H_2S \rightarrow FeS_2 + H_2$. This yields free energy and

a supply of hydrogen to act as an electron donor for the reduction of elemental sulfur: $H_2 + S \rightarrow H_2S$. Separating the hydrogen into protons and electrons across a simple membrane would create a proton motive force driving an ATPase that captured the free energy released in the high-energy phosphate bond of ATP. A system like this would need only three proteins: a hydrogenase (transferring electrons across the membrane), a sulfur reductase (catalyzing the formation of H_2S from S^0 and water), and an ATPase (to phosphorylate ADP). It would probably work, albeit very inefficiently, with just one, the ATPase, and a membrane. Bacteria that use iron and sulfur to generate energy are abundant today in anoxic environments.

- Heterotrophs. Modern organisms that consume organic compounds (heterotrophy) process them through a long series of intermediate reaction products, each requiring a different protein catalyst. These intermediates are not directly available in useful concentrations, because any that appears is quickly consumed. This need not have been the case in the early oceans, where, in the absence of living organisms, substances such as ribose or nucleotide triphosphates that were formed by purely chemical processes could accumulate. Once fairly efficient replicators had evolved, however, they would quickly increase in numbers, and by doing so would deplete the supply of dissolved nutrients. There would then be very strong selection for a creature with metabolism, even of the simplest sort; any creature that encoded a protein that acted as a very inefficient enzyme, converting some previously unexploited substance into the desired resource. This would represent the simplest possible metabolic pathway, with one step, a single enzyme that modifies a single substrate so as to produce energy, or a substance useful to the cell. The success of such creatures, however, would in turn deplete the substrate. As the substrate became limiting to growth, variants able to scavenge low concentrations of the substrate would spread; some of these variants would have two copies of the gene, thereby producing greater quantities of the enzyme. If the environment contains substances chemically similar to the substrate, selection would favor any type able to convert one of these substances into the substrate. This might be achieved by an enzyme similar to the one that the organism already possesses, given that the two substances involved are similar. Variants would arise in which one copy of the duplicated gene encoded a protein able to catalyze the reaction, however inefficiently, and selection would subsequently enhance its ability to do so. The organism now has a metabolic pathway with two substrates and two enzymes, whose end-product is the resource that the cell requires. The repetition of this process would eventually lead to the construction of complex metabolic pathways in which the initial substrate and the final product are separated by a long series of steps, in each of which a specific enzyme causes a simple modification of an intermediate metabolite.

• Aerobic respiration. Heterotrophs now live exclusively on organic compounds originally made by autotrophs. This can be accomplished anaerobically by fermentation. This route might have evolved early in the history of bacteria, but again the long series of intermediate reactions it involves makes it unlikely that it was a truly primitive system. Once the ocean had become oxygenated then aerobic metabolism could evolve, with the great advantage of using oxygen as a very strong electron acceptor. Heterotrophic bacteria are capable of utilizing a vast range of compounds, including such unlikely food sources as pesticides and petroleum.

Ancient metabolic machinery can leave traces in the rocks, because metabolic processes often discriminate between isotopes (photosynthesis preferentially uses light ^{12}C over heavier ^{13}C, for example) and thus alters the isotopic balance. These traces are difficult to read, however, except for the Great Oxidation Event, and the course of events in early bacterial evolution is still obscure.

The very oldest rocks date from about 3800 Mya, about 740 My after the formation of the Earth. At this time the intense meteor bombardment of the early Earth was slackening, the crust was consolidating and the cycle of deposition and erosion was starting up. At some time between 3800 Mya and 3000 Mya the first cells appeared, following the prior evolution of self-replicating molecules, probably RNA. These events have left no trace in the fossil record, however, and at this point the long trail that we have traced backwards in time fades out completely.

CHAPTER SUMMARY

	Clade	Coalescent (Mya)	Sibla taxon
5.1	Ourselves: *Homo sapiens*	0.2	*Homo erectus*
5.2	*Homo + Pan*	4	*Australopithecus africanus*
5.3	Hominoidea	21	*Proconsul*
5.4	Primates	44	*Plesiadapis*
5.5	Eutheria	120	*Eomaia*
5.6	Mammalia	210	*Hadrocodium*
5.7	Amniota	330	*Protoclepsydrops*
5.8	Tetrapoda	374	*Acanthostega*
5.9	Sarcopterygii	415	*Osteolepis*
5.10	Osteichthyes	420	*Cheirolepis*
5.11	Gnathostomata	414	*Climatius*
5.12	Chordata	430	*Myllokunmingia*
5.13	Deuterostomia	530	*Haikouella*
5.14	Bilateria	544	*Kimberella*
5.15	Metazoa	555	*Palaeophragmodictya*
5.16	Unikonta	740	(*Monosiga*)
5.17	Eukaryota	900	*Melanocyrillium*
5.18	Life	3240	*Archaeosphaeroides*

● FURTHER READING

Here are some pertinent further readings at the time of going to press. For relevant readings that have been released since publication, visit the book's Online Resource Centre at
www.oxfordtextbooks.co.uk/orc/bell_evolution/

Section 5.1   Finlayson, C. 2005. Biogeography and evolution of the genus *Homo*. *Trends in Ecology and Evolution* 20: 457–463.

Schwartz, J.H. and Tattersall, I. 2010. Fossil evidence for the origin of *Homo sapiens*. *Yearbook of Physical Anthropology* 53: 94–121.

Green, R.E. et al. 2006. Analysis of one million base pairs of Neanderthal DNA. *Nature* 444: 330–336.

Section 5.2   McHenry, H.M. and Coffing, K. 2000. *Australopithecus* to *Homo*: transformations in body and mind. *Annual Reviews of Anthropology* 29: 125–146.

Section 5.3   Wood, B. and Harrison, T. 2011. The evolutionary context of the first hominins. *Nature* 470: 347–352.

Section 5.4   Martin, R.D. 2003. Combing the primate record. *Nature* 422: 388–389.

Section 5.5   Springer, M.S., Murphy, W.J., Eizirick, E. and O'Brien, S.J. 2003. Placental mammal diversification and the Cretaceous–Tertiary boundary. *Proceedings of the National Academy of Sciences of the USA* 100: 1056–1061.

Section 5.6   Cifelli, R.L. and Davis, B.M. 2003. Marsupial origins. *Science* 302: 1899–1900.

Section 5.7   Reisz, R.R. and Muller, J. 2004. Molecular timescales and the fossil record: a palaeontological perspective. *Trends in Genetics* 20: 237–241.

Brusatte, S.L. et al. 2010. The origin and early radiation of dinosaurs. *Earth-Science Reviews* 101: 68–100.

Motani, R. 2009. The evolution of marine reptiles. *Evolution Education Outreach* 2: 224–235.

Section 5.8   Anderson, J.S. 2008. The origin(s) of modern amphibians. *Evolutionary Biology* 35: 231–247.

Section 5.9   Clack, J. 2006. The emergence of early tetrapods. *Palaeogeography, Palaeoclimatology, Palaeoecology* 232: 167–189.

Section 5.10   Friedman, M. and Sallan, L.C. 2012. Five hundred million years of extinction and recovery: a Phanerozoic survey of large-scale diversity patterns in fishes. *Palaeontology* 55: 707–742.

Section 5.11   Young, G.C. 2010. Placoderms (armored fish): dominant vertebrates of the Devonian period. *Annual Review of Earth and Planetary Sciences* 38: 523–550.

Section 5.12   Johanson, Z. 2010. Evolution of paired fins and the lateral somatic frontier. *Journal of Experimental Zoology* 314B: 347–352.

Section 5.13   Zamora, S., Rahman, I.A. and Smith, A.B. 2012. Plated Cambrian bilaterians reveal the earliest stages of echinoderm evolution. *PLoS ONE* 7: e38296.

Section 5.14   Conway Morris, S. 2006. Darwin's dilemma: the realities of the Cambrian 'explosion'. *Philosophical Transactions of the Royal Society of London B* 361: 1069–1083.

Xiao, S. and Laflamme, M. On the eve of animal radiation: phylogeny, ecology and evolution of the Ediacara biota. *Trends in Ecology and Evolution* 24: 31–40.

Section 5.15   Brooke, N.M. and Holland, P.W.H. 2003. The evolution of multicellularity and early animal genomes. *Current Opinion in Genetics and Development* 13: 599–603.

Section 5.16   Redecker, D., Kodner, R. and Graham, L.E. 2000. Glomalean fungi from the Ordovician. *Science* 289: 1920–1921.

Section 5.17   Knoll, A.H., Javaux, E.J., Hewitt, D. and Cohen, P. 2006. Eukaryotic organisms in Proterozoic oceans. *Philosophical Transactions of the Royal Society of London B* 361: 1023–1038.

Embley, T.M. and Martin, W. 2006. Eukaryotic evolution, changes and challenges. *Nature* 440: 623–630.

Section 5.18   Lopez-Garcia, P. et al. 2006. Ancient fossil record and early evolution (ca. 3.8 to 0.5 Ga). *Earth, Moon and Planets* 98: 247–290.

Schopf, J.W. 2006. Fossil evidence of Archaean life. *Philosophical Transactions of the Royal Society of London B* 361: 869–885.

● QUESTIONS

1. Describe how the age of rocks is estimated from their isotopic composition.

2. What is meant by saying that an extinct taxon is "ancestral" to a modern group? Explain the concept of a "sibla taxon."

3. Write a brief account of the history of our species in terms of major population movements.

4. Draw a phylogenetic tree showing the relationship among species of *Homo* and the relationship of *Homo* to *Australopithecus* and *Pan*. Indicate the main living and extinct taxa and their distribution in time.

5. Describe the history of the human lineage after its divergence from the lineage leading to chimpanzees.

6. Describe the general innovations of primates and the particular innovations of apes that made possible the evolution of human structural attributes such as upright posture and dexterity.

7. Draw a diagram to explain the relationships among the major lineages of mammals and write a brief account of the main events in mammalian evolution during the Mesozoic and Cenozoic.

8. The mammalian lineage diverged very early into Monotremata and the much larger clade of Metatheria + Eutheria. Likewise, Lepidosauria diverged very early into Rhynchocephalia and the much larger clade Squamata. To what extent do groups such as Monotremata and Rhynchocephalia preserve ancestral character states?

9. Illustrate the radiation of the main groups of amniotes by using a phylogenetic tree and a time-line.

10. Outline the series of sibla taxa from *Osteolepis* to *Plesiadapis* and list the main innovations they represent.

11. Give an account of the evolution of early tetrapods from aquatic ancestors.

12. Describe the radiation of marine vertebrates from the Silurian to the end of the Paleozoic. Use a diagram with a time-line to indicate the relationships between ostracoderms, placoderms, and the modern groups of cartilaginous, lobe-finned, and ray-finned fishes.

13. Give an account of the Cambrian radiation of the Bilateria. What do you think might have been the causes of this radiation? Why did so many groups appear so abruptly in the fossil record?

14. Describe what is known about the evolution of metazoans between their first appearance in the fossil record about 570 Mya and the Middle Cambrian. Why do you think that more than 2000 My elapsed between the evolution of cells and the evolution of multicellular eukaryotes?

15. Write an essay on the first two billion years of life on Earth.

16. What do you think is the most crucial fossil that has not yet been found?

You can find a fuller set of questions, which will be refreshed during the life of this edition, in the book's Online Resource Centre at www.oxfordtextbooks.co.uk/orc/bell_evolution/

PART 3
Origins

Every link in the chain of our ancestors traced in the previous chapter represented the modification of a pre-existing state: evolution is modification through descent. Each modification, however, must itself originate in some way. Before proceeding to examine the process of modification, therefore, we need to tackle the problem of origins. It is a problem that arises at every level of phylogeny. At the highest level, we need to explain the origin of the variation within species that leads to adaptation and divergence. At a somewhat lower level, the origin of the species themselves is the classical problem of origins that stimulated the development of evolutionary thought. At still lower levels, we must understand the origin of new ways of life, and, at the lowest level of all, the origin of life itself. This part of the book deals with origins at all three levels in the tree of life.

The Origin
of Variation

6

In every generation, mutation and recombination produce genetic variation; some variants reproduce faster than others and therefore increase in frequency. Variation and selection recur in every generation. There are therefore two theories of evolution: the first is that it is directed by variation, and the second that it is directed by selection.

Lamarckian evolution is appropriately directed modification. The first is Lamarck's theory of the inheritance of acquired characteristics, published in the year of Darwin's birth. According to this theory, individuals perceive the state of their environment during their lifetimes and respond appropriately to it. This may involve some trivial change, such as the thickening of part of the cuticle or the enlargement of a muscle, or a more profound reorganization, perhaps through the formation of new body cavities as the result of a change in the pattern of circulation of fluids. There is thus an inherently progressive principle in nature, causing a general increase in the level of organization. Selection may occur, but plays a subordinate role; adaptation to novel environments is caused primarily by the spontaneous appearance of appropriately-directed variation. We can translate this into modern terms by saying that the environment elicits favorable mutations.

So far as we know, this theory is wrong. It is not wrong as a matter of principle; indeed, it is an internally consistent and intellectually satisfying theory of evolutionary change. It is wrong as a matter of fact. No mechanism that would act as a specific directing principle to produce appropriate genetic variation has yet been identified. During the early years of the study of bacterial genetics a number of experiments were devised to investigate this issue. They involved challenging bacteria with a new and hostile environment, and then tracing the origin of the adaptations that evolved. In all cases, it was found that the mutations that conferred adaptation, and which increased in frequency through selection in the new environment, occurred before the environment changed. These results show that adaptation occurs through the selection of pre-existing variation, and not through the elicitation of appropriate variation.

Darwinian evolution is the selection of undirected variation. The remaining possibility is the theory of evolution through selection proposed by Charles Darwin and Alfred Russel Wallace. This does not require that mutations occur at random: on the contrary, it is well-known that some genes mutate more frequently than others, and that different sites within a gene may have different rates of mutation. Nor does it assert that mutations are not induced by the environment: it is equally well-known that many mutations are caused straightforwardly by physical agents such as ionizing radiation. The crucial point is that mutations are not *appropriately* induced by environmental factors. Suppose that we expose bacterial cultures, or populations of any organism, either to ultraviolet light or to EMS (ethyl methyl sulfonate, a mutagenic chemical). Both are toxic, and inhibit growth; both are also powerful mutagens, and in the appropriate doses will cause a range of mutations. It will not be found, however, that the mutations induced by ultraviolet light are specifically resistant to ultraviolet light, and not to EMS; nor that the mutations induced by EMS are specifically resistant to EMS, and not to ultraviolet light. Adaptation to either agent, or to any less dramatic environmental challenges, is through the selection of *undirected* variation. The direction (or lack of direction) taken by evolution is determined by the action of selection in each generation, and not by any inherent directional property of variation itself.

Origin and fate involve different processes. Hence, the origin of variation is distinct from the fate of variation. This divides the study of evolution into two parts. The first concerns the origin of variation, which is governed by processes such as mutation and recombination. The second concerns the fate of variation, which is governed by processes such as natural selection and genetic drift. The emphasis that we place on one or the other depends on the phylogenetic scale we are considering. Adaptation will ultimately depend on beneficial genetic alterations that enhance fitness in some set of conditions. Alterations that have only a small effect on phenotype occur so frequently that their occurrence can be taken for granted; in this case, selection governs the outcome of evolution. Alterations that have a large effect on phenotype occur very seldom and will have a profound effect on the outcome of evolution; in this case, mutation governs the outcome of evolution. These two extremes are sometimes called "microevolution" and "macroevolution." This is something like the difference between "weather" and "climate," in the sense that the terms refer to the same phenomena at different scales. The terms do not imply that different processes are involved. Potentially beneficial mutations will continually arise, whose frequency falls with their effect: mutations of larger effect are rarer. Selection cannot bring into being beneficial genetic alterations of any effect whatsoever, but once they appear will screen them, and their subsequent modifications, regardless of their effect.

The distinction between the origin and the fate of variation is clear and useful. It is not quite as absolute as it seems. Variation is produced by mutation and recombination, which together determine how readily and how far a lineage will respond to selection. The rate of response, however, might itself be selected. For example, lineages with a genetic predisposition to higher rates of mutation might be less fit (because most mutations are deleterious) or more fit (because they will give rise to the few superior variants that survive in the long term). We shall discuss these possibilities in the sections dealing with the selection of mutator alleles (Section 6.1.3) and the evolution of sex (Section 15.2.3). In the three chapters in this part of the book, we shall be concerned only with the origin of novelty at all genetic scales, from nucleotides to phyla.

6.1 Mutation is the source of genetic variation.

Mutation is a general term (it simply means "change") for any kind of difference between the genomes of parent and offspring. The least difference is a point mutation, the insertion of a different nucleotide at a given position in the genome of the offspring. Larger differences involve the deletion of a few existing nucleotides, or the insertion of a few new ones. Still larger differences may involve the complete deletion of entire genes, the insertion of complete genes, or the rearrangement of the order of genes along a chromosome. Whole chromosomes may be gained, or lost, or fused; the entire genome, even, may be duplicated. All these events, from the smallest to the largest, constitute mutations.

Mutation is inevitable. A mutation is an error in copying information from one genome to another. The probability that an error will be made depends on the resources that are allocated to detecting and correcting errors. This probability can never be reduced to zero, however, except in the impossible extreme of devoting an infinite amount of resources to the task. The text of this book, for example, contains about a million characters, which encode about the same information as the yeast genome. It has been very carefully checked, but one or two errors will nevertheless remain. Producing a million books independently (roughly equivalent to the lineage produced by 20 doublings of a single ancestral yeast cell) is bound to generate a very large number of errors. The occurrence of mutations does not need any special explanation; it is an inevitable byproduct of any system of copying.

Mutation can produce minor or major changes in structure. The snapdragon (*Antirrhinum*) is a common garden plant with showy flowers produced on a long spike-like inflorescence. The wild-type flower consists of four whorls: the outermost whorl bears five sepals; the next is the corolla, which has five lobes that are fused for part of their length to form a corolla tube; there then is a set of four stamens; and finally in the center of the flower the innermost whorl consists of two carpels bearing a style terminating in the stigma. The corolla is magenta in color, with a patch of yellow on the lower lip. This basic layout

can be modified in many ways, from the trivial to the fundamental, by mutations in different genes. Some of these mutations are illustrated in Figure 6.1.

- *nivea*. These flowers are albino because the production of anthocyanin pigment has been blocked. Loss-of-function mutations in several genes, often caused by transposon insertion, can lead to similar phenotypes.

- *pallida*. Mutations in genes that act late in the pathway leading to anthocyanin synthesis cause pale or streaky flowers. These mutations are also often due to transposon activity.

- *deficiens*. In this mutant the second whorl, as well as the first, consists of five separate sepals, while the third whorl bears five carpels joined together to form a ring of stigmatic tissue.

Figure 6.1 Floral mutants of snapdragon (*Antirrhinum*).

- *plena*. The third whorl bears petal-like structures that are fused with the normal petals in the second whorl, while the fourth whorl contains abnormal structures that resemble petals.

- *centroradialis*. The inflorescence becomes determinate: after 8 or 9 flowers have appeared along the spike it is terminated by a flower with unusual shape.

- *cycloidea*. The flowers of this mutant are more symmetrical than those of the wild type, because the three lower lobes of the corolla all resemble one another, imparting a slightly flattened shape to the flower.

Hence, mutations in genes controlling flower structure may have effects ranging from trivial alterations in pigmentation to fundamental modifications of inflorescence development and floral symmetry.

Mutation provides new selectable variation. The inevitability of mutation implies the continuous operation of selection. Moreover, the rate of mutation will limit the extent to which adaptation can be improved by selection. If there were no mutation selection would not occur. Conversely, mutation rates might be so high that selection is ineffective, if changes in gene frequency caused by selection are overwhelmed by countervailing changes caused by mutation. Consequently, the rate of mutation is a fundamental attribute of evolving lineages. It can be estimated in several ways.

- Observing the rate at which dominant genetic disorders appear in the families of unaffected parents. This is an underestimate, as most mutations will not cause disease.

- Propagating asexual lineages by choosing a single individual at random to be the parent of the whole of the next generation. This allows mutations to accumulate because the random choice of a single individual removes any possibility of selection.

- Comparing the sequences of pseudogenes in sister species that have diverged for a known period of time. Since these genes are not affected by selection, their divergence is caused by mutation alone.

- Simply sequencing whole genomes of lineages that have been isolated for a known period of time. This can be used to count mutations directly.

All these methods give estimates in the region of 10^{-9}–10^{-10} errors per base pair per replication. This is the fundamental rate of mutation that provides the basis for genetic variation and evolutionary change.

6.1.1 Populations often maintain high levels of genetic variation.

A character that takes two or more discrete states in a population is said to be polymorphic. If this variation is caused by alternative alleles it constitutes genetic polymorphism. Hair color and eye color are familiar human examples, especially in populations of European descent. Hair color (like skin color) is largely determined by melanin synthesis in epidermal cells, and in particular by the balance of eumelanin (black or brown pigmentation) and pheomelanin (yellow or red). Red hair color is strongly associated with recessive alleles of the melanocortin receptor gene *MC1R*. When the gene is activated by the ultraviolet component of sunlight it switches on eumelanin production, causing a visible darkening of the skin. Reduced activity of *MC1R* leads to red hair color, pale skin, and poor tanning. Visible polymorphisms arising from variation in pigmentation are common in other organisms, such as coat color in dogs or corolla color in flowers.

Classical and balance theories predict different levels of variation. Until quite recently, it was not clear whether polymorphism is exceptional, with most characters being uniform, or, conversely, whether many or even most characters are polymorphic. The classical point of view is that almost all mutations are deleterious, so that any given gene has a fully functional allele and a number of more or less severely deleterious alleles. Selection will act to eliminate the inferior alleles, so the population will be genetically almost uniform, each locus being represented by a single allele at high frequency. At very long intervals a beneficial mutation might arise and spread through the population, which would be polymorphic while the resident allele was being replaced. Apart from these rare and transitory episodes, however, the population is expected to be genetically almost uniform.

The contrary point of view was that alleles do not always have unconditionally deleterious (or beneficial) effects, but might instead complement one another so that the heterozygote is the fittest genotype. Essentially, the notion is that having two functional versions of a gene may extend the range of conditions in which an individual can flourish and thereby increase its fitness. It follows that heterozygotes will tend to increase in frequency through natural selection, so that both alleles will be conserved in the population. The population cannot consist exclusively of heterozygotes, since if this were the case 50% of the subsequent generation would be homozygotes, but neither allele can be eliminated and heterozygote advantage necessarily leads to the maintenance of allelic diversity. This is the balance theory of genetic polymorphism. The most dramatic example of heterozygote advantage was discovered 60 years ago in Africa (see also Section 14.6.1). Some people suffered from a type of anemia associated with a peculiar shape of their red blood cells, which are sickle-shaped rather than circular. This is a genetic disease caused by a single mutation in the gene that encodes the hemoglobin molecule. Detailed surveys showed that the disease was most prevalent in areas with a high incidence of subtertian malaria. Individuals who are homozygous for the normal allele are highly susceptible to malaria. Those who are homozygous for the sickling allele develop a severe and often fatal anemia. Heterozygotes have only a mild anemia and are less susceptible to malaria. In regions where malaria is endemic, the heterozygous individuals are more vigorous because their resistance to malaria more than counterbalances their anemia, and a genetic polymorphism for the normal and sickling alleles is actively maintained by balancing selection.

New technologies have revealed unexpectedly high levels of variation. So long as observations were restricted to obvious differences in appearance, only variation associated with some clear visual effect could be assessed. There are also many examples of polymorphism in morphological or physiological characters, but these examples do not definitively establish the prevalence of polymorphism because the genes involved constitute only a small part of the genome.

The situation changed in the 1960s when a new technology made it possible to measure the movement of a specific protein on a gel column in response to an electric current. A mutation that altered the amino acid sequence of a protein in such a way as to alter the net electrical charge of the protein would cause it to migrate at a different rate and thus occupy a different position on the gel after a given period of time. This new technique of gel electrophoresis immediately revealed that a large proportion of genes were polymorphic in amino acid sequence. Moreover, the procedure always underestimates the real level of polymorphism, because only certain kinds of mutation can be detected. Genetic variation was clearly more extensive than had previously been thought.

The situation changed again from the 1980s on when further advances in technology made it feasible to study the nucleotide sequence of DNA directly. The high fidelity of DNA replication ensures that any particular change is unlikely to occur when an individual reproduces, but the large number of nucleotides in the genome means that a change of some sort is quite likely to happen, especially when simple point mutation is supplemented by deletions, duplications and rearrangements of genes or parts of genes. Consequently, individuals will often differ, even in laboratory cultures of asexual organisms living in very simple conditions (Section 13.2.3). The variability of natural populations can be expressed as the probability that two randomly chosen individuals differ at a given nucleotide position. For human populations, this probability is about one in a thousand. (It is about the same, curiously enough, for wild yeast.) Even small genomes consist of millions of nucleotides, so any two unrelated individuals are likely to differ at thousands of sites. Another way of expressing the same statistic is that a length of one thousand nucleotides will include, on average, one or two variable sites.

Hence, the advent of genomics has settled the old dispute by proving that most genes are polymorphic at the level of nucleotide sequence. But how is all this variation maintained? Paradoxically, there seems to be too much variation for all of it, or even a large fraction of it, to be maintained by selection favoring heterozygotes. This is because heterozygotes will be maintained at high frequency through the elimination of homozygotes, either through reduced viability or reduced fertility. This process will necessarily reduce the average viability or fertility of the population, in proportion to the intensity of selection acting against homozygotes. This need have no demographic consequences, so long as only a few loci are involved. If hundreds or thousands of loci are involved, however, the demographic cost of eliminating homozygotes at all these loci simultaneously will destroy the capacity of the population to replace itself, and it will go extinct.

Neutral alleles drift in frequency by sampling error. Many mutations have little effect on fitness, especially synonymous mutations that do not alter the amino acid specified by their codon. Others will alter the structure of a protein without any appreciable effect on its activity. These neutral alleles will not be affected by natural selection, and will instead fluctuate randomly in frequency. When a neutral allele first appears, it may become extinct soon afterwards because the individual which bears it happens to die without reproducing; or, just as likely, it persists or even increases in frequency, purely by chance. This process of genetic drift is likely to create a good deal of fluctuating variability in the form of neutral alleles at intermediate frequencies in the population. This explains why synonymous polymorphisms are usually much more frequent than polymorphisms that involve a change in protein structure. As a rough rule of thumb, they are usually five to ten times as frequent. This reflects the loss of function that is usually caused by changing the sequence of a gene, because inferior variants are systematically eliminated by natural selection whereas neutral variants are invisible to selection.

Genetic drift does not itself contribute directly to adaptive evolution. It may contribute indirectly because some mutations that alter protein structure, although neutral in current conditions, may be beneficial when conditions change. The fluctuating variability of these neutral alleles represents a pool of potentially functional alleles able to contribute to future adaptation.

Mutations that actually reduce fitness are expected to be lost from the population. If their effect is slight, however, they may persist through genetic drift for some time; and, of course, they are continually being reintroduced by mutation. Instead of being completely eliminated, therefore, they will remain at some low frequency at which mutation is just balanced by selection. Again, this does not contribute to adaptive evolution, unless a mutation that is deleterious in some conditions is beneficial in others. In this case, even alleles that are deleterious in current conditions may contribute to future adaptation. Antibiotic resistance is (unfortunately) a familiar example. Some alleles confer resistance simply by slowing down growth and thereby reducing the toxicity of

antibiotics (such as penicillin) that target growing cells. In normal circumstances these slow-growing variants would be deleterious, but when antibiotics are administered their shortcomings become benefits and they spread rapidly through the population.

The neutral theory of genetic polymorphism has reconciled the classical and balance theories: there is extensive genetic variation (as the balance theory predicts), but most variants have little or no effect on fitness, so the population is effectively uniform with respect to fitness (as the classical theory requires).

This is not quite the end of the story. Heterozygote advantage is only one example of conditional fitness, in which the fitness of an allele in a given individual depends on which allele it is associated with. The fitness of an allele may also depend on whether it is expressed in a male or female individual, in a haploid or diploid individual, or in a sexual or vegetative individual. All of these situations may lead to the active maintenance of polymorphism by selection. Most pervasively of all, the fitness of an allele may depend on the conditions that an individual experiences during its lifetime. In a heterogeneous environment, ecological variation among sites may lead to local specialization and thus regional diversity. This kind of diversifying selection need have no baleful demographic consequences for the population and may support high levels of genetic polymorphism for functional alleles. It is discussed in Sections 14.2 and 16.1.

6.1.2 The rate of evolutionary change is limited by two scaled mutation rates.

The effect of mutation on the force of selection can best be appreciated by calculating two scaled mutation rates: the genomic mutation rate, which is scaled by genome size; and the mutation supply rate, which is scaled by population size.

The genomic mutation rate determines the mean fitness of the population. The rate at which mutations arise per locus is simply the fundamental rate multiplied by the number of base pairs per gene, which is about 2000; hence, the mutation rate per locus is about 2×10^{-6}. Likewise, the overall rate of mutation per genome is the fundamental rate u multiplied by the genome size G, which varies among species: the genomic mutation rate is thus $U = Gu$. Bacteria have small genomes with an overall mutation rate of about 0.001 per replication; eukaryotic microbes such as yeast have a value of about 0.01, and animals a value of about 0.1. Since many divisions occur in the animal germ line, this is equivalent to about one mutation per generation. Many of these mutations will occur in non-coding regions, so they will have little or no effect on fitness.

Hence, the offspring of animals like us may bear one or two new mutations that are not present in either of their parents. Since most mutations are deleterious, they will be less fit than their parents are. The genomic mutation rate thus governs the rate of degradation of fitness, which is made good by selection during the lifetime of the cohort of offspring. The effect of the mutation does not matter: more severe mutations will cause a greater depression of mean fitness, but are in the same proportion more effectively removed by selection.

The mutation supply rate determines the evolvability of the population. Mutation is two-edged: although most mutations are deleterious, mutation also provides the only source of new adaptive variation. A few mutations may be beneficial, depending on how well-adapted the population is. If the conditions of growth have recently changed the population may be rather poorly adapted, so that there is a good deal of opportunity for improvement; in other words, a relatively large fraction of mutations will be beneficial. If the population has lived for a long time in the same conditions it will have become well-adapted, and only a very small fraction of mutations will be beneficial. In either case, the amount of evolution that will occur will be limited by the number of mutations that occur, which will depend on the fundamental mutation rate u multiplied by the population size N: the mutation supply rate M is thus $M = Nu$. The mutation supply rate is the attribute of a population, rather than a species. It is governed largely by abundance, N.

Microbes are much more evolvable than animals and plants. Because microbes are more abundant (larger N) than larger organisms, they have a greater mutation supply rate. Now, it is generally true that the mass of particles in equal logarithmic size classes is independent of size. In other words, if we sampled a sufficiently large volume of seawater, the mass of bacteria (each weighing about 10^{-12} g) would be roughly equal to the mass of whales (each weighing about 10^6 g). But being so much smaller, there are so many more bacteria. In fact, of all the individuals in a sample of water or soil, about 99% are bacteria, about 99% of the remainder are eukaryotic microbes such as diatoms and ciliates, and multicellular organisms make up only the tiny number remaining. Obviously, microbes can be expected to evolve much more rapidly than animals or plants.

Mutation is usually deleterious but may sometimes be beneficial. The genomic rate of mutation creates a burden of poorly adapted genotypes which are removed by purifying selection, as described in Section 14.1. The mutation supply rate generates a few well-adapted genotypes, which spread through directional selection, as described in Section 14.2.

6.1.3 The mutation rate itself evolves.

The rate of mutation limits the rate at which evolution can occur; but it is not a fixed limit, like the rate of diffusion. The rate at which errors occur is governed by the fidelity with which DNA is copied through the mediation of proteins, but these proteins are themselves encoded by DNA. Mutations in genes that specify the proteins responsible for DNA replication and repair will themselves alter the mutation rate. Hence, the mutation rate must be an evolved feature of organisms, and not merely a physical limit.

Mutator genes can spread in asexual populations under stress. Mutator phenotypes causing a genome-wide elevation of the mutation rate by a factor of 10–1000 often appear in bacterial populations. They are usually caused by loss-of-function mutations in RNA polymerase genes. Most of the mutations they cause are deleterious, so they occur only at low frequency. When a population is stressed, however, these mutations often spread to high frequency. The reason is as follows. In a stressful environment, a successful beneficial mutation will sooner or later arise and spread. If mutator and non-mutator alleles are equally frequent, it will probably arise in the mutator strain, because this strain will have the greater mutation supply rate. Bacteria are asexual, so the mutator allele is linked to the beneficial mutation it has elicited; hence, as this mutation spreads, the mutator allele responsible must spread with it. The spread of a gene through the effect it produces at a linked locus is called "hitch-hiking."

It is more likely that the mutator allele occurs only at low frequency. In this case, the probability that a beneficial mutation arises in the mutator strain is the frequency of the strain multiplied by the elevation of the mutation rate that it causes. Thus, if a mutator causes a thousand-fold increase in mutation rate, it is almost certain to generate the first successful beneficial mutation if its frequency is 10^{-3} or more. The mutator allele will then spread through the population, by hitch-hiking with the beneficial mutation it has elicited, and in doing so will eliminate the normal, non-mutator allele. On the other hand, if its frequency is only 10^{-6}, the first beneficial mutation is almost certain to occur in the non-mutator strain, because this is so much more abundant. When this happens, it is the non-mutator allele that spreads, and the mutator allele is eliminated from the population. This principle is illustrated in Figure 6.2(a).

In other words, the probability that a mutator allele, with some defined effect on mutation rate, will spread through a population depends on its frequency: above a certain frequency it will tend to spread, and below this frequency it will usually be eliminated.

Now, once a mutator allele has become fixed, most of the subsequent mutations it generates will be deleterious. It will be linked with these too, and will consequently decline in frequency. Hence, mutator alleles should have a "saw-tooth" pattern of evolutionary dynamics in stressful environments, first spreading because they are linked to beneficial mutations and then declining in well-adapted populations.

These arguments have been conclusively demonstrated by experiments with bacteria in chemostats: mutator alleles decline if they are initially present below a certain threshold, and spread if they exceed it; one of these experiments is shown in Figure 6.2(b).

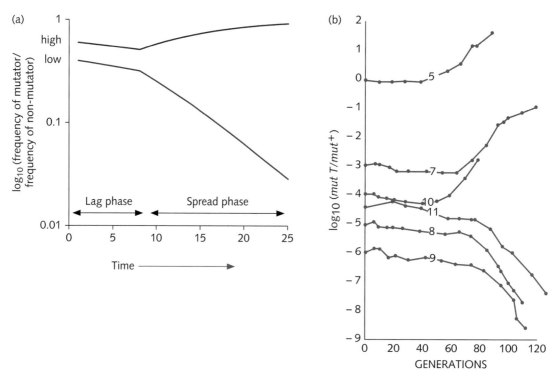

Figure 6.2 Dynamics of a mutator allele. (a) An allele that increases the mutation rate in an asexual population initially falls in frequency because it generates many deleterious mutations to which it is linked. Eventually a beneficial mutation is generated and spreads. It is likely to appear in a mutator lineage if this is sufficiently frequent, in which case the mutator allele spreads because it is linked with the beneficial mutation that it has generated. If the mutator lineage is rare the beneficial mutation is likely to arise in a normal lineage, and the mutator lineage is driven out of the population. Note that the y-axis is the logarithm of the ratio of frequency of mutator and normal types. (b) The experimental dynamics of mutator alleles introduced into laboratory populations of *E. coli* at different frequencies.

Reproduced from Chao, L. and Cox, E.C. 1983. Competition between high and low mutating strains of *Escherichia coli*. *Evolution* 37: 125–134 with permission from John Wiley & Sons Ltd.

These processes are also part of the normal dynamics of bacterial populations. A double screening with two antibiotics is a good way of isolating mutator alleles in the laboratory; and natural isolates of antibiotic-resistant bacteria are often mutator strains. The evolution of bacterial mutation rates through the spread of mutator alleles is therefore a matter of considerable interest from the point of view of human health.

In sexual populations the mutation rate is a compromise. Mutator alleles also occur in sexual organisms; since they are loss-of-function mutations, they arise quite frequently. However, they do not usually tend to spread. This is because they soon become unlinked, by genetic recombination, from any beneficial mutations they may cause. Hence, the mutation rate is chronically low. It is not zero, because detecting copying errors is costly, in terms of resource use or delayed replication. It is rather driven down to the value at which these costs just balance the advantage of avoiding deleterious mutations.

6.2 Genes are often transferred from distant relatives in bacteria.

DNA is a chemical substance; transcription, translation, and replication are chemical processes. DNA from one species can be introduced into cells of another species, where it will perform its normal role in protein synthesis, provided that the rest of the genomic apparatus (such as promoter sequences) is in place. This is how genetic engineering works. It is also how a lineage may abruptly acquire a new, fully functional gene that extends its ecological range. Horizontal transfer, or lateral transfer, represents a sort of macromutation that provides a lineage with a new gene without the need for a gradual process of mutation and selection.

Many bacteria have mosaic genomes. Any species of animal or plant will have a distinctive genome that closely resembles that of its sister species and is more different from that of less closely related species. Bacteria are not like that—not always, at least. Some bacteria will take up quite large pieces of DNA and incorporate these into their genome, thus acquiring new metabolic capabilities at a stroke. Some do so much more readily than others and consequently their genomes are a mixture of genes from different sources.

Many genes have been recently introduced into the genome of the enteric bacterium *E. coli* by lateral transfer. They can be detected because some of their features, such as nucleotide composition (GC versus AT base pairs), are different from those of resident genes. This is a very striking result. Suppose that we study some random isolate of *E. coli*—one obtained from your gut,

for example. On average, nearly 20% of its genome will have been recently acquired. These sequences have not been acquired all at once; they are the result of about 200 different transfers. Hence, the genomes of many bacteria are very dynamic: on average, an *E. coli* lineage will acquire new gene sequences by horizontal transfer at a rate of about 16 kb per My.

Plasmids can move genetic material between bacterial lineages. One way in which genes can move readily between bacterial lineages is through conjugation, a kind of one-sided sexual process in which DNA passes from one cell to another through a narrow cytoplasmic bridge. This process is directed by a special kind of conjugative plasmid, which thereby passes from the lineage it has infected to a previously uninfected lineage. Conjugative plasmids are genetic parasites, similar in some respects to viruses, except that they do not cause disease. As well as transferring a copy of their own genome, however, they will often transfer adjacent sections of DNA, which are then inserted into the chromosome of the recipient cell along with the plasmid.

Bacteria can acquire new genes directly from the environment. Dying bacteria release substantial amounts of DNA into the environment, so that soil contains about 1 μg/g and seawater about 1–100 μg/L DNA. Consequently, natural populations of bacteria are bathed in a weak solution of DNA, and this can be taken up and integrated into the chromosome by prophage-like particles ("gene transfer agents"), where it may occasionally provide genetic variation that contributes to adaptation. It is difficult to assess the importance of this kind of macromutation and I do not know of any conclusive evidence that it contributes to adaptation.

The cassette–integron system, on the other hand, is likely to provide an important source of new genes. The cassette is a small circular molecule comprising an open reading frame and a recognition sequence. The integron is a chromosomal region comprising the *intI*

gene encoding an integrase mediating the insertion of the cassette at the recombination site *attI* downstream of a promoter. Once the element has become integrated into the bacterial chromosome it will be expressed and replicated and thus may immediately provide a new metabolic function to the cell. Many kinds of elements such as these can be found in a cupful of soil, and in principle could kick-start adaptation to novel substrates or toxins. In practice, we do not yet know how important this process might be.

6.3 New genes arise by modification and recombination.

More complex organisms have more genes. In general, larger and more complex organisms have more genes. Yeast has about 6000 genes and we have about 22,000 genes. The relationship between genome size and individual complexity and integration is by no means straightforward, because a large part of the genome of many organisms does not encode proteins; the reasons for this are discussed at more length in Section 11.4. Nevertheless, anyone would expect that specifying a mammal, a fish or a fly would require more genes than specifying a bacterium, a mold, or an alga. Figure 6.3 shows that this is broadly true, and thereby introduces a fundamental question in evolutionary biology: how do new genes evolve?

The origin of new genes is the fundamental source of variation for evolutionary innovation. It is a fundamental question because the evolution of new kinds of organism clearly depends on the evolution of new kinds of gene. The evolution of seed plants from algae, or the evolution of animals from amoebas, requires a mechanism for generating new genes. Two kinds of process are responsible for the expansion of the genome through the addition of genes with new functions:

* modification of a gene by mutation and successive substitution
* recombination of parts of one or more genes to form a new chimeric gene.

Transformation and recombination are both necessary consequences of gene replication, and together

provide the fuel for the evolution of novel kinds of organisms.

6.3.1 New genes can arise by modification.

New versions of a gene evolve through mutation and selection. Mutation creates new allelic versions of a gene. Most are inferior to the currently prevailing type, and these lineages continually arise and die out. At last a superior allele may appear and spread through natural selection, eventually replacing the prevailing type. This may itself be replaced subsequently by an even better allele, so there is a stepwise improvement of the gene. This is a very common process that is often involved in adaptation to a changing environment. It will usually lead only to a modified version of the same gene, however; it is unlikely to generate an entirely new gene.

New genes evolve by duplication and divergence. So long as only a single copy of a gene is present it cannot evolve into a new kind of gene if it is essential for the normal functioning of the organism. Genes are quite often duplicated, however, in the course of replication. Once there are two copies of a gene, one can retain the ancestral state while the other becomes modified by mutation and selection so as to become more capable of performing some new function, such as metabolizing a substrate or regulating the expression of another gene.

When a gene has become duplicated one copy may be lost quickly to restore the original condition, as gene deletion is a common process. If both copies

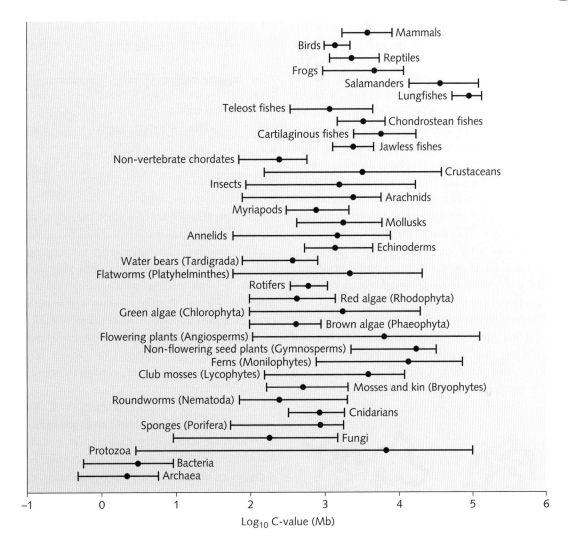

Figure 6.3 Genomes vary greatly in size—although not as greatly as you might have supposed.

Reprinted by permission from Macmillan Publishers Ltd: Nature Reviews Genetics. Gregory, T.R. 'Synergy between sequence and size in Large-scale genomics' Vol. 6, Iss. 9, pp. 699–708 © 2005.

persist, one may be inactivated by loss-of-function mutations, as these are neutral if a single copy of the gene is sufficient for normal function. This copy is then replicated as a pseudogene and gradually decays under mutation pressure, while the integrity of the other copy is maintained by purifying selection. In some cases, gene duplication is favorable because it increases the level of expression of the gene by doubling the rate of transcription. In this case, purifying selection protects the original sequence of both copies. Finally, duplication may be followed by divergence to create a new gene. This outcome is exceptional, but gene duplication is such a common process that it provides a continual source of new genes.

6.3.2 New genes can arise by recombination.

New genes may be new combinations of parts of old genes. New combinations often have new capabilities because of the modular nature of proteins and genes. Modular structures such as segmental appendages can diversify into a range of different variations on a theme (Section 9.1.1). A similar principle holds at the level of genes and proteins, and explains how new kinds of gene can arise.

Proteins have modular structure. Most proteins consist of a combination of domains, which are stable units capable of folding autonomously. The two main kinds of secondary structure, the α-helix and the β-sheet, can be linked into a motif, such as a β–α–β pattern in which two β-sheets are connected by an α-helix. Several such motifs linked together form a domain (Figure 6.4). Most eukaryotic proteins have several domains, each comprising about 50–200 amino acids. Each domain contributes not only a structural unit but also a range of potential capabilities to a protein. Hence it is not necessary (or usual) for novel proteins to evolve through step-by-step replacement of amino acids. Rather, substituting and rearranging modules can give rise to an indefinite number of fully functional proteins.

Genes have a (partly) modular structure. Genes and the elements that regulate them are also partly modular in organization. A typical gene may have a series of upstream transcription-factor-binding sites,

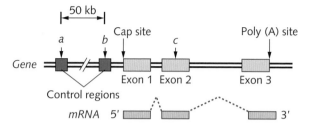

Figure 6.5 Genes have a (partly) modular structure.

a signal peptide sequence, a promoter element, and a series of exons and introns flanked by untranslated regions. Exons do not necessarily correspond to the coding sequences for modules, but they will often encode a large fraction of a module, or of several modules. Hence, shuffling exons will generate new proteins by rearranging modules (Figure 6.5). Adding or deleting or rearranging upstream regulatory sequences is also likely to have functional consequences by altering expression patterns.

Recombination of modules creates new functional genes. Thus, both proteins and the genes that encode them are partly modular in construction, formed of a series of units linked together. Each has a degree of autonomy and may therefore contribute a ready-made functional unit to the protein or gene that acquires it. New combinations of modules can arise in two ways.

- First, by failure of replication: slippage, deletion, translocation, or recombination. For example, recombination between nearby points on the chromosome may bring together elements that were previously far apart to form a new structure.

- Secondly, by illegitimate replication: transposition or retrotransposition. For example, the mRNA produced by a gene can be reverse-transcribed into a distant location in the genome through the action of a retroposon. This inserts a copy of the gene, or of part of the gene, into a new site where it has new neighbors. This element can then become associated with parts of neighboring genes to form a new compound gene.

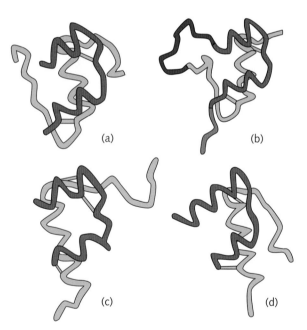

Figure 6.4 Proteins have modular structure. These diagrams illustrate the structure of the insulin peptide family, with the A domain in red, the B domain yellow, the C domain blue and the D domain green: (a) insulin; (b) IGI; (c) bombyxin-II; (d) human relaxin 2.

Reproduced from De Meyts, P. 2004. Insulin and its receptor: structure, function, and evolution. *BioEssays* 26 (12): 1351–1362. © 2004 Wiley Periodicals, Inc.

The modular structure of proteins and the liability of genes to be duplicated and recombined creates almost unlimited potential for new genes to arise, and explains how the large genomes of complex animals and plants have evolved from smaller and simpler predecessors.

6.4 New ways of life evolve when new genes arise.

In recent years it has become possible to examine the genetic basis of adaptation in great detail. This research has led to the discovery of new genes that arose only a few million years ago and has given concrete examples of how these genes have arisen through transformation and recombination.

6.4.1 A new antifreeze gene allows fish to live in Antarctic waters.

The seawater near the permanent ice of Antarctica can be as cold as –2 °C, because the dissolved salt lowers the freezing point below that of freshwater. Most fish cannot live at this temperature, because they would freeze solid. One group thrives in these frigid seas, however, where it has radiated into more than a hundred species. These are the notothenioids, which dominate the fauna of the shallow coastal bays where other fish are rare. They owe their success to the production of an antifreeze protein, present at high concentrations (about 35 mg/mL) in the blood, that binds to nascent ice crystals and prevents them from growing. It is a quite simple molecule consisting of 5–50 repeats of the triplet Thr–Ala–Ala, with a disaccharide attached to each Thr residue.

The antifreeze gene has evolved from a digestive enzyme gene. When the gene encoding the antifreeze protein was sequenced, the sequence immediately downstream of the gene was found to be very similar to the gene encoding trypsinogen, a digestive enzyme, in other teleosts. This suggested that the antifreeze gene had somehow evolved from the trypsinogen gene. The first exon of the trypsinogen gene encodes the signal peptide that governs the transport of the protein, and is almost identical to the first exon of the antifreeze gene. This is followed by an intron which is again almost identical in the two genes. At the boundary between this intron and the second exon occur the nine nucleotides specifying the Thr–Ala–Ala repeating unit of the antifreeze protein. The rest of the trypsinogen gene sequence does not occur in the antifreeze gene, except that the 200 or so nucleotides at the downstream end are almost identical in both genes.

These facts allow us to reconstruct (Figure 6.6) the evolution of the antifreeze gene from the trypsinogen gene of the ancestor of modern notothenioid fish.

- The upstream signal peptide sequence and intron were recruited from the ancestral gene and retained in the new antifreeze gene.

- The nine-nucleotide sequence encoding the Thr–Ala–Ala tripeptide was amplified by replication slippage or unequal crossing-over (both fairly common kinds of mutation) to make successively longer and more effective proteins.

- The entire central sequence of the evolving gene was deleted (another common type of mutation), resulting in a purer repetitive protein.

- A frameshift in the downstream sequence inherited from the trypsinogen gene created a stop codon at the end of the series of nine-nucleotide repeats, again stripping off a superfluous region of the protein, and resulting in an untranslated downstream sequence in the new gene.

This evolutionary sequence must have passed through intermediate forms in which both the amplified antifreeze sequences and the ancestral trypsinogen sequences existed in the same gene. This intermediate stage has now been identified in the giant Antarctic toothfish *Dissostichus mawsoni*, where the amplified Thr–Ala–Ala motif is present alongside the exons descending from the original gene.

Hence, a novel antifreeze gene evolved by a process involving the recruitment, amplification, and deletion of intron and exon sequences from parts of a digestive-enzyme gene, making possible for an ancestral notothenioid lineage to radiate into the previously uninhabitable coastal waters of the Antarctic.

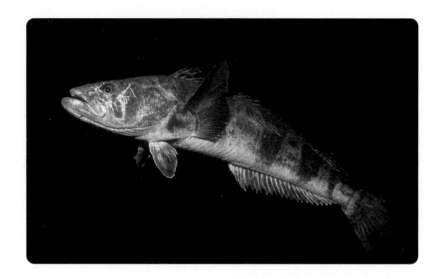

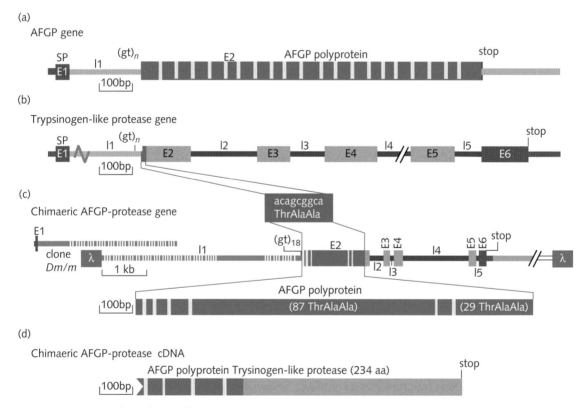

Figure 6.6 Reconstructing the evolution of the antifreeze gene from the trypsinogen gene of the ancestor of modern notothenioid fish.

Photograph © Rob Robbins. Diagram reprinted by permission from Macmillan Publishers Ltd: Cheng and Chen. Evolution of an antifreeze glycoprotein. *Nature* 401 (6752) © 1999.

The divergence of intron sequences between the antifreeze and trypsinogen genes can be used to calculate the time when they began to diverge, using calibrated molecular data from other teleosts. This gives an origin about 5–14 Mya, which is consistent with the 10–14 Mya when the southern oceans froze in the mid-Miocene.

Antifreeze has evolved independently in Arctic fish. Arctic fishes such as the northern cod *Boreogadus* have also evolved antifreeze proteins that enable them to live in water close to the ice margin. The proteins are remarkably similar, in both cases based on the Thr–Ala–Ala tripeptide repeat. The genes responsible, however, are not the same: the signal peptides, the codons, and the intron/exon boundaries are all quite different. This is a striking example of convergent evolution. Very similar proteins, serving the same function, have evolved in Arctic and Antarctic fish of distantly related groups from different ancestral genes.

6.4.2 Genes affecting male sexual behavior in *Drosophila* have evolved recently by duplication, exon shuffling, and retroposition.

Genes affecting sexual physiology and behavior often evolve very rapidly through sexual selection, as will be described in Section 15.4. Consequently, many examples of new genes involve sex, and here we give three examples of new genes affecting male sexual behavior in *Drosophila*.

Sphinx is a new gene affecting courtship behavior in Drosophila. *Sphinx* is a gene located on chromosome 4 of the fruit fly *Drosophila melanogaster*. It is expressed in the accessory gland, which secretes seminal fluid, so it is clearly a gene involved in sexual function. If the gene is knocked out, normal male–female courtship is unaffected. If males are isolated with other males, however, the *sphinx* knockouts display

much more vigorous courtship, going through all the normal stages short of copulation. This male–male courtship is so active that when several males are placed together, lines of courting males, each behind the other, may appear—sometimes these close to form a circle of endlessly moving male flies. This suggests that the function of the intact gene is to reduce male–male courtship, which is common in most *Drosophila* species but not in *D. melanogaster*.

This is clearly a gene that has a dramatic effect on courtship behavior (Figure 6.7). How did it evolve? The molecular data show that *sphinx* is a chimeric gene (hence the name: the Sphinx of ancient Egypt had a lion's body but a human head). A part of the ATP synthase F-chain gene from chromosome 2 was inserted into a new site on chromosome 4. This happened through retroposition: the mRNA of the gene was reverse-transcribed into its new site quite fortuitously through the action of a selfish genetic element (Section 11.4.2). This element was then combined with upstream sequences at this site, including a regulatory sequence, to form a new functional gene. This happened about 2 Mya.

Jingwei is a new testis-specific gene in Drosophila. Alcohol dehydrogenase is a very widespread enzyme that catalyzes the oxidation of alcohols to aldehydes, which are then transformed to acetate, which can be used as a nutrient. In the ancestral lineage of two African species of fruit flies, *Drosophila teissiri* and *D. yakuba*, the mRNA transcribed

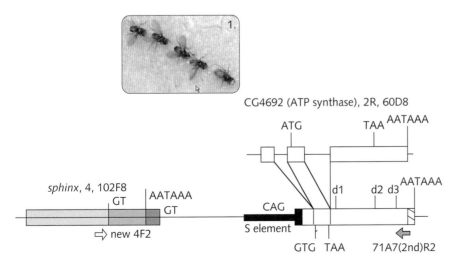

Figure 6.7 *Sphinx* affects courtship behavior in *Drosophila* and has evolved recently from ATP synthase.

Photograph reproduced from Dai et al. The evolution of courtship behaviors through the origination of a new gene in *Drosophila*. *PNAS* 105 (21): 7478–7483 © (2008) National Academy of Sciences, U.S.A. Diagram reproduced from Wang et al. Origin of *sphinx*, a young chimeric RNA gene in *Drosophila melanogaster*. *PNAS* 99 (7): 4448–4453 © (2002) National Academy of Sciences, U.S.A.

from the *Adh* gene that encodes the protein was reverse-transcribed into a new location in the genome. This retroposed copy was identical to the original, except that all the introns had been stripped out. It happened to be inserted close to a copy of a gene called yellow emperor (*ymp*), whose precise function is unknown but which presumably affects male sexual development because it is expressed in the testis. This copy, as it happened, resulted from a duplication of the original gene. This combined the first three exons of the *ymp* copy with the intron-less retroposed *Adh* sequence to form a new gene, which was named *jingwei*. The downstream exons of the *ymp* copy were not included in the new gene, and are now degenerating as a consequence. The new gene is thus a chimera, formed from elements of two quite different pre-existing genes (Figure 6.8). It inherited the regulatory mechanism of the ancestral *ymp* gene, as its expression remains specific to the testis. Its function is not known. It has evolved rapidly since the divergence of the two species, however, so must have been modified by quite strong selection since its first appearance. The original transposition event that led to the evolution of the new gene (calculated from the divergence between the ancestral and derived copies of the *Adh* gene) occurred about 2.5 Mya.

Sdic is a new gene affecting sperm motility in D. melanogaster. *Sdic* is a gene encoding a dynein protein active in the sperm tail. It has evolved very recently, because it is absent from the sister species of *D. melanogaster*, and must therefore have arisen and spread in the last 2–3 My. From its nucleotide sequence, it consists of parts contributed by two other genes, *AnnX* and *Cdic*. Intact copies of these two genes are situated on either side of the *Sdic* region itself on the X chromosome. In fact there are four copies of *Sdic* in this region, which have clearly arisen by two successive episodes of duplication. Finally, a retroposon has become inserted downstream of each copy of *Sdic*.

Some clever molecular detective work has discovered how *Sdic* arose and evolved.

- The first event must have been a duplication of the whole region embracing *AnnX*, *Cdic*, and the intervening sequences.
- Next, one copy of *AnnX* became fused to the neighboring copy of *Cdic* through the deletion

of sequences between them. This formed the new chimeric gene *Sdic* between an intact copy of *AnnX* downstream and one of *Cdic* upstream.

- At about the same time, a copy of a retroposon was inserted just downstream of *Sdic*.
- The promoter sequences of the new gene are recruited from non-coding *Cdic* exons.
- The entire *Sdic* gene, with retroposon, was then duplicated twice, presumably by non-homologous recombination.
- The four *Sdic* copies then gradually diverge by point mutation.

Thus, *Sdic* illustrates the formation of a new gene by a process involving duplication, exon shuffling, and retroposition (Figure 6.9).

6.4.3 The genome is a dynamic equilibrium between gene loss and gene gain.

The extensive variation in gene number among different kinds of organism is caused by the outcome of two opposed processes: the loss and gain of genes.

- Gene loss. Genes are continually lost by deletion, usually caused by recombination. In most circumstances this is deleterious, so selection preserves the normal complement of genes.
- Gene gain. The origin of new genes is the basis for fundamental adaptive changes. The main mechanisms involved in creating new genes are as follows.
  - **Duplication and divergence.** When a gene is erroneously replicated twice, the spare copy is free to acquire new functions without penalty, and occasionally does so.
  - **Retroposition.** A retroposed copy of an mRNA molecule, or part of one, is inserted at a new position in the genome.
  - **Transposition.** Integrating a mobile element into a pre-existing gene can create new functional capabilities.
  - **Exon shuffling.** Recombination or retroposition can bring together exons from different genes to create a hybrid structure.

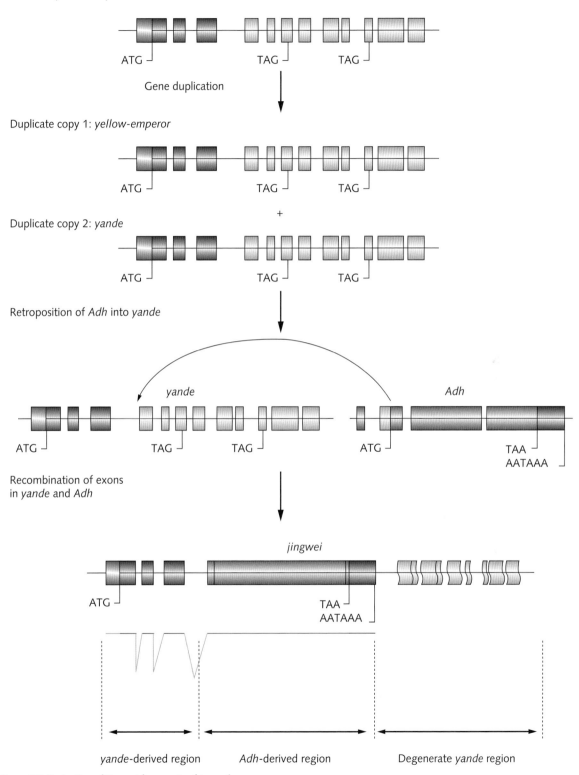

Figure 6.8 Derivation of *jingwei* from parts of two other genes.

1. Ancestral structure

2. First duplication event

3. Deletion, formation of new chimeric gene *Sdic* and insertion of retrotransposable element

4. Second duplication event; divergence

5. Duplication of the whole cluster

6. Divergence of the copies

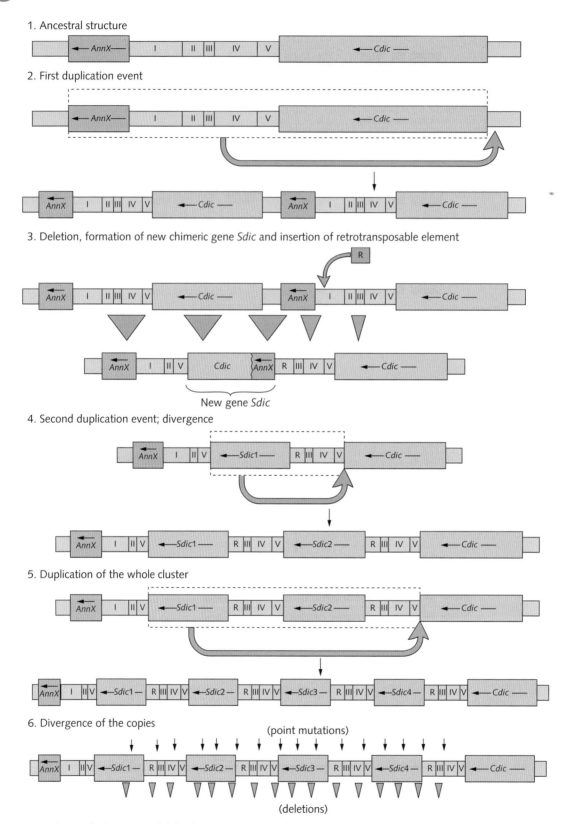

Figure 6.9 Evolution of *Sdic* in *Drosophila* by duplication, exon shuffling, and retroposition.

Reprinted from Ponce and Hartl. The evolution of the novel *Sdic* gene cluster in *Drosophila melanogaster*. *Gene* 376 (2). © 2006, with permission from Elsevier.

Duplication of an exon can create a new exon–intron structure.

- **Lateral transfer.** The movement of genes between distantly related lineages can provide new adaptive potential in a single step.

All of these processes have been documented; some are more important in bacteria, others in eukaryotes. All are rare, but over time they cause revolutions in genome structure.

● CHAPTER SUMMARY

Lamarckian evolution is appropriately directed modification.
Darwinian evolution is the selection of undirected variation.
Origin and fate involve different processes.

Mutation is the source of genetic variation

- – *Mutation is inevitable.*
- – *Mutation can produce minor or major changes in structure.*
- – *Mutation provides new selectable variation.*

- Populations often maintain high levels of genetic variation.
  - – *Classical and balance theories predict different levels of variation.*
  - – *New technologies have revealed unexpectedly high levels of variation.*
  - – *Neutral alleles drift in frequency by sampling error.*

- The rate of evolutionary change is limited by two scaled mutation rates.
  - – *The genomic mutation rate determines the mean fitness of the population.*
  - – *The mutation supply rate determines the evolvability of the population.*
  - – *Microbes are much more evolvable than animals and plants.*
  - – *Mutation is usually deleterious but may sometimes be beneficial.*

- The mutation rate itself evolves.
  - – *Mutator genes can spread in asexual populations under stress.*
  - – *In sexual populations the mutation rate is a compromise.*

Genes are often transferred from distant relatives in bacteria.

- – *Many bacteria have mosaic genomes.*
- – *Plasmids can move genetic material between bacterial lineages.*
- – *Bacteria can acquire new genes directly from the environment.*

New genes arise by modification and recombination.

- – *More complex organisms have more genes.*
- – *The origin of new genes is the fundamental source of variation for evolutionary innovation.*

- New genes can arise by modification.
  - – *New versions of a gene evolve through mutation and selection.*
  - – *New genes evolve by duplication and divergence.*

- New genes can arise by recombination.
  - – *New genes may be new combinations of parts of old genes.*
  - – *Proteins have modular structure.*

- *Genes have a (partly) modular structure.*
- *Recombination of modules creates new functional genes.*

New ways of life evolve when new genes arise.

- A new antifreeze gene allows fish to live in Antarctic waters.
  - *The antifreeze gene has evolved from a digestive enzyme gene.*
  - *Antifreeze has evolved independently in Arctic fish.*

- Genes affecting male sexual behavior in *Drosophila* have evolved recently by duplication, exon shuffling. and retroposition.
  - *Sphinx is a new gene affecting courtship behavior in* Drosophila.
  - *Jingwei is a new testis-specific gene in* Drosophila.
  - *Sdic is a new gene affecting sperm motility in* D. melanogaster.

- The genome is a dynamic equilibrium between gene loss and gene gain.

● FURTHER READING

Here are some pertinent further readings at the time of going to press. For relevant readings that have been released since publication, visit the book's Online Resource Centre at **www.oxfordtextbooks.co.uk/orc/bell_evolution/**

Lewontin, R.C. 1974. *The Genetic Basis of Evolutionary Change*. Columbia University Press, New York

Section 6.1 Sniegowski, P.D., Gerrish, P.J., Johnson, T. and Shaver, A. 2000. The evolution of mutation rates: separating causes from consequences. *BioEssays* 22: 1057–1066.

Section 6.2 Boto, L. 2010. Horizontal gene transfer in evolution: facts and challenges. *Proceedings of the Royal Society of London B* 277: 819–827.

Section 6.3 Long, M. 2001. Evolution of novel genes. *Current Opinion in Genetics and Development* 11: 673–680.

Section 6.4 Cheng, C-H.C. and Chen, L. 1999. Evolution of an antifreeze glycoprotein. *Nature* 401: 443–444.

Long, M., Betran, E., Thornton, K. and Wang, W. 2003. The origin of new genes: glimpses from the young and old. *Nature Reviews Genetics* 4: 865–875.

Dai, H., Chen, Y., Chen, S. et al. 2008. The evolution of courtship behaviors through the origination of a new gene in *Drosophila*. *Proceedings of the National Academy of Sciences of the USA* 105: 7478–7483.

● QUESTIONS

1. Discuss the role of variation in Lamarckian and Darwinian theories of evolution. How would you distinguish the two by experiment?

2. Define "macroevolution" and "microevolution." Is there a fundamental difference between them?

3. If you were asked to estimate the mutation rate in a laboratory population of fruit flies, how would you proceed?

4. How do deleterious mutations, neutral mutations, and beneficial mutations contribute to the genetic variation of a population?

5. Define the mutation supply rate and the genomic mutation rate. How do they affect the process of evolution?

6. How does the mutation rate evolve in asexual and sexual populations?

7. What are the consequences of lateral gene transfer in bacteria?

8. What are the main ways in which new genes can arise from the modification of old genes?

9. Explain how the evolution of genes and proteins is affected by their modular structure.

10. Write an essay on the evolution of antifreeze proteins in polar fish.

11. Describe how novel genes affecting male mating behavior have evolved in *Drosophila*.

12. What are the main genetic mechanisms that can lead to the appearance of new genes?

You can find a fuller set of questions, which will be refreshed during the life of this edition, in the book's Online Resource Centre at **www.oxfordtextbooks.co.uk/orc/bell_evolution/**

7 The Origin of Species

One of the most obvious attributes of living organisms is that many of them, especially large and familiar organisms like animals and plants, usually fall into rather sharply demarcated categories called species. It was the attempt to explain this fact, indeed, that gave the first impetus to evolutionary biology. This chapter is about how lineages that acquire new genes, in the ways described in the previous chapter, evolve into new species.

Continued selection produces new species. Lineages always tend to change because they adapt to new conditions of growth or evolve different ways of mating. In time they may become sufficiently different from their ancestor to be recognized as a distinct species—although this is a human and subjective decision rather than a sharp natural boundary, as we shall see. There are three ways in which this might occur.

- In the first place, a single lineage might become so extensively modified that it constitutes a new species, distinct from its ancestor. The outcome is the replacement of one species by another.

- Secondly, a lineage might become divided into two (for example, by an external barrier such as a large river), only one of which became modified. The outcome is a new species, with the ancestral species persisting.

- Finally, when a lineage is divided in two both may become divergently modified. The outcome is the evolution of two new species, with the disappearance of the ancestor.

This classification is logical rather than real—in practice, most lineages will often become divided into several incompletely separated descendant lineages, which then diverge to different extents and may proceed to evolve into distinct species or (much more often) not. It is no more than a classification intended to aid thought. The crucial factor is whether lineages continue to exchange alleles through mating, because it is only when this exchange is blocked that new species can evolve.

7.1 Diversification is the natural tendency of lineages.

The natural tendency of life is to diversify, because of the cumulative nature of genetic change over time (Section 2.3). Any two lineages that do not exchange genes will almost certainly diverge. There are three reasons for this.

- The first is that neutral substitutions will occur entirely at random. Indeed, the extent of divergence at neutral sites can be used to date the coalescence of the two lineages.

- The second is that beneficial mutations arise in a random order that will usually be different in every lineage. If the effect of a mutant allele depends on the state of other loci it will alter the relative fitness of other potential mutations. Hence, the fixation of a particular mutation will make it more likely that some mutations will subsequently be fixed, and less likely that others will be. Since this applies to every successive beneficial mutation, independently evolving lineages may acquire different combinations of alleles, even when becoming adapted to similar ways of life.

- The third is that lineages may become adapted to different ways of life, consuming different food, for example, or tolerating different physical conditions. This will almost always involve mutations in different genes, and is therefore likely to cause substantial phenotypic divergence. This is the process that leads to the emergence of lineages with distinctively different morphological, physiological and behavioral characteristics that we recognize as species.

In the last chapter we described how new genes arise through processes such as mutation, duplication, and recombination. In this chapter we shall describe how the new genes of independently evolving lineages lead to the origin of new species.

7.1.1 Asexual lineages can diversify indefinitely but usually become specialized to a limited number of discrete ways of life.

In asexual organisms lacking horizontal gene transfer all lineages are independent, and are free to diverge indefinitely, potentially giving rise to a wide range of specialized types adapted to different ways of life. Whether or not this will happen depends on *opportunity* and *cost*.

- The opportunity for adaptation depends on how "lumpy" the environment is. If natural environments offer an unlimited number of potential ways of life, then a very large number of types may evolve, each differing only slightly from many others. On the other hand, the number of ways of life may be limited; there is, for example, a limited number of sugars that can be fermented by yeasts, and only a limited number of animal hosts that can be infected by parasitic worms. In this case, we would expect that a correspondingly limited range of specialized types will evolve.

- The cost of adaptation is the degree to which specialization to a particular way of life reduces success in following other ways of life. Human occupations provide simple analogies: no-one is likely to be successful both as a jockey and as

a sumo wrestler, for example. If there is a steep cost of adaptation then we expect the evolution of many specialized types. Conversely, if there is little or no cost we expect that a few generalists will be able to span the whole range of available ways of life, so there will be much less diversity.

Hence, the inevitable tendency for diversification is channeled into the diversity of distinctively different kinds that we observe in nature through the number of exclusive ways of life that are available. The remaining factor is time. A very ancient clade may now exploit all the opportunities available to it, whereas a new clade will not yet have diversified very far.

Asexual dandelions have undergone a miniature adaptive radiation. Dandelions are very familiar weeds of suburban lawns, and are often found in waste places in town and country. Their history is fascinating. The ancestors of the modern species evolved in the region between the western Mediterranean and the Himalaya, where they still live. They are quite conventional diploid, sexual plants with outcrossing and normal meiosis. Their descendants, which occupy most of the north temperate and boreal zone, are very different. Most are polyploid and completely asexual. The dandelions on your lawn have showy yellow flowers, like those in Figure 7.1, but they seldom produce normal

Figure 7.1 An asexual population: a field of dandelions, *Taraxacum officinale*.

pollen and their seeds develop without fertilization. Moreover, a sexual plant may fertilize an asexual plant, but the progeny are then all asexual. The consequence has been a great radiation of asexual lineages, which often differ slightly and are named as separate species. There are currently some 2000 or so named species in the dandelion genus, *Taraxacum*, the great majority of which are asexual. They are all similar in appearance, however: small perennial herbs arising from a basal rosette and producing the characteristic wind-dispersed fruits. The loss of sexuality has allowed asexual lineages to undergo a miniature adaptive radiation into a large number of forms, which although named as species are in fact no more than slightly differentiated clones.

Bdelloid rotifers have evolved into strongly differentiated types. Bdelloid rotifers are not nearly as familiar as dandelions, but they are even more common and widely distributed. They are minute animals a fraction of a millimeter in size that are found everywhere in temporary bodies of water, from rock pools to roof gutters. If you take a pinch of moss anywhere from Angola to Antarctica, squeeze it into a watch glass and examine the result under a microscope, you will find small transparent creatures that loop along like miniature leeches. They are the only large group of completely asexual animals: they produce large eggs that hatch into small versions of the adults (Figure 7.2), but no males or sperm have

ever been observed. Their nearest relatives are the monogonont rotifers, which are abundant in lakes and ponds. Monogononts are sexual, and it is not surprising that they are easily classified into a few hundred species each with distinctive characteristics. Bdelloids can be classified just as easily, however, and it would be impossible to say, merely from the descriptions of the species, which group was sexual and which asexual. This illustrates how clades become adapted in the long term to a restricted number of ecological niches.

Bacteria have very high levels of diversity lumped into ecotypes. Bacteria have only a restricted range of morphology, and classical methods identify only a few thousand different kinds. There are not many more named bacteria than there are named dandelions. In the last 20 years the application of modern methods of automated phenotypic profiling and gene sequencing has revealed a vast cryptic diversity. A single cupful of soil harbors tens of thousands of kinds of bacteria, judging from genomic divergence, and worldwide there are probably a billion or more different kinds. We are only just beginning to understand and catalogue this diversity, but it appears to be organized into ecologically distinct clusters that have been called ecotypes. The evidence for this is that strains that are phenotypically similar, because they use similar carbon substrates, for example, or infect the same host, also tend to be genetically similar. Thus, genetic surveys usually identify clusters of genotypes with similar ecological properties. Like rotifers, lineages of bacteria seem to be arranged into more or less discrete kinds, each following a distinctive way of life.

7.1.2 Sex hinders divergence.

Species and gender are complementary attributes of sexual eukaryotes. Eukaryotes are sexual organisms. The exchange of genes among lineages occurs in bacteria, but the complete mingling and recombination of genomes by a process involving gamete fusion and meiotic reduction is a unique character of eukaryotes. Their sexual cycle is governed by the rules of fusion, dictating who can mate with whom. There are two basic rules that apply very broadly: two individuals can mate and produce viable offspring if they are of

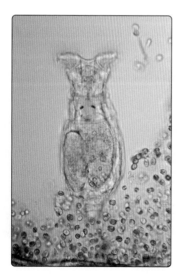

Figure 7.2 An asexual individual: the rotifer *Rotaria rotatoria*.
©istock/NNehring

like species and unlike gender. Consequently, species and gender are necessary and universal consequences of sexuality.

This is a sexual definition of species: a species is the set of lineages which exchange genes among themselves while being isolated from all other lineages. It is necessarily a relativistic definition, because the rate of genetic exchange among lineages depends on the remoteness of their common ancestor. Dogs and diatoms diverged about 1000 Mya and never exchange genes; dogs and wolves diverged about 0.01 Mya and will exchange genes whenever the opportunity occurs. Between extremes like these, any intermediate state is possible, and all intermediate states do in fact occur. In any familiar group, such as fish or frogs, lay people will almost always agree with competent naturalists about the range of species that exists, and about which species each individual belongs to. Indigenous people whose livelihood depends on a close knowledge of the local animals and plants classify them in almost exactly the same way as professional taxonomists. Nevertheless, there are usually a few individuals that do not fit into the general scheme, because they are hybrids. Pumpkinseed and bluegill sunfish are familiar and distinctive species of centrarchid fish in North American lakes and rivers; but hybrids are occasionally formed between them. Roach and bream are equally distinctive and equally abundant species of cyprinid fishes in European freshwaters; they occasionally hybridize too. Fish and other eukaryotes really can be grouped into sexually isolated species with distinctive characteristics, but their isolation is seldom absolute. This is the natural consequence of evolutionary divergence: species are only a strongly marked stage in the divergence of differently specialized lineages.

Sexual isolation permits ecological divergence whereas recombination prevents divergence. Any two lineages that are sexually isolated from one another will evolve independently and will therefore tend to diverge from one another, just as asexual lineages of organisms like bdelloid rotifers or bacteria will tend to diverge, to the limits set by opportunity and cost. The converse is also true: lineages that are not sexually isolated will diverge only with great difficulty, if at all. The reason is that random mating

followed by genetic recombination rapidly breaks up combinations of genes, and thereby prevents any particular well-adapted combination from spreading in the population.

This principle will operate whenever the performance of a structure depends on the interaction between its parts. This is a concept that is very familiar to us through the design of devices. You may cut down a tree with an axe (a device consisting of a sharp wedge attached to a long handle) or a saw (a thin toothed blade with a grip). Combining the features of the two would leave you with ineffective tools such as a thin blade attached to a long handle, or a hand-held wedge. A key must fit its lock to be effective; given a pair of locks, exchanging their keys will destroy the function of both. Biological examples are just as easy to find. Insects with long tongues (such as bumble bees) are attracted to flowers with nectaries hidden in long corollas (such as touch-me-nots), whereas insects with short tongues (such as flies) are attracted to flowers with open and easily accessible nectaries (such as daisies). Combining the two would be equally disastrous for flowers and insects. At a more fundamental level, any highly integrated system such as an organ, a skeleton, a metabolic network, or a developmental pathway is likely to be severely disrupted if its parts are combined randomly with those of another.

Lineages in sexual populations are continually fusing and recombining. They cannot readily diverge like asexual lineages because any specialized adaptation that requires a particular combination of interacting parts (genes, or gene products such as enzymes or bones) cannot for long be maintained whole. Sex is a powerful brake on diversification. Conversely, populations will become divergently specialized only to the extent that they are sexually isolated.

Species are the stage at which sexual isolation is sufficient to permit substantial divergence. Whether or not two populations evolve into differently specialized types depends, therefore, on the relative strength of two processes. One is natural selection favoring certain combinations of character states, each well adapted to a particular way of life. In each generation, those combinations that enhance the fitness of individuals will tend to increase in frequency. The other is genetic recombination, which randomizes combinations of genes and thereby breaks down any combinations of character states. In each generation, any combination that has increased in frequency through selection will be destroyed by recombination. Hence, the extent of divergence that will evolve depends on the balance between selection and recombination.

Imagine a population of mice living on an island where there is grassland and forest: they eat the seeds of the grasses and nuts from the trees. These food items require different kinds of processing: grass seeds are easily harvested and digested but are low in protein; nuts are high in protein but are much more difficult to process and digest. The difference between them creates the potential for two kinds of mouse to evolve: one specialized to garner large quantities of grass seed as rapidly as possible, using weak jaws and rapid gut throughput, the other specialized to harvest carefully a lesser quantity of nuts, which require strong jaws and a more powerful digestive system. This potential is unlikely to be realized if the mice form a single large population in which individuals mate at random. The functionally appropriate combinations of behavioral, morphological and physiological characteristics favored by selection will be erased in every generation by recombination. Now suppose that the island consists of two large areas of land joined by a narrow neck. What difference will this make? It will make no difference if both parts of the island have both forest and grassland. Each will simply constitute a smaller copy of the same island. If one is mostly forest and the other largely grassland, however, two specialized types may evolve because the forest mice and the grassland mice will tend to become divergently specialized to eat nuts and seeds respectively, while the two seldom meet and mate because contact is limited by the narrow neck of land joining the two parts of the island. If the ecological difference between the two parts of the island is sufficiently great, and if the neck of land joining them is sufficiently narrow, then the two specialized types are likely to evolve into recognizably distinct species.

In this imaginary example, any intermediate situation may occur in the balance between forest and grassland or the rate of migration between the two halves of the island. Hence, any intermediate degree of differentiation between the populations of mice living in the different halves of the island may also occur. At the extreme, they will come to constitute different species, but with somewhat less ecological difference or somewhat more genetic contact evolve into slightly different varieties or races. Hence, species mark an important stage at which very low levels of sexual contact permit clear ecological divergence, but all kinds of intermediate stages are also to be expected. This has the important implication that we can trace the origin of species along an unbroken series of intermediate stages that demonstrate how biological diversity evolves.

7.2 Populations that are permanently separated may diverge through drift or selection.

Some species consist of very similar individuals, or have a continuous range of variation without any marked geographical variation or clear division into different kinds. Maple trees, monarch butterflies, starlings, and herrings are familiar examples. Other species have two or more distinct kinds of individual (apart from any male–female dimorphism). The most general term for a distinct kind that is not a species is a *variety*. There are several other terms in common use, however, some nearly equivalent and others with some specialized meanings. A *morph* is a type that is distinguished by a single genetic difference, or a few simple differences, which may involve cryptic characters such as alleles at enzyme-coding loci; this is a term used mainly by population geneticists. A *cultivar* is a variety that has been deliberately produced by artificial selection. An *ecotype* is a variety with a restricted ecological distribution. A *race* is a variety with a restricted geographical distribution. A *subspecies* is a race so strongly and consistently differentiated from others that it is given a formal taxonomic name. The range of terms itself indicates that there are levels of differentiation below as well as above the rank of species.

7.2.1 Species may vary continuously over their geographical distribution.

Populations from opposite ends of the distribution of a species may be very different, without any sharp break marking the boundary between one kind and another. For example, there is a general tendency (with many exceptions) for body size to increase from the tropics to the poles among related species, or among populations of the same species. This gradual shift in character state is called a cline. One example among many is the tendency for body size and wing length to increase from south to north in European populations of the fruit fly *Drosophila subobscura*. It is a particularly interesting example because the species was introduced into North America in about 1980 and spread rapidly throughout temperate regions. Ten years afterwards, there was no difference between southern and northern populations; but after twenty years a south–north cline in wing length that mirrored the European cline had become established. Clearly, clinal patterns of variation can evolve very rapidly.

7.2.2 Races are geographically restricted varieties.

Other species are divided into more or less sharply differentiated types living in different regions. Organisms with feeble powers of dispersal are likely to evolve distinct kinds in different parts of their range, either through divergent specialization to different conditions or simply because historical differences become magnified over time. The salamander *Ensatina eschscholtzii* (Figure 7.3) occupies the belt of territory surrounding the Central Valley of California. Individuals move only about 20 m on average from where they were born. Its appearance is extremely variable, ranging from a uniform dun coloration to a pattern of yellow or pink patches on a black background. Each color type occupies a distinct region, and is quite sharply separated from its neighbors. The most likely explanation of this variation is that each type imitates a locally abundant toxic species, such as the rough-skinned newt *Taricha torosa*. The

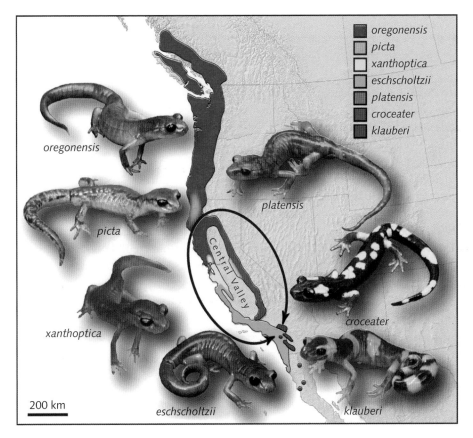

Figure 7.3 A variable species: the salamander *Ensatina eschscholtzii* in western North America.

Image kindly supplied by Dr Thomas J. Devitt.

distinction between neighboring races would then be maintained by strong natural selection, despite frequent interbreeding at the margin of the range of each type. The exception occurs at the southern end of the distribution, where types that are very different in appearance scarcely interbreed at all. In this case, there is a clockwise gradation around the ring of populations circling the Central Valley, with races that differ only slightly in appearance and readily interbreed replacing one another in sequence, until races of very different appearance that are almost fully isolated meet at the extremes.

Large gulls have a similar pattern of variation, on a larger geographical scale, despite their much greater powers of dispersal. Herring gulls of broadly similar appearance are found through the northern hemisphere. Throughout their vast range, from North Africa to the Arctic, they differ mainly in body size and in the darkness of their dorsal plumage, which varies from silver-gray to black. More than twenty different kinds have been described, some differing only slightly and frequently interbreeding whereas others are named as distinct species. They are a good example of a cluster of diverging types that range from local varieties to sexually isolated populations of different appearance that can be given different species names.

Clines and races are extreme cases of the geographical distribution of genetic variation. Most broadly distributed species are intermediate between these extremes; there is a fairly smooth increase in genetic variation over rather large distances, interrupted by boundaries between distinctively different kinds. Human populations are an example of this common pattern (see Section 4.1). Genotype frequencies change gradually over large distances, driven by the influences of history (who got there first), migration (who arrived subsequently) and selection (which phenotypes flourish). This is interrupted by occasional sharp differences at continental boundaries, especially for ancient and isolated groups such as the Aborigines of Australia. The phenotypic differences between human populations in different parts of the world are quite marked; the genetic differences are rather small, because of our common ancestry from the bands that moved out of Africa in recent times.

7.3 Divergent selection leads to more or less strongly marked varieties.

Populations living far apart will readily diverge because they are necessarily isolated from one another, so that combinations of genes are seldom broken up by recombination. Conversely, we would not expect distinct varieties to be preserved within the same region, within which individuals or their offspring can freely disperse. This is because we expect dispersal to be more effective than selection in changing gene frequencies. If selection is very strong, this is not necessarily true. There are many examples of distinct varieties that are found in close proximity because they are adapted to different conditions of life.

7.3.1 Varieties are readily produced by artificial selection.

Plant breeders routinely produce varieties by deliberately selecting individuals with desirable qualities. *Brassica oleracea* is a tall biennial plant found on the sea cliffs of southern and western Europe. It has thick fleshy leaves that store nutrients, and had become a table vegetable by Roman times. Selective breeding to exaggerate one structure or another of the plant has since given rise to the familiar vegetables illustrated in Figure 7.4: kale (more leafy but otherwise similar to the ancestral form), cabbage (spherical cluster of immature leaves), broccoli (flower heads), cauliflower (cluster of aborted floral meristems), Brussels sprouts (edible buds), kohlrabi (swollen stem), and others. Clearly, many species have the potential to evolve into a range of morphologically distinct varieties within short periods of time.

7.3.2 Several ecotypes of a species may live in the same geographical area.

Divergent adaptation generates distinct imperfectly isolated groups within species. Littorina is a common

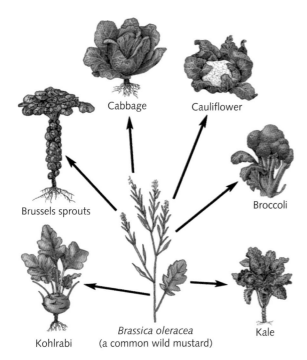

Figure 7.4 A recent diversification: useful crops derived from wild mustard, *Brassica oleracea*.

From Cain et al. *Discover Biology*, third edition © 2007 by W.W. Norton & Company, Inc. 2000, 200 by Sinauer Associates, Inc. Used by permission of W. W. Norton & Company, Inc.

in wave-stressed sites, and a thick-shelled type with a narrow aperture where crabs are abundant. How these two types are distributed on the shore depends on local circumstances. In northern England the thin-shelled form is found mainly in crevices at the extreme upper end of the tidal zone, with the thick-shelled form living among boulders in the Fucus zone of the mid-shore, as shown in Figure 7.5. In northwest Spain the thin-shelled type lives low on the shore, where it can shelter in mussel beds, while the thick-shelled form lives in the barnacle zone higher on the shore, where it is much more vulnerable to crab attack. On the Baltic coast of Sweden the tidal range is small and the two types are horizontally rather than vertically distributed, the thin-shelled form living on rocky headlands exposed to severe wave action while the thick-shelled form is found in the intervening bays, where crabs are more abundant. In each case, two

snail of the intertidal zone on both sides of the Atlantic. There are two chief threats to its existence. One is physical: wave action will tear the snail from its hold on the rocks and hurl it against the shore. The other is biological: crabs can break the shell with their powerful pincers. No single shell design will provide protection against both. A broad foot is required to grip the rock, but this requires a wide aperture in the shell that offers purchase for the powerful chelae of a crab. Consequently, there are two successful kinds of shell. The first is a high-spired, thin-walled kind with a broad aperture able to resist wave action. The second is a low-spired, thick-walled kind with a narrow aperture able to resist crab predation. Intermediate kinds would be unsuccessful anywhere; a thin-walled type with a narrow aperture, for example, would be unsuccessful on exposed shores because it could not cling tightly to the rock, while it would be equally unsuccessful if it lived where crabs could easily crack it open.

Hence, there are often two varieties of a species of *Littorina*: a thin-walled type with a broad aperture

Figure 7.5 Ecotypes *of Littorina saxatilis*. The thin-shelled morph (H) is found in crevices on the cliff face; the thick-shelled morph (M) lives among the boulders lower down the shore.

Images kindly supplied by Dr Sophie Webster.

differently specialized varieties live in close proximity to one another.

The two varieties are not species, because they readily interbreed, but their distinctive appearance is nevertheless maintained even though the two varieties may be separated by only a few meters of distance on the shore. This situation is only found, however, in species with direct development, with young that develop within the egg mass, glued to a rock, and crawl away as miniature adults. This greatly restricts dispersal and thereby restricts sexual contact between varieties, allowing natural selection to maintain the difference between them. Other species of *Littorina* have planktonic larvae that can settle on any site on the shore, so that the population is completely mixed in every generation. These species do not have ecotypes. They may nevertheless respond to local conditions of growth through development: individuals settling on the upper shore are able to detect the presence of crabs, and grow thicker shells as a result. Hence, genetically based ecotypes can evolve when there is some degree of sexual isolation, whereas phenotypes may be modified developmentally when there is complete mixing.

Parasites may have varieties specialized for infecting particular hosts. The cuckoo is a nest parasite that lays its eggs in the nests of other species. Some hosts are able to identify cuckoo eggs, and will tip them out of the nest. Consequently, cuckoos have evolved to mimic the eggs of their hosts. Different host species, however, lay eggs of different size and appearance,

so a female cuckoo must lay an egg that specifically resembles those already in the nest, as illustrated in Figure 7.6. Consequently, there are several varieties of cuckoo, each corresponding to a particular host species whose eggs it mimics. So far as is known, there is little if any sexual isolation between these varieties, which cannot be distinguished by random nuclear or mitochondrial DNA sequences. Consequently, the mimicry is not very exact, but it is sufficiently close to make it less likely that an egg will be ejected from the nest by the legitimate parents.

Parasites that are able to exploit a range of hosts often have varieties specialized for attacking particular host species. For example, aphids are plant parasites that feed on the phloem of their hosts. The common pea aphid *Acrosiphon pisum* feeds on a range of legumes, including alfalfa and clover. One variety grows better on alfalfa, and is much less successful if it is moved to clover; another is a clover specialist that grows poorly on alfalfa. The distinction between the two is maintained by this strong divergent selection and by partial sexual isolation caused by females choosing to lay eggs on the same host species they grew on (see Section 7.4.1). In this case, sexual isolation has been strong enough to permit substantial genetic divergence between the two varieties.

7.3.3 Strongly marked varieties are often named as subspecies.

When two or more forms are strongly and consistently distinguishable, they are often given different

Figure 7.6 "Host races" of cuckoo *Cuculus canorus* mimic the eggs of the species they parasitize.

Photograph of cuckoo courtesy of GabrielBuissart. This file is licensed under the Creative Commons Attribution-Share Alike 3.0 Unported license.
Photograph of eggs courtesy of Dr Marcel Honza.

names, even when they are still regarded as belonging to the same species. Varieties are often named in horticulture, such as the *Antirrhinum* varieties described previously (Section 6.1). They are often the outcome of mutations in one or two genes. The least inclusive clade of wild animals and plants that is given a formal name is the subspecies, which is designated by adding a third name to the usual Linnean binomial. The house mouse *Mus musculus*, for example, has three subspecies that occupy different regions of the world: *Mus musculus musculus* lives in eastern Europe and northern Asia; *Mus musculus domesticus* is found in western Europe, the Levant, and northern Africa (and was introduced to the Americas); and *Mus musculus castaneus* occupies south-eastern Asia. The three types, which diverged about 350,000 years ago, are similar in appearance and usually interbreed when they come into contact. There are some fixed genetic differences between them, however, and when *musculus* and *domesticus* interbreed in central Europe the male offspring are sterile (see Section 7.4.2). These are lineages in the process of forming distinct species, a process that may or may not proceed to completion.

Our lice have evolved with our clothing. We are the only animal that clothes itself. Clothing is an artificial body covering, and the amount of clothing we wear depends on climate—obviously, more is required in colder regions. This creates an opportunity for another kind of animal: lice. Our nearest relatives, the chimpanzees, are infected by lice, which live in their fur and suck their blood. We are also infected by a louse that has shared our fate since the chimp–human divergence about 6 Mya. Indeed, molecular phylogeny

of our lice shows that they diverged from chimp lice at about this time. As our lineage evolved, our lice encountered a problem: no fur. They could only retreat to the permanently furred area of the head, where they have remained ever since. (Our pubic lice also inhabit a permanently furred area but have a different ancestry.) The invention of clothes, however, gave the lice back their original habitat. They could shelter in this artificial fur, making forays from time to time to take their blood meal. Head and body lice are illustrated in Figure 7.7. Their divergence has had two interesting consequences.

- The first is that we can use the molecular divergence of head lice and body lice to date the origin of clothing. The same cladistic techniques used to date the divergence of chimp lice and human lice can be used to date the divergence between head lice and body lice. They show that the lineage of body lice emerged about 100,000 years ago, around the time that we first moved out of Africa. Clothes are seldom preserved in ancient settlement sites, but the evidence of the lice is that full body covering was invented by some human groups at about this time.

- The second is the degree of divergence between head and body lice. How different are head lice and body lice? They are morphologically very similar, although head lice tend to be smaller. The molecular data suggest that there are no fixed differences between them, so that neither forms a distinct clade. There is even some experimental evidence showing that lineages with the characteristic features of body lice can evolve from head lice within a few generations. One very detailed

Figure 7.7 The human louse has evolved two specialized varieties: the body louse (left) and the head louse (right).

molecular analysis detected separate Eurasian and American clades, the American clades consisting of head lice alone whereas the Eurasian clade consisted of head and body lice, neither forming a distinct clade of its own.

What are we to make of this? Head lice and body lice might be assigned to separate species, since they can be recognized from their morphology and live in different places. Or they might be named as subspecies, since the morphological differences are slight and there is clear evidence of interbreeding. Or they might be recognized merely as races because they are too similar to warrant formal names. Or they might not be recognized as any taxonomic entity at all, because they do not constitute monophyletic clades. All these possibilities can be defended because evolution does not respect the boundaries that we try to impose on its products. Formal names are useful when we want to refer to a particular kind of organism, but they do not reflect any real discontinuity in nature. Lineages continually tend to separate, through divergent specialization, and at the same time they always tend to cohere, through mating and recombination. The nomination of a species represents no more than a stage in this process, at which divergence begins to predominate over coherence. This is the signature of the evolutionary origin of different kinds of organism, and leads us to expect that all intermediate stages between varieties and strongly differentiated species will be found.

7.3.4 New species are continually evolving.

In this case, the formation of new species is not a rare and special event that we can never hope to witness. We should be able to find new species routinely emerging in communities of animals and plants, so that speciation can be studied directly like any other natural process. In fact, we can study it in our own backyard. The recent deglaciation of the northern part of North America opened up new areas to animals and plants that flooded in from the south and promptly began to diversify in their new territory. Some are in the process of forming new species.

Two nascent species of stickleback have evolved in postglacial lakes. The three-spined stickleback,

Gasterosteus aculeatus, is a small fish likely to be familiar to anyone living in the northern hemisphere. It has a very characteristic appearance, with bands of bony amour around its trunk and long, sharp spines sticking upward from its back and outward from its pelvic girdle. These protect it against attacks by predatory fish such as trout and perch. They are sexually dimorphic, with brightly colored males that construct nests, where they inseminate visiting females individually, and which they use to nurture and defend the young. Sticklebacks live in the sea, but readily invade freshwater and are found in streams and lakes throughout Canada, the United States, Europe, Russia, and Japan. When the ice retreated, sticklebacks began to move into the rivers and lakes of the new postglacial landscape. As they did so, they evolved in response to their new conditions of life, as we shall describe in Section 14.2.1. They also began to diversify.

One of the small bodies of water that sticklebacks colonized is Paxton Lake, in British Columbia. Like all lakes, Paxton Lake has two physically distinct zones: an inshore littoral zone where rooted plants can grow, and a deeper offshore pelagic zone. The sticklebacks in these two zones are morphologically and ecologically different, as is evident from Figure 7.8. The limnetic form that lives in the pelagic zone is slim and long-bodied, with many finely-divided gill rakers that efficiently sieve out small planktonic organisms, whereas the benthic form that lives in the littoral zone is stout-bodied, long-jawed

Figure 7.8 Benthic (above) and limnetic (below) sticklebacks (*Gasterosteus aculeatus*) from Paxton Lake, British Columbia. Both fish are gravid females.

Photo courtesy of Todd Hatfield.

and feeds on relatively large prey such as amphipods and worms. These two forms tend to breed at different sites in the lake, and although hybrids are produced they tend to be poorly adapted either to benthic or to limnetic habitats.

Mate-choice experiments in the laboratory have shown that females prefer to mate with males of the same type, apparently as a consequence of differences in size and coloration. In the field, both types nest in the littoral zone, the benthics among macrophytes and the limnetics mainly in shallower unvegetated areas. Mate choice in sticklebacks is strongly influenced by the nuptial coloration of the male, which evolved in the common ancestor of the clade. The ancestral state is a black throat, which has evolved into a vivid red in *G. aculeatus*. This is used both as a threat display towards other males and as an enticement for females: females prefer intensely red-throated males, and males with redder throats receive more eggs. They also pay a price, as red throats are easily detected by visual predators; the conflict between survival and sexual success is discussed further in Section 15.4.

Benthic and limnetic forms of *G. aculeatus* represent intermediate stages in speciation. The only other named species in the genus, *G. wheatlandi*, is a slightly more advanced stage. It has small bright green males with black-spotted throats. It occupies the same areas in estuaries but does not interbreed with *G. aculeatus*, which indeed displaces *G. wheatlandi* males from their nests. Sticklebacks show how speciation is a continuous process, from the modifications that evolve when marine forms enter freshwater, through the diversification of ecologically specialized types, to the emergence of sexually isolated species. All these stages have occurred quite recently, since deglaciation about 0.01 Mya.

There has been a modest radiation of whitefish in northern Eurasia and North America. Whitefish in the genus *Coregonus* are another characteristic fish of the north temperate and subarctic zones. They are different from sticklebacks in almost all respects. They are much larger (up to several kg) shoaling fish that are broadcast spawners with no sexual dimorphism and no parental care. Nevertheless, whitefish populations often include a smaller, plankton-eating type and a larger, benthic, bottom-

Figure 7.9 Benthic and limnetic forms of lake whitefish (*Coregonus clupeaformis*).

Reproduced with permission from Jeukens, J. and Bernatchez, L. 2012. Regulatory versus coding signatures of natural selection in a candidate gene involved in the adaptive divergence of whitefish species pairs (*Coregonus* spp.). *Ecology and Evolution* 2 (1): 258–271. doi: 10.1002/ece3.52.

feeding type, which are illustrated in Figure 7.9. Thus, similar ecological specializations evolve in whitefish and sticklebacks, despite their very different ways of life. Moreover, they have evolved repeatedly. Independent radiations of large benthic and small limnetic forms are commonplace in North American and European whitefish.

Whitefish neither construct nests nor care for their young. Male and female are alike; the eggs are externally fertilized during spawning on shoals in late fall, are deposited on the bottom and hatch in the following spring. Ecotypes may spawn at different times or in different places, but there appears to be much less opportunity for sexual selection than in sticklebacks. Nevertheless, there is good evidence from mtDNA surveys that benthic and limnetic types in Yukon lakes are sexually isolated. Moreover, hybrids between the whitefish ecotypes suffer higher embryonic mortality, equivalent to a selection coefficient of about 0.2–0.4. This may be attributable to incompatibility between genes governing development time, resulting in the asynchronous emergence of hybrid larvae. The whitefish ecotypes are well on their way to becoming distinct species.

Remarkably, many other fish that have succeeded in colonizing postglacial lakes have undergone parallel radiations. For example, four morphs of Arctic char (*Salvelinus alpinus*) have been described from an isolated volcanic lake in Iceland: a limnetic planktivore, a large and a small benthic type, and a large piscivore. Smelt (*Osmerus mordax*) have evolved anadromous and freshwater forms, with lake populations evolving sympatrically into limnetic and benthic ecotypes. North American lakes may support several kinds of benthic and limnetic round whitefish *Prosopium* with different diets that are named as species but may be comparable to ecotypes in *Coregonus*. In short, the occupation of new lakes by the few species of fish capable of expanding into a recently deglaciated landscape has often been followed by the evolution of trophically specialized benthic and limnetic types. The details of this process vary from species to species, but the general principle of divergent specialization into ecologically distinct benthic and limnetic types has occurred independently in many lineages.

Emerging species can be seen all around us among familiar local organisms. Rain forests, tropical lakes, and coral reefs support a multitude of species and are justly regarded as demonstrating in the clearest way the generation of new kinds through natural selection. But we do not have to travel to remote and exotic places to see speciation in action. The lakes and forests of North America and Europe provide striking examples of speciation in progress, in the familiar kinds of animals and plants that we see all around us.

7.3.5 Closely related species often hybridize.

The gradual emergence of new species implies that hybridization may be very common when it can occur. There is nothing inevitable or irreversible about the divergence of lineages. Two isolated or divergent lineages may for long continue to be sexually compatible, and will cross more or less frequently whether they are named as races, subspecies or species. The outcome depends on how frequently they mate together. If the lineages are connected through cross-mating more strongly than they are separated by divergent selection, they will merge into a single lineage and speciation will proceed no further. On the other hand, if mating is restricted or infrequent then more or less distinct species can be recognized despite the production of many hybrids in each generation.

Nascent species of sticklebacks can collapse into a single lineage. Enos Lake is a small body of water, a few hectares in extent, on a creek that flows into the sea near the southeastern corner of Vancouver Island.

It was one of the localities from which the distinct benthic and limnetic types of three-spine sticklebacks were recorded in the 1970s. By the late 1980s the distinction between the two was somewhat less than before; by the late 1990s it had disappeared altogether. There is now only a single type of stickleback in the lake, representing the highly variable population resulting from hybridization between what were previously distinct kinds. The reason for this collapse is not known for certain, although it coincided with the introduction of an exotic crayfish that may have altered the ecology of the benthic zone. At all events, it provides a rare directly observed example of the tentative nature of the early stages of species formation.

Species identity can be preserved despite frequent hybridization. Conversely, lineages may continue to retain distinctive ecological and morphological characteristics despite frequent hybridization, if divergent selection is sufficiently strong. Ducks fall into easily recognizable species, for example, yet at least half of them produce hybrids in nature. The species remain separate because the hybrids usually do not develop normally or are sterile with either parent. Many other examples could be cited: occasional hybridization between closely-related species is the rule, not the exception. Species can remain distinct, however, despite hybridization, provided that an equally powerful countervailing process of selection favors the parental types.

The fire-bellied toad, *Bombina bombina*, is found in lowland areas of eastern Europe. Its name comes from the red and black mottling of the belly; when attacked by a predator it will roll over to expose this highly conspicuous pattern, which warns the predator that its prey is best left alone (lick one of these toads and you will begin to sneeze uncontrollably—or so I have been told). It breeds in ponds; the males have a large vocal sac and the mating chorus of a good-sized population can be heard a mile or more away. The closely-related yellow-bellied toad *Bombina variegata* is similar in appearance, with the same anti-predator behavior but a yellow-and-black belly. It occurs in more mountainous districts to the west and breeds in puddles, where the males, lacking a vocal sac, emit a more modest chirp during the mating season. The two are clearly distinct, as you can see from Figure 7.10, and have long been named as separate species. Nonetheless, they have also long

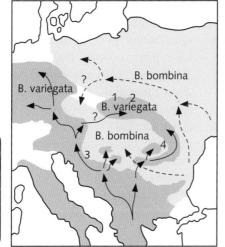

Figure 7.10 The intermingled distribution of the fire-bellied toad (*Bombina bombina*, upper right) and the yellow-bellied toad (*Bombina variegata*, lower left) in Europe. Hybrids are formed where they meet. The arrows on the map are possible migration routes from southern refugia after deglaciation.

Photographs kindly supplied by Dr Boris Timofeev. Map reproduced by permission of the author from Arntzen, J.W. 1978. Some hypotheses on postglacial migrations of the fire-bellied toad, *Bombina bombina* (Linnaeus) and the yellow-bellied toad, *Bombina variegata* (Linnaeus). *Journal of Biogeography* 5: 339–345.

been known to hybridize quite frequently. The junction between the two species is sometimes a zone a few kilometers in width across which one species smoothly replaces the other through a continuous series of hybrid populations with varying frequencies of alleles from either parent. At other sites it is more of a mosaic where sites dominated by one species or the other can be found close together. Individuals from either species mate freely with the other when they meet, so the hybrid zone must be stabilized by the balance between dispersal between breeding sites (which will tend to broaden the zone) and the inferiority of hybrid offspring (which will tend to narrow it). Both contribute to maintaining species identity in *Bombina*. Each species tends to prefer its own breeding site, pond or puddle. At the same time, the mortality of hybrid tadpoles is about ten times that of pure-bred tadpoles. Consequently, the two species remain distinct in most areas rather than collapsing into a single hybrid swarm.

7.4 Species are recognized when ecologically distinctive forms become sexually isolated.

These examples show us how lineages can diverge through a series of stages that we call, for convenience, varieties, races, subspecies and species. These human categories are signposts along a continuous process of divergence that can be studied like any other natural process. Consequently, we can investigate how lineages become permanently separated and distinct. Varieties and races readily melt back to the ancestral type; subspecies may be somewhat more permanent; species may hybridize, but often mark the beginning of an irreversible separation between lineages and the emergence of permanent diversification. How does this come about?

New species may arise through divergent sexual selection or through divergent natural selection. Broadly speaking, there are two principal routes for the evolution of distinct species, corresponding to the two fundamental events of the sexual cycle.

- The first is that species formation is the direct response to divergent **sexual selection**; the indirect response is ecological specialization, and isolation is maintained primarily by the failure of gametes to fuse successfully (this has been called "prezygotic" isolation). I shall call this "sexual speciation."

- The second is that species formation is the direct response to divergent **natural selection**; the indirect response is sexual incompatibility, and genetic isolation is maintained primarily by the failure of the zygote to develop or to complete meiosis successfully ("postzygotic" isolation). This has been called "ecological speciation."

These categories are not mutually exclusive: there are clear examples of both processes. Moreover, they are both quantitative: all degrees of ecological divergence and sexual isolation can be found.

7.4.1 Lineages that do not meet cannot mate.

Once two sister lineages are unable to mate they are free to diverge, and will evolve into different races, subspecies and species as time goes by. The simplest and perhaps the most general reason for failing to mate is that they live in different places and never meet.

Populations separated by a physical barrier will tend to diverge over time. Southern California and Nevada were occupied by a large lake (Lake Manly) and extensive river systems only a few hundred thousand years ago. In the mid-Pleistocene the climate became more arid and the lakes and rivers began to shrink. As they shrank, populations of fish such as pupfish in the genus *Cyprinodon* became restricted to smaller and smaller bodies of water. When the landscape had reached its present condition of arid scrubland and desert, surviving populations had often become restricted to isolated stream systems or even to small clusters of springs and pools in valley floors.

Being completely isolated from one another by an impassable barrier of hot dry rock and sand these populations were free to diverge and as a result have often evolved into distinctive forms. About 30 species of *Cyprinodon* have been described from the region, together with many other forms named as subspecies, most of them found in only a few localities. *Cyprinodon diabolis* is an extreme example: this species comprises only a few hundred individuals living in an isolated pool in southern Nevada, shown in Figure 7.11, where it has diverged from the neighboring populations of *Cyprinodon nevadensis* within the last few tens of thousands of years. Despite their distinctive appearance, these isolated species of pupfish will usually mate successfully with one another when they are brought into contact in the field or in the laboratory.

Populations separated by a physical barrier may become unable to mate with one another. If the range of a species becomes divided by impassable barriers the populations on either side of the barrier will begin to diverge. The formation of a land bridge between North and South America in the Pliocene (about 3 Mya) is an example on a large geographical scale. It permitted terrestrial biotas to mingle, so that carnivores and ungulates moved south, for

Figure 7.11 The desert pupfish *Cyprinodon diabolis* is found only in this geothermal pool in the Amargosa Desert of Nevada.

example, while opossums and armadillos moved north. At the same time it split marine species into permanently isolated populations on the Atlantic and Pacific sides of the land bridge. These are now slowly evolving into distinct species because they can no longer mate with one another. For example, there are about 20 species of snapping shrimps in the genus *Alpheus* living in the shallow coastal waters on the Pacific coast of Panama. Most of them occupy different kinds of site along the shore. There are 20 corresponding species on the Caribbean coast (Figure 7.12), each living in the same kind of site and looking so similar to its counterpart on the other side of the isthmus that the two are often given the same name. Individuals from different coasts do not mate, however, even when brought together in the same aquarium, although individuals from the same coast mate freely. This shows that sexual isolation has already occurred. Moreover, there are fixed genetic differences between the two members of a pair, showing that their divergence has begun, although the morphological and ecological differences between them are as yet very slight. Species pairs living in deeper water or on offshore islands are the most

Figure 7.12 The snapping shrimp species *Alpheus bouveri* (Atlantic) and *Alpheus javieri* (Pacific) have diverged from an ancestral population split 6 Mya by the closure of the Isthmus of Panama.

Photographs of snapping shrimps © Dr. Arthur Anker.

divergent, because they have been separated since the isthmus first began to rise about 15 Mya. The least divergent pairs live in coastal mangrove forests, which became isolated later, when the isthmus finally closed about 3 Mya. In these shrimps, geographical isolation has led to almost complete sexual divergence within a few My, whereas ecological divergence is slight because species pairs have continued to occupy the same habitats.

Barriers to mating that evolved when populations were apart may keep them isolated when they meet again. Species and subspecies that have diverged recently in isolation are in a state of flux. If they are brought back into contact they may hybridize readily, so that two distinctive forms are lost by merging into a single descendant lineage. This seems to have happened repeatedly in the *Cyprinodon* community of southwestern North America as the result of shifting hydrological conditions. On the other hand, genes affecting sexual behavior and development may have become altered enough to prevent hybridization, in which case the lineages are permanently isolated and will continue to diverge. This has happened in *Alpheus* and other marine animals on opposite sides of the Isthmus of Panama.

The ultimate reason for sexual divergence is the coevolution of male and female attributes. Males and females have identical interests only in strictly monogamous species. Otherwise, the most successful males might be those who mated with many females, for example, whereas the most successful females might be those who curtailed this behavior by insisting their mates help to rear their offspring. The evolution of sexual interactions is described in Chapter 15. Now, the success of male attributes depends to some extent on the current state of the female population (how they choose mates), while females are likewise affected by the state of the male population (how they evade or manipulate female choice). Consequently, any isolated population will evolve in a particular way depending on the combinations of mutations affecting sexual behavior that arise and spread. It is unlikely that two populations isolated from one another will follow the same trajectory. Hence, if they subsequently come into contact again their sexual behavior will be somewhat different, and this may be enough to prevent an individual from either population mating with an individual from the other.

The simplest situation is found in marine invertebrates such as sponges and sea-urchins that simply cast their gametes into the water column. Successful fertilization is mediated by the interaction of one or more binding proteins on the surface of the sperm with receptors on the surface of the egg. This binding-receptor complex evolves very rapidly—as much as 50 times faster than genes not involved in mating—and always differs between species, even closely related species. In animals with copulation and internal fertilization, the shape of the genitalia may determine whether or not mating is successful. In many insect groups, for example, males have morphologically complex genitalia whose structure is highly species-specific. Mating may also depend on chemical cues such as pheromones, which also differ between sister species. Hence, when a male fly meets a female, the outcome will depend on a very complex sequence of interactions involving chemical and behavioral cues, genital complementarity, the fate of sperm in the female reproductive tract, and the ability of sperm to bind to and penetrate the egg membrane. All of these attributes may evolve along different paths in long-isolated populations, so when these are brought back into contact they may well be unable to mate with one another.

Sexual isolation may evolve if groups of individuals do not meet despite living in the same area. Hence, speciation will often require the prior separation of populations whose mating systems diverge in isolation, such that they seldom interbreed when they again come into contact. This may occur on a large geographical scale, such as the many examples of species that have evolved on remote oceanic islands or in isolated lakes. It may also occur on much smaller geographical scales, however, provided that there is some barrier to mixing.

- Spatial separation can be on a large scale, such as the many examples of speciation on islands or in lakes. It may also be on a small scale, such as different spawning beds in the same lake. Four morphs of whitefish occur in Lake Femund, in Norway. They are all given the same species name, *Coregonus laveratus*, which is applied to a wide variety of closely related fish in northern Europe. The four morphs are specialized for different diets, and consequently differ in the position of the mouth, the length of the lower jaw and the number of gill rakers. They spawn in different places: in deep water, in shallow water, in bays, and in the river feeding the lake. The difference in spawning site permits a modest divergence, both in genotype and in morphology.

- Temporal separation requires different breeding seasons, such as insects that feed on several host plants. The pea aphid is a small insect that feeds on legumes by stabbing the leaves and sucking out the sap. It attacks dozens of plant species, but individuals collected from one species are often poorly adapted to others. The aphids collected from alfalfa, for example, do not grow well on red clover, and vice versa. Moreover, the alfalfa aphids are genetically distinct from the red clover aphids. Aphids living on different host plants will hybridize readily, but the extent of hybridization varies with the degree of ecological specialization: specialized types that are largely restricted to a particular host plant species seldom hybridize successfully with other specialized types. Since their host plants often grow in the same area, different specialized lineages of the pea aphid often live very close to one another. In principle they might never meet, if they were always dispersed from one plant to another of the same species. In practice there is a fair amount of mingling, and this prevents the complete sexual isolation of races restricted to a particular host plant species. In the past, aphids found on a particular host were named as separate species, so there were dozens of species within the pea aphid genus. They were very difficult to distinguish by morphology, however, and most have since been demoted to subspecies. Whether or not they have a formal name is not important. They illustrate the fact that a radiation of specialized types may occur on the same plot of land, although the extent of specialization will usually be restricted by encounters between differently specialized individuals, resulting in a diverse flock of races that represent an intermediate stage in the formation of completely isolated species.

Whitefish and pea aphids illustrate the principle that lineages will diverge, become ecologically distinct and evolve into separate species depending on the degree of sexual isolation between them. When this is more or less absolute, because it is enforced by geographical and geological barriers that can seldom if ever be crossed, the divergence of sexual systems will eventually lead to the emergence of new species. When the barriers are more porous, admitting more or less frequent mating among types living in the same place, divergence occurs more slowly and the outcome is a range of ecologically specialized types on the borderline of speciation.

7.4.2 Long-isolated lineages may not mate successfully when they meet.

In practice, lineages that are isolated for a few thousand generations may mate perfectly well when they meet again, because the genes that control mating have not greatly diverged. This is the normal situation; the formation of permanently separated species is an exceptional event. Even so, they may have diverged in other ways, for example by becoming specialized for different ways of life, by following different developmental pathways, or by accumulating somewhat different complements of chromosomes. Any of these processes may obstruct hybridization and the fusion of lineages, not because hybridization does not occur, but because the hybrids are not successful. If they are inviable or sterile then the parental lineages are completely isolated, no matter how freely they may mate.

Hybrids may be inviable or sexually sterile. Any hybrid individual from a cross between two sexual lineages must first develop to maturity and then itself produce gametes and mate successfully. Hence, there are two barriers to hybridization between long-separated lineages.

- The first is the failure to develop successfully through mitosis. The two haploid genomes that contribute to the zygote may not be capable of directing the development of a successful diploid individual. Hybridization will then be opposed by natural selection.

- The second is the failure to complete the sexual cycle through meiosis. The hybrid diploid genome may not be capable of segregating successfully into functional haploid gametes. In this case, hybridization will be opposed by sexual selection.

These two processes can be combined: for example, hybrids might develop with abnormal mating behavior that will prevent fertilization by either parent.

Hybrid inferiority is caused by genetic incompatibility. When a gene is expressed, the effect it has on the fitness of the individual will depend on the environment. The key to understanding why hybrids often have lower fitness than either parent is that the genome is a part—often a very important part—of the environment. This is because the expression of a gene depends on a complex web of regulation governed by other loci, while its effect on the fitness of the individual which bears it depends on how its product interacts with the products of many other loci. A rather simple-minded cultural analogy might involve two societies: in one, workers make either light bulbs or electrical sockets, whereas in the other they make either candles or candlesticks. Both societies are adequately illuminated. A combination of workers, however, would produce either light bulbs with candlesticks, or candles with electrical sockets; neither would produce much light.

When two lineages have long been separate their genomes will have diverged, and consequently any locus in one lineage will have become adapted to the genome within which it has evolved. That is, the alleles that predominate at this locus will have spread because they interact favorably with alleles at other loci that predominate in the genome of the lineage. They will not necessarily interact favorably with the different set of alleles that have spread in the other lineage. For the great majority of loci this will not matter, either because allelic effects are not sensitive to genetic background or because similar alleles have become fixed in both lineages. A very small number of loci whose effects are incompatible, however, might be sufficient to raise an almost insuperable barrier to successful hybridization.

Hybrids may not develop normally. In the first place, hybrids may not be able to grow or develop normally, and either die or give rise to defective adults. The swordtail (*Xiphophorus helleri*) is a popular aquarium fish with a silvery body and an extended ventral tail lobe, the sword. The platyfish (*X. maculatus*) is a related species that has black spots on its back and sides caused by pigmented cells called melanophores.

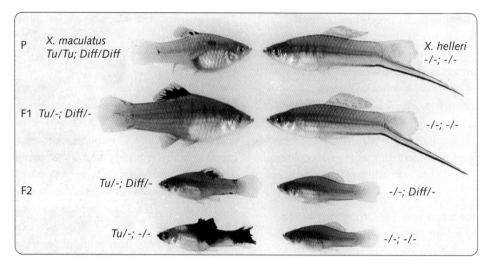

Figure 7.13 Species of platyfish are kept separate by the inviability of hybrids. (a) P, the parental generation: *Xiphophorus maculatus* female and *X. helleri* male. (b) Hybrid (F1) female, which displays a benign nevus-like melanotic lesion, is backcrossed to an *X. helleri* male. (c) F2, backcross progeny segregating the oncogenic *Tu* allele and the tumor suppressor *Diff* allele. Fish that carry one *Tu* allele but do not have any copies of *Diff* (bottom left) have malignant melanomas.

The two species can be crossed, producing hybrids with a large number of black spots. When these hybrids are in turn crossed with *X. helleri*, half the progeny are unspotted and half have a range of phenotypes which includes an invasive malignant melanoma—they are killed by skin cancer. This pattern of inheritance, which is illustrated in Figure 7.13, is attributable to two loci, a sex-linked tumor (*Tu*) locus and an autosomal suppressor *R* locus. *X. maculatus* carries alleles at *Tu* that specify melanophore formation and alleles at *R* that regulate its expression, whereas *X. helleri* lacks both melanophore-producing and suppressor alleles. The hybrids have a copy of *Tu* from their *X. maculatus* parent and are heterozygous at the *R* locus, so they develop an increased level of spotting. When back-crossed to the *X. helleri* parent, half the offspring inherit the melanophore-producing allele at *Tu* but lack its suppressor at *R*. Consequently, they suffer from severe and often fatal melanoma. This effectively separates the two species because gene exchange between them cannot be perpetuated. This is an extreme example of how the interaction between genes—in this case the *Tu* gene and its suppressor—can give rise to severely crippled phenotypes. Less dramatic cases have been found in other pairs of closely related species in a variety of organisms.

Hybrids may not reproduce normally. Hybrids may develop normally yet fail to produce functional gametes. This is again likely to be attributable to maladaptive gene interactions, in this case between the genes that govern meiosis. It may also be caused by gross changes in chromosome structure, if they hinder the orderly pairing and separation of homologues. Centric fusion is one example: two chromosomes which both have centromeres near their tips can fuse to form a single chromosome whose centromere is then in the middle. This need not be severely debilitating, because in heterozygotes the two separate chromosomes from one parent may be able to pair more or less normally with the single fused chromosome from the other. The outlook is much worse for hybrids between populations in which different centric fusions have occurred. Suppose that in one population chromosomes A and B have fused (forming AB) while C remains separate, whereas in the other A and C have fused (forming AC) while B remains separate. Hybrids may develop normally, but when chromosomes form pairs at the onset of meiosis AB from one parent will tend to pair both with AC and with B from the other, while AC will tend to pair both with AB and with C. The outcome is likely to be complex and often unbalanced multivalents which cannot segregate normally, which prevents the formation of functional gametes.

Centric fusion is quite common in mammals (including humans) and is responsible for the bewildering complexity of populations of the European house mouse, *Mus musculus domesticus*. The standard type has a diploid complement of 40 chromosomes, all with centromeres near their tips. Other types have fewer chromosomes because of centric fusions. As these have occurred independently in different localities there is a network of hybrid zones throughout western Europe and North Africa. Each of the dozens of local races is isolated from its neighbors by the partial or complete sterility of hybrids arising from improper chromosome segregation during meiosis. At a larger geographical scale, the subspecies *domesticus* is replaced by the subspecies *musculus* in eastern Europe and northern Asia, and by the subspecies *castaneus* in southern Asia (Section 7.3.3). These hybridize to a varying degree where their distributions overlap, but remain distinct because hybrid males are impaired or sterile, partly because of differences in chromosome structure and partly because of specific interactions between divergent genes. Finally, other wild mice are named as different species, such as the short-tailed mouse *Mus spretus*. Hybrids are only rarely formed in nature, and hybrid males are completely sterile. The backcross between males of either parental species and hybrid females sometimes yields fertile males, however, so the incompatibility is likely to be caused by a few interacting loci. Thus, the house mouse illustrates the whole spectrum of differentiation from local races to distinct species, with hybrid sterility at each level caused by the divergence in isolation of chromosome structure or gene sequence.

7.5 Rapid speciation gives rise to swarms of sister species.

Hawaiian crickets have evolved rapidly through sexual selection on male song. Species of cricket in the genus *Laupula* look very similar and all eat a variety of forest plants in Hawaii (Section 1.2.3). They are not ecologically distinctive in any obvious way. The males have different courtship songs, however, and females prefer the song of their own lineage. This necessarily leads to sexual isolation between lineages, and thereby to the emergence of species which are not yet ecologically specialized. The phylogenetic tree of the species in the genus can be estimated from gene sequences, and the age of the islands in the group is known from geological evidence. Combining the two gives the rate at which new species have evolved over the last few million years. This turns out to be 1–4 species per My.

Pacific tree snails illustrate rapid speciation and rapid extinction. About 120 species of tree snails in the family Partulidae live on the volcanic islands of the tropical Pacific. These islands often have very rugged terrain, because the volcanic landscape is easily carved by rain and weather into steep-sided valleys that the snails can cross only with difficulty. For example, on the small island of Moorea, which is no more than 10 km across, there were until recently about a dozen species of the single genus *Partula*, whose diversity of form is illustrated in Figure 7.14. They are closely related and most likely diverged from one another on the island. They occupy somewhat different but overlapping sites, such as drier versus wetter conditions, or low shrubs versus the branches of trees.

The shell of any snail is coiled either in a left-hand or a right-hand spiral. *Partula* is unusual in that some species are right-coiled and others left-coiled. This leads to an unusual complication in mating. Like most other land snails, it is hermaphroditic, and mates both as a male, transferring sperm via a penis, or as a female, receiving sperm into a cloaca. A right-coiled snail, acting as male, will probe with its penis for the cloaca of its partner, acting as female (later, they change roles). It is likely to succeed only if its partner is also right-coiled, and mating between individuals of opposite coil is usually unsuccessful. Thus, the direction of shell coiling sets up a rather unusual sort of barrier to mating. Some of the species of *Partula* are left-handed and some right-handed, which partly explains why they hybridize in some combinations, but not in all. Even when hybrids are formed, however, they may have abnormal genitalia, presumably because of genetic incompatibilities.

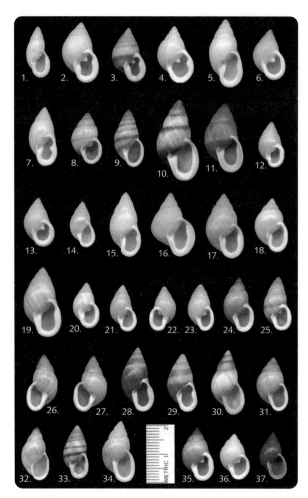

Figure 7.14 The diversity of form among the tree snails (*Partula*) of Moorea, a small island a few kilometers north-west of Tahiti.

Composite scan by Bill Frank and Harry G. Lee MD, FLS, Jacksonville FL, USA

In short, the tree snails of Moorea were a swarm of weakly divergent lineages that have become partially isolated because neighboring populations have become ecologically specialized or have evolved the opposite pattern of shell coiling. They represented an early stage in species formation that took place since the origin of the island about 2 Mya.

The use of the past tense is deliberate. Most of these species have become extinct since they were first studied in the early twentieth century. This followed from the well-intentioned introduction of the giant African snail *Achatina* as a food for the inhabitants. This was so successful that the introduced snails proliferated to such an extent that they threatened the crops. The solution was to introduce a predatory snail, *Engladina*, to control *Achatina*. This was entirely successful, but having wiped out

Achatina, *Engladina* then began to eat the native *Partula* species. Most are now extinct in the wild.

Relaxing cytogenetic constraints promotes speciation. A few organisms lack the distinct centromeres that usually control chromosome pairing in meiosis. Instead, their chromosomes have a diffuse centromeric property that enables them to pair along their whole length. This has the interesting consequence that when a chromosome breaks (not an uncommon event) both fragments may be stably transmitted, whereas in other organisms the fragment lacking a centromere would be lost and the progeny would be inviable. The outcome is that a wide range of types can be generated quite rapidly through the fission and fusion of chromosomes and chromosome fragments. In principle these can hybridize readily, although in practice any considerable variation in chromosome number is likely to lead to difficulties during meiosis. Consequently, groups with this kind of chromosome are liable to radiate quite readily into a wide range of races and species. Sedges belonging to the genus *Carex*, illustrated in Figure 7.15, are a good example. These are small inconspicuous rush-like plants of the northern hemisphere. There is an enormous range of chromosome numbers in the genus, from 6 to 60 or more, and speciation often involves changes in chromosome structure and number. Moreover, many species are self-compatible and some may be mostly selfing. Hybrids are rather readily formed, although their success is often low and the fertility of hybrid progeny decreases with the difference in chromosome number between the parents. Because they are freed from one of the usual constraints that hinder the

Figure 7.15 A clump of sedges: *Carex prasina*.

diversification of sexual lineages they are extremely diverse: more than 2000 species have been described within this single genus, and if you live north of 35 N latitude, in North America, Europe, or northern Asia, you are likely to encounter dozens of species of *Carex* in a day's ramble in forest, marsh, or moor. Many species look similar and grow in similar places. Others are ecologically quite distinctive, however, especially those that grow in lowland marshy places and those that grow in upland forests. They are older than the stickleback and whitefish races (most date back at least 0.1–1 Mya), but those occupying the forests and wetlands of Canada and northern Europe have established themselves there much more recently, through dispersal from the south since deglaciation. The swarm of sedges found in the forests and bogs of the northern hemisphere show how lineages can separate, diversify, and form full species over short periods of time, subsequently intermingling as their ranges shift.

7.6 Diversity is the dynamic equilibrium between the origin and extinction of species.

There is no fixed catalogue of species in the world; new species are continually being added and old ones disappearing forever. Most of this chapter has described how new species come into being, but the counterpart is that species also become extinct.

7.6.1 All species eventually become extinct.

It is a familiar observation that rare things vanish. This is because the few remaining individuals of a species may all become sick, or die accidentally, or simply fail to meet. It is a less familiar fact that all species become rare from time to time. There are two reasons for this.

- The first is that resources may diminish or enemies increase. The English sparrow, *Passer domesticus*, was the most common bird in London fifty years ago, but it is now so uncommon that its survival is in question. It expanded from the countryside into urban sites in the nineteenth century because it could feed on spilt grain. When markets were relocated in the later part of the twentieth century its population shrank with its food base. At the same time, the moorhen, *Gallinula chloropus*, was a common bird of riversides, streams, and canals, but withered after the introduction of an effective semi-aquatic predator from North America, the mink. Both of these examples involve human intervention (it is easily documented) but the lesson is perfectly general: species fluctuate in abundance widely over time and often become rare.

- The second arises from the stochastic nature of survival and reproduction. If you take $100 to the casino then you will sooner or later lose it all, whether or not the house has an advantage. The reason is very simple: so long as you have money you can continue to play, but on the first occasion that you run out of money you will be obliged to leave. A lineage may persist for a long period of time, but if it fluctuates randomly in abundance it must sooner or later become rare. If it then becomes extinct it cannot recover.

We are accustomed to the animals and plants we see around us, and tend to assume that they have always been common. They have not. Natural communities are in a continual state of flux, species often become rare, no matter how common they have previously been, and while rare they are likely to become completely extinct.

7.6.2 Species persist for about 1–10 My on average.

The longevity of species can be estimated either from fossils or from phylogenies.

- Estimates from fossils are based on the first and last appearances of morphologically distinctive types in the fossil record. Any given type may have evolved before the first fossil yet discovered, however, and may have persisted after the last, so fossils provide a minimal estimate of species longevity. For marine invertebrates, this is about 4 My as a rough average; for terrestrial plants it is about 1 My for herbs, 4 My for shrubs, and 8 My for trees.

- Estimates from phylogenies are based on estimates of the time to the most recent common ancestor for extant taxa. Because this necessarily omits extinct sister taxa, it will provide a maximal estimate of species longevity. For marine invertebrates this is about 13 My on average; for terrestrial arthropods 9 My; for vertebrates 4.3 My; and for plants 7.5 My.

These estimates are reasonably consistent, given the very different evidence each is based on: as a loose generalization, most species persist for 1–10 My before becoming extinct.

7.6.3 Some species persist for very long periods of time.

Some clades are producing new species very rapidly: the average age of African Rift Lake cichlids or Hawaiian silverswords, for example, is 0.1–0.5 My. Conversely, some species are extremely durable. It is difficult to be sure, for the only direct evidence comes from the morphological similarity of fossils. Nevertheless, the extreme conservatism of form is very impressive in groups such as cycads (little apparent change over the last 25–35 My), gingko trees (Figure 7.16; 100 My), and clam shrimps (200 My). These cases illustrate the general rule that once a lineage has become very well adapted to a way of life then mutations that would improve its adaptedness are exceedingly rare, so the lineage persists with little or no change for as long as that way

of life will support populations. Cultural analogies come readily to mind: you may be reading this with the aid of technology that did not exist twenty years ago, but if you are peeling an apple at the same time you will be using the sharpened metal blade technology that was invented some 5000 years ago, when the first urban civilizations appeared in Egypt and China.

7.6.4 Extinction is balanced by speciation in the long term.

Despite the longevity of some species, all become extinct sooner or later, so far as we know. The origin of new species must therefore balance the loss of old, at least in the long term. There are two reasons for this, both arising from the fact that many species are specialized for a particular way of life.

The first is that there are many combinations of conditions, created by the physical properties of environments, which require particular specializations. This is easy to understand for extreme environments, such as the boiling hot acid springs of Yellowstone Park or the bitter cold of the Antarctic shore. It applies with equal force, however, to much milder sites such as a saltmarsh or an alpine meadow. In each case a species will flourish only if it possesses a certain suite of adaptations. Moreover, no species can be successful over the whole range of conditions, because adaptations that reinforce success in (say) a

(a)

(b)

(c)

Figure 7.16 A living fossil: the gingko tree (left). The leaves of *Gingkoites huttoni* from the Jurassic (170 Mya) (above, right) are similar to those of the living *Gingko biloba* (below).

hot spring will be unnecessary and expensive luxuries that guarantee failure in a normal pond a few hundred meters away. Hence, if a species specialized for some such way of life should become extinct, its disappearance opens up an opportunity for a new species to exploit.

The second reason is more subtle. Species become adapted to their partners and enemies as well as to the physical conditions of life (Chapter 17). When a species becomes extinct the event will close some ecological opportunities and open others. In many parts of the world, for example, we have fished cod almost to the point of extinction. If the Atlantic cod, *Gadus morhua*, should disappear completely, its obligate parasites would of course disappear with it. Less specific enemies, such as seals, would suffer more or less severely. Its competitors, such as other groundfish and squid, would be likely to benefit. These short-term changes in community composition would in turn benefit at second hand other organisms (such as parasites of squid, or the crabs that would otherwise have been eaten by the seals) and harm others (such as the snails that are now less common because crabs are more abundant). In the longer term, this cascade of consequences opens up new ways of life that are difficult to predict.

The more or less fixed range of opportunities presented by the physical environment and the endlessly shifting set of opportunities presented by the biotic environment imply that any extinction creates a vacant space that will sooner or later be occupied by a new lineage capable of exploiting it.

7.6.5 Mass extinctions are caused by rare events of large magnitude.

There is one important qualification that needs to be added to the general principle that, in the long run, speciation balances extinction. This is that "the long run" may be a very long run indeed. Extinction can occur very quickly, sometimes within a few generations; new species always take much longer than this to evolve. There is a fundamental imbalance between the timescales of extinction and speciation. This is made dramatically clear on those few occasions in Earth history when a large fraction—sometimes the majority—of species are wiped out by some great convulsion.

Mass extinctions have punctuated the history of life. There is a very general rule that governs how the environment changes over time: events of small magnitude are much more frequent than events of large magnitude. Clouds pass across the sun every day; forest fires occur every few years; severe ice storms or devastating hurricanes every few decades; earthquakes may be centuries apart; and in recent times extensive glaciations have retreated and returned after tens or hundreds of thousands of years. On even longer timescales, catastrophes of vast extent depopulate the entire Earth. These are the mass extinctions. There have been at least three since the metazoan radiation of the Cambrian. The best-known is the end-Cretaceous event when an asteroid strike put an end to well-known groups such as dinosaurs, and ammonites about 65 Mya. The most disastrous occurred much earlier, at the end of the Paleozoic about 250 Mya. The end-Permian event destroyed an entire marine biota, or reduced it to tatters, clearing the stage for a completely different range of species that today dominate the seas of the world.

Most species became extinct during the end-Permian event. It has not yet been established for certain exactly what happened towards the end of the Permian. It seems likely that there were two events a few million years apart that contributed to the catastrophe, although it was the end-Permian event that was decisive. The most likely cause was the vast outpouring of volcanic magma in what is now Siberia, covering a million km^2 of land, but other events may also have contributed. There is good (although not conclusive) evidence that oceans became anoxic. What is not in doubt is that within a very short space of time many major groups disappeared completely and that about 90% of all species became extinct. Some groups that had played a major part in Paleozoic biotas were lost, although some, such as trilobites, had been declining in diversity for some time previously. Coral reefs disappeared from the oceans for 10 My. Graptolites, and eurypterids (sea-scorpions) vanished. Among fish, the acanthodians and placoderms did not survive. Crinoids and brachiopods were very severely pruned, although a few species managed to struggle across the boundary. On land, at least half of all families of tetrapods were lost, with the cynodonts (mammal ancestors; Section 5.7) and archosaurs among the few survivors. Eight orders of insects disappeared.

Evolution restored biodiversity in a different form. The Great Dying of the end-Permian depopulated the oceans and the continents for 10–30 My, until diversity had begun to recover by the end of the Triassic. By the end of the Jurassic marine biodiversity was as great as it had been during the Paleozoic. Although the range of diversity had by then recovered, however, the nature of this diversity had permanently and irreversibly altered. The typical Paleozoic marine fauna of sessile filter-feeding animals such as brachiopods, phoronids, crinoids, and bryozoans had been replaced by more actively foraging groups such as gastropods and starfish. The continuous operation of evolutionary diversification at length filled up all the ways of life that had been vacated, but it did not fill them up with the same groups, or even with ecologically similar representatives of unrelated groups. Graptolites

and trilobites have gone forever; crinoids and brachiopods remain ecologically marginal; sea-urchins and clams have instead proliferated. Evolution is an historical process whose course depends on initial conditions. It is therefore not reversible, and a mass extinction permanently resets the composition and balance of natural communities—until the next mass extinction.

To put the end-Permian event into perspective, imagine an event in the near future that destroys every person, every bird, every spider, and every conifer within a few thousand years. Only a few species of mammals and grasses survive, let us say, although other groups such as frogs and ferns are little affected. The world would be changed completely and forever. The lost species would eventually be replaced, although only after a lapse of some millions of years. But—replaced by what? We cannot yet predict what will happen when a mass extinction presses the reset button. At a time when we ourselves are the likely cause of the next great extinction event, this thought should give us pause.

● CHAPTER SUMMARY

Continued selection produces new species.

Diversification is the natural tendency of lineages.

- Asexual lineages can diversify indefinitely but usually become specialized to a limited number of discrete ways of life.
  - *Asexual dandelions have undergone a miniature adaptive radiation.*
  - *Bdelloid rotifers have evolved into strongly differentiated types.*
  - *Bacteria have very high levels of diversity lumped into ecotypes.*
- Sex hinders divergence.
  - *Species and gender are complementary attributes of sexual eukaryotes.*
  - *Sexual isolation permits ecological divergence whereas recombination prevents divergence.*
  - *Species are the stage at which sexual isolation is sufficient to permit substantial divergence.*

Populations that are permanently separated may diverge through drift or selection.

- Species may vary continuously over their geographical distribution.
- Races are geographically restricted varieties.
  - *Other species are divided into more or less sharply differentiated types living in different regions.*

Divergent selection leads to more or less strongly marked varieties.

- Varieties are readily produced by artificial selection.

- Several ecotypes of a species may live in the same geographical area.
  - *Divergent adaptation generates distinct imperfectly isolated groups within species.*
  - *Parasites may have varieties specialized for infecting particular hosts.*

- Strongly marked varieties are often named as subspecies.
  - *Our lice have evolved with our clothing.*

- New species are continually evolving.
  - *Two nascent species of stickleback have evolved in postglacial lakes.*
  - *There has been a modest radiation of whitefish in northern Eurasia and North America.*
  - *Emerging species can be seen all around us among familiar local organisms.*

- Closely related species often hybridize.
  - *Nascent species of sticklebacks can collapse into a single lineage.*
  - *Species identity can be preserved despite frequent hybridization.*

Species are recognized when ecologically distinctive forms become sexually isolated.
  - *New species may arise through divergent sexual selection or through divergent natural selection.*

- Lineages that do not meet cannot mate.
  - *Populations separated by a physical barrier will tend to diverge over time.*
  - *Populations separated by a physical barrier may become unable to mate with one another.*
  - *Barriers to mating that evolved when populations were apart may keep them isolated when they meet again.*
  - *Sexual isolation may evolve if groups of individuals do not meet despite living in the same area.*

- Long-isolated lineages may not mate successfully when they meet.
  - *Hybrids may be inviable or sexually sterile.*
  - *Hybrid inferiority is caused by genetic incompatibility.*
  - *Hybrids may not develop normally.*
  - *Hybrids may not reproduce normally.*

Rapid speciation gives rise to swarms of sister species.
  - *Hawaiian crickets have evolved rapidly through sexual selection on male song.*
  - *Pacific tree snails illustrate rapid speciation and rapid extinction.*
  - *Relaxing cytogenetic constraints promotes speciation.*

Diversity is the dynamic equilibrium between the origin and extinction of species.

- All species eventually become extinct.

- Species persist for about 1–10 My on average.

- Some species persist for very long periods of time.

- Extinction is balanced by speciation in the long term.

- Mass extinctions are caused by rare events of large magnitude.
  - *Mass extinctions have punctuated the history of life.*
  - *Most species became extinct during the end-Permian event.*
  - *Evolution restored biodiversity in a different form.*

● FURTHER READING

Here are some pertinent further readings at the time of going to press. For relevant readings that have been released since publication, visit the book's Online Resource Centre at
www.oxfordtextbooks.co.uk/orc/bell_evolution/

Rieseberg, L.H. and Willis, J.H. 2007. Plant speciation. *Science* 317: 910–914.

Section 7.1   Barraclough, T.G., Birky, C.W. and Burt, A. 2003. Diversification in sexual and asexual organisms. *Evolution* 57: 2166–2172.

Fontaneto, D., Herniou, E.A., Boschetti, C., Caprioli, M., Ricci, C. and Barraclough, T.G. 2007. Independently evolving species in asexual bdelloid rotifers. *PLoS Biology* 5: 0914–0920.

Section 7.2   Pereira, R.J. and Wake, D.B. 2009. Genetic leakage after adaptive and non-adaptive divergence in the *Ensatina eschscholtzii* ring species. *Evolution* 63: 2288–2301.

Section 7.3   Johannesson, K. 2003. Evolution in *Littorina*: ecology matters. *Journal of Sea Research* 49: 107–117.

Guénet, J-L. and Bonhomme, F. 2003. Wild mice: an ever-increasing contribution to a popular mammalian model. *Trends in Genetics* 19: 24–31.

Kittler, R., Kayser, M. and Stoneking, M. 2003. Molecular evolution of *Pediculus humanus* and the origin of clothing. *Current Biology* 13: 1414–1417.

McKinnon, J.S. and Rundle, H.D. 2002. Speciation in nature: the three-spine stickleback model systems. *Trends in Ecology and Evolution* 17: 480–488.

Bernatchez, L., Chouinard, A. and Lu, G. 1998. Integrating molecular genetics and ecology in studies of adaptive radiation: whitefish, *Coregonus* sp., as a case study. *Biological Journal of the Linnean Society* 68: 173–194.

Section 7.4   Orr, H.A. and Presgraves, D.C. 2000. Speciation by postzygotic isolation: forces, genes and molecules. *BioEssays* 22: 1085–1094.

Section 7.5   Mendelson, T.C. and Shaw, K.L. 2005. Rapid speciation in an arthropod. *Nature* 433: 375–376.

Cowie, R.H. 1992. Evolution and extinction of Partulidae, endemic Pacific island land snails. *Philosophical Transactions of the Royal Society of London B* 335: 167–191.

Section 7.6   Erwin, D.H. 1990. The end-Permian mass extinction. *Annual Review of Ecology and Systematics* 21: 69–91.

Erwin, D.H., Bowring, S.A. and Yugan, J. 2002. End-Permian mass extinctions: a review. In *Catastrophic Events and Mass Extinctions: Impacts and Beyond*, edited by C. Koeberl and K.G. MacLeod, Geological Society of America Special Paper 356, 363–383. Geological Society of America, Boulder, CO.

Sahney, S. and Benton, M. 2008. Recovery from the most profound mass extinction of all time. *Proceedings of the Royal Society of London B* 275: 759–765.

● QUESTIONS

1.   Explain why extremely narrow specialists and extremely broad generalists are unlikely to evolve. Illustrate your answer using economic analogies.

2.   Compare and contrast the diversification of specialized lineages in sexual and asexual organisms.

3.   Argue for and against the proposition that the species is the sole distinctive unit of natural variation, and come to a reasoned conclusion.

4.   Describe with examples how ecologically specialized varieties can evolve and be maintained within a single species.

5.   What ecological processes might explain the distribution of thin-shelled and thick-shelled forms of the same species of snail on a rocky shore?

6. How did the loss of most body hair and the subsequent invention of clothing affect the evolution of ectoparasites of humans?

7. Describe the evolution of open-water (limnetic) and bottom-feeding (benthic) types in freshwater fish.

8. How might hybridization contribute to species breakdown and species formation?

9. Explain how long-continued geographical isolation might lead to species formation.

10. How might one species split into two while occupying the same geographical area?

11. Why are hybrids between species often inferior in vigor to either parent?

12. Describe the ecological and genetic factors that have led to rapid diversification in present-day animals and plants.

13. Why do species become extinct?

14. Provide a reasonable estimate of the average longevity of a species and its variation among species. Why do some species persist unchanged for very long periods of time?

15. How have mass extinctions changed the composition of the world's biota?

16. Are human activities likely to lead to mass extinction?

You can find a fuller set of questions, which will be refreshed during the life of this edition, in the book's Online Resource Centre at **www.oxfordtextbooks.co.uk/orc/bell_evolution/**

8 The Origin of Innovation

How do fundamentally different kinds of organisms evolve?

Sister species are usually very similar because they are closely related. In most cases they look similar and have similar ways of life. In time they will diverge, and each may give rise to new species which diverge further. This cumulation of differences in independently evolving lineages will gradually lead them further and further apart. Any expanding clade will show this tendency. So long as phenotypic divergence is more or less uniform there will be a continuous spectrum of variation among species within the clade, all of which can be recognized as examples of the same kind of organism. From time to time, however, something more remarkable will happen: one sister lineage becomes distinctively specialized and eventually gives rise to a more or less extensive sister clade that can be recognized as a different kind of organism.

Even more than in the case of species the recognition of a different kind of organism is subjective. We can say whether two organisms belong to different clades, at some specified depth of branching, because "clade" is a natural category, accurately marking the outcome of a natural process. We cannot unambiguously assert that two organisms belong to different kinds, because "kind" is an artificial category based on human judgment. It often corresponds with one of the higher ranks—orders, classes, and phyla—of conventional Linnean taxonomy. Even when these are strictly monophyletic, the subdivision of a clade into a limited set of named sister clades is always a matter of human judgment. Despite being subjective, however, it focuses attention on the sharp phenotypic discontinuities between certain clades. Structures such as the water vascular system of echinoderms, the foot of mollusks, or the flower of angiosperms occur in all the members of a particular clade and nowhere else. These are the characters that we described in Chapter 4 as innovations.

Innovation is triggered by new ecological opportunities or by genetic revolutions. We distinguished between implementing innovations, which are themselves the characters responsible for the distinctiveness of a clade, and potentiating innovations, which make it possible for new suites of adaptive characters to arise.

- Implementing innovations evolve when new ecological opportunities arise. From time to time, new conditions of life become available. If no extant group is well-adapted to these conditions, selection will favor the evolution of a new range of specialized types. This may arise when a few groups colonize a remote and isolated site, such as the amphipods of Lake Baikal or the moa of New Zealand (Section 9.3). A more mundane and much more frequent opportunity is provided by rare or recurrent events creating conditions that are not fully exploited by other species, such as a retreating glacial front or a new saltmarsh.

- Potentiating innovations evolve through fundamental genomic changes. From time to time, new genes, combinations of genes and consortia of genes evolve that can form the basis of innovations that permit new kinds of organism to evolve. The amniote egg, cellulose and the extracellular matrix, for example, are characters that enable existing conditions to be more efficiently exploited by new kinds of organism.

In either case the distinctive signature of innovation is that there is often no clear series of intermediate forms linking a clade which possesses an innovation and its sister clade which lacks it. The evolution of new kinds of organism therefore involves two kinds of process: the origin of innovations, and the loss of intermediate forms.

Modification, duplication, and fusion are the routes to innovation. Three kinds of genetic process might contribute to the evolution of an innovation.

• Modification. The accumulation of change through the selection of successive beneficial mutations can result in a complex, integrated structure quite different from the homologous structure in the sister lineage. This will usually occur gradually over long periods of time.

• Duplication. If there is only one version of an essential gene it cannot readily be modified because this would severely impair its normal function. Once it is duplicated, however, the copy may diverge, and acquire new functions, while the ancestral version continues to support its original function. This may occur quite rapidly because the usual constraint on modification, that the ancestral function be maintained, is relaxed.

• Fusion. The partial or complete fusion of genomes or cells belonging to long-separated lineages creates a new composite organism. This is free to evolve in new and unexpected directions. This can happen almost instantaneously, on the very rare occasions that fusion results in a viable individual.

All three processes are very common events. Modification always accompanies divergence; the duplication of some locus in the genome is likely to occur in most generations; and in many bacterial lineages, at least, fusion is common because they will incorporate foreign genetic material quite readily. In most cases modification will be minor, duplicate copies will disappear, and fusion products will be inviable. Once in a while, however, a new kind of organism will appear, flourish, and give rise to a distinctive new clade.

All three processes also correspond to similar processes involved in the origin of new genes. They are simply operating on larger scales: more time for modification, greater scope of duplication, more extensive fusion. There is no essential difference in the processes that fuel the evolution of genes, species and kinds: variation and selection. It is the ecological and genetic scales on which these processes occur that determine the difference in the magnitude of their outcomes.

8.1 New structures evolve from old structures.

In previous chapters we have emphasized that new genes evolve from pre-existing genes—however rearranged and modified—while new species likewise evolve from common ancestors. The same principle continues to operate for new kinds of organism, no matter how different they may seem. The innovations that mark the distinctive nature of a clade have without exception evolved by the modification of a pre-existing ancestral state. Some innovations evolve quite readily, especially when similar phenotypes can be produced by a variety of genetic mechanisms. At the other extreme, some innovations are unique, evolving only once in the history of life despite the long-continued existence of structures that could have served as starting points. Multicellularity and C4 photosynthesis are examples of innovations that

have evolved repeatedly; feathers are an example of an innovation that has evolved only once.

8.1.1 Multicellular organisms evolve from unicellular ancestors through a series of intermediate forms.

The Volvocales is a remarkable group of green algae that provides clear examples of the transition from unicellular organisms, though colonial organisms with definite form but no cellular differentiation, to multicellular organisms with different kinds of cell. The main stages in this transition are illustrated in Figure 8.1. We shall use them to illustrate the main features of how complex, integrated organisms evolve from much simpler predecessors.

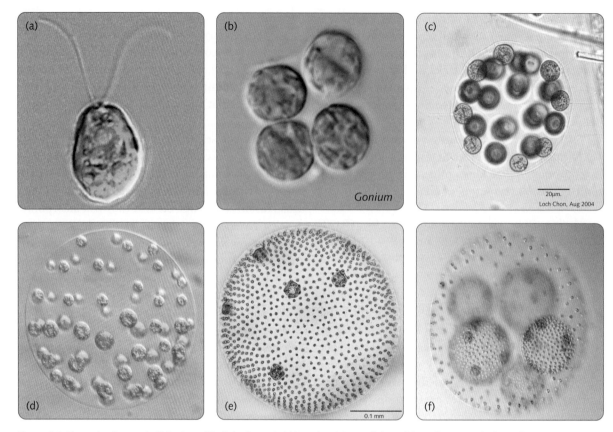

Figure 8.1 The series from unicellular to multicellular forms in Volvocales: (a) unicellular: *Chlamydomonas*; (b) few cells: *Gonium*; (c) many undifferentiated cells: *Eudorina*; (d) many cells, soma-germ distinction: *Pleodorina*; (e) large colony with pronounced soma–germ division of labor: *Volvox*; (f) three generations in a single *Volvox* colony.

(b) provided by Peter A. Siver, Chrysophytes LLC. Image (c) is taken from *The Algal Web* with permission from Dr John Kinross. Images (d) and (f) kindly provided by Prof. Dr. Armin Hallmann, University of Bielefeld. Image e) is © Steffen Clauss.

- Some members of the Volvocales are unicells, all of which reproduce by producing unicellular offspring. For example, *Chlamydomonas* is a normal unicellular chlorophyte that reproduces by an equal fission. The family to which it belongs, however, contains a unique range of colonial and multicellular forms.

- Some are colonies, with distinct individuality, in which all cells reproduce by producing multicellular offspring. The smallest and simplest forms, like *Basichlamys*, *Gonium*, and *Pandorina*, are regular colonies consisting of 4–32 cells, each of which is similar in appearance to *Chlamydomonas*. Each cell in the colony reproduces by producing a miniature colony of the same form, which swims through the water propelled by flagella.

- Others are colonies with two kinds of cells, somatic cells and germ cells. Only the germ cells reproduce, producing multicellular offspring with the same division of labor. Larger species of *Eudorina* and *Pleodorina*, with 32–128 cells, show the first indications of a division of labor between a sterile caste of somatic cells and a smaller number of germ cells, which alone reproduce. This tendency is taken further in *Volvox*, which has a spherical body made up of 1000 or more cells. Most are small, flagellated somatic cells, which line the surface of the sphere and are responsible for photosynthesis and locomotion. A few are much larger, immotile germ cells, which develop inside the parental spheroid into miniature *Volvox* (themselves with germ cells) before being released.

This illustrates the primary division of labor among cells within the body of a multicellular individual, between somatic cells responsible for vegetative function, and germ cells that are responsible for reproduction. You may ask how selection can favor the evolution of sterile cells that never reproduce, as it will normally favor the greatest possible rate of replication. The reason is that dividing labor maximizes the productivity of the individual as a whole, so that the output of the germ line more than compensates for the sterility of the soma. Hence, the alleles that govern this division of labor will be replicated faster than alleles that direct the development of an equivalent number of unicells.

A few major genes direct development. An asymmetric division early in colony growth creates the distinction between small somatic cells and much larger germ cells, or gonidia. The gonidia develop inside the parental spheroid into miniature *Volvox* (themselves with gonidia) before being released. This crucial step in constructing a differentiated adult involves three developmental genes.

- The crucial asymmetrical division is regulated by *gls*, and if this is inactive all cells develop as somatic cells and the adult is sterile.

- Following this division, *regA* directs the expression of somatic genes in the small cells.

- *lag* directs the expression of gonidial genes in the large cells.

RegA mutants at first develop normally but the somatic cells eventually resorb their flagella and redifferentiate as gonidia, so the colony consists entirely of small germ cells. *RegA* appears to regulate the expression of nuclear genes encoding chloroplast proteins and thereby keeps the somatic cells small by preventing them from synthesizing chloroplast material. *RegA* mutants which also have defects in colony matrix synthesis fall apart and can be cultured as unicells. Hence, the range of forms between unicells and differentiated multicellular organisms can be reconstructed in this group by using the appropriate combinations of mutations.

The multicellular body evolves through a series of intermediate forms. Transforming a unicell such as *Chlamydomonas* into a complex multicellular organism such as *Volvox* requires several major changes in organization.

- The growing clone must be held together. This is achieved by incomplete cell division, which leaves cytoplasmic bridges between daughter cells, and by the transformation of cell walls into an extracellular matrix.

- Cell differentiation requires loss of reproductive ability by somatic cells and loss of motility by germ cells.

- There must be a regular developmental program with a fixed endpoint. This involves limiting the number of cells in the colony; performing an inversion of the growing embryo (in a fashion resembling gastrulation in animals) so that the flagella of the somatic cells face outward; and determining the fate of somatic and germ cell lineages through an asymmetrical cell division.

These changes are most unlikely to happen all at once. By mapping the characters required for multicellularity onto a phylogenetic tree for the Volvocales estimated from chloroplast gene sequences, as shown in Figure 8.2, we can trace how multicellularity has evolved step by step in this group.

The closest unicellular relatives of *Volvox* all reproduce by dividing several times within the same cell wall, producing a group of identical daughter cells held physically in close proximity to one another. Hence, the evolution of multicellularity is more likely to happen in Volvocales than in groups with strictly binary fission.

- The first step in the transition to multicellularity is the transformation of the cell wall into an ECM. Again, this is more likely to occur in unicellular Volvocales than in other algae, because they possess a cell wall in three layers. In some species, the middle layer is expanded and gelatinous, resembling a rudimentary ECM. At about the same time, the number of cell divisions during development was fixed, regulating the number of cells in the adult colony.

- The second step was the retention of cytoplasmic connections between cells, through incomplete cell division. This permits morphogenetic

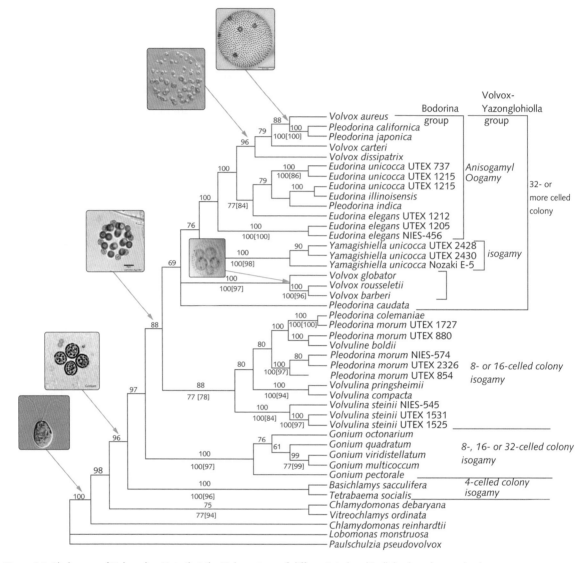

Figure 8.2 Phylogeny of Volvocales. Note that the *Volvox* stage of differentiated multicellular form has evolved at least twice in this group.

movements during development, and communication between cells.

- The third step was the inversion of the growing individual during development, which was made possible by the prior formation of cytoplasmic connections between cells. The ECM was greatly expanded at about the same time, allowing greater flexibility in development.

- The fourth step was the restriction of reproduction to a particular group of cells, creating a sterile caste of somatic cells.

- The final step was the increased specialization of the germ cells, which become immotile and move into the interior of the cell, through a developmental program involving an asymmetric cell division that partitions somatic from germinal lineages.

Hence, the Volvocales provide an excellent example of how a major transition, from unicells to differentiated multicellular individuals, can evolve through a stepwise sequence of changes, each of which can be illustrated in extant species.

Multicellularity has evolved independently in several lineages. If you look at the phylogeny of the Volvocales in more detail, however, it quite clearly shows that multicellularity has not evolved through a uniform trend leading towards increased complexity of organization. There are two main complications.

- The first is that natural selection does not enforce a uniform tendency towards higher levels of organization. Characters can be lost, as well as gained. Lineages with expanded ECM and advanced division of labor have sometimes evolved towards a simpler colonial structure.

- Secondly, multicellularity has evolved independently in several lineages. *Volvox* turns out to be a grade of construction, rather than a monophyletic clade. The *Volvox* grade can be reached by different paths, only one of which has involved a developmental program with asymmetric cell division during development separating soma from germ.

8.1.2 C4 photosynthesis is a novel mechanism that evolves from the C3 state.

One of the most abundant proteins on Earth is an enzyme, ribulose-1,5-bisphosphate carboxylase/oxygenase (always called Rubisco for short) which can make up 30% of all the soluble protein in a plant. It is abundant for two reasons. The first is that it is critical in photosynthesis: it catalyzes the addition of CO_2 to ribulose phosphate to make 3-phosphoglycerate, which is then transformed into sugar. This is the crucial reaction by which plants and algae convert atmospheric CO_2 into organic carbon compounds that supports most of life on Earth. The second is that it is very inefficient. Rubisco works about a hundred times more slowly than most other enzymes, and therefore has to be synthesized in great quantity to be effective. Even worse, it is readily reversible: if exposed to O_2 it will catalyze the oxidation of ribulose phosphate to phosphoglycolate, which is metabolically useless and has to be recycled to recover its carbon with release of CO_2—a process called photorespiration because it is akin to the respiration of animals. Rubisco binds to CO_2 more strongly than to O_2, but since O_2 is more than 500 times as abundant as CO_2 in the atmosphere, respiration can compete with photosynthesis for the active site of the enzyme. This is a dramatic example of imperfect adaptation, perhaps stemming from the evolution of Rubisco some 2800 Mya in cyanobacteria when the atmosphere had very high levels of CO_2 and very low levels of O_2.

Rubisco is particularly inefficient in hot or dry places, so the plants that grow there will tend to evolve adaptations that make photosynthesis more efficient and that reduce the rate of photorespiration. The only way of doing this is to pack as much CO_2 as possible close to the active site of Rubisco. This is done by creating a compartment for the Rubisco, then carrying CO_2 into it, so that the local concentration saturates the active site and elevates the rate of photosynthesis. A useful way of envisaging the process is that it recycles the CO_2 that would otherwise be lost in respiration back into photosynthesis. Only about 3% of plants have evolved this capacity, but they are disproportionately important in the global economy: the largest group is the grasses, which includes many of our most important crop plants, such as wheat and maize, and which collectively contribute about a quarter of global plant production. Almost all of them share two features, which are illustrated in Figure 8.3. The first is the formation of a kennel for Rubisco. This is the "bundle sheath," a prominent sheath of cells surrounding each leaf vein. The second is a reaction sequence to pump CO_2 into these sheath cells. The crucial enzyme is PEP carboxylase, which fixes CO_2 into 4-carbon acids in mesophyll cells that are then transported into the bundle sheath cells where they release CO_2 within their chloroplasts, in close proximity to Rubisco. The consequence is that Rubisco can be saturated with CO_2 even when the leaf stomata are shut down because it is hot and dry. This CO_2 is then fixed into 3-carbon acids as in other plants. It is this ability to pump CO_2 around Rubisco that distinguishes "C4" from normal "C3" plants.

C4 photosynthesis evolves by the modification of pre-existing C3 mechanisms. Most of the enzymes which drive the C4 shuttle (such as PEP carboxylase) are present in C3 species, where they support other functions. It is not the enzymes themselves so much as their expression in particular tissues that drives the C4 system. These enzymes are modified in the normal

(a)

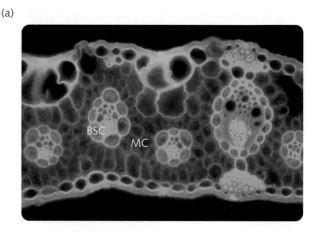

(b)

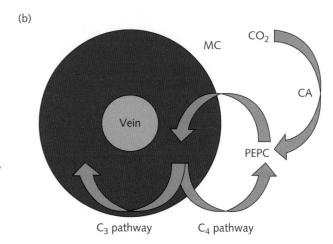

Figure 8.3 C4 photosynthesis. (a) Vertical section through a leaf to show the typical anatomy of a C4 plant: BSC marks a bundle sheath cell and MC marks a mesophyll cell. (b) Schematic diagram of the conversions taking place in the different compartments of the leaf in C4 plants. CA, carbonic anhydrase; PEPC, PEP carboxylase.

(a) reproduced from P.W. Hattersley, in Watson, L., and Dallwitz, M.J. (1992 onwards) The grass genera of the world: descriptions, illustrations, identification, and information retrieval; including synonyms, morphology, anatomy, physiology, phytochemistry, cytology, classification, pathogens, world and local distribution, and references. Version: 5 February 2014. http://delta-intkey.com. (b) reprinted from Christin et al. Can phylogenetics identify C4 origins and reversals? *Trends in Ecology and Evolution* 25 (7). © 2010, with permission from Elsevier.

way, by the substitution of beneficial mutations, to optimize their activity in C4: for example, the substitution of serine at position 780 (of the maize sequence) in PEP carboxylase is essential for C4 function, along with other changes.

Rubisco evolves within the C4 environment. Rubisco has a very complex tertiary structure, and its activity may be modified by changes far from the active site. The switch from C3 to C4 photosynthesis alters the optimal configuration of the molecule and has led to a succession of substitutions that improve its specificity for CO_2. Many of these substitutions have occurred independently in different C4 lineages.

C4 photosynthesis evolves through a series of intermediate states. In C3 plants, photosynthesis is carried out in the mesophyll cells of the leaf. In C4 plants there is a division of labor between mesophyll and bundle sheath cells. Intermediate stages would have only a partial C4 cycle, or an incomplete partition of enzymes between the two tissues. These intermediate stages not only exist, but may be demonstrated by related species within the same genus. For example, *Flaveria* ("goldentops," a genus in the daisy family Asteraceae) has not only C3 and C4 species but also species representing intermediate stages. Figure 8.4 shows the procession of stages in the evolutionary history of the genus.

(1) The most deeply branching species (such as *F. pringlei*) have the usual C3 system, which is therefore the ancestral character state for the genus. There is no bundle sheath distinct from the mesophyll, and no biochemical mechanism to decarboxylate C4 compounds.

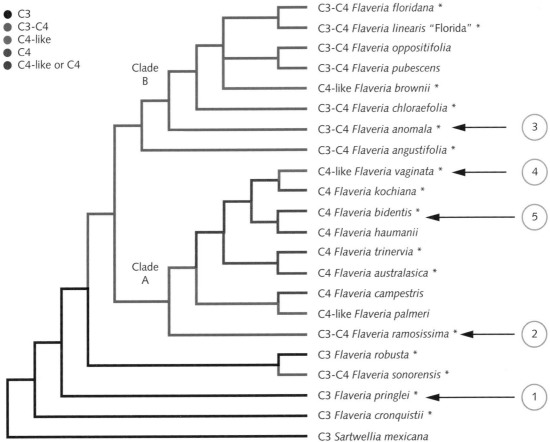

Figure 8.4 Evolution of C4 photosynthesis in *Flaveria*. The numbers correspond to the stages described in the text.

Photograph Courtesy of Forest & Kim Starr. This file is licensed under the Creative Commons Attribution-Share Alike 3.0 Unported license. Diagram reproduced from McKown and Dengler. 2007. Key innovations in the evolution of Kranz anatomy and C4 vein pattern in Flaveria (Asteraceae). *American Journal of Botany* 94 (3), with permission from Dr Athena McKown.

(2) The first stage in the evolution of the C4 system is shown by species (such as *F. ramosissima*) that show some differentiation of mesophyll and bundle sheath but have only a rudimentary C4 pathway.

(3) The next stage is further differentiation of mesophyll and bundle sheath, with a partial C4 cycle, as in *F. anomala*.

(4) Other species, such as *F. vaginata*, have a full C4 cycle but have not yet fully segregated the photosynthesis enzymes between mesophyll and bundle sheath.

(5) Finally, species such as *F. bidentis* have a fully functional C4 system.

These morphological stages are paralleled by intermediate biochemical stages. The decarboxylation of the 4-carbon oxaloacetic acid generated by PEP carboxylase is accomplished by NADP malic enzyme, which converts it to malate. C3 and C4 species have different versions of this enzyme, with intermediate species having a mixture of the two, together with an intermediate form. Hence, the C4 phenotype evolves through a series of morphological and biochemical intermediates.

There are several different outcomes of selection for C4 function. The NADP malic enzyme system of *Flaveria* is only one of the routes to C4 photosynthesis. There are two other decarboxylation enzymes that can be recruited: NAD malic enzyme (oxaloacetic acid is converted to aspartate) and PEP carboxykinase (forms PEP, which returns to the mesophyll). Moreover, there are many kinds of leaf anatomy able to support C4 photosynthesis. An extreme version is to create compartments for PEP carboxylase and Rubisco within the same cell, leading to a single-cell version of C4 photosynthesis.

The transition from normal C3 to C4 photosynthesis is a biochemically well-understood process that neatly illustrates several general principles: exaptation (C4 systems evolve from prior C3 systems), interaction (the structure of Rubisco evolves as a consequence), gradualism (intermediate stages connect C3 ancestors with C4 descendants), and historicity (similar phenotypes evolve along different genetic routes).

8.1.3 Feathers have evolved from epidermal structures of dinosaurs.

The C3–C4 and unicellular–multicellular transitions are also examples of innovations that evolve quite readily, at least in some groups. Other innovations, despite some pre-existing tendency, evolve very seldom, and some are unique. Feathers are a good example.

Feathers evolved from keratin structures in the epidermis of amniotes. Amphibians have a naked skin, whereas other tetrapods develop structures that project above the epidermis, providing protection or insulation or a range of other benefits that enhance adaptation to a completely terrestrial way of life. Reptiles have scales or scutes, mammals have hair, and birds have feathers. These are all epidermal structures made of the protein keratin, in contrast to the collagen fibers of the underlying dermis. Feathers are the most complex. The simplest feather-like structure would be a hollow pointed cylinder of keratin, growing from a follicle produced by the invagination of the epidermis into the underlying dermis (hairs grow as solid columns of keratin from a similar kind of follicle). The development of feathers suggests how they evolved from this simple origin, as illustrated in Figure 8.5. Several similar cylinders arising from the same base would produce a fluffy structure like a down feather. Alternatively, the branching of a central axis of growth, the feather vane, would result in a palm-leaf structure; branching of the branches and then a repeated branching of these sub-branches yields a broad flat surface whose elements are stuck together like Velcro. If this structure is slightly asymmetrical, so that its surface is curved with the central axis displaced to one side then it will generate lift when moved up and down—a wing, in fact. Among living organisms, feathers are known only from birds, and it is their complex asymmetrical wing feathers that enable them to fly. They are unmistakably an innovation; how did they evolve?

Feathers evolved from projecting epidermal fibers in dinosaurs. Until very recently it would not have been possible to provide a convincing answer to this question. Feathers are preserved only in very unusual

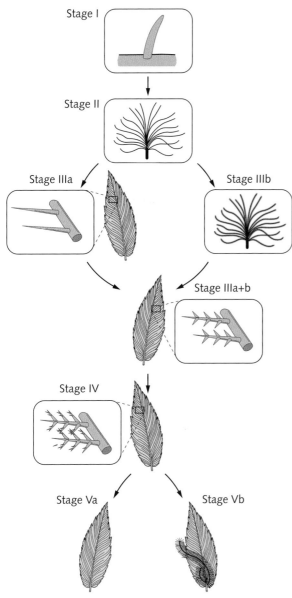

Stage I

Stage II

Stage IIIa

Stage IIIb

Stage IIIa+b

Stage IV

Stage Va

Stage Vb

Figure 8.5 Stages in the evolution of feathers from a simple cone.
Redrawn with permission from Prum, R. O. (1999). 'Development and Evolutionary Origin of Feathers', *Journal of Experimental Zoology*. 285: 291-306

transition between reptiles and birds. In the last fifteen years, however, an extraordinary series of feathered dinosaurs has filled in most of the gaps. Most have been found in Liaoning in northeastern China, where they were quickly buried by volcanic ash. Some of these have only filamentous feathers, which cannot have contributed in any way to flight and which must therefore have served some other function. The most likely possibility is that they provided thermal insulation, like the hair of mammals. Others have vaned feathers, but were too large to fly, had short forelimbs, and bore feathers only in positions (such as the tail) where they could not have formed a wing. A few had the potential for flight. *Microraptor* had vaned feathers on forelimbs, hindwings, and tail; it could certainly have glided, and might have flown if it could have flapped its limbs fast enough, although it still lacked the enlarged sternum that the wing muscles are attached to in modern birds. *Archaeopteryx* comes close to the end of this transition, lacking only a horny toothless beak and the reduction of the tail to a stump to transform it to a modern bird.

These recent discoveries have provided the series of intermediate forms between dinosaurs and birds that fills in the history of this transition. They have also shown that there was nothing inevitable about the transition. Dinosaurs seem to have had a predisposition to express feather-like structures in their skin, and did so for a very long while before these structures were co-opted for gliding and flight. There is some evidence that even dinosaurs such as the gigantic *T. rex* may have been feathered, at least as juveniles. It was only when aberrant versions of a warm or waterproof down were expressed in much smaller animals that any approach to flight became possible. Feathers are another example of exaptation: the recruitment of a structure evolved for mundane reasons that by chance creates unanticipated potential for evolving a completely new way of life.

circumstances—rapid burial in anoxic sediments that give rise to fine-grained laminated rocks. The famous late Jurassic (about 150 Mya) fossil *Archaeopteryx* has been known for over a century as an intermediate stage between archosaurs and birds, possessing bird-like characters such as feathers but also crocodilian features such as an elongated (and feathered) tail and toothed jaws. It was about the size of a blue jay. For many years it stood almost alone as a witness of the

The transitions from unicellular to multicellular form, from C3 to C4 photosynthesis, and from scales to feathers illustrate some of the fundamental features of cumulative change driven by natural selection: some initial degree of exaptation, a series of intermediate stages, and a range of outcomes. Much more modest episodes of evolution, such as the experimental evolution of amide metabolism (Section 13.5.1) share the same set of features. Thus, innovations that constitute major shifts in morphology or physiology can evolve through the same mechanism that produces adaptation over short periods of time in the field or the laboratory.

8.2 Duplication provides an opportunity for functional divergence.

Duplication offers the possibility of retaining the ancestral state of one copy of a structure while the other copy is modified to carry out some different activity. This principle applies at every level of structure, from genes through cells and organs to whole bodies.

- Colonies. The Portuguese Man O'War *Physalia* is a fearsome marine animal trailing poisonous tentacles many meters long suspended from a float that is blown by the wind across the surface of the sea. It can inflict painful or even fatal injuries on swimmers who come into contact with it, and is dangerous even when stranded on the beach. The tentacles are poisonous because they bear cnidocytes, the characteristic stinging cells of all cnidarians such as jellyfish, sea anemones, corals, and hydras. Like corals, its body is made up of many interconnected units called zooids, each of which has its own complement of organs. Unlike corals, the zooids have become specialized for different tasks: feeding zooids, reproductive zooids, protective zooids, and the zooids that form the float. Its modular construction has made it possible for a compound organism to evolve by the division of labor among its component parts.

- Individuals. In an organism such as *Volvox* the duplication of cells creates the potential for some to differentiate as soma and others as germ. In larger multicellular organisms, such as seaweeds, plants, fungi, and animals, the somatic cells may themselves become further differentiated for capturing and processing food, respiring, excreting, reproducing, and so forth.

- Genomes. Whole genomes can be duplicated by a division of the chromosomes without the usual division of the cell, or by hybridization between related species. Bread wheat, for example, is a hexaploid formed by successive hybridization among three diploid species of wild grasses.

- Genes. Two copies of a gene are readily created by errors occurring during replication, such as unequal crossing-over.

At any level, duplication creates redundancy. Redundant structures are free to diverge because a novel function can evolve without damaging the original function. This simple principle supplies the most important source of innovation in evolution.

8.2.1 Gene duplication can lead to novel capabilities.

New genes can evolve from simpler predecessors through gene duplication. This happens quite often through crossing-over during meiosis between chromosomes that are not correctly aligned, leading to a homologue with two copies of a gene while the other homologue has none. The duplicated copies are then closely linked, in the same region of the chromosome. When this happens repeatedly, a string of several copies of an ancestral gene can be produced.

As only a single copy is necessary for normal function in the ancestor, a duplicate is redundant, and often deteriorates through mutation until it becomes inactive and is eventually lost. However, it also has the opportunity to evolve new functions, because it can be modified without any loss of the original function, which is still supported by the ancestral copy of the gene. Hence, duplication followed by divergence is an important route for the evolution of new kinds of gene.

Visual pigments originate through gene duplication. We are visual animals who know the world around us largely through our eyes. At a glance, we can recognize the size, direction of movement, velocity, shape, and color of an object, and this information is passed to the brain for processing within a few milliseconds of perceiving it. This involves a great deal of morphological infrastructure (eyes and their associated tissues) and physiological processes (conduction and interpretation of nerve impulses). The initial event in vision, however, is always the absorption of a photon by a pigment molecule, which sets off the cascade of events resulting in an internal representation of the external world.

The visual pigments of animals are the opsins, which span the cell membrane of the rod and cone cells of the retina. Each opsin molecule is bound to a non-protein molecule called retinal, derived from vitamin A. Retinal changes its shape when it absorbs a photon of light, and in doing so it changes the shape of its associated opsin. This change is transmitted to another protein, setting off a signal cascade that eventually results in the propagation of a nervous impulse. This will enable an organism to respond to the difference between light and dark. Coupled with an eye, it will enable it to form a monochromatic image of its surroundings.

The rod cells of the retina contain rhodopsin, an extremely sensitive molecule that makes it possible to detect movement and shape in dim light. The cone cells contain different kinds of opsin that are less sensitive and work best in bright light. A small change in the structure of the opsin molecule alters how retinal is bound and thereby alters its sensitivity to light of different wavelengths, causing it to absorb light more strongly at one wavelength than another. Hence, an organism bearing two kinds of opsin can distinguish between two wavelengths of light by receiving impulses from two different kinds of receptor cell.

Vertebrates have four kinds of cone opsin, which absorb violet, blue, green, and red light. This is the basis of color vision. They are encoded by very similar genes that have clearly arisen by duplication, with subsequent slight modification of a small number of amino acids in one of the transmembrane domains of the protein creating differences in absorption. They diverged very early in chordate evolution, probably in the Cambrian; surprisingly, the dim-vision pigment rhodopsin appears to be the most recently-evolved molecule, as shown in Figure 8.6. Natural selection alters the visual capabilities of animals through the ratio of rods to cones or through the range of available pigments. Nocturnal mammals may have 90% or more rods, while diurnal mammals have 90% cones. The number of functional cone opsins is quite variable, because of the frequent degeneration of opsin genes in circumstances where color vision is unnecessary. Most mammals, for example, have only two types, and therefore only dichromatic vision, because mammals have been small nocturnal animals for most of their history (Section 5.6). Lemurs and tarsiers are examples. By contrast, catarrhine primates (Old World monkeys, apes, and humans; Section 4.4) have three, because of a duplication of the gene encoding the long-wavelength opsin on the X-chromosome; this gives us trichromatic vision, which is useful to diurnal animals foraging most of the time in bright light. Platyrrhines (New World monkeys) have an odd intermediate position: they usually have three allelic versions of a long-wavelength gene, so that individuals may have either dichromatic or trichromatic vision depending on the combination of alleles they inherit. Hence, gene duplication introduces potential variation that is molded by selection into different visual systems adapted to particular ways of life.

(a)

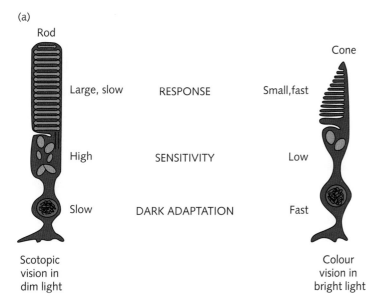

Rod

Large, slow RESPONSE Small, fast

High SENSITIVITY Low

Slow DARK ADAPTATION Fast

Cone

Scotopic vision in dim light

Colour vision in bright light

Figure 8.6 Rod and cone photoreceptors in vertebrates. The phylogenetic tree shows the radiation of the rod opsin (rhodopsin) and the four cone opsins. The deepest split is between the long-wavelength (red) cone opsin and all others; rhodopsin is a recently derived pigment. RH1, rod opsin (rhodopsin); RH2, rod-opsin-like cone opsin (green); SWS1, shortwave-sensitive cone opsin 1 (UV-violet); SWS2, shortwave-sensitive cone opsin 2 (blue); LWS, longwave-sensitive cone opsin (red).

(a) Reproduced from Shichida and Matsuyama. Evolution of opsins and phototransduction. (b) Reproduced from Collin, S.P., et al. The evolution of early vertebrate photoreceptors. Both from *Philosophical Transactions B* 364 (1531) © 2009, The Royal Society.

(b)

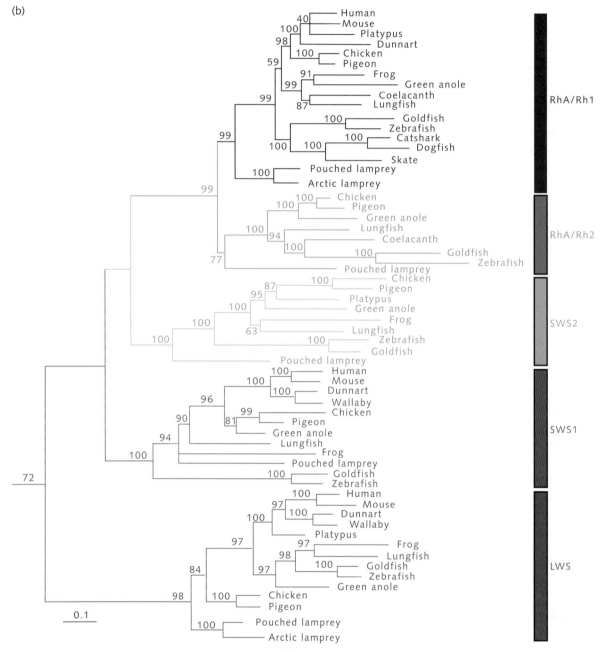

Figure 8.6 *(Continued)*

The visual pigments are themselves only a small part of the diversification of opsin proteins. The pineal "eye" of reptiles, for example, is a light-sensitive organ containing another variety of opsin called parietopsin capable of discriminating between different wavelengths of light. More generally, there is a large range of non-visual but light-sensitive opsins that regulate circadian rhythms and other physiological processes that respond to light. All in all, the opsin family comprises about a thousand different proteins, all originating through divergent selection on duplicate copies of ancestral genes. As for the ancestral genes themselves, they evolved, as all new genes do, from old genes: the opsins used as light-intercepting pigments in animal eyes have evolved from similar proteins in bacteria, where photon capture is used to generate energy.

Whole-genome duplication can lead to adaptive radiation. The yeast radiation has involved gene gain and loss, including whole-genome duplication. In the lineage leading to baker's yeast, *Saccharomyces cerevisiae*, the whole genome was duplicated, followed by the piecemeal loss of about 90% of the duplicated genes. In lab stocks, large segmental duplications arise spontaneously at a rate of about 10^{-10} per division. Individually duplicated genes can be generated by transposition. Moreover, experimental lines often show high levels of aneuploidy. The yeast genome, which is the best-known of all eukaryote genomes, is in a perpetual state of flux.

8.3 The fusion of dissimilar cells has led to new kinds of eukaryote.

The loss of the cell wall gave the earliest eukaryotes the ability to engulf and digest bacterial prey. This unlocked the potential for a new way of life as a predator, but it also created risks. In some cases, the prey might evade digestion and proliferate within the cell that had eaten it as a parasite. Very rarely indeed the two might evolve a symbiotic relationship and thereafter form a compound organism in which both entities were perpetuated. All modern eukaryotes constitute one such ancient partnership, with a heterotrophic bacterium whose descendants are mitochondria. Many have a second partnership with an autotrophic bacterium whose descendants are chloroplasts. The eukaryotes are a novel kind of organism originating some 2000 Mya after the appearance of the first cells through the fusion of dissimilar partners.

Endosymbiosis occasionally evolves through the survival of prey. *Pelomyxa* is a very large amoeboid organism living in the mud at the bottom of ponds—it was originally found in the Elephant Pond at Oxford, where the carcasses of large animals were left to decompose until their skeletons could be prepared for the University Museum. It has no mitochondria, but a dense cellular population of symbiotic bacteria. *Paulinella* belongs to a group of amoebae that build a protective test. Most feed by engulfing bacteria with pseudopodia sent out from an opening in the test, but *Paulinella* has a resident cyanobacterium that functions as an actively photosynthetic symbiont. Even metazoans can evolve such partnerships—some corals, hydras, and flatworms have endosymbiotic algae, and subsist largely on the carbon they fix. These examples show that symbiotic associations between eukaryotes and bacteria evolve quite readily.

Indeed, this process can sometimes be observed in the laboratory. Two French biologists observed a remarkable instance of the evolution of a novel partnership when their cultures of *Amoeba* became accidentally infested by an unknown bacterium able to multiply inside the host cells. The bacterium was initially pathogenic, killing most of the amoebas it infected. The bacteria released by the lysis of the dead host would then infect other amoebas. A few hosts survived this infection, though retaining a residual population of bacteria in their cytoplasm. When these amoebas divided, both daughters inherited an intracellular population of bacteria, so the bacteria were necessarily vertically transmitted with their host. The evolution of a specific relationship between the amoebas and the bacteria could be demonstrated by reciprocal transplant experiments: bacteria extracted from the perennially infected host lines were pathogenic in naive hosts, and unevolved bacteria were pathogenic to evolved hosts whose resident bacteria had been cleared out by antibiotics. Moreover, in some lines this relationship became obligate after about a hundred host generations—the amoebas were no longer able to grow successfully without the bacteria. This shows how it is possible to reconstruct the initial stages of the evolution of highly-modified bacterial endosymbionts such as chloroplasts and mitochondria in the laboratory.

Eukaryotes incorporate two radical innovations originating by fusion. The most recent common ancestor of modern eukaryotes had a single organelle, the mitochondrion. It is the site of respiration, where ATP, the energy currency of the cell, is generated by transferring electrons to oxygen. This is a very important process indeed; the turnover of ATP in the mitochondria of organisms like us is roughly equal to our body weight per day. Some modern groups, such as *Giardia*, the "beaver-fever" pathogen you may acquire if you drink from northern streams that have drained through a beaver pond, have no mitochondria, or only highly degenerate derived structures. These groups are all internal parasites that have lost the mitochondria of their ancestor. The origin of mitochondria from bacteria is clearly indicated by two universal characteristics. The first is a separate genome that resembles a bacterial genome, although it is much smaller. The second is the enclosure of the mitochondrion within two sets of membranes, one representing the original bacterial membrane and the other the vacuolar membrane of the host cell. The metabolism and replication of eukaryote cells therefore still bear the stamp of their origin. In the first place, the mitochondria replicate independently of the nucleus and without using the machinery of centromeres and microtubules. Secondly, proteins that need to cross the membrane system separating the host cytoplasm from the mitochondrion must normally be tagged with a transit peptide which is then removed once the journey is completed. Hence, eukaryote cells are compound organisms that rely on the cooperation between partners.

Partners may not have the same interests. The nuclear and mitochondrial genomes have a common interest, the propagation of the cell or individual. More precisely: genes in either the nucleus or the mitochondrion that enhance the propagation of the cell or the individual will tend to spread as their lineage replaces others with less effective genes. This argument is flawed, however. For the sake of simplicity, consider the mitochondria of a unicellular organism such as a yeast or a diatom. Suppose that there were an allele of a mitochondrial gene that increased the rate of replication of the mitochondrial genome while reducing the contribution that the mitochondrion makes (through oxidative metabolism) to the wellbeing of the cell as a whole. This allele will tend to spread through the population of mitochondria within the cell (because it replicates faster than average), whereas it will tend to be eliminated by competition between cells (because cells with a high frequency of this allele replicate more slowly than average). Conversely, any element in the nuclear genome that can destroy the mitochondrial genome will tend to spread (because its lineage will increase more rapidly without the necessity for supporting mitochondria) provided that mitochondria are not needed (because metabolism is anaerobic). Hence, the interests of independent replicators may differ, even when they are components of the same organism (see Section 11.4).

The conflict of interest between replicators within the cell is bound to hazard their joint wellbeing. It will be partly resolved by any process (such as genetic recombination) that shifts genes from one compartment to the other. This movement is almost entirely from the mitochondrial genome to the nuclear genome. Hence, the mitochondrial genome tends to become much smaller than that of its bacterial ancestor. In most eukaryotes it encodes little more than the set of proteins responsible for the generation of high-energy phosphate bonds used in respiration. Parasites may not need even this residual function, and the whole of their mitochondrial genome is then transferred to the nucleus.

Mitochondria all descend from a single bacterial ancestor. All eukaryotes have mitochondria, except for a few unicellular parasites that have secondarily lost their mitochondria. Consequently, the most recent common ancestor of Eukaryota already possessed a mitochondrion that had evolved from a captured bacterium, probably a planktonic marine α-proteobacterium. The original invasion of the ancestral host cell—whose nature is still unknown—and the subsequent series of invasions that have occurred during the radiation of eukaryotes are sketched in Figure 8.7.

Plastids have evolved along many routes. Plastids have a much more complicated ancestry than mitochondria. The chloroplasts of green algae have descended directly from a cyanobacterial endosymbiont because they have two membranes. All land plants—mosses, ferns, and seed plants—have inherited this chloroplast. The plastids of red seaweeds have a similar structure and probably descended from the same ancestral cell. Hence, photosynthetic eukaryotes all descend from a single ancestral partnership between a eukaryotic cell (with mitochondria) and a cyanobacterium.

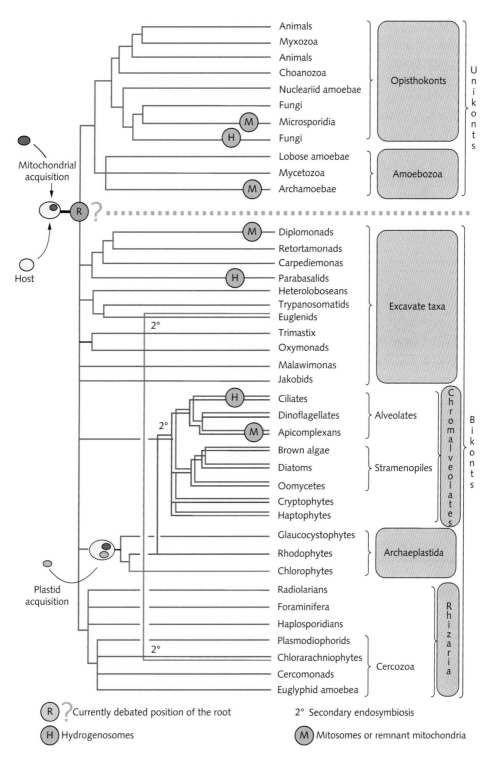

Figure 8.7 Acquisition of mitochondria and plastids by eukaryote lineages. The original compound cell resulting from the fusion of a host cell with the mitochondrial ancestor is the root lineage (R) of eukaryotes, whose position is not yet conclusively determined (hence the ?). H and M indicate the presence of hydrogenosomes and mitosomes, which are reduced versions of mitochondria typical of fermenting organisms living in anoxic environments or inside living cells. Doubled branches indicate secondary (2o) endosymbioses.

Other kinds of algae are much more complicated, because other photosynthetic eukaryotes bear plastids with three or four membranes and remnants of other genomes. This is because they originated as partnerships between two eukaryotes, one of which already had a partnership with a cyanobacterium. The main lines of evolution of plastid-bearing eukaryotes are illustrated in Figure 8.8.

Euglenids are motile green flagellates that can be found in any pond whose nuclear genome resembles a trypanosome—the organism that causes sleeping sickness in humans—while their plastid genome is

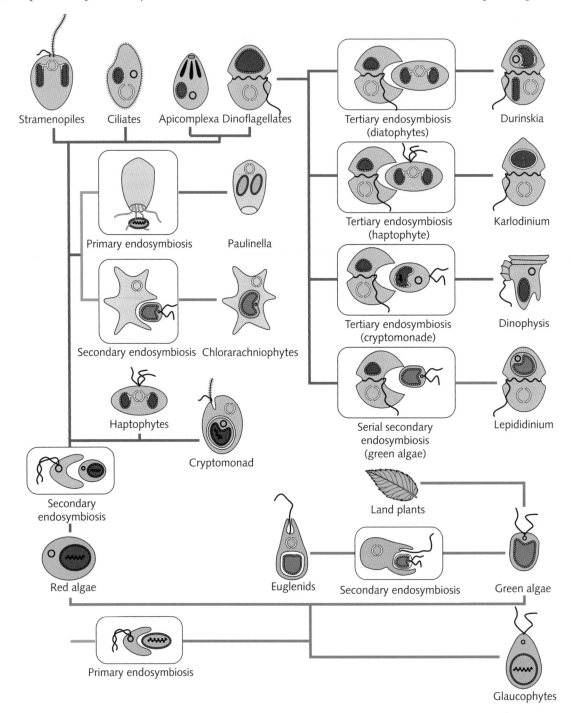

Figure 8.8 Radiation of plastid-bearing eukaryotes.

Reproduced from Keeling, P.J. The endosymbiotic origin, diversification and fate of plastids. *Philosophical Transactions B* 365 (1541) © 2010, The Royal Society.

like that of a green alga. Their ancestors were predators, consuming algal cells, whose prey resisted digestion, and were retained in the cell as photosynthetic partners that eventually evolved into organelles. Dinoflagellates are very abundant unicellular marine organisms which have acquired their plastid from a red alga. Cryptomonads are another group of unicellular planktonic organisms that not only have a red algal chloroplast but also still retain a rudiment of the ancestral nucleus with a few chromosomes.

The outcome of this sequence of fusions is that many species of phytoplankton are built like Russian dolls. Their chloroplasts may have as many as four membranes—the cytoplasmic membrane of the host cell, the plasma membrane of the eukaryote endosymbiont (and sometimes a remnant of its nucleus), and the two membranes of the original cyanobacterium. Familiar and ecologically important organisms such as diatoms have evolved through serial endosymbiosis: the origin of a new kind through fusion followed by modification.

8.4 The origin of life

Life has three fundamental characteristics: individuals are able to replicate, populations are variable, and lineages evolve. The origin of life and the very early stages of the evolution of living organisms will never be known for sure, because they occurred long ago and involved molecules that could leave no fossil record. We do know that life appeared quite soon once liquid oceans were formed, because cellular organisms have left physical and chemical traces from very early times (Section 5.18). How these first cells evolved, and what their predecessors were, cannot be directly demonstrated. What we could do instead is to construct a continuous series of intermediate stages between organic molecules and simple self-replicating cells as a demonstration of how life might have evolved from non-living precursors. This has not yet been achieved in full. Some crucial intermediate steps have been validated in the laboratory: in particular, the existence of truly self-replicating nucleic acid molecules and the formation and replication of membrane-bound protocells.

8.4.1 Protocells arise spontaneously and are readily selected.

Compartments are necessary for natural selection to act on phenotypes. The key property of a cell is that it retains the physical link between a chemical product and the genetic system that encoded it. It is only when this link between genotype and phenotype exists that natural selection will drive the evolution of more powerful and effective metabolic systems. Without cells, any products encoded by a replicator would simply diffuse away, without supplying any advantage to the replicators. As in any society, producers must have possession of the fruits of their labor to gain any advantage from them.

This principle can be restated in a more powerful way: should compartments able to replicate ever arise they would rapidly replace all diffusive, non-compartmental systems because any improvements are transmitted to progeny. Compartmental systems evolve rapidly because they are readily selected, whereas diffusive systems may scarcely evolve at all.

Cell-sized hollow spheres are formed spontaneously by fatty acids in water. The cells of modern organisms are bounded by complex membranes based on a phospholipid bilayer that incorporates a wide range of channels, pumps and pores that regulate their traffic with the external environment. The first cells must have been far simpler, forming spontaneously by a chemical process. Moreover, they must have had some inherent ability to grow and divide. Although these seem to be very burdensome conditions at first sight they are satisfied by fatty acids. Fatty acids are very stable molecules that spontaneously form cell-sized microspheres in water. The microspheres have two layers, with hydrophobic groups on the inside and hydrophilic groups outside. Molecules are continually exchanged between the layers, so that any external supply of fatty acids leads to growth because new molecules continually enter the outside layer, which then re-equilibrates with the inside layer. Hence, the microsphere grows in size. The structure is dynamically unstable, so that after growing for a while it buds off a smaller sphere, which in turn begins to grow.

8.4.2 Protocells could house the first selectable self-replicating systems.

Fatty acid bilayers are readily permeable to small charged molecules. Microspheres will continue to increase in numbers, given a supply of fatty acid molecules. They are quite inert and are not living in any sense. However, they are quite permeable; small charged molecules, even nucleotides, readily pass through the membrane to the interior of the microsphere. This gives them the potential for assembling through chemical processes alone a pool of molecules that could provide the raw materials for metabolism. Indeed, this potential has been exploited by biochemists who have developed microspheres for isolating biochemical reactions in order to improve the yield of a product by employing evolutionary technology. Hence, it is most likely that the first cells were heterotrophic, using the pool of organic molecules dissolved in seawater as the basis for metabolism and replication.

The first replicators were probably small RNA molecules. In modern cells, DNA encodes protein enzymes that drive metabolism. The very earliest chemical systems capable of self-replication cannot have worked like this, because even if they could appear spontaneously (a most unlikely event) the proteins would diffuse away from the DNA encoding them. Hence, these early systems must have combined the properties of a replicator with that of an enzyme catalyzing replication. The only substance we know of with these properties is ribonucleic acid, RNA. This is universally used in cellular organisms as a carrier of genetic information, in the form of mRNA; it is also used as the genomic nucleic acid in some viruses. Its versatility is based on the tendency of single-stranded RNA to fold into complicated tertiary structures which form binding sites like those of protein enzymes. RNA enzymes, or ribozymes, are capable of catalyzing chemical reactions, including those involved in nucleic acid replication.

Recall Sol Spiegelman's Qβ experiment, where a population of RNA molecules rapidly evolved novel properties (Section 2.5.1). This demonstrates the tendency of a replicating system to evolve, even when it is merely a chemical. These RNA molecules are not truly self-replicating, however, because they require the presence of a protein replicase which they do not themselves encode. A truly self-replicating system requires that the RNA itself acts as its own replicase. This is a very difficult problem in chemistry that was not solved until 2009, when the sequence of the first truly self-replicating RNA system was announced. It is not a very likely candidate for the molecule ancestral to all later self-replicators, because it is not very evolvable—random changes do not seem to lead to progressive improvement through natural selection. Nevertheless, it clearly demonstrates that truly self-replicating chemical substances do exist.

8.4.3 Evidence for the chemical formation of the first self-replicators is still incomplete.

Short RNA molecules can form spontaneously. We do not know how many self-replicating RNA polymers there are, but all must be quite long, consisting of a few dozen nucleotides at least. Hence, their initial formation requires that the rate of polymerization must exceed the rate at which the growing chain is disrupted by physical and chemical forces. This is unlikely to occur in solution, where mononucleotides are likely to be too dilute to add one unit to the chain before it is shortened. It might easily occur on charged surfaces such as clay particles, however, where the local concentration of nucleotides can be much higher than in solution. In the laboratory, short chains of a dozen or so nucleotides can be formed in this way. They are too short to function as self-replicators, regardless of sequence. This is an important gap in our knowledge of how the first self-replicators appeared.

The main kinds of organic molecules are readily formed by physical processes. The formation of organic molecules from inorganic predecessors presents no difficulty. It has been known for fifty years that electric discharge in a reducing atmosphere containing carbon dioxide, ammonia and water vapor will produce a wide range of substances including sugars, amino acids and nucleotides. The simple experiment that demonstrated this striking fact is illustrated in Figure 8.9. All the basic molecules necessary for life can be produced by physical and chemical processes that were active in the early Earth system. The pathway to RNA has not yet been traced conclusively, however, because it has been unexpectedly difficult to find how the pyramidine nucleobases cytosine and uracil can

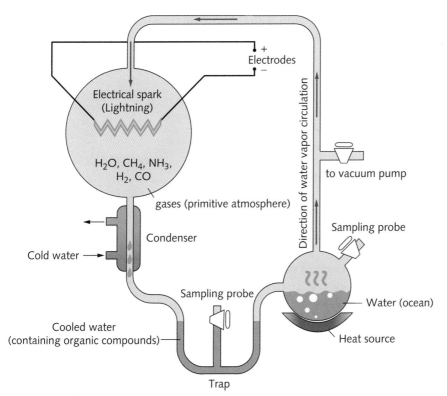

Figure 8.9 The Miller–Urey experiment demonstrating that a range of organic substances are formed spontaneously under early-Earth conditions.

Courtesy of Yassine Mrabet. This file is licensed under the Creative Commons Attribution-Share Alike 3.0 Unported license.

be bonded with ribose. Recent advances in chemistry have shown how this might occur, but we still lack a well-established route to the chemical synthesis of nucleotides in the environment of the early Earth.

Origins are discovered from two kinds of evidence: the fossil record (such as the origin of terrestrial vertebrates or flying birds) or the range of variation among living kinds (such as the origin of multicellularity or C4 photosynthesis). Neither is available for the origin of life, or even of cellular organisms, nor will it ever be. The success of life itself has long since consumed all the signs of its very early stages. What we can do is to demonstrate that there is no obstacle to believing that living cells could originate from indisputably non-living predecessors. Some of the pieces of this puzzle have been discovered; others are still missing. Those that have been found indicate that the bridge from chemistry to biology will eventually be built.

● CHAPTER SUMMARY

Innovation is triggered by new ecological opportunities or by genetic revolutions. Modification, duplication, and fusion are the routes to innovation.

New structures evolve from old structures.

- Multicellular organisms evolve from unicellular ancestors through a series of intermediate forms.
  - *A few major genes direct development.*
  - *The multicellular body evolves through a series of intermediate forms.*
  - *Multicellularity has evolved independently in several lineages.*

- C4 photosynthesis is a novel mechanism that evolves from the C3 state.
  - *C4 photosynthesis evolves by the modification of pre-existing C3 mechanisms.*
  - *Rubisco evolves within the C4 environment.*
  - *C4 photosynthesis evolves through a series of intermediate states.*
  - *There are several different outcomes of selection for C4 function.*
- Feathers have evolved from epidermal structures of dinosaurs.
  - *Feathers evolved from keratin structures in the epidermis of amniotes.*
  - *Feathers evolved from projecting epidermal fibers in dinosaurs.*

Duplication provides an opportunity for functional divergence.

- Gene duplication can lead to novel capabilities.
  - *Visual pigments originate through gene duplication.*
  - *Whole-genome duplication can lead to adaptive radiation.*

The fusion of dissimilar cells has led to new kinds of eukaryote.

  - *Endosymbiosis occasionally evolves through the survival of prey.*
  - *Eukaryotes incorporate two radical innovations originating by fusion.*
  - *Partners may not have the same interests.*
  - *Mitochondria all descend from a single bacterial ancestor.*
  - *Plastids have evolved along many routes.*

The origin of life.

- Protocells arise spontaneously and are readily selected.
- Protocells could house the first selectable self-replicating systems.
  - *Fatty acid bilayers are readily permeable to small charged molecules.*
  - *The first replicators were probably small RNA molecules.*
- Evidence for the chemical formation of the first self-replicators is still incomplete.
  - *Short RNA molecules can form spontaneously.*
  - *The main kinds of organic molecules are readily formed by physical processes.*

● FURTHER READING

Here are some pertinent further readings at the time of going to press. For relevant readings that have been released since publication, visit the book's Online Resource Centre at **www.oxfordtextbooks.co.uk/orc/bell_evolution/**

Section 8.1   Herron, M.D. and Michod, R.E. 2007. Evolution of complexity in the volvocine algae: transitions in individuality through Darwin's eye. *Evolution* 62: 436–451.

Christin, P-A., Freckleton, R.P. and Osborne, C.P. 2010. Can phylogenetics identify C4 origins and reversals? *Trends in Ecology and Evolution* 25: 403–409.

Fucheng, Z., Zhonghe, Z. and Dyke, G. 2006. Feathers and 'feather-like' integumentary structures in Liaoning birds and dinosaurs. *Geological Journal* 41: 395–404. (See also: Rubin, J. 2010. Paleobiology and the origins of avian flight. *Proceedings of the National Academy of Sciences of the USA* 107: 2733-2734.)

Section 8.2   Trezise, A.E.O. and Collin, S.P. 2005. Opsins: evolution in waiting. *Current Biology* 15: R794–R796.

Section 8.3   Keeling, P.J. 2010. The endosymbiotic origin, diversification and fate of plastids. *Philosophical transactions of the Royal Society of London B* 365: 729–748.

Falkowski, P.G., Katz, M.E., Knoll, A.H. et al. 2004. The evolution of modern eukaryotic phytoplankton. *Science* 305: 354–360.

Gould, S.B., Waller, R.F. and McFadden, G.I. 2008. Plastid evolution. *Annual Reviews of Plant Biology* 59: 491–517.

Lane, C.E. and Archibald, J.M. 2008. The eukaryotic tree of life: endosymbiosis takes its TOL. *Trends in Ecology and Evolution* 23: 268–275.

Section 8.4 Robertson, M.P. and Joyce, G.F. 2010. The origins of the RNA world. *Cold Spring Harbor Perspectives in Biology* LXXIV doi: 10.1101/cshperspect.a003608

Mansy, S.S. and Szostak, J.W. 2010. Reconstructing the emergence of cellular life through the synthesis of model protocells. *Cold Spring Harbor Perspectives in Biology* LXXIV doi: 10.1101/sqb.2009.74.014.

QUESTIONS

1. What is meant by "a different kind of organism?" How different is different?

2. Compare and contrast the innovations that might evolve in response to (a) a change in the environment, and (b) a change in the genome.

3. Describe the main kinds of structural change that can lead to innovation, and give an example of each.

4. Write an essay on the evolution of multicellularity in Volvocales. Is this group a good model for the evolution of multicellularity in other groups, such as animals and plants?

5. Explain why the division of labor between germ cells and somatic cells may be advantageous. Given your explanation, why are there so many unicellular organisms?

6. Give a brief account of C4 photosynthesis and identify the conditions in which it is superior to C3 photosynthesis. How does the C4 system evolve from an ancestral C3 state?

7. Explain how feathers could have evolved from simpler structures in animals unable to fly, through a series of intermediate states each of which had an advantage of some sort.

8. Why have feathers evolved only once, whereas multicellularity and C4 photosynthesis have evolved independently in many lineages?

9. Write an essay explaining how duplication can lead to the evolution of novelty.

10. Describe the early stages in the evolution of an intracellular symbiont, using examples from contemporary organisms.

11. How do we know that mitochondria and plastids are derived from bacterial endosymbionts?

12. Write an essay on the evolution of the eukaryote cell as a compound structure.

13. Identify the main steps that must be verified experimentally to provide a plausible route for the origin of life. Which have been demonstrated so far? What are the main gaps in our current understanding?

14. Life is characterized by metabolism and replication. Which came first?

You can find a fuller set of questions, which will be refreshed during the life of this edition, in the book's Online Resource Centre at **www.oxfordtextbooks.co.uk/orc/bell_evolution/**

PART 4
Adaptation

The bodies of living organisms are very complex, highly integrated structures. How they can have come about through some natural process is the central problem of evolutionary theory. Moreover, the body plan of organisms varies in relation to their way of life. It varies on a large scale, for example between marine and terrestrial organisms, but it also varies on a small scale, for example between species of fish with different diets. This is the problem of adaptation: to explain how bodies become appropriately constructed for a particular way of life. It can be expressed in three contexts. The first concerns the body itself, and the principles that govern the pattern of adaptation and the degree of specialization. The second concerns the development of the body: how the pathway from a unicellular egg to a multicellular adult has itself evolved. Finally, the pathway of development is governed by genes, and underlying the evolution of bodies and their development is the evolution of the genome. This part of the book is about adaptation in the context of bodies, the way in which they develop, and the genomes they bear and transmit.

Adaptation and Evolved Design

9

In Chapter 4 we explored the long list of innovations involved in the evolution of humans from remote single-celled ancestors. These included clearly functional features such as jaws and differentiated teeth, or embryonic development involving an amnion or placenta, which contribute to success in new ways of life. They also include features such as radial cleavage and a single posterior flagellum which have likewise been inherited by the members of large clades from their common ancestor, but whose utility is not as clear. This chapter describes how both functional utility and ancestry contribute to adaptation and the evolved design of organisms.

Corresponding features of related species are homologous. Evolved structures, unlike fabricated structures, are constrained by their origin as modifications of an ancestral state. Hence, structures in two species may be similar in form, not because they operate in a similar fashion, but because their common ancestor had the corresponding structure. Characters that are similar among species because they are inherited from the common ancestor of the group are said to be homologous. In the terminology of Chapter 3, shared primitive and shared derived character states are homologous among the members of a clade. The wing of a gull and the wing of a goose, for example, are homologous because their most recent common ancestor was winged.

This way of describing homologous characters, although conventional, is not completely satisfactory, because it is not wings that are transmitted from generation to generation, but rather copies of the genes responsible for wing development. Homology is therefore an attribute of genes, and refers only indirectly to characters. Two species descending from a common ancestor may express similar states of a given character because that character is encoded in each descendant by a copy of the same gene or genes borne by the ancestor. When we refer to "homologous characters," we are using shorthand for this long-winded but more precise phrasing.

Figure 9.1(a) illustrates the concept of homology for an imaginary case in which the cuticle of an insect can become melanized as the result of a mutation in either of two genes. The character state "melanized cuticle" is homologous among species with a melanized common ancestor, that is, species in a clade with "melanized cuticle" as a shared character. In this case, melanization evolved once, as the consequence of a mutation in the lineage leading to the most recent common ancestor of clade X. This illustrates a situation in which a character evolves in a lineage and fails to evolve in its sister lineage.

Figure 9.1(b) shows a different distribution of melanism, arising from independent mutations in different lines of descent. Melanized cuticle is homologous only among members of the same clade. This is not affected by the secondary loss of melanization in other members of the clade. Melanization is not homologous among species belonging to different clades (clades X and Y in the figure), in which the altered character state has been caused by mutations in different genes. Moreover, it would not be homologous even if the *same* mutation had occurred independently in the ancestry of these clades. However, the character "cuticle" (or equivalently the character state "cuticle present") is homologous among all the species in the diagram.

(a)

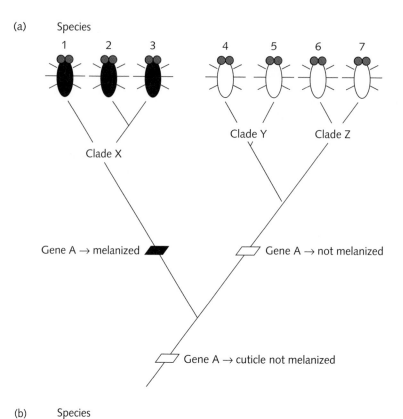

(b)

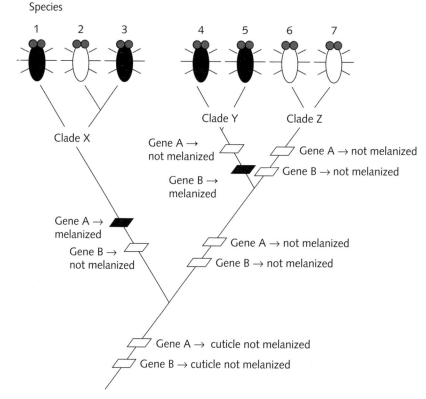

Figure 9.1 Homology. (a) Melanized cuticle is homologous in species 1, 2, and 3 because the condition was inherited by each species from the most recent common ancestor of clade X. (b) Mutations in two genes, A and B, can cause melanization of the cuticle. Gene A mutates in the common ancestor of clade X and gene B in the common ancestor of clade Y. There is a back-mutation suppressing melanization in species 2 of clade X. The character state "melanized cuticle" is homologous in species 1 and 3. It is also homologous in species 4 and 5. It is not homologous in any comparison of species in clade X with species in clade Y.

The degree of homology varies with phylogenetic level. In practice, we often do not know much about the genetic basis of characters, and assume that a shared character has a common genetic basis among the members of a monophyletic group. Granted this assumption, which can be tested by experimentation when necessary, any shared ancestral character or shared derived character is homologous among the common ancestor of the group and all its descendants. As an example, consider the beak of a blue jay.

- The beak of the blue jay (*Cyanocitta cristata*) and the beak of its sister species, Steller's jay (*C. stelleri*), are homologous structures because both are inherited from a common ancestor that lived about 5 Mya. In this case, the particular state of the beak, a stout, pointed type capable of dealing with a wide range of food items, is homologous in the two species.

- The beak of the blue jay and the beak of (say) an albatross or an eagle are also homologous, because the beak is a shared derived character of all modern birds (Neornithes), and the genes responsible for its formation are assumed to have been transmitted from the most recent common ancestor of the group, which lived about 120 Mya. Here, it is the general state of the beak, as a toothless prolongation of upper and lower jaws covered with a horny bill, that is homologous among members of the clade.

- The beak of a blue jay and the beak of a hawksbill turtle are not homologous in the same sense, because birds and turtles evolved independently from different non-beaked ancestors. However, both are modifications of the upper and lower jaws, which are homologous structures, because they are shared primitive characters of Amniota. The formation of the beak is presumably regulated in part by the same genes in birds and turtles (those responsible for the general structure of the jaws) and in part by different genes, or different mutations (those responsible specifically for the development of the jaws into a beak).

- Finally, the beak of a blue jay and the mandibles of a dragonfly are not homologous. They are not modifications of a shared ancestral character, since jaws evolved from branchial bars in the ancestors of gnathostomes, whereas mandibles evolved from segmental limbs in the ancestors of arthropods. The development of branchial bars and segmental limbs is governed by different sets of genes. Non-homologous characters are sometimes said to be analogous.

Hence, the central principle of independent evolution leads to the recognition of homology at different phylogenetic levels. The particular form of the stout, pointed beaks of sister species of jays is homologous; the beak itself is homologous in all birds; and the jaws which may be modified to form a beak are homologous in gnathostomes. Below a certain level of comparison it is difficult to recognize any homology at all, as in the mouthparts of gnathostomes and arthropods. Even at this deep level, however, the main mechanisms of animal development tend to be evolutionarily conserved. Although it is often useful to make a sharp distinction between homologous and non-homologous characters (as we shall do), we should recognize that these are two ends of a continuum—in between, homology simply makes a smaller contribution to character state in more distantly related organisms.

Homology expresses the ancestral constraint on the evolution of utility. Homologous structures evolve from the same ancestral state and thereby reflect that state to some degree in their current state. In the next two sections we shall first describe the evolution of utility and then explain how utility can be constrained by ancestry.

9.1 Utility favors optimal design.

The correspondence between the design of an organism and its way of life is the single most striking observation that has motivated the development of biology. It is a signal that the utility of a structure may induce its evolution in particular circumstances irrespective of the ancestral state of a lineage. The utility of evolved structures is expressed at all phylogenetic levels, from the general body plans of phyla to the idiosyncratic features of individual species.

9.1.1 The fundamental features of body plans are effective adaptations to particular ways of life.

The shared derived characters of large groups, such as placentation or jointed limbs, indicate the history and relationships of these groups. They are far from being arbitrary phylogenetic markers, however: in most cases, they are very effective adaptations to particular ways of life.

Metazoan body plans are adaptive. The presence of a body cavity, its structure and its mode of development were very prominent themes in working out the phylogeny of Metazoa in the days of classical morphology-based systematics. They are often unreliable guides, however, because body cavities play important roles in the lives of animals, and are therefore readily molded into different shapes and sizes by natural selection.

For example, consider how a body cavity affects the ability of a simple worm-shaped animal to move in a viscous medium such as mud or sand. Its locomotion is powered by muscle, but muscles alone cannot produce movement. This is because a muscle can forcibly contract, but cannot forcibly extend. Hence, it must be re-extended by the contraction of another muscle, to which it is linked by a skeleton.

A very simple skeleton can be made with a body wall of two muscle layers, circular and longitudinal, surrounding a fluid-filled interior. Contracting the circular muscles makes the body longer and thinner, because the fluid is incompressible, and hence its volume must be conserved. Similarly, contracting the longitudinal muscles makes the body shorter and fatter. Neither action will be effective by itself, but since the circular muscles re-extend the longitudinal muscles, and vice versa, their coordinated action can produce directed movement. Imagine a large worm-like animal lying in soil or sediment with the longitudinal muscles contracted, as shown in Figure 9.2. It now relaxes the longitudinal muscles and contracts the circular muscles in the forepart of its body. The consequence is that the body is extended forward, against the resistance provided by the thickened hind region.

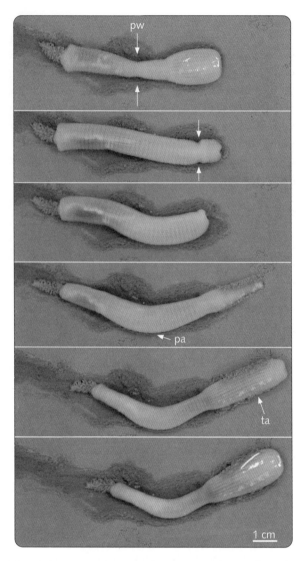

Figure 9.2 Locomotion in a simple unsegmented worm: forward movement produced by successive contraction of circular and longitudinal muscles in a priapulid burrowing in sediment. pa, penetration anchor; pw, peristaltic wave; ta, terminal anchor.

Republished with permission of the Geological Society of America, from Vannier et al. Priapulid worms: pioneer horizontal burrowers at the Precambrian–Cambrian boundary. *Geology* 38 (8) © 2010; permission conveyed through Copyright Clearance Center, Inc.

The animal now relaxes the circular muscles and contracts the longitudinal muscles to shorten and thicken the forepart of the body, while doing the reverse in the hind region. This will draw the rear part of the body forward, against the resistance of the plug formed by the thickened fore region. In this way, cyclical contraction and relaxation of circular and longitudinal muscles can produce forceful burrowing in soil or sediment.

In a simple burrowing worm like this, the coelom or an equivalent body cavity provides a hydrostatic skeleton that allows circular and longitudinal muscles to oppose one another. It is a very important innovation of Bilateria, because it unlocked the food available in marine sediments, which are inaccessible to flatworms that can only crawl on the surface. At the same time, it is only useful to relatively large animals, 1 cm or more in length, such as the priapulid worm illustrated in Figure 9.2. Much smaller organisms, 1 mm or less in length, cannot push aside sand or clay particles, and instead crawl between them. Hence, small animals like rotifers or kinorhynchs cannot use a coelom for burrowing, and either lack a body cavity or have a highly modified version.

Animals such as arthropods and tetrapods use limbs for locomotion. The fundamental principle remains the same, except that the skeletal elements used to oppose muscles are rigid. This allows a greater variety of movements, in which hinged levers can be used for running, digging, swimming, or flying. In small animals like insects the limb muscles are small enough to be accommodated inside stiff-walled tubes. In larger animals with more bulky muscles this would not work as well, because a larger tube would either be much more massive, increasing the inertia of the limb, or would have thin walls that would be likely to buckle or fracture. Above a certain size, the muscles must be placed outside a relatively thin, strong tube, as they are in tetrapods.

Earthworms use a hydrostatic skeleton to burrow through soil, a much more resistant medium than marine sediments. They are able to do this because their coelom is divided into many semi-independent compartments, the segments, by transverse walls. Segmentation permits a greater degree of coordination of the muscles, with several trains of contraction and relaxation traveling simultaneously along the body, as shown in Figure 9.3. This permits the worm to thrust more powerfully into the soil.

Earthworms also have setae, thin extensions of the cuticle, on each segment. These grip the soil in segments that are thickened by contracting the longitudinal muscles, providing a more secure anchor to thrust against. In arthropods the segments bear a much greater diversity of appendages. In a shrimp or an insect, for example, different segments may bear antennae, mouthparts,

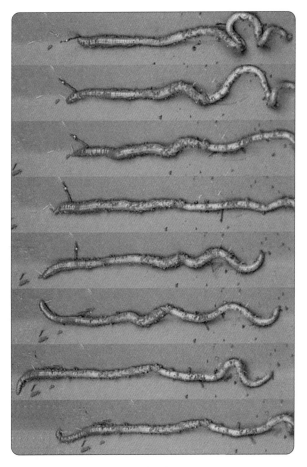

Figure 9.3 Locomotion in a segmented worm: several simultaneous waves of contraction drive forcible burrowing through soil in earthworms.

Image taken from <www.practicalbiology.org>, an online teaching resource from the Nuffield Foundation and the Society of Biology.

walking legs, gills or wings. This allows the regional specialization of the body to evolve, through the developmental program that governs the structure of each segment. Modifications of this program underlie the characteristic body plans of crustaceans, centipedes, spiders, winged insects, and other kinds of arthropod. Hence, the modular construction of segmented animals leads to the evolution of a great range of forms adapted to different ways of life.

Green plant body plans are also adaptive. The shared derived characters of land plants include fluid-conducting vessels and roots (Tracheophyta), and a seed containing nutrient reserves to support the growth of the young plant (Spermatophyta). These are adaptations to terrestrial conditions, where tall plants can monopolize the supply of light. Seaweeds can grow to a very large size without needing a stiff skeleton because they are supported by the water. Land plants such as mosses must remain small because they are unable to extend their bodies upwards against gravity. Vascular plants such as ferns, conifers, and flowering plants use thin, stiff tubes of cellulose to resist gravity and to transport fluids to the top of the plant. Bundles of tubes, embedded in a ground substance such as lignin, allow trees to grow up to 30 m in height. Specialized roots anchor the plant and enable it to tap into underground sources of water and nutrients more effectively. The seed evolves as a spore supplied with nutriment for the embryo that increases the likelihood of successful germination in harsh conditions.

Coeloms, limbs, and segmentation are all features of the fundamental body plans of large groups of animals. Likewise, vessels, roots, and seeds are diagnostic for large groups of plants. All are clearly functional features that enable the members of each group to follow particular ways of life that would otherwise be impracticable. They are often modified to produce strikingly different variations on the same underlying theme, such as the creeping foot of snails and the suckered arms of squid among the mollusks. The close fit of form to function is the most striking property of living organisms.

9.1.2 Small details of structure can be important adaptations.

The utility of the fundamental features of body plans can be traced down to the most insignificant details of structure. There are very many examples of highly specialized structures that play an important part in a particular way of life. We will give just two examples of minor modifications with clear adaptive consequences.

The extraordinary jaw of Epibulus *is precisely adapted to catch prey.* The jaws of many teleosts can be shot forward to catch prey by the contraction of muscles attached to hinged skeletal elements. Imagine four rods joined with one another by movable joints to form a square, a four-bar linkage. The contraction of a muscle running diagonally between two corners of the square will pull it into a much narrower parallelogram,

as shown in Figure 9.4(a). If the square has a fixed base, this movement will protrude one corner of the parallelogram forward. This is used to catch agile prey that might escape if the fish simply swam towards

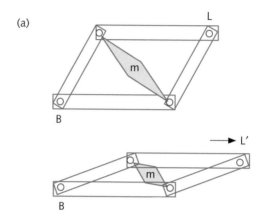

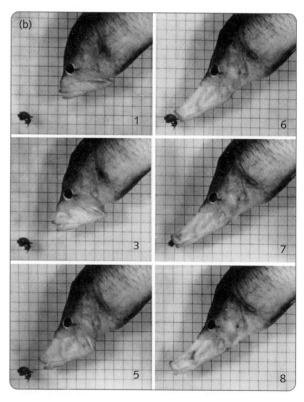

Figure 9.4 (a) A four-bar linkage. Contraction of the muscle m deforms the quadrilateral, producing an extension in one direction from L to L', given a fixed base B. (b) Slingjaw wrasse *Epibulus* capturing a small crustacean. Total time from 1 to 8 was 0.04 seconds.

Part (b) reproduced from Westneat and Wainwright. Feeding mechanism of *Epibulus insidiator* (Labridae; Teleostei): evolution of a novel functional system. *Journal of Morphology* 202: 129–150 © 1989 Wiley-Liss, Inc.

it. The slingjaw wrasse, *Epibulus insidiator*, has the unique ability to protrude its jaws forward in less than 40 ms by a distance greater than the length of its head (Figure 9.4(b)). This has been accomplished by adding two ligaments to form a novel six-bar linkage with the ability to capture shrimps and small fish at a distance where they would normally be safe.

The Y-skeleton of Polyxenus *is precisely adapted for running on smooth surfaces.* Most millipedes are rather stout-bodied animals that can burrow in leaf litter by means of the powerful thrust that their many pairs of legs can provide. *Polyxenus* has only about a dozen pairs of legs, and is a swift runner that has

the remarkable ability to move in narrow crevices and to walk upside-down on any surface, however smooth—even a mirror. It has evolved this ability through three modifications of the usual myriapod body plan. The first is a very light, non-calcified cuticle. The second is a flat plate of chitin next to the claw at the extremity of each leg, which can be pressed to the surface and acts like a suction cup when the claw cannot take hold. The third is a rigid lattice at the base of each limb provided by thin bars of chitin, especially the Y-shaped bar on the anterior face of each leg, which is shown in Figure 9.5. This gives a solid base for muscles that produce a large angle of swing and a powerful grip on the surface.

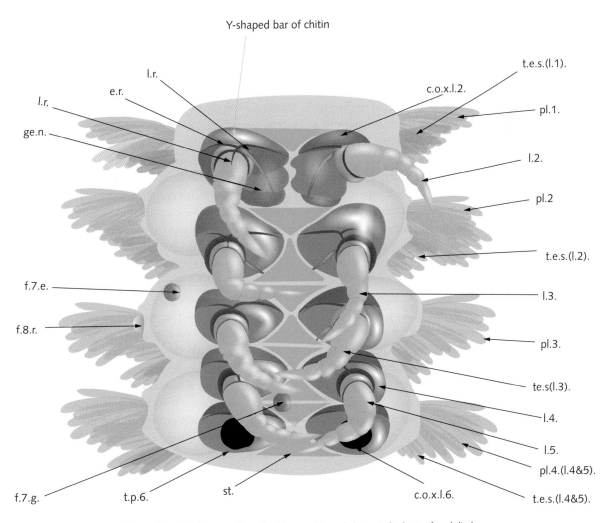

Figure 9.5 Ventral view of the millipede *Polyxenus*. Note the Y-shaped bar of chitin at the base of each limb.

Reproduced from Manton, S.M. 1956. The evolution of arthropodan locomotory mechanisms. Part 5. The structure, habits and evolution of the pselaphognatha (diplopoda). *Zoological Journal of the Linnean Society* 43 (290): 153–187 © 2008, John Wiley and Sons.

This is what enables the animal to run swiftly upside-down on ceilings.

This Y-skeleton is the unique shared derived character for the group to which *Polyxenus* belongs, and could be interpreted as a mere developmental idiosyncrasy that is useful for purposes of classification. In fact, it is an essential component in a suite of adaptations that enables *Polyxenus* to exploit a habitat that no other millipede can use.

9.1.3 The design of organisms is often closely fitted to their conditions of life by natural selection.

Within limits imposed by the range of materials that can be synthesized, the structure of organisms is continually honed by natural selection. Any variant fractionally better than the average of the population will tend to spread, and consequently we expect that most individuals will be closely adapted to their conditions of life. This should apply, not only to the presence or absence of characters, but also to the precise values of character states, such as the length of a limb or the sugar concentration of nectar. This is the optimization theory of adaptation: we should be able to calculate exactly how we expect organisms to be constructed, in order to maximize their fitness.

Neither machines nor organisms can be perfect, however. This is because most machines must be capable of performing more than one function, and different functions often interfere with one another. Suppose we wish to make a lever capable of moving a heavy load as swiftly as possible. Figure 9.6 illustrates the interference of strong and rapid motion. If we place the fulcrum close to the point of application of force, the load will be moved swiftly but weakly; if we place it far away, the load will be moved strongly but slowly. Any one task is performed well at the expense of performing others less well. A lever capable of moving heavy loads, whether a crowbar or a mole's forelimb, will not be capable of imparting rapid motion, like a catapult or a deer's forelimb. This is called *functional interference*. It means that we must decide what combination of strength and swiftness maximizes the effectiveness of the lever for some useful task. The result will be a design that is optimal for this particular task, at the cost of being ineffective for other tasks. This constraint bears even more severely

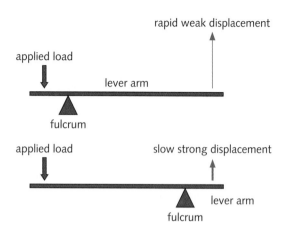

Figure 9.6 The position of the fulcrum relative to the applied load governs the work done by a simple lever.

on organisms than on machines, because an organism must be capable of carrying out all the tasks necessary for survival, growth and reproduction.

Optimal structures are compromises. Functional interference means that structures cannot be perfect, even within the limits of the materials they are composed of. They must instead represent compromises between different functional requirements, which maximize some overall measure of performance at the cost of falling short in particular features that contribute to performance. This applies equally to evolved and to fabricated structures.

The simplest source of functional interference is between quality and quantity. This is always imposed by a fixed limit on the time, energy or material available for use. Suppose that we are constructing cans to hold food, using a fixed quantity of metal, and want to know how thick the wall of the can should be. Thicker cans will be more robust. Very thin cans will be very fragile, so a moderately thick wall is necessary. A quite modest thickness will be adequate for most purposes, however, and adding more metal will not greatly improve the robustness of the can. Adding metal, however, will always decrease the number of cans that can be manufactured from a given quantity of metal. The optimal solution is found by multiplying the number of cans that can be made by their robustness, which produces a hump-shaped curve whose maximum value is the best compromise between producing more cans and producing stronger cans. This is not the strongest possible can design, nor is it the design that yields the most cans. It is the optimal design, the best can that can be made, given the constraints that we have imposed.

A more subtle source of functional interference is presented by the shape of the can. The can must hold the greatest possible volume of produce. Given the wall thickness, and thus the weight of metal in each can, should we make the can tall and thin, or short and thick? The answer is the optimal ratio of height to diameter. Elementary geometry enables us to calculate the volume enclosed by a cylinder of given height and diameter, holding its total surface area constant. This is shown in Figure 9.7. We can see that the optimal shape for a can is for the height to be equal to the diameter.

These simple arguments illustrate a powerful way of designing structures. It applies equally to cans of beans and to snail shells, so we can use it to interpret the design of either manufactured structures like cans or evolved structures like snail shells. We must use it with caution, however, because it is only successful if we have correctly specified the quantity that is to be maximized. For manufactured structures, this is the utility of the finished product, since this will be maximized by human agents. For evolved structures, it is the fitness of the individual, since this will be maximized by natural selection. In both cases, other considerations may intervene. If you measure a sample of cans in the supermarket, you will find that most are either longer and thinner, or shorter and fatter, than the optimal shape that we have predicted. There are therefore considerations other than economy of material that manufacturers use in designing cans. If you measure a large number of species of snail, you will find exactly the same thing—most species are either short and squat, or long and slender. This is presumably because fitness is more strongly influenced by other factors than by economy of material. Optimization is a valuable theory of evolved design to the extent that the criterion that we have chosen to maximize reflects overall fitness.

Bones have evolved to resist the stress they normally experience. Consider how a long bone, such as the femur, is designed in order to resist the stresses that it will ordinarily experience. This is an engineering problem, which we must first simplify in order to get a solution. A simple long bone can be visualized as a tube enclosed by a bone shell (of given diameter) containing marrow (of given thickness). It must resist a bending stress of some magnitude, for example that experienced when jumping, which will tend to deform the bone. A hollow tube is a good basic design, because tubes are stiffer than solid rods of the same length and mass, which is why the frame of your bicycle is made of hollow tubes rather than solid rods. We could make a very thick shell, with little marrow. This would be very strong, but would also be expensive to build. Or, we could make a much cheaper, thin shell, but this would be likely to break under stress. The optimization problem is then to find the thickness of bone that is just sufficient to resist an applied force that tends to bend the bone.

The physical attributes of bone are well known, so calculating the optimal thickness of a long bone is a simple engineering problem, like calculating the thickness of a beam designed to lift a given weight. It can be expressed in terms of the ratio of the internal diameter to the external diameter, which would be 1 for a solid bone and close to 0 for a very thin shell. The answer is 0.63. This can now be compared with real bones, which in fact vary between about 0.5 and 0.7.

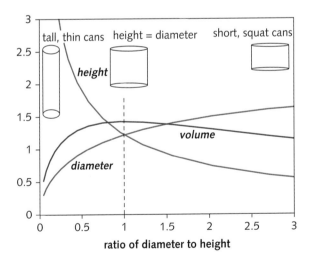

Figure 9.7 Optimal can design. The shape that maximizes volume for a fixed wall thickness and total weight of metal has height equal to diameter. The detailed working is as follows: area $A = 2\pi r^2 + 2\pi rh$. Hence constraint $dA/dr = 0$, i.e. $dh/dr = -2 - (h/r)$. Volume $V = \pi r^2 h$; $dV/dr = 2\pi rh + \pi r^2(dh/dr)$, hence $dV/dr = 0$ when $dh/dr = -2h/r$. $dA/dr = 0$ and $dV/dr = 0$ when $-2h/r = -2 - (h/r)$. Hence, optimal shape is $h = 2r$, i.e. height = diameter.

The design of membranes balances the gain and loss of gases. A slightly more complicated problem is presented by walls or membranes that must be porous, in order to admit substances into a cell or organ. For example, some species of tree have much larger leaves than others, while leaves may also differ in size according to their position on the tree. We can try to calculate how the optimal size of leaves varies with the conditions of growth. The main function of a tree leaf is to perform photosynthesis, so it must permit carbon dioxide to enter the leaf tissue through the stomata. This will necessarily allow water vapor to leave, however, which may reduce the efficiency of photosynthesis and expose the plant to drought stress. A reasonable approach— although not the only one—is to find the leaf size that maximizes water-use efficiency, which is the amount of carbon fixed by photosynthesis per unit of water lost.

This is quite a complex problem, because both photosynthesis and water loss will depend on environmental variables such as air temperature, relative humidity, light intensity, and wind speed. The central principle, however, is that the rates at which carbon dioxide, water vapor and heat are exchanged between the leaf and the atmosphere are all proportional to the ratio of wind speed to leaf size. Hence, large leaves have more resistance to heat and mass transfer, which may increase water-use efficiency (by reducing water loss) or reduce it (by reducing carbon dioxide uptake). The result is that large leaves are more efficient only when light intensity is low. Hence, we expect sun leaves (exposed in the forest canopy) to be small, whereas shade leaves (growing further down, where there is less light) should be large, and this is broadly true. Moreover, the range of light intensities over which water-use efficiency increases with leaf size is much greater at high temperatures than at low temperatures. Hence, we expect plants in cold regions to have small leaves, and tropical plants to have larger leaves, which again is broadly true. In this way, we can interpret leaves of optimal size evolving so as to maximize water-use efficiency in different conditions.

Eggshells present a similar problem. Like leaves, they have many minute pores through which gases can diffuse. If the shell has too many pores, the embryo will dry out; if it has too few pores, the embryo will suffocate. We can calculate the optimal permeability of the shell, which turns out to depend on size. Larger eggs should be more permeable than small

eggs, mostly because it takes longer for oxygen to diffuse into the interior of a large egg. Hence, we expect ostrich eggs to be more permeable than songbird eggs, and can even predict precisely how permeability should increase with size. This prediction is consistent with the data, as illustrated in Figure 9.8, so this feature of eggshells can be interpreted as an optimal compromise between respiration and water economy.

Optimality is a powerful method of interpreting adaptation, because it is capable of predicting precisely the structures that should evolve, provided that the sources of functional interference have been correctly identified. We shall meet it again in different contexts in later sections. It is not a complete theory of adaptation, however, because it assumes that the optimal character state can always evolve through natural selection. This is a plausible assumption for characters such as leaf shape, because there will be abundant variation in most species on which selection can act. It is unlikely to be equally true for more fundamental features of organisms, such as the appendages that develop on the different segments of an arthropod. Centipedes and beetles may both evolve as rapid predators, and will consequently share some attributes, such as sharp mandibles and running limbs. In other respects they will be utterly different, because they are modified versions of different ancestors. Hence, a complete theory of adaptation must take into account ancestry as well as utility.

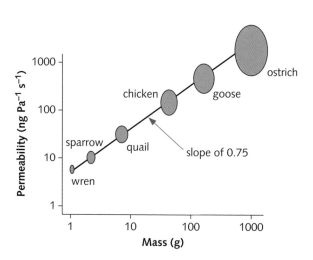

Figure 9.8 The permeability of eggshells increases with egg size in a predictable way.

9.2 The optimal design of evolved structures is constrained by ancestry.

Suppose that a few fish are introduced into a large, remote, fishless lake. They might be chub, for example, that would normally live in small streams and rivers. How will they adapt to their new home? This depends on the tension between two processes.

- The first is the fundamentally conservative process of reproduction. Offspring closely resemble their parents, and so the fish population many generations in the future will be very similar to the original immigrants, in the absence of any other process.

- The second is the modifying and diversifying tendency of variation and selection. Many generations hence, the population may have become better adapted to living in a large lake rather than a small stream, for example by evolving into a larger bottom-feeding form resembling a whitefish. Perhaps it might also have branched into specialized types resembling sunfish (littoral forms picking small invertebrates from vegetation), catfish (benthic foragers with flattened bodies feeding on large invertebrates), or minnows (small plankton-eaters).

Hence the outcome of evolution will depend on the extent to which selection is able to overcome the inertia of merely replicating the ancestral state.

9.2.1 Adaptation is constrained by ancestry.

Adaptation is generally a process of sequential modification driven by selection, so that the fully adapted state is reached through the cumulation of unit genetic changes (Section 2.3). Consequently, adaptation is likely to occur only if there is a genetic route from the ancestral condition to the adapted condition, along which each successive change is advantageous. A particular kind of adaptation might be able to evolve from one ancestral state, but not from another. Thus, tetrapods could evolve from lobe-finned fish with monaxial fins and accessory air-breathing organs through a series of intermediate forms, each succeeding because of their improved ability to exploit terrestrial resources (Sections 4.9 and 5.9). Tetrapods could not evolve from ray-finned fish, because this would require a large, abrupt and simultaneous modification of gills and fins that is extremely unlikely to happen. This is an example of an ancestral constraint.

The ancestral constraint is lack of appropriate variation. Selection will drive adaptation quite readily when the ancestral state is free to vary in a particular direction, giving rise, for example, to slightly modified fin skeletons or larger lungs in lobe-finned fish. Conversely, if this kind of variation does not occur, then selection will be ineffective, and adaptation and diversification are unlikely to occur. This is not to say that ray-finned fish are not variable at all, but rather that they show little if any variation in the required direction: any slight modification of a rayed fin, for example, would be very unlikely to confer the ability to walk on land, however feebly. Lack of appropriate variation will cause the ancestral state to be preserved. There are two main reasons for a lack of appropriate variation.

- The ancestral state may be resistant to change because some basic feature of the organism must be retained in its current form. That is, some feature crucial to adaptation cannot be slightly altered without reducing fitness. The highly permeable skin of amphibians provides an example. Many groups of tetrapods have returned to the sea because appropriate changes in their physiology and reproduction have evolved through natural selection. Turtles, ichthyosaurs and whales are examples. There are no marine amphibians, however, because their skin is very sensitive to salt water (Section 4.8). Resistant variants would require fundamental changes in character state, and consequently they arise very rarely, if ever, and selection cannot operate effectively. Hence, the ancestral state of amphibians has prevented any marine form from evolving in the entire 375 My history of the group.

- The ancestral state may also be resistant to change if the course of development cannot readily be modified. Amphibians again provide an example. The egg, as well as the skin, is sensitive to salt water, and no amphibian has evolved the ability to breed in the sea. Moreover, most amphibians have an aquatic larval stage, and hence are limited to damp habitats with standing water. This does not completely prevent amphibians from living in arid regions; some can rear their young without standing water (like the red-backed salamander of eastern North American forests), and others are able to live in arid land (like the spadefoot toads of the south-western North American deserts). It does, however, explain why the characteristic tetrapods of dry places are amniotes like mice and lizards, rather than amphibians like frogs and salamanders.

Ancestral constraint may not be absolute. Ancestry constrains the evolution of particular characters and thereby adaptation to particular ways of life. It is seldom a complete obstacle—it hinders adaptation rather than preventing it completely. Some groups of ray-finned fish, for example, have become amphibious despite the limitations imposed by their ancestral state. The most advanced are the mudskippers (Figure 9.9), which forage above the water line on tropical shores, feeding on insects and small crabs. They have thickened gill lamellae that prevent their gill filaments from collapsing in air, enabling them to breathe out of water. The lining of the mouth is also highly vascularized, enabling them to breathe by gulping air. They have a robust pectoral girdle with strong shoulder muscles that enable them to extend the pectoral fin and then thrust backwards, propelling them on land rather like a person moving with crutches. There are several other ray-finned fish that have evolved a variety of ways to breathe in air and move on land, but all of them are restricted to a narrow strip of territory just beyond the margin of the water, and none have become capable of a permanently terrestrial way of life.

The first lineage to evolve may exclude others. When an innovation first appears, it enables the lineage possessing it to follow a new way of life where it has no competitors. Selection will cause the structure involved to improve over time, until a high level of adaptation is reached. If another lineage living in the same place subsequently evolves a similar innovation, it will at first be imperfectly adapted, and is unlikely to be an effective competitor. Hence, the prior occupation of some way of life may obstruct adaptation in other groups, even if they are not constrained by ancestry, because of the sequential nature of evolution, from the

Figure 9.9 The mudskipper *Periophthalmus* is the most specialized terrestrial form so far evolved in the actinopterygian (ray-finned) clade; by contrast the terrestrial forms that have evolved in the sarcopterygian (lobe-finned) clade include frogs, lizards, birds, dogs, and people.

original primitive state to higher levels of adaptation. This is similar to the tendency of the first successful company in the field to monopolize a particular business sector, for which there are many examples, from automobiles to computer software. It is a reasonable explanation of cases where the first step in adaptation to a particular way of life can be taken, but fails to progress in the presence of an established and highly adapted competitor.

For example, the forelimbs of bipedal archosaurs were available to become specialized for flight through modifications of the limb bones and the evolution of feathers to form an effective flight membrane. There are many living species of lizards (lepidosaurs) able to run fast on long hind legs, such as basilisks and the frilled lizards of Australia. There are others (the "flying dragons") that are capable of gliding, by using a flight membrane of skin stretched over expanded ribs. Neither innovation has led anywhere close to powered flight, already monopolized among diurnal vertebrates by birds.

9.2.2 Primitive characters are more highly constrained.

The degree of variation that can occur in a character, without being inviable, will depend on how this character affects the overall functioning of the individual. In general, primitive characters that evolve early in the history of a group are likely to be more highly constrained.

- Ancestral constraint will be least severe for characters that can be modified separately, without changing the state of other characters. The color of a patch of feathers, the number of gill rakers, and the shape of leaves, for example, are recently evolved features that often distinguish sister species of birds, fish, and trees. They can usually be modified without causing or requiring changes in other parts of the body. Hence, recent derived characters are usually free to vary, will be only lightly constrained by ancestry, and are likely to respond to selection in the near future, as they have just done in the past.

- Ancestral constraint will be most severe for characters that cannot be modified without extensive changes in other characters. Features that evolved long ago are likely to be highly resistant to change, because all the features that evolved subsequently must conform to them. Events in early development, such as gastrulation, have little or no meaningful variation, since any slight deviation from the normal course results in a dead or deformed embryo. Hence, primitive characters are severely constrained by ancestry and are unlikely to respond to selection.

Ancestral constraint obstructs the independent evolution of similar characters in different lineages. Hence, the severity of ancestral constraint is reflected by the frequency of independent evolution. We can express this through the concept of homology. Figure 9.10 shows two imaginary phylogenies with different balances of homologous and non-homologous characters involved in adaptation to two ways of life. In both cases, sister species express homologous characters. In one case, sister taxa at the next lower level are unlike, showing that the change in character state occurred independently in the ancestry of the two sets of sister species. In the other case, sister taxa at the next lower level are alike, because they inherit the change of character state that had already occurred in their most recent common ancestor. This represents fewer independent origins of each altered character state, and thus more severe constraint on adaptation. Hence, the severity of the constraint attributable to ancestry is reflected by the phylogenetic level at which sister taxa are alike. For example, a patch of red feathers on the

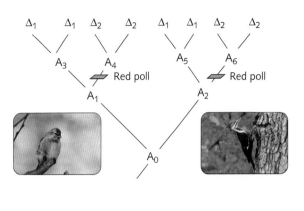

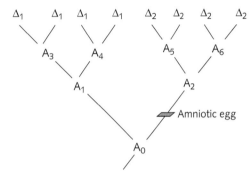

Figure 9.10 Homology and ancestral constraint. The ancestor at each node is designated A_x, and the two possible character states in the terminal taxa are Δ_1 and Δ_2. In the lower phylogeny, the genetic change causing the altered character state (amniote egg) Δ_2 occurred in the lineage leading to A_2, and Δ_2 is homologous in all species; likewise, Δ_1 is always homologous. In the upper phylogeny, a genetic change causing Δ_2 (patch of red feathers on head, i.e. red poll) occurred independently in the lineages A_1–A_4 and A_2–A_6, so Δ_2 in the descendants of A_4 is non-homologous with Δ_2 in the descendants of A_6; likewise, Δ_1 is homologous in sister species only.
Pileated woodpecker courtesy of Joshlaymon. This file is licensed under the Creative Commons Attribution-Share Alike 3.0 Unported license. *Carduelis flammea* courtesy of Cephas. This file is licensed under the Creative Commons Attribution-Share Alike 3.0 Unported license.

crown of the head has evolved independently in many lineages of birds. This character is variable, and hence is only weakly constrained by ancestry and readily responds to selection. On the other hand, all birds hatch from shelled eggs; no birds give birth to live young or have an aquatic larva. This is a highly constrained character that has never been fundamentally modified by selection in the whole history of the group.

All organisms share a small set of precursor metabolites. At an even more fundamental level, ancestral constraints on the operation of all living organisms are reflected by the uniformity of metabolism (Section 1.1). A great variety of energy and carbon sources can be harnessed by bacteria, enabling them to follow a correspondingly wide range of ways of life. Once energy and carbon have been acquired by the cell, however, the pathways that lead to the synthesis of biological compounds are remarkably similar in all bacteria, and indeed in all organisms.

Any food source is processed by the cell to yield monomers such as amino acids and nucleosides that are assembled into the characteristic macromolecules such as proteins and amino acids. These monomers are synthesized from precursor metabolites into which the food source is chemically transformed. The range of precursor metabolites, however, is much more restricted than the range of food sources. Whether the cell is taking in carbon dioxide, glucose, DDT, ethanol, or any one of a thousand other possibilities, it will transform it into one of just 12 precursor metabolites. Here is the complete list:

Glucose-6-phosphate

Fructose-6-phosphate

Pentose-5-phosphate

Erythrose-4-phosphate

Glyceraldehyde-3-phosphate

3-phosphoglycerate

Phosphoenolpyruvate

Pyruvate

Acetylcoenzyme A

Oxaloacetate

α-Ketoglutarate

Succinyl CoA

The shortness of this list is in strong contrast to the many fundamentally different energy-generating systems used by bacteria, and to the thousands of substrates that heterotrophic bacteria can consume. Moreover, these precursor metabolites are almost always generated by one of just three reaction sequences: glucose catabolism, pentose generation, and the tricarboxylic acid cycle (Krebs cycle). Furthermore, once monomers have been synthesized they are assembled into macromolecules in even fewer ways. DNA and RNA synthesis from purines and pyrimidines, for example, is very similar in all cells, bacterial or eukaryotic.

This is why a biochemistry textbook that describes the elementary processes of metabolism will probably not bother to mention the organisms where the crucial research was carried out. It doesn't matter. As the great French microbiologist Jacob Monod once remarked, "Anything that is true for *E. coli* is also true for elephants, but more so."

It is also why biotechnology works. We can move genes in and out of very distantly related organisms, such as bacteria and tomatoes, without wondering whether they will work in their novel genomic environment, because the basic processes of cell chemistry are very similar, even in organisms that are outwardly so different.

Hence, the biochemical pathways responsible for metabolism are highly conserved in all cells, despite the great range of substrates they are capable of dealing with. This is the legacy of their descent from Archaean bacteria.

9.3 Evolutionary engineering combines utility and ancestry.

The evolutionary tension between utility and ancestry is responsible for two characteristic features of evolved designs.

- Divergence and adaptive radiation. Homologous structures may be modified for different ways of life in species belonging to the same monophyletic group. This corresponds to the implementing innovations of Chapter 4, where bird beaks were given as an example (Figure 4.2). Structures that are adapted for different ways of life may be superficially dissimilar, but are nevertheless modified versions of the same ancestral structure. The beak is a variable structure that is readily selected for enhanced performance to filter plankton from water, tear apart the flesh of prey, crack thick-husked fruit, and so forth. It thereby contributes to the adaptive radiation of birds.

- Convergence and parallel evolution. Conversely, non-homologous structures may be independently modified for similar ways of life in different groups, by virtue of some potentiating innovation in the clade that includes these groups. A familiar example is the evolution of wings in tetrapods and insects, made possible by the regional specialization of the body in Bilateria. The same applies to the loss of structures: within the clade Tetrapoda, for example, limbless forms such as caecilians, amphisbaenids, and typhlopid snakes have independently evolved for a permanently subterranean way of life (Figure 9.11).

Adaptive radiation and parallel evolution document the course of evolution by showing how contemporary species bear modified versions of ancestral characters. They provide examples of cases where the ancestral constraint has been overcome: different structures have evolved, despite sharing the same ancestral state, or similar structures have evolved from very different ancestral states. In other cases, the ancestral constraint is stronger, and adaptation fails to occur at all.

- Failure of divergence: some kinds of structure evolve from the ancestral state of a particular group, whereas others do not.
- Failure of convergence: some groups adapt to a particular way of life, whereas others do not.

The four possible outcomes of convergent and divergent selection are shown schematically in Figure 9.12. These provide the characteristic features of evolutionary engineering, and we shall examine them one by one in the next four sections.

9.3.1 Divergence leads to adaptive radiation.

Three major adaptive radiations of vertebrates were described in Chapter 4: placental mammals, including the parallel radiations of Afrotheria and Laurasiatheria, marsupials, and birds. Many large groups, such as insects, mollusks, and flowering plants, show even more extensive radiations. The cases that most clearly illustrate evolutionary principles, however, are those that arise when a relatively small group occupies an extensive but isolated site, where, in the absence of well-adapted competitors, it adapts to many ways of life that are elsewhere occupied by quite different groups. There are many examples of island radiations, including the cichlid fishes of Lake Victoria, Darwin's finches on the Galapagos Islands, and the silversword plants of Hawaii. Here, we describe two less familiar but equally illuminating examples of how ecological opportunities are rapidly filled when well-adapted competitors are absent.

The Baikal amphipods have radiated into an extraordinary range of forms. Lake Baikal is a vast, ancient, isolated lake in eastern Siberia. It has a volume of 23,000 km³ and a maximum depth of 1620 m; for comparison, Lake Superior has a volume of 12,100 km³ and a maximum depth of

Figure 9.11 Convergent evolution of limbless tetrapods from limbed ancestors. (a) *Gymnophis* (caecilian: Amphibia, Caecilia). (b) *Amphisbaena* (amphisbaenid: Lepidosauria, Amphisbaenidae). (c) *Anguis* (lizard: Lepidosauria, Anguidae). (d) *Typhlops* (snake: Lepidosauria, Typhlopidae).

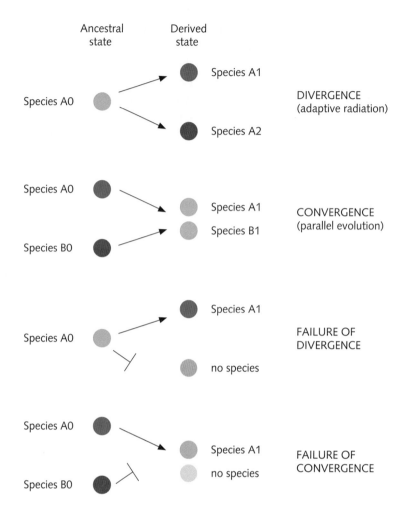

Figure 9.12 Convergent and divergent adaptation. Circles represent species (or clades); colors represent character states.

406 m. Baikal holds as much water as all the Great Lakes combined. Lake Tanganyika is nearly as large and as deep, but animals and plants are confined to a shallow oxygenated layer at the surface; Baikal is a cold lake that is oxygenated at all depths. Lake Superior is a young, postglacial lake, a few thousands of years old; Tanganyika is much older, dating from about 12 Mya; Baikal is older still, probably forming 20–25 Mya. It is almost a world unto itself, a separate biosphere where a unique community of animals has evolved. About 1800 species of aquatic animals have been described from Baikal; more than half of them, about 1000 species, have evolved in the lake, and are found nowhere else in the world.

The isolation of Baikal in central Asia has led to an impoverished fauna. With the exception of one group, there are barely two dozen species of fish in the lake—scarcely as many as in a pond in Ontario or Oxfordshire. The exception makes the crucial point: there are 27 species of gobies, or sculpins, that are found nowhere else. In other lakes, these are a minor component of the fish community. In Baikal, with few competitors, they have evolved so as to occupy ways of life that are elsewhere occupied by many different kinds of fish.

Baikal also has few kinds of crustaceans. In particular, the crayfish and opossum shrimps of comparable lakes in North America, or the freshwater crabs of the African Rift Valley lakes, are absent. This ecological gap has been filled by an extraordinary radiation of amphipods belonging to a single family, the Gammaridae, known as freshwater shrimps, or scud. This is a widespread but rather inconspicuous group of freshwater and marine crustaceans, normally small, laterally flattened animals a few millimeters long living under stones near the edge of the water, creeping on the bottom and feeding on algae and detritus. They are evolving in Baikal to exploit the opportunities left open by the absence of well-adapted competitors.

One group consists mainly of small, smooth-bodied forms that burrow in sand and sediment, consuming detritus. Another includes stout-bodied, armored forms with numerous spines and processes that forage in the littoral zone, consuming invertebrates. Some are large scavengers up to six inches long, capable of eating dead fish. Others are predators on other invertebrates. Some have become adapted to the abyssal depths of the lake, eyeless forms without pigment. There are parasitic species that live on sponges, or even on other amphipods. There is even an abundant pelagic species that lives in the open water of the lake. Some idea of the extent of this radiation is given in Figure 9.13.

The Baikal amphipods are a dramatic example of the power of natural selection to drive divergent

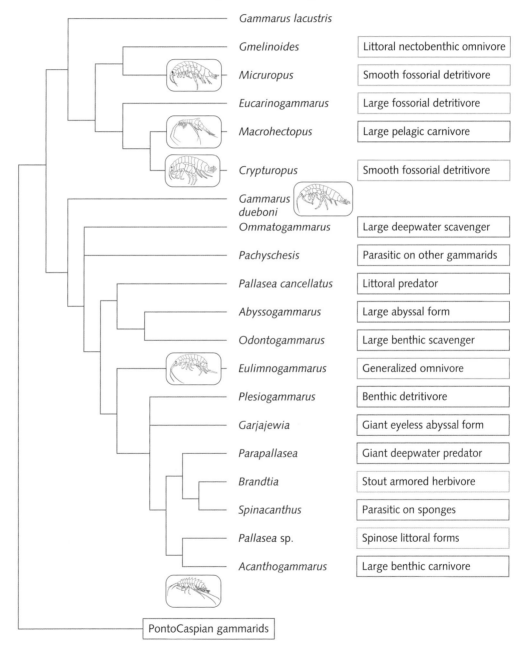

Figure 9.13 Radiation of gammarids (freshwater shrimps) in Lake Baikal.

Reproduced from McDonald, K.S., Yampolsky, L. and Duffy, J.E. (2005) 'Molecular and morphological evolution of the amphipod radiation of Lake Baikal', *Molecular Phylogenetics and Evolution* 35; 323-343 Elsevier specialization to different ways of life, once the ecological opportunity is presented.

Flightless birds evolved into a clade of browsers in New Zealand. New Zealand is so isolated in the Southern Ocean that it was the last considerable area of habitable land colonized by humans, who arrived about 700 years ago. It is an ancient fragment of land that separated from the main Australia/Antarctica land mass about 80 Mya. Its isolation prevented many modern groups, including mammals, from reaching it, while its age provided the time required for the evolution of a distinctive flora and fauna.

When the first human bands arrived, they found no mammals other than a few bats. What they did find were many kinds of large flightless birds, belonging to a group (the Ratites) elsewhere represented by ostriches, emus, rheas, and cassowaries. These were

the moa. They were exterminated within a hundred years or so, but their subfossil remains are sufficiently abundant and well-preserved to give us a clear idea of their structure and ways of life. They even still contain traces of DNA that can be used to reconstruct their phylogeny.

The radiation of the moa is shown in Figure 9.14. It seems to have been associated with the cooling trend towards the end of the Miocene, about 10 Mya. In the absence of ungulates and other herbivorous mammals, they adapted to browsing ways of life that were elsewhere already occupied by deer and similar mammals. The ancestral form lived on the edge of the forest in high-altitude grasslands. Its descendants became specialized for eating berries, or tough leaves, or even twigs and branches. At the same time, they evolved a wide range of body sizes, from relatively small birds about 15 kg in weight to giants of more

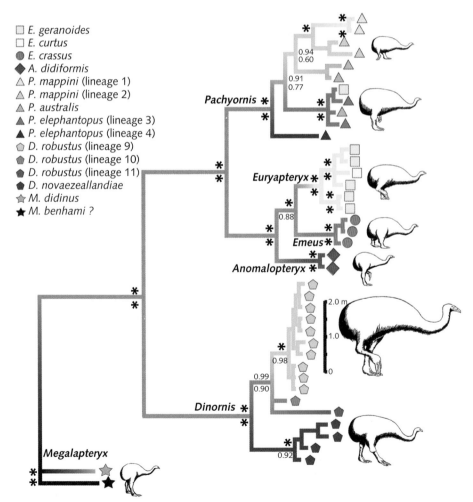

Figure 9.14 Radiation of the moa of New Zealand.

Reproduced from Baker et al. Reconstructing the tempo and mode of evolution in an extinct clade of birds with ancient DNA: the giant moas of New Zealand. *PNAS* 102 (23) © 2005 National Academy of Sciences, U.S.A.

than 200 kg. Moas are a good example of how extensive divergence in structure and lifestyle can evolve over a few million years, in the absence of previously adapted competitors.

9.3.2 Divergence often fails to evolve because of ancestral constraint.

The divergent specialization of homologous structures is a strong signal of adaptation through natural selection acting to modify an ancestral state. There is no guarantee that it will succeed, however, and we expect to find many cases in which adaptation fails to occur because selection cannot drive incremental improvement from a particular ancestral state. This would lead to noticeable disparity among clades in the extent of diversification. In Chapter 5 we came across many examples in which a clade comprised only a few species whereas its sister clade comprised many thousands of species.

- Monotremes and therians
- Birds and crocodiles
- Lepidosaurs and rhynchocephalians
- Teleosts and holosteans

In these cases, one taxon possesses an innovation that enables it to adapt to many ways of life, whereas the other does not. Several groups of birds have independently evolved nectarivory, but there are no nectarivorous crocodiles, for an obvious reason: it is unlikely that a large aquatic fish-eater would evolve into a small nectar-sipper through a series of functional intermediate forms. A less obvious but particularly revealing situation is when a group diversifies widely but nevertheless fails to adapt to particular ways of life.

Marsupials have not evolved into fully-adapted aquatic forms. If another group has previously become specialized for a particular way of life it might be difficult for competitive types to evolve, because in the initial stages of adaptation they will be inferior to their already adapted relatives. We have compared the radiation of marsupials, for example, with that of placental mammals (Section 5.6). In most cases, they have adapted to similar ways of life—burrowing, eating ants and termites, browsing, and so forth. There

are some exceptions, however, and in particular there are no fully-adapted aquatic marsupials comparable with whales, manatees, and seals. This might be an ancestral constraint, since the marsupial pouch must be a handicap in swimming. Nevertheless, there is an aquatic possum (*Chironectes*), living along rivers in South America, that has overcome this difficulty by evolving a muscular closure. (Remarkably, there is a pouch in males, too: it encloses the scrotum, streamlining the body.) The only placental mammals native to Australia, indeed, are those which can readily disperse to a distant land mass: cetaceans and bats. The absence of fully aquatic and flying marsupials is most simply explained by the prior adaptation to these ways of life by placental mammals.

Insects have not evolved into large terrestrial forms. Insects have become adapted to practically all the ways of life available to terrestrial animals between a millimeter and a centimeter in size. Some are smaller, tiny parasitic wasps little more than a tenth of a millimeter long. A few are larger; the largest are the Goliath beetles and walking sticks that may weigh as much as a mouse. Large insects are the exception, however, and very few approach the size of the average frog, lizard, or mammal. The main reason for this is that insects breathe through pores in their cuticle, the tracheae. These are long narrow tubes through which oxygen diffuses from the air. The rate of diffusion is proportional to the length of the tube, and is inadequate to support aerobic metabolism when the tube is more than a few millimeters in length. There are some ways of palliating this—pumping air by inflating the abdomen, for example, or simply increasing the number of tracheae. The reliance on diffusion, however, is a fundamental limit to insect size. It may not be a coincidence that much larger insects evolved in the Carboniferous, including dragonflies with a wingspan of a foot or more, when oxygen levels were much higher than they are now.

An insect might well respond that we have got this the wrong way round: a body plan based on lungs and pumped blood means that tetrapods cannot evolve very small size, and are thereby excluded from many of the ways of life followed by insects. The very smallest tetrapods (currently the New Guinea frog *Paedophryne*) are much bigger than the average insect, and tetrapods the size of a small wasp or fly could not function at all, since there would be no space for the circulatory system they depend on. Whichever point of view you take, however, the conclusion is the same: insects and tetrapods are both restricted to a characteristic range of sizes because both are limited by ancestral body plans that cannot be switched by any process of gradual modification.

9.3.3 Convergence leads to parallel evolution.

Different groups may independently evolve similar adaptations to a given way of life. We illustrated adaptive radiation in Chapter 4 in terms of innovations, ancestral characters that contribute to diversification. We can also recognize "themes" in evolved designs, such as the flipper-like forelimbs of marine tetrapods such as whales, seals, manatees, plesiosaurs, and ichthyosaurs. Each case represents a variation on this theme, a shared derived character (or set of characters) associated with a particular way of life that evolves independently several times within a clade. A few innovations are unique, such as the water vascular system of echinoderms, but in most cases innovations become themes on a broader phylogenetic scale. Warm blood is an important innovation for the clade Mammalia, for example, but it is a theme for the clade Tetrapoda because it evolved independently in mammals and birds.

- Variations on a theme may evolve from homologous characters in the ancestral lineages whose structure and development is encoded by copies of the same genes. An example would be the long snout, sticky tongue and loss of teeth found in mammals that eat ants or termites, a set of correlated characters that has evolved independently in anteaters, pangolins, aardvarks and echidnas within the clade Mammalia (Section 4.5).

- Among less closely-related lineages, variations on a theme may evolve from non-homologous characters whose structure and development are encoded by copies of different genes. The wings of birds and bats (within the clade Amniota) are homologous because both have evolved as modifications of the tetrapod forelimb. The wings of flying vertebrates and flying insects (within the clade Bilateria) are non-homologous because insect wings have evolved as modifications of

segmental appendages derived from outgrowths of the body wall.

This is a useful but not (as we have already emphasized) an absolute distinction. Vertebrate and insect wings are certainly quite different, but similar signaling molecules are involved in their development: for example, the product of a "hedgehog"-like gene is involved in determining the antero-posterior axis of both the chick wing bud and the fly imaginal wing disc.

It is possible to make a distinction between the convergence of clearly homologous characters in closely-related organisms ("parallel evolution") and the convergence of non-homologous characters in distantly-related organisms ("convergent evolution"). Because this is a matter of degree, we refer to both as parallel evolution: the vital point is that similar structures may evolve from different starting points, but always retain the signature of the ancestral state.

Parallel evolution is commonplace in vertebrates. We described several examples of parallel evolution at different phylogenetic levels among vertebrates in Chapter 4.

- Leaf-eating in primates (4.4).
- Modes of placentation in mammals (4.5).
- Spiny armor in rodents (porcupines), insectivores (hedgehogs), afrotherians (tenrecs), and monotremes (echidnas) (4.6).
- Ways of life such as leaf-eating, browsing, and gnawing in placental mammals and marsupials (4.6).
- Antifreeze and electrolocation in fish (4.10).
- Electrogenesis in cartilaginous and bony fish (4.11).
- Placentation in sharks and mammals (4.11).

In each case, we can use phylogenetic analysis to trace the independent routes to adaptation. Figure 9.15 shows a current phylogeny of electric fish, showing that the capacity to generate and detect electric currents has evolved several times, probably about eight times altogether. It is characteristic of benthic ambush predators, especially those hunting in turbid water. In all cases the basic unit involved is a modified muscle cell, and electric organs are all quite similar, although they may be located in different parts

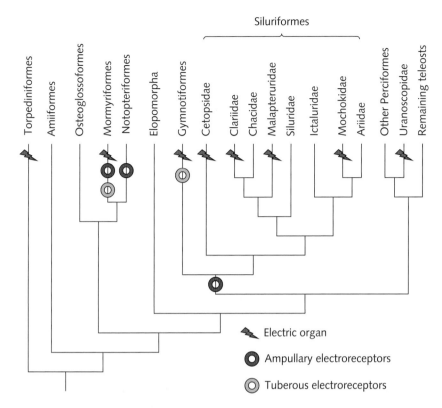

Figure 9.15 Independent evolution of electric organs in fish.

of the body. Moreover, very similar changes in the sequence and expression of a sodium channel gene are associated with electrogenesis in the two main weakly electric groups, the mormyriform fishes of Africa and the gymnotid fishes of South America. It seems likely that most of the lineages of electric fish have evolved in similar ways, even to the extent of incorporating similar changes in homologous genes.

There are many other striking cases of similar structures that have evolved independently in several groups.

- Billed snouts in ducks, platypus, and hadrosaurs (duck-billed dinosaurs).
- Narrow, tusked snouts in fish-eaters such as crocodiles, phytosaurs, gars, and pike.
- A bivalve shell in mollusks, brachiopods, and ostracods.
- The succulent body form of desert plants such as cacti and euphorbs.
- Silk production in spiders, silk moths, and caddis flies.
- Halteres in Diptera and Strepsiptera.

Similar ways of life have also evolved independently in different groups.

- Insectivory among plants living in acidic, nitrogen-poor bogs.
- Parasitic vines such as Loranthaceae and Viscaceae.
- Very large plankton-sievers such as baleen whales and whale sharks.
- Echolocation in bats, oilbirds, and cave swiftlets.
- Parasitic plants lacking chlorophyll.
- Hovering nectar-feeding in humming-birds, honeycreepers, sunbirds, and sphinx moths.

Eyes evolve independently but are controlled by the same regulatory genes. At deeper phylogenetic levels, ancestral states become increasingly different, giving rise to more severe constraints, and clear examples of convergent evolution become less frequent. Similar adaptations by widely-separated lineages do occur, however, and illustrate well the persistence of fundamental homologies in the development of the characters involved.

Simple eye-spots consisting of a patch of pigment on a membrane occur in many eukaryotes, including many groups of unicellular protists. An eye capable of collecting and focusing light is found in five metazoan clades: Cnidaria, Panarthropoda, Annelida, Mollusca, and Chordata. Their eyes are very diverse, but the range of possible eye structures is limited by simple physical constraints. An image can be formed by shadows (pinhole eyes), by refraction (eyes with lenses), or by reflection (eyes with mirrors). The eye itself may be a single unit, or may be a compound structure made up of many simple units. Consequently, there are $2^3 = 8$ kinds of eye, all of which are found in animals (Figure 9.16). Hence, selection has been able to generate the complete range of basic morphological structures.

On the other hand, eyes detect only a fraction of the full range of wavelengths found in sunlight. This is because eyes evolved in marine animals, and water filters out most of the wavelengths in sunlight. The limited range of the eyes of terrestrial animals thus represents a general ancestral constraint, although receptors for ultraviolet light have independently evolved in some insects and vertebrates.

The details of eye structure are tailored by natural selection for different ways of life. The ability to resolve an image is about 10,000 times greater in eagles than in flatworms. Animals living in the dark depths of the ocean, on the other hand, have eyes that are about 100,000 times more sensitive to light than the eyes of an eagle.

We have already met the camera-type lensed eyes of vertebrates and cephalopods as examples of the independent evolution of complex structures (Section 1.3.4). Despite their similarity, they develop in quite different ways: vertebrate eyes develop from the neural plate, inducing the overlying epithelium to form a lens, whereas cephalopod eyes develop by the

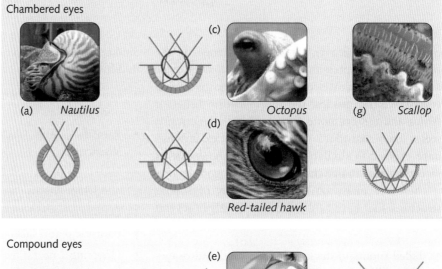

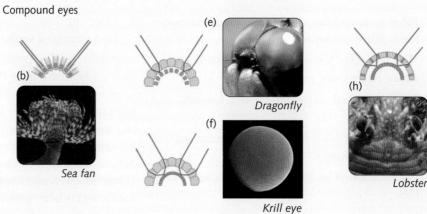

Figure 9.16 The range of possible eye designs. The diagrams illustrate simple and compound eyes using shadows (A and B), refraction (C to F), and reflection (G and H).

Reproduced from Fernald, R.D. 2006. Casting a genetic light on the evolution of eyes. *Science* 313 (5795): 1914–1918.

infolding of a sheet of epidermal cells. This leads to morphological differences in the structure of retina and cornea (cephalopods have no cornea).

There is also an important cellular difference between them. All eyes use a membrane-bound opsin pigment to absorb light, but it is packaged in two ways. Vertebrate eyes are ciliary eyes, in which the optical pigment is packed into modified cilia, whereas cephalopods have rhabdomeric eyes in which it is located in microvillar extensions of the cell membrane. The compound eyes of arthropods are a very different kind of rhabdomeric eye made up of repeated units, and clearly represent another independent evolution of image-forming eyes. Ciliary and rhabdomeric eyes use different opsins and have different signal-transduction mechanisms to translate the interception of photons into a nerve impulse. Eyes are a good example of a complex structure evolving independently in all three major bilaterian lineages: deuterostomes, lophotrochozoans, and ecdysozoans.

Nevertheless, the development of eyes shows remarkable parallels in all three lineages. In particular, there is the stunning demonstration that expressing the *Pax6* gene from mice induces eye development in *Drosophila*, and the corresponding gene from *Drosophila* induces the formation of eye structures in frogs. Clearly, one of the master regulatory genes involved in eye development is conserved in all Bilateria. Indeed, *Pax6* genes from squid, tunicates, and amphioxus are all capable of inducing eye development in *Drosophila*.

Many of the genes governing eye development in flies and frogs are quite different, of course. The involvement of the same regulatory genes, however, shows that very different kinds of animal eyes are independent modifications of a common ancestral state. This is a good example of the graded series of homologies that connects structures with similar function in distantly related organisms. Very different outcomes still bear the stamp of common ancestry.

Very small aquatic animals have similar adaptations for swimming. Instead of asking how similar structures evolve in different groups, we can instead ask how different groups adapt to similar ways of life. A good example is provided by very small aquatic metazoans, less than 1 mm in length, which evolve towards a common body plan because of the physical forces that govern motion through a fluid.

We are large organisms, and this strongly influences how we view the world. The motion of large fast animals such as sharks and whales through water is limited almost entirely by pressure drag, the difference in pressure between the leading and the trailing ends of the body (Section 1.3.1). There is a second source of drag, however, created by the friction of the surface of the body against the water. The pressure drag is created by inertia, the tendency of a heavy body to resist acceleration but, once impelled, to continue in motion. The friction drag is caused by viscosity, the tendency of the fluid to resist motion. The key to understanding how swimming organisms are constructed is that the ratio of inertial forces to viscous forces depends on body size. For large fast swimmers inertial forces predominate, so an intermittent powerful thrust and a streamlined body maximize velocity, leading to a similar conformation of the body in fish, whales, ichthyosaurs or squid. For small swimmers, however, less than a millimeter in length, viscous forces predominate and inertia is negligible. This leads to profound differences in how force is supplied and how bodies are constructed.

Suppose that you are swimming breaststroke in a pool, making a power stroke with arms and legs and then gliding before the next stroke. Clearly, the glide could take you for several meters, over ten or twenty seconds. This is because you are a large organism where a past action will continue to impel your body through inertia. Now suppose you are a very small organism, a bacterium, swimming in the same pool. How far will you travel before you stop, and how long will this take? The answer is that you will travel less than 1 Å (the diameter of a hydrogen atom) and stop in about 3/1000 of a second. To understand this situation, imagine that you are swimming in tar, unable to glide because the medium is too viscous.

Consequently, small swimming organisms must be constructed in a particular way: they cannot function simply as scale models of large organisms. In particular, they cannot use a powerful intermittent propeller, such as a tail fin powered by muscles. They must instead use a continuously acting low-power propeller, so that all distance gained is due to present activity and none to past activity. This low-power propeller is the eukaryote flagellum. A flagellum is a flexible oar. Ordinary stiff oars would not work, if you were submerged, because they would produce reciprocal motion—the recovery stroke would take you as far backward as the power stroke takes you forward.

There are two ways of getting round this. One is to use rotary motion, like a corkscrew. The other is to use sinusoidal motion, like an eel. Both techniques are employed by unicellular organisms, but there is a very sharp phylogenetic divide: swimming bacteria use corkscrews, whereas swimming algae use an eel-like motion. This is a good example of an ancestral constraint: bacteria have not evolved the sinusoidal flagellum of eukaryotes because it is formed of an extension of the cytoskeleton, which bacteria do not possess.

The thrust developed by a flagellum is proportional to its length, and in principle a small metazoan could be propelled by a sufficiently long flagellum. In practice this would not work, because a very long flagellum would be distorted by water currents, and would be unable to produce directional motion. An equivalent force can be generated, however, by the coordinated beat of many short flagella, which we call cilia. Consequently, small aquatic organisms such as ciliates, rotifers and the larvae of marine invertebrates are all propelled by bands of cilia (Figure 9.17). In larger organisms cilia are only effective if they are somehow grouped together, as in the comb rows of ctenophores. For the most part, however, ciliary locomotion is impracticable in animals more than about 1 mm in size, and is replaced by swimming or paddling powered by muscles.

9.3.4 Convergence often fails to evolve because of ancestral constraint.

Parallel evolution is an eloquent witness of evolution, because it shows how superficially similar structures have descended through the modification of different ancestral states. The reverse of the coin, however, is that structures may fail to evolve because the ancestral state of a group prevents, or at least seriously hinders, its members from evolving in some directions.

Angiosperms and insects have failed to become fully marine. Terrestrial communities are dominated by flowering plants and insects, which are the most abundant and diverse types of plants and animals in

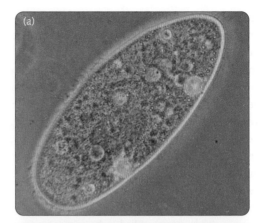
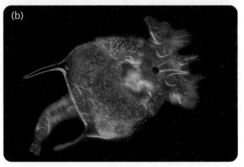
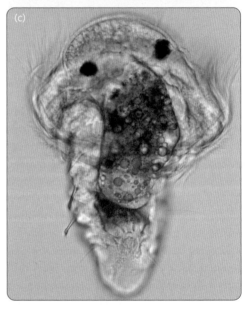

Figure 9.17 Three minute aquatic organisms from different major groups, all propelled by bands of cilia. (a) Ciliate; (b) Rotifer; (c) Trochophore larva of annelid.

most parts of the world. Flowering plants appeared in the Cretaceous, about 125 Mya; insects became abundant in the Carboniferous, over 300 Mya. Uncountable billions of flowering plants and insects have lived in close proximity to the ocean for millions of years, and yet have failed to make more than the most modest advance towards a fully marine way of life. There are examples in the other direction, too: echinoderms and chaetognaths are just two examples of abundant marine groups that have failed to spread into fresh water. The presence of an ecological opportunity is no guarantee that adaptation will occur, even in very abundant groups over very long periods of time.

Animals have not evolved cellulose-based skeletons. Multicellular organisms are supported and protected by skeletons of strong, stiff material. Metals, ceramics, and plastics would be useful, but cannot be made in conditions tolerated by carbon-based organisms. Instead, two other ways of supporting multicellular bodies have evolved.

- The first is to deposit a mineral, usually calcium carbonate, over the body surface to form a protective layer, as in corals and snails. This shields the vulnerable body behind a thick wall. The draw-back is that it prevents growth, so corals grow as colonies of zooids each building a new chamber, whereas snails must continually create new chambers of the shell for the growing body to inhabit.

- The second is to construct a composite material from fibrils embedded in a ground substance, the method used to fabricate fiberglass or carbon fiber. This is more complicated and expensive, but can yield light, tough and flexible materials.

The basal split between bikonts and unikonts is accompanied by the almost exclusive use of different structural fibers by these two lineages: cellulose in bikonts and chitin in unikonts. These are the most abundant biological molecules on Earth, making up about 50% of all biological production.

The basic chain of cellulose is a polymer of glucose. Bundles of chains are linked by hydrogen bonds to form microfibrils embedded in the cell wall. When these are oriented parallel to the long axis of the cell,

cell division produces stiff fibers capable of supporting plant stems. Cellulose is found in the cell walls of multicellular algae (rhodophytes, phaeophytes, and chlorophytes), but often as only a minor component. Algae often use polymers of other sugars, especially mannose and xylose. These are not as strong, nor as chemically stable, as cellulose, but for organisms supported by seawater this is not a crucial factor. Cellulose provides structural support for bryophytes, the first land plants, but it loses stiffness when wet and, by itself, will not support tall stems. Vascular plants waterproof their cellulose with pectin, hemicellulose, and, above all, lignin. This extraordinary substance is a polymer (of three different alcohols), but has no regular structure—the basic units have no set repeating pattern. This makes it highly resistant to enzymatic attack. Indeed, bacteria have never evolved to metabolize lignin, an almost unprecedented failure in bacterial history. Instead, it falls only to brute-force attack by peroxides secreted by basidiomycete fungi, after which the fragments can be mopped up by bacteria. The combination of cellulose and lignin, which we know as wood, is the implementing innovation that underlies the evolution of trees.

Animals and fungi, by contrast, use chitin rather than cellulose as a structural fiber (Section 4.16). Where did cellulose and chitin come from? Cellulose is produced by many bacteria. If you leave a liquid culture of *Pseudomonas* undisturbed on the lab bench for two or three days, for example, you will often observe that a thick film has formed on the surface. This is caused by selection for mutants which extrude cellulose fibrils that glue the cells together, giving them priority for access to atmospheric oxygen (the evolutionary dynamics of this behavior are described in Section 16.1.2). Chitin is not known to be used as a structural fiber by bacteria, but nitrogen-fixing rhizobial bacteria that populate the root nodules of legumes produce a four-sugar version that is involved in host-plant recognition. Hence, the potential to produce cellulose and chitin was already present in the bacterial ancestors of eukaryotes.

The restriction of cellulose and chitin to bikonts and unikonts respectively is not perfectly complete. Surprisingly, cellulose is found in tunicates, a rather late-diverging lineage of animals. The explanation is that tunicates seem to have acquired the ability to synthesize cellulose by horizontal transfer of a gene from a bacterium, so the exception proves the rule. This apart, the two major structural proteins used by multicellular organisms are a good example of the pervasive constraint of ancestry.

9.3.5 Unique adaptations are examples of failure of convergence.

Convergence always creates difficulties in classifying organisms, because similar structures may be evolved by unrelated lineages. The most desirable characters for classifying organisms, therefore, are unique shared derived characters, which evolve only once. Some of these are also witnesses of powerful ancestral constraint. If a large, diverse and abundant group is characterized by a shared derived character that occurs nowhere else, this demonstrates the retention of an ancestral character by a wide range of successful forms. If the character is clearly useful, it also demonstrates the inability of other groups to evolve the same character.

Nematocysts are unique to cnidarians. The tentacles of cnidarians contain a unique specialized cell, the nematocyst, that unleashes a poisoned harpoon when it is triggered by contact (Figure 9.18(a)). This is a very effective prey-catching device capable of immobilizing large prey—large jellyfish,

indeed, can even be dangerous for bathers who bump into them. It has evolved only once. There is, indeed, a very similar structure in the "polar capsules" of myxozoans, an enigmatic group of unicellular parasites living inside the cells of marine invertebrates. Molecular evidence has recently shown that myxozoans are closely related to cnidarians—that they are, in fact, a highly reduced parasitic jellyfish. This neatly confirms the unique origin of the cnidarian nematocyst.

Stereom is unique to echinoderms. In contrast to the ancestral constraint on the use of structures involving cellulose, lignin, chitin, cartilage, and bone, calcification has evolved many times in unicellular and multicellular organisms, including coccolithophores, foraminiferans, rhodophytes, charophytes, corals, bryozoans, mollusks, and echinoderms. It is a low-cost way of producing a strong but heavy and inflexible skeleton. Even in this case, however, the manner in which calcium carbonate is mobilized has a distinctive phylogenetic signature. The skeleton of echinoderms, for example, is made up of elements (such as spines) that seem to be single crystals of calcite. In life, each calcite block of the skeleton is penetrated by an intricate series of narrow canals, forming a "stereom" (Figure 9.18(b)). This unique structure, inherited from Cambrian ancestors, makes it easy to identify fossil echinoderms, even when their outward appearance is different from modern

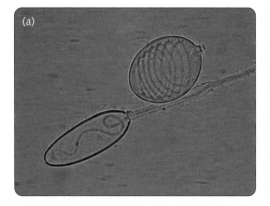
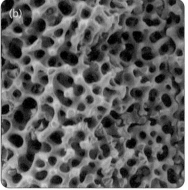

Figure 9.18 Unique adaptations. (a) Nematocyst of Cnidaria. In the undischarged nematocyst, the shaft is coiled inside the cell; when stimulated the shaft is forcibly everted, penetrates the prey, and delivers a dose of toxin. (b) Stereom of Echinodermata. The skeleton of echinoderms is made up of calcite blocks with a characteristic foam-like structure.

Nematocyst image courtesy of Teresa Carrette, James Cook University. Stereom image © 1994-2014 by the Regents of the University of California, all rights reserved.

forms. Why it should have evolved has not yet been satisfactorily explained, however: the stereom is an example of an innovation that clearly reflects ancestry, but whose function is unknown.

9.4 Simple ancestral states are connected to complex descendant states—or vice versa—by a series of intermediate states.

The geological history of whales (Sections 1.3.2 and 3.1.3) and tetrapods (Section 5.1.9) shows how ancestor and descendant states are connected by a chain of intermediates. This is a very general rule that applies even to very complex, highly integrated structures. Their evolution can be traced step by step from simpler precursors; and likewise, if they cease to be functional, their gradual decay can be charted through time.

9.4.1 Complex structures evolve through a series of intermediates, each of which is favored by selection in some environment.

Many motile unicellular algae have an eyespot, consisting of a patch of pigment on a membrane. It is a very simple kind of eye indeed: it has no cornea, no iris, no lens, no retina and no nerve. It cannot form an image, nor even detect movement. All it can do is to detect light. The light is absorbed by the pigment, which produces a signal across the membrane, which triggers a change in behavior—the flagella alter their beat so as to propel the cell towards the light. Even the simplest eye may play an important part in the life of an organism.

All grades of eye structure are found among mollusks. Other kinds of eye are much more complex, of course, and it is not obvious how they can have evolved by a process of gradual modification. However, we do not have to rely on imagination to show that the transition from the simplest to the most complex can be traced through a series of intermediate forms. We can instead point to these series among extant organisms such as mollusks.

- Some mollusks have no trace of eyes. Monoplacophorans are bizarre segmented mollusks whose only living representative closely resembles its early Paleozoic ancestors. It lives in the lightless abyssal depths of the ocean, where an eye would serve no function, and has presumably never evolved. Aplacophorans, which are sluggish wormlike mollusks living on the sea bed, are also primitively blind. Scaphopods are the tusk-shaped shelled mollusks that burrow in marine sands and sediments; since they burrow head-down they need no eyes, but they probably evolved from eyed ancestors, so that their lack of eyes is derived. This is certainly true for blind mollusks whose nearest relatives are sighted, but which themselves live in environments where vision is not required. *Janthina* is a good example; it is a pelagic marine snail that is buoyed up by a cluster of bubbles and drifts with wind and tide, eating jellyfish whenever it bumps into them. Although it lives in the open, it has no ability either to pursue its prey or to flee from its predators, and therefore no need to see them either.

- There are several opisthobranch mollusks that also eat jellyfish, but are able to swim clumsily by undulating their whole body, and these often have very simple eyes. Perhaps the simplest is found in the nudibranch *Phylliroe:* merely a patch of pigment granules underlying an unmodified epidermis, and closely appressed to the pedal ganglion. It is no more complicated than the eyespot of an alga, and its usefulness is likewise restricted to telling the difference between light and dark. In other nudibranchs and pteropods the eye is a distinct vesicle embedded in the dermis, but is otherwise no more complex or versatile. In such cases the extreme simplicity of the eye is primitive, but simple eyes, like complete blindness, may also evolve through degeneration. Freshwater snails that live in caves, for example, often have eyes that are no more than pigmented vesicles sunk in the dermis, but these are clearly derived from the much more complex eyes of their ancestors in streams and lakes.

- In some pteropods, a simple patch of pigment, or a pigment-bearing vesicle, is protected by a modified epidermal layer that represents the first approach to a cornea. *Styliola* has an eye of this kind, in which the pigmented vesicle is supplied with a single nerve cell. At this point, then, we have an eye with a very limited functional range but with three components: a device for collected light, or at least permitting it to enter, a device for absorbing light, and a device for transmitting the information. In *Styliola*, these components are extremely simple. More sophisticated eyes evolve by one of two routes: as an innervated patch in a pit or as an innervated vesicle sunk in the skin.

- The eyes of limpets are open cups, consisting of a sheet of epidermal cells whose distal ends are pigmented and supplied with an optic nerve. This sheet is infolded to form a depression with a broad opening. Bivalves have eyes of this sort, often arranged in long rows so that they are able to detect the movement of a shadow overhead, triggering the closure of the valves. In other archaeogastropods the opening is narrower, and the internal space may be partly filled with a translucent gelatinous material. The pigmented zone of the epidermal cells begins to resemble a retina. The most advanced eyes of this kind are found in nautiloids, which are active pelagic carnivores able to detect and pursue prey. Their eyes have no lens, but the opening of the optic cup has become so restricted that the eye can function like a pinhole camera, capable of forming a distinct image.

- In many prosobranchs and opisthobranchs the eye is a multicellular vesicle embedded in the skin in which cells in different regions become specialized for different functions. The interior is filled with fluid, and may contain a hyaline body that represents a crude lens, capable of collecting light although not able to form an image. The cells on the outer surface of the vesicle are thinner, to admit light; those on the inner surface are heavily pigmented and constitute a retina. The eyes of polyplacophorans (chitons) and pulmonates (land snails) are basically of this type, although pulmonate eyes are more highly organized than the others. Heteropods have quite complicated eyes, in which light enters through a transparent cornea, is focused by a nearly spherical lens, and is then absorbed by a complex, layered retina. These are naked pelagic mollusks that pursue active prey.

- The most advanced eyes belong to another group of active pelagic mollusks, the cephalopods. Their eyes are in many respects astonishingly similar to those of vertebrates. They include the full range of structures: a transparent cornea, an iris diaphragm, a lens that can be adjusted by ciliary muscles, and a retina in which light-sensitive cells, their pigmented ends directed outwards, are innervated distally by nerve fibers that are collected into an optic nerve leading to a large brain.

The cephalopod eye raises the same problem as the vertebrate eye: how could so complex a structure have evolved gradually through a series of intermediate types? In mollusks, however, the mystery dissolves, because a complete series of intermediates exists, from the simplest pigment patch to the perfected optical instrument, as illustrated in Figure 9.19. The degree of sophistication of the eye is related to the

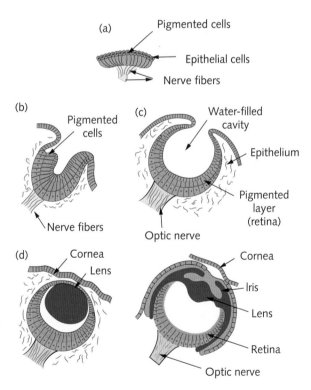

Figure 9.19 Series of eyes of increasing complexity found in mollusks.

way of life of the organism. Simple eyes are found in forms that have little need for any but the most limited vision; complex image-forming eyes are borne by active pelagic carnivores. There is no need to suppose that the most complex and highly integrated structures evolve by any process other than the cumulation of adaptive change through continued selection.

9.4.2 Vestigial structures chart the effect of removing selection.

A new way of life may lead to the elaboration of structures that are newly functional; but it may also lead to the decay of structures that are newly redundant.

Whales retain vestiges of hindlimbs. Back to whales. A large fast-swimming tetrapod propelled by a tail fin needs forelimbs (modified as paddles) as control surfaces, but has no need of hindlimbs. Consequently, selection will favor variants with reduced hindlimbs, because they have less drag and can swim more efficiently, and this will eventually lead to the complete loss of hindlimbs. The pelvic girdle is now unnecessary, and will likewise tend to be lost. This is likely to be a slow process, because it is governed chiefly by the lack of selection against mutations that cause the failure of the pelvic girdle to develop correctly. As these slowly accumulate, the girdle will degenerate, but it will persist for many generations despite its lack of current utility. Hence, whales still express a rudiment of the pelvic girdle of their distant terrestrial ancestors. Vestigial structures of this sort are an eloquent witness of the ancestral state of lineages that have adopted a novel way of life.

Troglobytes retain vestiges of eyes. Cave systems are often colonized by freshwater organisms such as shrimps, salamanders, and fish which live in perpetual darkness. Eyes are useless in caves, and loss-of-function mutations will accumulate because they are not eliminated by purifying selection. Consequently, most troglobytes—the animals you will find in deep sunless caves—are blind, although they still retain traces of the eyes their ancestors once possessed. In some cases, such as the characin fish *Astyanax mexicanus* shown in Figure 9.20, there is a pigmented surface-dwelling form with fully

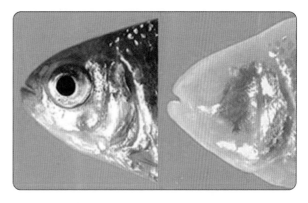

Figure 9.20 Fully sighted surface variety and unpigmented blind variety of the Mexican cave fish *Astyanax mexicanus*.
Image kindly supplied by Professor William Jeffery.

functional eyes and a blind unpigmented form living in the dark interior of caves.

Parasites are often secondarily simplified. Evolution does not always favor increased complexity and integration. Parasites exploit the resources already gathered by their hosts and so can dispense with many of the structures that free-living organisms require—eyes, limbs, and even guts. Obligate internal parasites such as tapeworms and acanthocephalans have lost many of the basic structural features of their free-living ancestors, to the point where their relationships are almost completely obscured. *Sacculina*, for example, is a rather structureless bag of tissue found attached to marine mollusks, but has a nauplius larva that identifies it as a copepod.

9.4.3 A major change in lifestyle evolves through a series of intermediate stages.

We have seen that whales, which are so highly adapted to a fully marine way of life, evolved from terrestrial artiodactyls via a series of intermediate forms (Section 3.1.3). We can trace these intermediate stages in the evolution of an aquatic lifestyle among living mammals.

All grades of aquatic specialization are found among mammals. Several lineages of mammals have become adapted for aquatic life, but the difficulties they have faced are formidable. Most terrestrial mammals can swim (giraffes are an exception, apparently), but movement relies on a weak paddling motion, body heat is quickly lost when the fur is wetted, and foraging is limited by the risk of drowning. In order to evolve into aquatic animals, a lineage must first pass through an intermediate stage in which these difficulties are overcome by modifying the structures that are adapted to terrestrial life. This intermediate stage is represented by semi-aquatic mammals such as mink, otters, and beavers. They live in or by the water, but often return to the land to eat, sleep, mate, and give birth. Semi-aquatic mammals have a characteristic suite of features.

- They live in fresh water. While fully aquatic mammals are mostly marine (river dolphins and manatees are exceptions), semi-aquatic mammals are almost always found in rivers and lakes.
- They have dense, non-wettable fur. This provides a light body covering for living on land, with sufficient insulation to work for long periods in cold water.
- They have modifications for aquatic locomotion, such as webbed feet or a long, propulsive tail.

These adaptations are quite inefficient. An animal swimming at the surface is hampered by wave drag, its kinetic energy dissipated into the potential energy of waves. A cumbersome layer of fur is not suitable for deep diving. Paddling is an inefficient means of locomotion because of the water resistance encountered by the recovery stroke. In fully aquatic mammals these difficulties are overcome by swimming fully submerged, with less drag. They have a layer of blubber, rather than fur, which provides insulation and allows the body to be streamlined. They use hydrofoils to swim, such as the tail flukes of whales. They also have physiological adaptations such as salt-secreting glands that allow them to exploit the greater possibilities of marine life.

Figure 9.21 summarizes adaptation in aquatic and semi-aquatic mammals in terms of the cost of transport (the work done in moving the body through water) and swimming speed. It shows the continuum between the weakest paddlers, such as mink, and the large, fast, efficient whales. The transition between semi-aquatic and fully aquatic mammals is somewhere between sea otters and seals.

The series of intermediate forms between terrestrial and fully aquatic mammals is documented in the fossil history of whales (Section 3.1.3). We can recognize the same stages in the evolution of

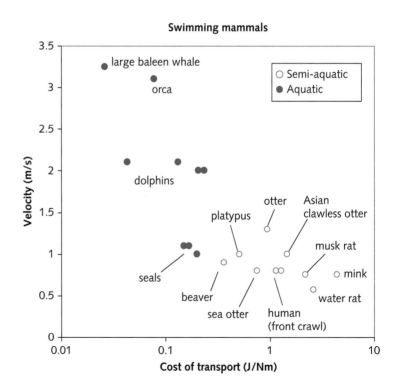

Figure 9.21 Energetics of swimming in aquatic mammals.

Based on data from Fish. 2000. *Physiological and Biochemical Zoology* 73: 683–698, Table 2.

aquatic mammals among living forms, and moreover use them to analyze the morphological and physiological changes involved in the transition to aquatic life. Despite the radical changes involved in re-engineering a terrestrial body for aquatic life, both fossil and living forms show us that the necessary succession of intermediate forms is in fact possible and does occur.

● CHAPTER SUMMARY

Corresponding features of related species are homologous.
The degree of homology varies with phylogenetic level.
Homology expresses the ancestral constraint on the evolution of utility.

Utility favors optimal design.

- The fundamental features of body plans are effective adaptations to particular ways of life.
  - *Metazoan body plans are adaptive.*
  - *Green plant body plans are also adaptive.*
- Small details of structure can be important adaptations.
  - *The extraordinary jaw of* Epibulus *is precisely adapted to catch prey.*
  - *The Y-skeleton of* Polyxenus *is precisely adapted for running on smooth surfaces.*
- The design of organisms is often closely fitted to their conditions of life by natural selection.
  - *Optimal structures are compromises.*
  - *Bones have evolved to resist the stress they normally experience.*
  - *The design of membranes balances the gain and loss of gases.*

The optimal design of evolved structures is constrained by ancestry.

- Adaptation is constrained by ancestry.
  - *The ancestral constraint is lack of appropriate variation.*
  - *Ancestral constraint may not be absolute.*
  - *The first lineage to evolve may exclude others.*
- Primitive characters are more highly constrained.
  - *Ancestral constraint obstructs the independent evolution of similar characters in different lineages.*
  - *All organisms share a small set of precursor metabolites.*

Evolutionary engineering combines utility and ancestry.

- Divergence leads to adaptive radiation.
  - *The Baikal amphipods have radiated into an extraordinary range of forms.*
  - *Flightless birds evolved into a clade of browsers in New Zealand.*
- Divergence often fails to evolve because of ancestral constraint.
  - *Marsupials have not evolved into fully-adapted aquatic forms.*
  - *Insects have not evolved into large terrestrial forms.*
- Convergence leads to parallel evolution.
  - *Different groups may independently evolve similar adaptations to a given way of life.*
  - *Parallel evolution is commonplace in vertebrates.*

- *Eyes evolve independently but are controlled by the same regulatory genes.*
- *Very small aquatic animals have similar adaptations for swimming.*

- Convergence often fails to evolve because of ancestral constraint.
  - *Angiosperms and insects have failed to become fully marine.*
  - *Animals have not evolved cellulose-based skeletons.*

- Unique adaptations are examples of failure of convergence.
  - *Nematocysts are unique to cnidarians.*
  - *Stereom is unique to echinoderms.*

Simple ancestral states are connected to complex descendant states—or vice versa—by a series of intermediate states.

- Complex structures evolve through a series of intermediates, each of which is favored by selection in some environment.
  - *All grades of eye structure are found among mollusks.*

- Vestigial structures chart the effect of removing selection.
  - *Whales retain vestiges of hindlimbs.*
  - *Troglobytes retain vestiges of eyes.*
  - *Parasites are often secondarily simplified.*

- A major change in lifestyle evolves through a series of intermediate stages.
  - *All grades of aquatic specialization are found among mammals.*

● FURTHER READING

Here are some pertinent further readings at the time of going to press. For relevant readings that have been released since publication, visit the book's Online Resource Centre at **www.oxfordtextbooks.co.uk/orc/bell_evolution/**

Section 9.1   Cain, A.J.1964. The perfection of animals. *Viewpoints in Biology* 3: 36–63 (see also *Biological Journal of the Linnean Society* 36: 3–29).

Westoby, M. and Wright, I.J. 2008. Land-plant ecology on the basis of functional traits. *Trends in Ecology and Evolution* 21: 261–268.

Section 9.2   Gould, S.J. and Lewontin, R.C. 1979. The spandrels of San Marco and the Panglossian paradigm: a critique of the adaptationist programme. *Proceedings of the Royal Society of London B* 205: 581–598.

Section 9.3   Sherbakov, D.Y. 1999. Molecular phylogenetic studies on the origin of biodiversity in Lake Baikal. *Trends in Ecology and Evolution* 14: 92–95.

Baker, A.J. et al. 2005. Reconstructing the tempo and mode of evolution in an extinct clade of birds with ancient DNA: the giant moas of New Zealand. *Proceedings of the National Academy of Sciences of the USA* 102: 8257–8262.

Fernald, R.D. 2006. Casting a genetic light on the evolution of eyes. *Science* 313: 1914–1918.

Section 9.4   Land, M.F. and Fernald, R.D. 1992. The evolution of eyes. *Annual Reviews of Neuroscience* 15: 1–29.

Fish, F.E. 2000. Biomechanics and energetics in aquatic and semiaquatic mammals: platypus to whale. *Physiological and Biochemical Zoology* 73: 683–698.

● QUESTIONS

1. The major features of extensive clades, such as body cavities or roots, can be interpreted as: (a) accurate indicators of ancestry, or (b) adaptations to particular ways of life. Attempt to reconcile these two points of view.

2. Explain the concept of optimality, giving examples of situations in which two functions cannot be maximized simultaneously.

3. As a practical exercise, visit your local supermarket and measure the height and diameter of canned foods. Write an account of your findings and evaluate how they can be related to optimality theory.

4. As a practical exercise, visit your local museum and measure the height and diameter of gastropod shells. Write an account of your findings and evaluate how they can be related to optimality theory.

5. Once you have performed (3) and (4), compare and contrast the morphology of food cans and snail shells.

6. Discuss the statement that: "The limb of a horse is homologous to the limb of a frog but not homologous to the limb of a grasshopper."

7. Write an essay on adaptive radiation, with examples from animals and plants.

8. Identify the conditions that will promote or obstruct adaptive radiation when a species is introduced into a new region.

9. To what extent does the ancestral character state affect the subsequent evolution of that character in a clade?

10. Why are there many more endemic species in Lake Baikal than in Lake Superior?

11. Describe and account for the failure of a major group to evolve specialization to a particular environment.

12. What is convergent evolution? Explain why it occurs, with examples from as wide a range of organisms as possible.

13. Contrast adaptive radiation and convergent evolution.

14. Write an essay on the evolution of eyes that evaluates the contributions of adaptive radiation and convergent evolution.

15. Explain how the form of minute aquatic organisms is governed by hydrodynamic principles.

16. There are more species of insects and flowering plants than of any other multicellular terrestrial organisms. Why have they failed to occupy marine environments?

17. A unique adaptation is one that occurs only in a single clade. List as many as you can think of. Do the clades that you have mentioned have anything in common?

18. Why have vestigial structures such as the hindlimbs of whales not disappeared completely?

19. Identify some vestigial structures in human anatomy.

20. Describe the intermediate series of forms leading to the evolution of aquatic mammals from terrestrial ancestors.

You can find a fuller set of questions, which will be refreshed during the life of this edition, in the book's Online Resource Centre at **www.oxfordtextbooks.co.uk/orc/bell_evolution/**

10

Evolving Bodies

Of all the major transitions that have punctuated the history of life, the appearance of complex multicellular organisms is perhaps the most striking. After 2000 My of bacteria with very simple organization, a new kind of organism evolved, powered by mitochondria and capable of forming complex bodies. Once the ancient bacterial limits had been surpassed—probably only once—many new ways of life become possible. Multicellular organisms, moreover, have features such as differentiation and death that are the consequences of new evolutionary processes.

10.1 Multicellularity and differentiation give rise to complex individuals.

There are a few multicellular bacteria that produce structures bearing spores or specialized vegetative cells. This is the limit of bacterial ambition, however. By contrast, several lineages of eukaryotes have independently evolved into complex multicellular organisms with dozens of divergently specialized cells arranged in tissues and organs.

is no physiological reason that a lineage should not continue to reproduce indefinitely by vegetative budding, but in practice none are known to do so. There must instead be an evolutionary reason. At the outset of thinking about the development of multicellular organisms, we are forced to think in evolutionary terms.

10.1.1 The body develops by the proliferation and diversification of cell lineages.

All multicellular organisms develop from single cells. Many multicellular organisms are able to reproduce vegetatively, a new individual developing from a large mass of parental tissue. For example, beech saplings often begin life as shoots arising from the root system of an existing tree. New individuals in corals and similar colonial animals develop as buds from the sides of existing individuals. A worm that reproduces by vegetative fission is illustrated in Figure 10.1. In all of these cases, however, the lineage is sooner or later reduced to a single cell. This can be a fertilized egg, but it can also be an asexual spore. At all events, it is always a single cell. There

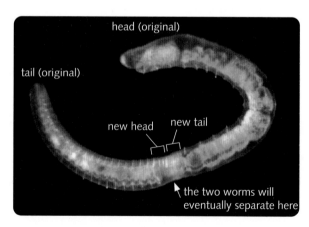

Figure 10.1 Vegetative reproduction in an annelid worm.
Image kindly supplied by Dr Alexa Bely.

A mass of tissue that is destined to develop into a new individual will contain many thousands of cells. From an evolutionary point of view, each will have two interests that are opposed to one another. On the one hand, it has an interest in belonging to an individual whose coherence and integrity makes it more likely that it—the cell—will be replicated. On the other hand, it has an interest in replicating as much as possible within the mass of tissue so that it will be over-represented in the ensuing individual, and hence more likely to pass to its descendants, even if this in some degree compromises the coherence and integrity of the individual. To put this a little more precisely, an allele that refrains from over-replication will benefit from belonging to a superior kind of individual, whereas an allele that directs a higher rate of replication would replace the more slowly replicating cell lineage during the expansion of the mass of tissue that will form that individual. Hence, there is a fundamental tension between altruistic and selfish cell lineages during the development of any kind of multicellular individual. The evolution of complex, highly integrated individuals depends on the resolution of this conflict.

The passage of the lineage, sooner or later, through a single cell resolves this conflict because it transfers selection from being competition among cell lineages within an individual to being competition among individuals. When every individual develops from a single cell, those developing with genotypes that include selfish, over-replicating alleles will fail to form successful individuals, and hence will be eliminated by selection. This explains why all complex, highly integrated individuals belong to lineages that pass through the stage of a single cell, where any defects are immediately exposed to selection.

The bacterial cell wall obstructs the evolution of multicellularity. Many bacteria form filaments of cells and the fossil record makes it clear that this habit evolved very early. In most cases the filaments are undifferentiated, although cyanobacterial filaments may include specialized nitrogen-fixing cells. Myxobacteria are more ambitious and can form slugs with a distinction of somatic and germ cells. This is the pinnacle of bacterial multicellularity,

however, and in a billion years of Proterozoic evolution the bacterial equivalents of plants and animals failed to evolve. This is an example of an ancestral constraint, which is largely attributable to the peptidoglycan cell wall of bacteria that did not allow cell–cell communication or a cytoskeleton to evolve.

The developing body is a diversifying clone of cells. A dividing egg gives rise to a clone of cells that remain physically associated with one another. It is possible to trace each cell lineage forward in time, and to identify the final form of each of its members. Alternatively, any differentiated cell in the adult body can be traced backward in time until the coalescent has been reached—which is, of course, the egg itself. For example, Figure 10.2 shows the complete branching diagram for cell fate in the nematode *C. elegans*, which is extensively used in the study of development. There is a close parallel with the phylogeny of a population of asexual microbes (Chapter 3). It is, in fact, the phylogeny of the clone of cells descending from a single-cell ancestor, but differs from that of a clone of unicellular organisms in two respects.

- First, there is very rapid phenotypic diversification. Very different kinds of cell can emerge within a few divisions.
- Secondly, this diversification follows a fixed schedule. The eggs of a given species all go through the same transformations in the same order.

The diversification of the clone of cells descending from an egg is clearly directed by physiological rather than evolutionary processes, and these physiological processes are in turn governed by instructions encoded by regulatory genes. These instructions can be modified by mutation, however, and the range of body plans we see in different organisms requires an evolutionary explanation.

10.1.2 The fundamental division of labor in multicellular organisms is between somatic cells and germ cells.

The evolution of multicellular organisms, so difficult within the constraints of the body plan of

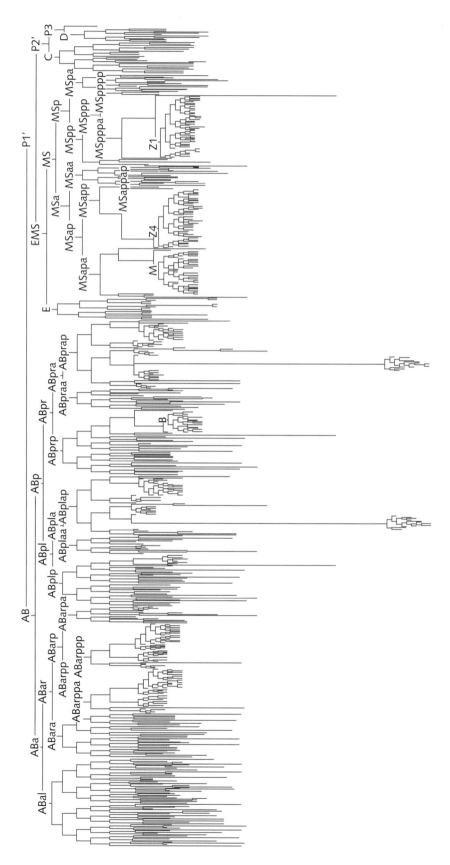

Figure 10.2 The fate of cell lineages during the development of the nematode *Caenorhabditis elegans*.

bacteria, readily and repeatedly occurs once this constraint is relaxed in eukaryotes. The Volvocales show how multicellular organisms with a division of labor between somatic cells and germ cells can evolve on more than one occasion through a series of intermediate forms (Section 8.1.1).

The distinction between soma and germ is fundamental and primary. The simplest multicellular organisms always express this distinction. Somatic cells do not themselves give rise to new individuals, but rather aid the germ cells to reproduce, and thereby enhance the reproductive capacity of the differentiated individual beyond that of an individual that consisted of germ cells only. The germ line is thus potentially immortal, insofar as its descendants give rise to the continuing lineage. The soma itself is dispensable, and terminates with the individual. The soma–germ distinction, therefore, introduces the phenomenon of natural death as a necessary part of the life cycle.

Larger multicellular eukaryotes with somatic division of labor as well as a soma–germ distinction have evolved in five groups: brown seaweeds (Phaeophyta), red seaweeds (Rhodophyta), green plants (Chlorophyta, including land plants), Fungi, and Metazoa. All repeat the patterns we have seen in Volvocales on a larger phylogenetic scale. Multicellular lineages have reverted to unicellularity in Fungi (yeasts, in several unrelated groups) and Metazoa (Myxozoa), for example. Within each group, except Metazoa, multicellularity has evolved more than once: the genetic basis of multicellularity in Volvocales, for example, is quite different from that in land plants.

The multicellular body is a good example of a major innovation that has evolved many times, by different routes, which in some revealing cases—such as the Volvocales—we can trace through its intermediate stages.

10.1.3 The germ line is often set aside early in development to protect it against invasion by selfish cell lineages.

The germ line is vulnerable to invasion by selfish cell lineages. Thinking about vegetative reproduction (Section 10.1.1) alerts us to a fundamental tension in the development of a multicellular individual. Mutant alleles that direct somatic cells to re-differentiate as germ cells may be selected, even though they threaten the integrity of the individual. The regA mutants of *Volvox* (Section 8.1.1) are an example of this; they would fail in most natural conditions because they give rise to colonies that are no longer motile. There are few similar cases in organisms like these, because the members of the colony are very closely related, so there is little opportunity for selfish mutants to arise.

Cellular slime molds have a completely different mode of development. For most of the time, they live as unicellular amoebas in the soil, consuming bacteria. When food runs short, they aggregate to form a compound slug-like organism that differentiates into a long stalk bearing a structure containing spores; this rises above the soil surface in order to disperse the spores. The amoebas that differentiate into spores propagate their genomes, whereas those that form stalk cells are necessarily sterile. It is easy to isolate lineages that preferentially move towards the spore-forming region of the slug, thereby ensuring their own propagation. They are less effectively dispersed, however, as the stalk is suppressed or even absent. Hence, the body plan of slime molds, based on the distinction between stalk (soma) and spore (germ), can be maintained only by selection acting among individuals, in the face of selection acting among cell lineages within individuals. Slime molds are particularly vulnerable to invasion by selfish cell lineages, as illustrated in Figure 10.3, because their body is formed by the aggregation of cells that are not necessarily closely related.

The bodies of large multicellular individuals are likewise vulnerable to selfish cell lineages, arising through somatic mutation, which could disrupt development by re-entering the germ line. In most Bilateria this cannot occur, because a certain cell lineage is set aside early in development as the sole source of germ cells (and of no other kind of cell). In our bodies, it is impossible for a kidney cell or a brain cell to re-differentiate as a gamete, for example.

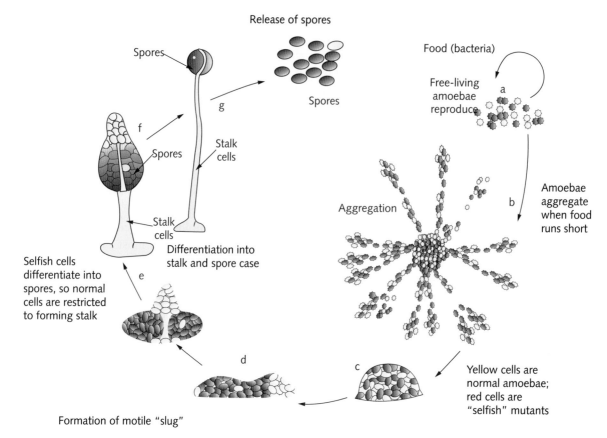

Figure 10.3 Life cycle of a slime mold to show how the development of a multicellular organism can be exploited and disrupted by selfish somatic lineages.

Reprinted by permission from Macmillan Publishers Ltd: Kessin, R.H. Evolutionary biology: cooperation can be dangerous. *Nature* 408 (6815). © 2000.

10.2 Major clades have distinctive body plans.

Developmental systems that control cell proliferation evolve in conditions that favor multicellular organisms. Once they are in place, the process of development can be modified gradually so as to generate more complex and more highly integrated bodies.

10.2.1 In most multicellular organisms, a range of cell types is produced by secondary somatic differentiation.

In organisms as simple as *Volvox* or slime molds there are only two kinds of cells, germ cells and somatic cells. In other multicellular organisms, different lineages of somatic cells follow separate developmental pathways to form a range of cell

types, each of which is specialized to perform a different activity.

In general, larger organisms have more distinct kinds of cell. The complexity of organisms over a wide range of size is illustrated in Figure 10.4. This is an example of the general principle, well known to economists, that labor tends to be more finely divided in larger communities. Thus, people living in the countryside used to undertake most routine tasks themselves, whereas in a town most people would make a living by specializing in a single task, such as baking, carpentry or weaving. In larger towns there will be more distinct kinds of occupation, because even a highly specialized task, such as making walking sticks or teaching evolutionary biology, will provide the whole of the income

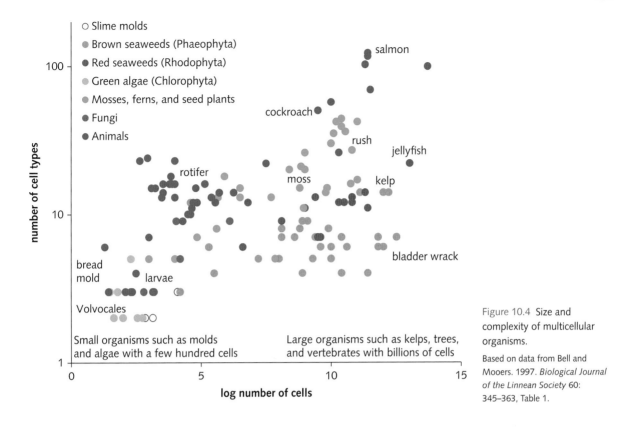

Figure 10.4 Size and complexity of multicellular organisms.

Based on data from Bell and Mooers. 1997. *Biological Journal of the Linnean Society* 60: 345–363, Table 1.

of one or a few people. In the same way, tropical regions with high rainfall can support large communities with many species, while there are few species in boreal forest and tundra. In both cases, this is a *competitive* division of labor, brought about because competition drives people into occupations that are currently under-staffed, or provides an advantage for species using resources that are currently under-utilized.

Within a commercial organization there is a *cooperative* division of labor that is designed to increase the profitability of the enterprise as a whole. Each person may share in all the tasks necessary to run a very small enterprise, such as a corner store, while a large corporation will employ people for a wide range of specialized tasks. This is because the revenue of the corporation is so large that it is worthwhile to employ someone whose sole task, such as sweeping the floor or producing advertising copy, would occupy only a very small fraction of the time of each employee. The differentiation of many kinds of somatic cells is a biological example of the cooperative division of

labor, which evolves when a diversity of cell types increases the rate of reproduction that the individual can achieve.

In a small organism, a task that represents only a small fraction of total activity will have to be carried out by the body as a whole, whereas in larger organisms it can be performed by a small number of dedicated cells. Hence, organisms follow the same general rule as organizations, and tend to become more complex as they grow larger.

It is obvious from Figure 10.4, however, that size is only one factor. At any given size, different kinds of organism display very different levels of complexity. Large brown seaweeds, such as kelps, may be tens of meters in length, and weigh several kilograms, but the frond, and the holdfast that attaches it to the rock, have few specialized cell types. Land plants have more kinds of cell, because they can exploit the soil through specialized roots, resist gravity and conduct fluids through systems of vessels, and intercept light with complex leaves. Animals have the most diversified bodies because the motility provided by muscle and nerve leads to opportunities for many other specialized cell types involved in locomotion, sensing, ingestion, digestion and so forth. The complexity of bodies has evolved in response to the way in which selection has molded them for different opportunities.

10.2.2 Cell division is strictly regulated to avoid cancer.

In a population of unicellular organisms, the capacity to grow and divide as rapidly as possible is positively selected. This does not usually apply to the cells forming the body of a plant or animal, because the somatic cells of a multicellular organism normally become terminally differentiated and thereafter do not divide, or produce only other cells of the same tissue.

This constraint on division can be breached, however, by several kinds of genetic damage. When this happens, the outcome is the uncontrolled growth of a somatic lineage as a tumor. In some organisms (such as plants) this is localized and causes little damage. In others (such as animals), cells are able to move around the body, and the tumor can endanger the integrity of the individual.

A tumor is a population of cells that are dividing in spite of the normal controls, and it will therefore evolve through selection for variants that are even more successful in evading these controls and hence replicate more rapidly. A tumor is thus an evolving population that becomes steadily better adapted to its environment, the body of the individual. In doing so, it often disrupts the normal function of other tissues, and may thereby cause the death of the individual. The tumor also dies, because tumors are not transmitted from individual to individual. This is a good example of an evolutionary process in which natural selection causes a progressive increase in fitness (of the tumor cells), but fails to provide any eventual benefit. Evolution does not necessarily result in long-term improvement.

Tumor cells are normally evolutionary dead ends, because they die with their host. Canine transmissible venereal sarcoma (CTVS) is an exception: it is a cell line transmitted among dogs from individual to individual by sexual contact. It is a striking instance of a unicellular parasitic mammal that originated on a single occasion, through the insertion of a LINE retrotransposable element upstream of the *c-MYC* oncogene. It has subsequently spread worldwide.

10.2.3 Each animal phylum has a distinctive body plan.

The earliest animals, however, were not active and motile: they were filter-feeding organisms firmly attached to the substrate, and incapable of locomotion, at least as adults. Their modern representatives, whose lineage dates back to the dawn of metazoan evolution some 600 Mya, are the sponges. With the aid of molecular data, we can reconstruct some of the stages in their evolution.

The body plan of a sponge is (in the simplest terms) a cylinder containing chambers lined with flagellated cells, the choanocytes. The constant beating of the flagella draws water from outside the sponge through the chambers, to be expelled upwards through the interior of the cylinder. This creates a draught—like the chimney of a fireplace—that helps to maintain the flow of water. The bacteria and other small organisms in the water are trapped by the mucous collar of the choanocytes and digested to nourish the sponge individual.

The choanocytes themselves resemble a group of unicellular organisms, the choanoflagellates, which are quite common in marine and freshwater ecosystems. Choanoflagellates often form colonies, which sometimes show a division of labor between soma and germ, rather like *Volvox*. Moreover, the molecular data has confirmed that choanoflagellates are the sister group of metazoans. Hence, sponges can be interpreted as modified colonial choanoflagellates with large bodies and a limited degree of differentiation among somatic cells (see Section 4.16).

The key innovation that enables more complex body plans to evolve is gastrulation, the growth of a cell layer from the surface to the interior of a developing embryo. Gastrulation leads to the distinction between inside and outside that is the basis of all other animal body plans. It is very ancient: fossilized embryos resembling gastrulas have been found in deposits dating back to the lower Cambrian (see Section 5.15). The outside/inside distinction creates two layers of cells, ectoderm and endoderm, that can then become specialized for different kinds of task—the ectoderm to regulate the flow of material and information across the body surface, and the endoderm for internal processing. It also creates the context for the evolution of novel cell types, muscle and nerve, to enable the body to respond to external signals. The simplest organism of this sort is a hollow cylinder comprising a blind gut, the enteron, with a food collection apparatus, such as tentacles. This is the body plan of the Cnidaria (see Section 4.15).

The earliest bilaterian animals would have been little different, except that they were able to crawl. They have left no fossils that we have yet found, but may have resembled acoel flatworms. These are like simple cnidarians lying on their side and moving by cilia. This does not necessarily mean that modern acoels are the little-modified descendants of these animals, but rather that they represent the same grade of construction, linking a sessile tentacle-feeder with a slow crawler. Directional locomotion implies a preferred posterior-anterior body axis, and the placement of sensory structures at the front—the beginning of the development of a head, seen in its most rudimentary form in acoels and similar animals. Crawling also implies a dorsal-ventral axis, with locomotory structures such as cilia and limbs below and respiratory and protective structures above. With the establishment of these fundamental body axes, the potential for more elaborate body plans has been established (see Section 4.14).

10.2.4 Repeated structures give rise to modular organization.

A ciliary creeper is a weak animal that cannot develop enough force to penetrate sediments, and hence cannot use the resources they harbor. Forceful burrowing requires a body cavity, the coelom, to enable the circular and longitudinal muscles of the body wall to provide powerful thrust (Section 9.1.1). Simple coelomate worms appear early in the geological history of animals, and are still common today. We have also seen how the power of the coelom can be enhanced by dividing it into many semi-autonomous compartments, the segments.

Segmentation is a very important feature of animal body plans because it enables different regions of the body to evolve more or less independently of one another. Any change in the development of a simple worm with an undivided body will inevitably affect all regions of the body, whereas a change in

the fate of a particular segment could, in principle, be confined to that segment. The modification of the number, nature and disposition of repeated parts is the key to the evolution of complex body plans in animals such as arthropods and chordates.

- The simplest modification is change in number: more segments in myriapods (such as centipedes and millipedes), more phalanges in limbs (such as the flippers of ichthyosaurs), or more vertebrae (such as snakes). There are many examples in other organisms, such as the number of leaves borne along the stem in green plants.

- Repeated structures may also change in state. Initially similar structures borne on different segments can be independently modified to serve different ends. The limbs formed as outgrowths of the body wall of arthropods have become modified along the body of crustaceans such as shrimps into sensory structures (such as antennae), mouthparts (such as mandibles), locomotory structures (limbs), respiratory structures (the gill-bearing parts of limbs), and other structures that aid in food handling, reproduction and escape from predators. The functional specialization of segments as the basis for the radiation of arthropod body plans is illustrated in Figure 10.5. The pattern of specialization represents a diversification of homologous structures (specified by the same set of genes) within the body of an individual. Likewise, mammal teeth are a good example of how the regional specialization of different parts can produce adaptation to different ways of life.

- Homologous parts may be modified differently in different lineages. Crustaceans represent one pattern of differentiation in arthropods. Many others have evolved. Myriapods have a long succession of walking legs. Chelicerates (spiders, scorpions, and related forms) have only four pairs of walking legs, and no antennae. Insects have three pairs of walking legs, and also wings. Extinct groups realize many other possibilities. It is this pattern of differentiation that underlies the divergence of major groups of arthropods.

Hence, when the body plan is modular one module can be modified without affecting others. This

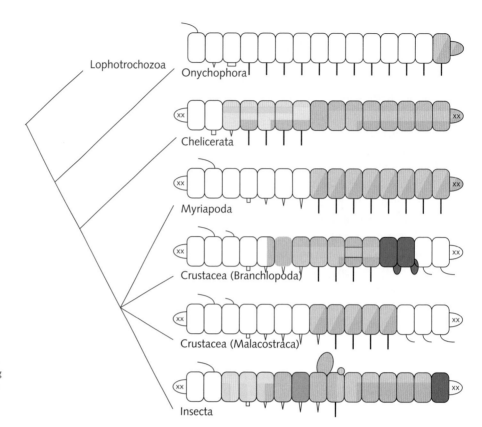

Figure 10.5 The functional specialization of segments underlies the adaptive radiation of body plans in arthropods.

Knoll, H. and Carroll, S.B. 1999. Early animal evolution: emerging views from comparative biology and geology. Science 284: 2129–2137.

reduces functional interference (Section 9.1.3) and makes it easier to evolve novel structures. New kinds of animals can evolve as different combinations of similar types of structure.

10.2.5 Life cycles often involve radical changes of form.

The body plan evolves as an adaptation to a particular way of life. The life cycle, however, may span several ways of life, and an organism will then evolve several body plans. This will happen when selection acts on body plans in relation to

- size, in which case there may be a succession of two or more bodies separated by metamorphosis;
- vegetative function, with a succession of individuals separated by reproduction;
- sex, with a succession of individuals separated by sexual fusion and fission.

Life cycles with these three characteristics are illustrated in Figure 10.6.

Animals that live at several scales develop indirectly. Organisms like us develop by the gradual transformation of the embryo into the adult form. In many other animals, development involves two quite different body plans (Figure 10.6), with the embryo being transformed abruptly into the adult by a process of metamorphosis (literally, "change of shape"). A familiar example is the transformation of a maggot into a fly, or a caterpillar into a butterfly, within the pupa. This process is initiated by pulses of the hormone ecdysone, and results in the degeneration of larval structures, the differentiation of adult structures from cell clusters within the larva, such as imaginal discs, the partial destruction and reformation of the larval nervous system, and the formation of male or female gonads and genital system. This complex series of events is governed by changes in the expression of some hundreds of genes.

One important reason for the evolution of indirect life cycles is that miniature versions of the adult may not be viable. The adult starfish, for example, is a large benthic animal with stereom, water vascular system and other features that cannot be

(a)

(b)

(c)

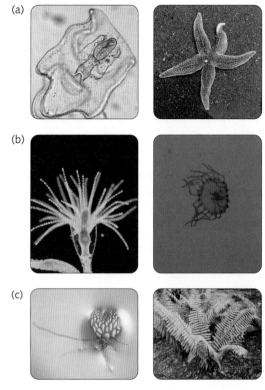

Figure 10.6 Radical changes of form during the life cycle. (a) Metamorphosis: larval (about 1 mm) and adult (about 10 cm) starfish, *Asterias*. (b) Alternation of phases: polyp (about 1 mm) and medusa (about 1 cm) of a jellyfish, *Obelia*. (c) Alternation of generations: gametophyte (about 1 mm) and sporophyte (about 1 m) of a fern, *Dryopteris*.

built on very small scales. The smallest known starfish is the Australian species *Patiriella parvivipara*, which is about 1 cm across. The larva of a starfish is much smaller, and must function as a minute free-swimming organism capable of feeding and growing in the sea. Hence, it is propelled by bands of cilia in the same way as other small aquatic metazoans (Section 7.3.3). An embryo like this cannot be transformed part-by-part into the adult, so a small group of set-aside cells develops into the adult form, the rest of the larval tissue being discarded at metamorphosis.

Vegetative and sexual phases of the life cycle may be separated. Cnidaria and some other animals, such as trematodes (parasitic flatworms), also have life cycles involving a succession of different kinds of individual. In these cases, however, the individuals are separated by reproduction. The classic example is the alternation of polyp and medusa in the life cycle of jellyfish (Scyphozoa) and other cnidarians (Section 4.15). Here, the polyp is a benthic individual that proliferates asexually, whereas the medusa is a pelagic swimming individual that produces sexual gametes. A life cycle like this exploits two very different sources of food, and utilizes an actively motile individual to find sexual partners.

Sexual organisms have a fundamental division between fusing and splitting phases. All sexual organisms must alternate between a haploid phase and a diploid phase. The haploid phase produces gametes, which fuse to form a diploid individual. The diploid individual produces spores that go through meiosis to reconstitute the haploid state. This is hard to discern in organisms like us, where the haploid stage is represented only by the gametes and their immediate precursor cells. It is much more obvious in organisms like ferns and seaweeds, where the haploid (gametophyte) and diploid (sporophyte) individuals live independently and are often very different in form. The familiar fern plant is a sporophyte that bears large green fronds, and reproduces by microscopic diploid spores borne on the underside of the fronds. These spores are dispersed by wind and germinate in the soil to develop into minute flattened haploid gametophytes. These produce gametes that fuse to form diploid spores that close the life cycle by developing into fern plants.

The key to understanding this life cycle is the divergent biology of gamete and spore. Gametes are specialized to fuse, and so the gametophytes produced by a fern plant should be small and clustered close to those produced by other individuals; they benefit from proximity. Spores are specialized to grow, so they will compete for resources. They will profit from being widely scattered, so that offspring compete as little as possible among themselves or with their parent. This is achieved by releasing them from large upright stems.

The life cycle may involve a complex succession of forms linked by growth and reproduction. Combining indirect development, a succession of phases and an alternation of sexual generations can produce very complex life cycles. For example, Figure 10.7 shows the main stages in the life cycle of a digenic trematode (the liver fluke), involving a succession of asexual individuals between sexual phases.

• The adult *fluke* is an internal parasite of verte-brates. It is a flattened, ribbon-like animal a few centimeters in length that infests the liver of ver-tebrates, including humans. It is hermaphroditic, producing eggs and sperm that fuse to form a *zygote*. This cleaves to form a large somatic cell and a small germinative cell. The somatic cells divide to form the tissues of the miracidium larva.

• The *miracidium* is liberated from the host into water and searches for a different kind of host, usually a snail. It bores into its new host and develops into a hollow ball of cells, the sporosac, with the germinative cells in the interior.

• The *sporosac* absorbs nutriment from the host and produces asexual embryos called redia. These escape from the sporosac and proliferate within the host.

• After several generations of *redia*, a different spe-cialized stage, the cercaria, develops. This leaves the host and lives briefly as a free-living organism in the water.

• The *cercaria* is consumed by the final vertebrate host, either directly or as its encysted form, the *metacercaria*.

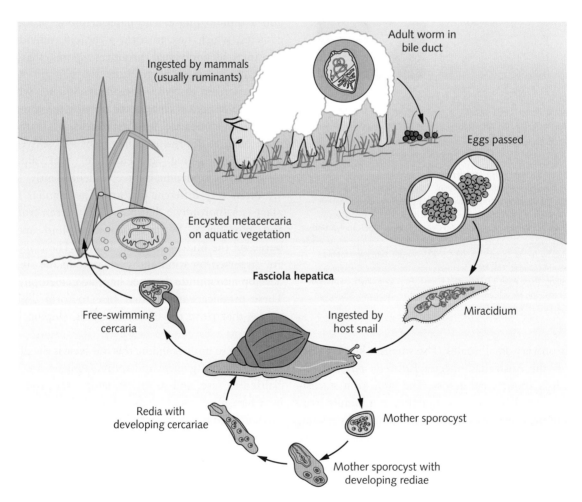

Figure 10.7 The fantastic voyage: the life cycle of a trematode, the liver fluke.

Image kindly supplied by Dr. Kimberly Bates.

- The cercaria then develops into the adult fluke, to close the life cycle.

What is the purpose of this fantastic voyage? There is no physiological or genetic need for development to follow such a tortuous route. Instead, this complicated life cycle is a legacy of the long evolutionary history of trematodes, adding stage to stage in order to connect one adult phase to the next through a succession of environments and host tissues.

10.3 The evolution of tetrapod body plans is based on a group of multiply duplicated genes.

There is an obvious but fundamental difficulty in the development of a complex multicellular organism: it proliferates from a single cell, yet comes to consist of many different kinds of cell. This is clearly not the result of genetic mutation, like the evolution of specialized cell types in a population of microbes. It is brought about because the development of each cell type is governed by the expression of a particular set of genes, and therefore requires a mechanism for switching on some genes and switching off others in different cell lineages within the developing body.

10.3.1 Regulatory genes control the development of multicellular organisms.

Gene expression is regulated by proteins that bind to DNA. The developmental switches are regulatory genes coding for proteins that bind to DNA and thereby either prevent or promote its transcription. This ability is associated with particular shapes assumed by the folded protein, which are determined by its amino acid sequence. Some regulatory genes control the expression of other regulatory genes, leading to a very complex system of checks and balances that guides the orderly development of the embryo. The evolution of this regulatory network is clearly of the first importance in explaining the evolution of complex body plans. As we shall see, regulatory genes are often among the first to be modified, even in microbial populations exposed to new agents of natural selection in the laboratory.

The homeodomain is the DNA-binding motif of many genes that regulate development. One important kind of structure responsible for DNA binding is the homeodomain, a sequence of 60 amino acids forming three helices that fit into the groove of the DNA double helix. The homeodomain is encoded by the 180-nucleotide homeobox region in the regulatory gene. The particular sequence of amino acids in the homeodomain determines which nucleotide sequences in the DNA will be most strongly bound. As a whole, however, the homeodomain region is strongly conserved among metazoans. In some genes, the homeodomain region is very similar even in animals as distantly related as mice and flies.

Homeodomain genes do not merely regulate other genes; they switch on or off whole sets of genes, including other regulatory genes. Hence, mutations in homeodomain genes can have very large effects on phenotype by altering developmental pathways. This is a crucial fact in understanding how mutation can provide the variation necessary to fuel the evolution of novel body plans.

Homeotic mutants reveal the identity of major genes involved in governing body plans. A particular set of homeobox genes is usually closely linked and arranged in a fixed order along a chromosome. Mutations in these genes give rise to homeotic mutants, which express inappropriate structures in specific regions of the body, as illustrated in Figure 10.8. These genes are the *Hox* genes. They are the best-known homeobox genes. *Hox* genes share three remarkable features.

- Genetic collinearity: they occur as a single cluster in which they are arranged in a fixed order.
- Morphological collinearity: the order of expression along the front-to-rear body axis is the same as the order of the genes along the chromosome.
- Temporal collinearity: the order of expression in time also corresponds to their order on the chromosome, genes closer to the 3′ end being expressed first.

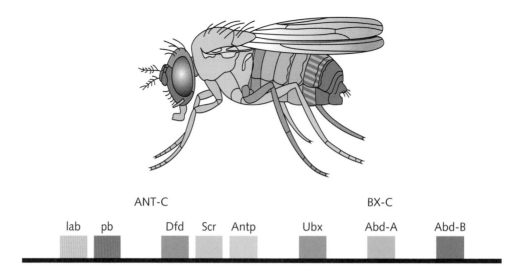

Figure 10.8 The *Hox* cluster governs basic features of the body plan in *Drosophila*. The collinear *Hox* genes and their regions of expression along the body of the adult fly.

Hox genes help to govern development in all Bilateria, and are especially well known for their role in determining segment fate in arthropods. The diversification of arthropod body plans, through the modification of the fate of structures developing on particular segments, is governed in large part by the regions within which particular *Hox* genes are expressed, and by the period of development in which they are expressed.

10.3.2 The evolution of metazoan body plans involves modification of the genes regulating development.

The Hox *cluster evolved from simpler predecessors.* Genes that encode DNA-binding proteins did not suddenly appear in the ancestors of metazoans. They are present in all organisms, including bacteria. The *Hox* cluster itself descends from non-*Hox* genes that have been identified in sponges and even in the unicellular sister group of Metazoa, the choanoflagellates. New genes do not arise abruptly and without predecessors; they are the modified descendants of genes that served other functions in ancestral lineages. The history of *Hox* genes is sketched in Figure 10.9.

Genes homologous to the *Hox* genes of Bilateria are expressed during the development of

anthozoans (sea anemones and corals). They are not collinear, however, and are scattered throughout the genome. They do not show definite anterior-posterior patterns of expression. It is not clear whether *Hox* genes are present in other classes of Cnidaria. No *Hox* genes have been found in Porifera or Placozoa.

Bilateria have larger suites of Hox *genes.* The *Hox* suite is conserved among Bilateria, although gene content and organization varies among major groups, as

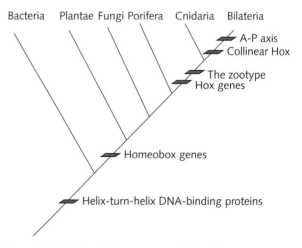

Figure 10.9 The *Hox* cluster genes were derived from genes present in the unicellular ancestor of Metazoa.

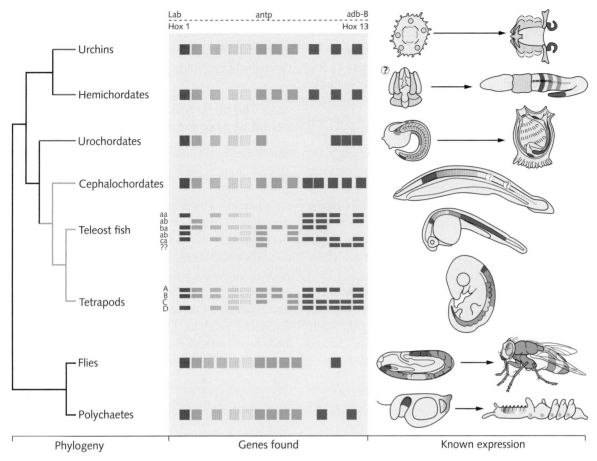

Figure 10.10 The evolution of the *Hox* cluster underlies the radiation of body plans in Metazoa.

Swalla, B.J. 2006. Building divergent body plans with similar genetic pathways. Heredity 97: 235–243.

shown in Figure 10.10. Hence, the *Hox* suite evolved in the most recent common ancestor of Bilateria from the genes already existing in Cnidaria, giving rise to the characteristic body axes of bilaterian animals. This would have been directly selected, because it provides the basis for directional locomotion. At the same time—although this could not have been directly selected—it would provide the potential for much greater diversification in body plans.

These facts give a clear picture of the genetic basis for the evolution of complex body plans in animals. The *Hox* genes of Bilateria descend from precursors that still govern development in Cnidaria, which were in turn derived from non-*Hox* regulatory genes in the ancestors of Metazoa.

The Hox *cluster is a gene family that has evolved by duplication and divergence.* The origin of the *Hox* suite is inferred to be a small group of *ProtoHox* genes in the metazoan ancestor. This gave rise, through complete duplication, to the *Hox* and *ParaHox* suites. The *ParaHox* genes comprise three or four developmental genes related to the *Hox* genes, but they are usually not linked, and do not give rise to homeotic mutants. The *Hox* cluster itself arose through duplication of genes in the original set to form a linked set of eight or more genes. The acoel flatworms (Section 5.14) lack many of the attributes of other Bilateria, and have a small *Hox* cluster of three or four genes, which may represent a situation close to the ancestral state.

Chordates generally have four *Hox* clusters, each with slightly different gene content. The individual genes correspond closely to homologous genes in other animals, such as the fruit fly, *Drosophila*. Hence, the evolution of vertebrates has involved two rounds of duplication of the entire cluster. The genome of amphioxus, inferred on other grounds to be the sister group of the vertebrates (Section 5.12), has a single *Hox* cluster, showing that this is the ancestral state for the vertebrate lineage.

In other groups of animals, *Hox* genes may be lost or gained, as indicated in Figure 10.10. For example, mollusks have rather scattered and incomplete *Hox* suites. This may be associated with their unusual body plans, which are far from the norm for Bilateria. Ray-finned fishes (Actinopterygii, Section 5.10) have six *Hox* clusters, due to a third whole-cluster duplication event early in their history. Hence, the diversification of body plans within the Bilateria is associated with the modification of the *Hox* cluster.

The mode of origin of new *Hox* genes, by duplication and divergence, implies that they are homologous (Section 7.2). (Copies of a duplicated gene in related species are often said to be "paralogous," but this is only a special case of being homologous.) Hence, their ancestry can be traced using the same methods that we would apply to species or asexual lineages (Section 4.3). This yields a phylogenetic tree (Figure 10.11) showing how the *Hox* genes and their associated developmental pathways have become diversified over time.

10.3.3 The phylotypic stage of a group is the period of peak expression of the *Hox* genes.

One of the most famous diagrams from the early years of evolutionary biology showed how embryos of a group of animals resemble each other more closely than the adults do. This is not true for the earliest stages of development, during which embryos of brooded species (such as birds) may be very different from those of non-brooded embryos (such as most fish). Rather, there is a stage, usually a little after gastrulation, at which the embryos of a large monophyletic group have a strong resemblance to one another, despite their subsequent divergence as adults. This reflects their evolution from a common ancestor (Section 1.4).

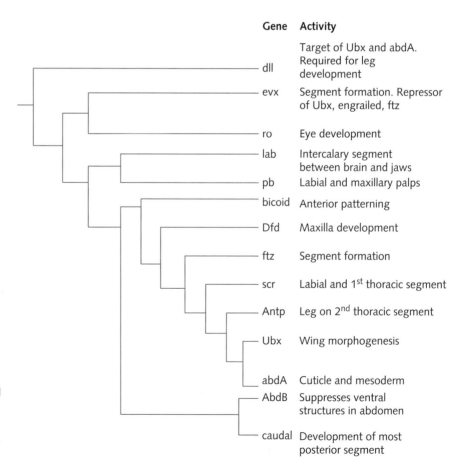

Gene	Activity
dll	Target of Ubx and abdA. Required for leg development
evx	Segment formation. Repressor of Ubx, engrailed, ftz
ro	Eye development
lab	Intercalary segment between brain and jaws
pb	Labial and maxillary palps
bicoid	Anterior patterning
Dfd	Maxilla development
ftz	Segment formation
scr	Labial and 1st thoracic segment
Antp	Leg on 2nd thoracic segment
Ubx	Wing morphogenesis
abdA	Cuticle and mesoderm
AbdB	Suppresses ventral structures in abdomen
caudal	Development of most posterior segment

Figure 10.11 Duplication and divergence of the *Hox* suite (and others) generates a repertoire of development genes.

Based on data from Ryan et al. 2007. *PLoS One* 2 (1): e153. doi:10.1370.

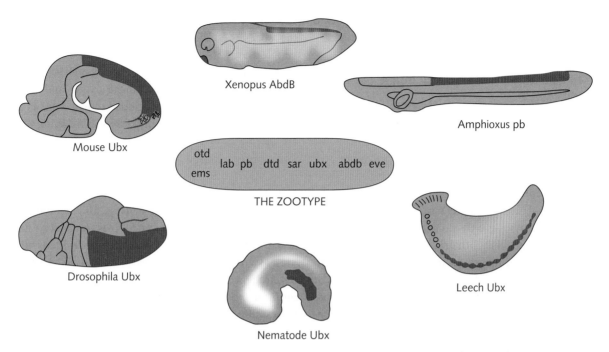

Figure 10.12 The zootype is the anterior–posterior pattern of expression of the *Hox* genes (and some others) that determine the relative position of structures in the developing embryo. The phylotypic stage of a particular group corresponds to the period when the zootype is most clearly expressed.

Reprinted by permission from Macmillan Publishers Ltd: Slack et al. The zootype and the phylotypic stage. *Nature* 361 (6412). © 1993.

The point during development at which the embryos of a monophyletic group most strongly resemble one another has been called the "phylotypic stage." In vertebrates, the phylotypic stage is the pharyngula, at which point the basic features of the vertebrate body plan have appeared as cell condensations but are not yet fully differentiated: notochord, dorsal neural tube, segmented musculature, and post-anal tail.

The anterior–posterior pattern of expression of the *Hox* genes (and some others) that determines the relative position of structures in the developing embryo leads to the characteristic body plan of a given phylum, which has been called the "zootype."

The phylotypic stage of a particular group corresponds to the period when the zootype is most clearly expressed (Figure 10.12). This ties the intuition of the early evolutionary biologists to the latest discoveries of molecular developmental biology.

The *Hox* family is a good example of a set of linked genes, crucial to the evolution of animal body axes, that has evolved by duplication and divergence from a simpler and less extensive ancestral set. The new knowledge of the genetics of development has illuminated the way in which basic body plans evolve, showing how more complex organisms can evolve from simpler ancestors through modification and selection.

10.4 The life history of an individual is its schedule of survival and reproduction.

Larvae and embryos are distinct stages in the succession of morphological forms through which an individual develops. In a more abstract view, development can be seen as a succession of episodes in each of which an individual has two fates: it survives, or not; and it reproduces, or not. This is its life history, shorn of all morphological or physiological detail. By virtue of its simplicity, it can be used to address general issues, such as when individuals should begin to reproduce, and why they should die.

10.4.1 The amount of reproduction is limited by its cost.

A simplistic view of natural selection is that it tends to maximize the rate of reproduction, because the genotypes of individuals that reproduce more will increase in frequency in the next generation. Individuals should therefore produce as many offspring as possible. This is not quite correct, for two reasons. The first is obvious: only those individuals that survive can produce any offspring at all. The second is more subtle, and hinges on the principle that evolution is driven by the differential success of lineages. If an allele predisposes an individual to produce more offspring, it will spread only if those offspring themselves survive to produce offspring. Hence, alleles that direct an increased rate of reproduction will spread only if two conditions are satisfied.

- The first is that producing more offspring does not entail a disproportionate reduction in the parent's chance of surviving to reproduce.
- The second is that a greater quantity of offspring can be produced without a disproportionate decline in their quality—their chance of surviving to reproduce.

These assumptions might be justified in exceptional circumstances. A few individuals arriving on a remote island where there are no predators or competitors may find a superabundance of resources sufficient for all their needs, and for their offspring's needs. Even in this extreme case, however, the population will then tend to increase until resources become scarce. When this happens, any resources that an individual spends on producing more offspring must necessarily be withdrawn from providing for its own survival, or from providing for its offspring's survival. The rate of reproduction is then curtailed by the cost of reproduction—the loss of quality, of the individual or of its offspring, that is entailed by producing more offspring.

An island of sheep. Soay is a small, remote island off the west coast of Scotland. Its name means "the island

of sheep," because it was home to a breed resembling the sheep first brought to Scotland by Neolithic people about 5000 years ago. The sheep were moved to the neighboring island of Hirta in the 1930s, when the few human inhabitants moved to the mainland, and since then they have continued to live there as an unmanaged herd. Far out in the North Atlantic, they are often exposed to severe weather, and after a wet and stormy winter a large fraction of the population is likely to die of starvation brought about by a combination of exposure, food shortage brought about by high population density, and debilitation caused by a parasitic nematode. It is unusual to find a self-contained population of large mammals that are easy to capture and release, and so an intensive research program was set up in the mid-1980s to monitor most of the individuals in the herd through their entire lives.

Each ewe gives birth to one or two lambs in late April or early May, or fails to reproduce. Carrying a developing fetus through the winter is a heavy metabolic burden for the mother, and may thereby impose a high cost of reproduction. This has been studied in detail by the scientific survey.

The probability that a female will survive from one breeding season to the next depends on its size: larger individuals are more likely to survive. At any given size, survival also depends on reproduction. In years when there is heavy pressure on the food supply, so that many sheep die of starvation, females that do not reproduce are more likely to survive than those which do, and females that produce a single lamb are more likely to survive than those which produce twins. These differences are greatest for the smallest ewes, and least for the heaviest. Hence, there is a survival cost that is most severe when the metabolic burden of pregnancy is greatest—in famine years, and among small females.

The survival of the lambs also depends on their size: again, larger individuals are more likely to survive. The heaviest lambs are twice as likely to survive their first year of life as the smallest. Larger mothers tend to produce larger lambs, so that this cost too bears most heavily on smaller females, with lower stores of resources.

In this way, the limited resources that are available to a mature ewe limit the number of surviving offspring she is able to produce. By devoting more resources to reproduction, she will be able to produce larger and more vigorous offspring, but only at the cost of putting her own life at risk.

10.4.2 Reproductive allocation is optimized by selection.

Because reproduction is costly, natural selection will not maximize the total production of offspring. Instead, it will lead to the evolution of optimal—not maximal—rates of reproduction in the same way that competing demands lead to the evolution of optimal structures (Section 9.1). The optimal quantity of reproduction at any given age will depend on how the survival of the parent is affected by the effort and risk involved in reproducing. The optimal distribution of these resources among offspring will depend on how the investment in each offspring affects its fate.

More means worse. The simplest tradeoff is between the number and size of offspring. A newborn is more likely to survive (and hence to reproduce) if it is supplied with a large reserve of the resources necessary for its own development and survival. Its parent, however, commands only a limited quantity of resources. It follows that producing high-quality offspring necessarily entails producing few offspring. In other words, the greatest number of surviving offspring, able themselves to reproduce, will generally not be achieved by producing as many offspring as possible, but rather by producing an intermediate number that maximizes the number of grand-offspring. Hence, the evolution of reproduction, like the evolution of structure, will be governed by a principle of optimality.

This means that the evolution of life histories can be treated as though it were a branch of engineering. A good example is the evolution of clutch size: how many eggs should a bird lay? This has been intensively studied in the great tit, *Parus major*, a small insectivorous bird, in woodland near Oxford, England. The tits are hole-nesting birds that are easily persuaded to build nests in artificial nest-boxes, allowing almost the whole population to be monitored. The number of eggs laid by a female in a single clutch varies between 2 and 15, with a mode at 9 eggs. Why do they lay fewer than they could? The reason is that the parents have only a limited ability to supply their nestlings with flies and caterpillars. Hence, nestlings in larger clutches each receive a lower ration, fledge at a smaller size, and are more likely to die in their first winter. The fate of nestlings is followed by putting light, coded aluminum bands on their legs in the nest. The leg-bands are easily identified in birds recaptured the following year, which allows the researchers to calculate the rate of survival. We can now use some simple algebra to calculate the optimal clutch size that will evolve under natural selection.

First, we calculate the probability of survival (S) of a nestling as a function of the size of the clutch (B) in which it was laid. This shows that survival decreases with clutch size in a roughly linear fashion, so $S = C - aB$, where C is the maximal probability of survival in very small clutches, and a is the rate at which survival decreases as clutch size increases. Now, the total number of offspring surviving from the clutch is the number of eggs (B) multiplied by the probability of survival of the nestling hatching from each egg (S), or $w = SB$. Since we know how nestling survival is affected by clutch size, we can substitute for S to obtain $w = B(C - aB)$. The optimal value of B is the value that maximizes w. This is calculated by differentiation: $dw/dB = C - 2aB$. Setting this to zero identifies the maximum value of B, $B^* = C/2a$. The two parameters are estimated from the survey data as $C = 0.177$ and $a = 0.0098$. Hence, the optimal number of eggs per clutch is $B^* = 9.03$, which is very close to the observed modal value (Figure 10.13). This is a very pretty example of how the theory of natural selection can be used to predict the value of a character, as an optimal compromise between producing too many offspring and producing too few.

Reproduction can be risky. Guppies are small tropical fish that are attacked by a range of predators, including larger fish. The females bear live young, and so become heavily pregnant, with large bellies that impede rapid swimming. The males have evolved bright coloration to attract the females (see Chapter 15), including highly conspicuous orange, yellow and red spots that are prominently displayed during courtship. Both pregnancy and courtship are risky. Colorful males are more likely to be attacked by predators, and heavily pregnant females are less likely to escape an attack. Hence, individuals who invest more in reproduction are less

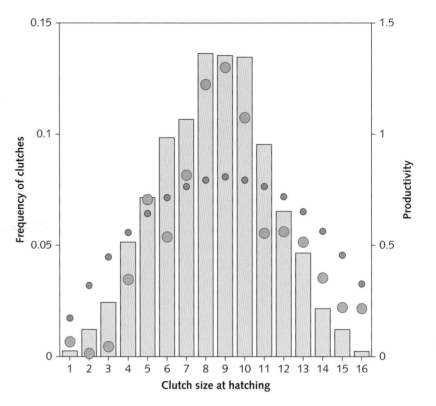

Figure 10.13 Optimal clutch size in the great tit. The histogram bars are observed values, a nearly Normal frequency distribution with mean 8.48 and standard deviation 1.75. The large open circles show the actual productivity of clutches, in terms of survival to three months of age. This curve is also unimodal and approximately Normal, with a mode at 9 eggs. The small filled circles give the productivity as calculated from the survival regression, as $B(C - aB)$, yielding an optimal value of $C/2a$ = 9.03 eggs.

likely to survive. The optimal reproductive strategy therefore depends on the pattern and severity of predation. In clear streams, males tend to be less highly ornamented, because colorful males are most visible to predators. Where there are large predators, females tend to reproduce when they are younger and smaller, even though they produce fewer offspring, because deferring reproduction until they are larger would entail a risk of being attacked before they can reproduce at all.

These examples show how the cost of reproduction not only limits the quantity of reproduction, but also governs how resources are allocated between reproduction and survival, or between a few large or many small offspring.

10.4.3 Selection favors an optimal schedule of reproduction.

At any given stage in life, the effort devoted to reproduction will be optimized by the balance between benefit and cost. The life history extends from birth to death, however, and the costs of reproduction will likewise extend through time. Present reproduction not only incurs present costs, but also incurs future costs: an increased effort at any given age may permanently weaken an individual, reducing their fecundity or survival for the rest of their life. Not only each episode of reproduction, but also the whole schedule of lifetime reproduction, will evolve towards an optimal compromise.

All-or-nothing situations lead to the evolution of suicidal reproduction. Pacific salmon (*Oncorhynchus*) migrate as smolts from the streams in which they were born to the depths of the North Pacific Ocean, two thousand kilometers or more away. They then swim all the way back to their natal stream to spawn. This was a very hazardous journey even before the first gill-net was set on the Fraser River; few would survive it, and the probability of surviving a second round trip to the Pacific and back would be very low indeed. An adult salmon that has reached the spawning stream has therefore nothing to gain by restraint, and should commit all its reserves to reproduction, rather than holding some back in a futile attempt to survive to breed for a second time. The suicidal reproduction of these fish is the consequence of a survival cost that makes any attempt to reproduce at all nearly certain to be fatal.

Atlantic salmon (*Salmo*) undertake a similar but somewhat less arduous migration to the north-west Atlantic; they spawn with great vigor, and many die, but a few survive to repeat the voyage and spawn for a second time. Sea trout move into coastal waters and back, without venturing into the open sea, and routinely survive for two or more spawning seasons.

Exclusively freshwater salmonids such as lake trout, and many related fish such as chars and graylings, make at most very limited journeys along the lakeshore to find their spawning grounds. They often survive to spawn for many years. Post-reproductive survival in this group of fish thus appears to be governed by a prospective cost of reproduction, the risk entailed by the spawning migration.

The life history will evolve when the intensity or the direction of selection varies with age. Guppies have less dramatic life histories, but they likewise evolve towards an optimal schedule of reproduction. For example, populations living in different places may be attacked by different kinds of predator. Pregnant females are at risk if they can be attacked by large predatory fish, such as the cichlid *Crenicichla*; their offspring are more at risk if they can be attacked by smaller predatory fish, such as *Rivulus*, a killifish. This will influence the age and size at which females reproduce, and the number and size of their offspring,

In Trinidad, guppies live in small streams in wet tropical forest, interrupted by waterfalls. Few fish can pass above the falls. *Rivulus* can travel short distances overland (by flipping and wriggling on wet soil), but *Crenicichla* is restricted to downstream sites. The evolution of the pattern of reproduction in female guppies has been worked out by field experiments in which guppies are transplanted from downstream sites, where *Crenicichla* attacks the adults, to upstream sites, where *Rivulus* is the only predator. The effect of transplantation is thus to increase the risk for young guppies, while reducing the risk for adults. Consequently, the population at the transplant site evolves to produce a smaller number of larger offspring in each brood. This change occurs very rapidly—within a dozen generations or so—because of the strength of natural selection acting on the basic determinants of fitness, survival and fecundity.

10.4.4 Senescence is the correlated response to selection for early vigor.

The life of an individual terminates with death, of course; but how does death come about? The simplest idea would be that it happens by accident. Every year, for example, some thousands of people die from choking. Suppose this were the only cause of death. It would then occur more or less independently of age (with the exception of young children, perhaps). Sooner or later, everyone would die from choking, but the chance of this happening would be the same at any age, and the old would be no more likely to die than the young.

The rate of mortality increases with age. Everyday experience tells us that this is incorrect. Old people are generally less vigorous in many ways, and in particular they are more likely to die than young people. In other words, the rate of mortality increases with age. This increase in the probability of dying, beyond a certain age, is called senescence. Senescence creates a natural limit to the lifespan. Someone with a healthy lifestyle, free of genetic diseases, may easily live to 85 years of age, and a few fortunate individuals live to be centenarians. The number of people living to 120 years of age is extremely small, however, and there is no reliable record of anyone ever reaching the age of 150. We also know, from everyday experience, that organisms as different as dogs and trees follow the same pattern: if they live long enough, then before they die they become increasingly frail. Senescence is a universal feature of multicellular organisms.

At first sight, this seems puzzling. Natural selection favors greater fitness, and hence should favor any extension of lifespan, so why should we not all live healthy lives until we are killed by an accident? The reason hinges on the concept of the cost of reproduction. The logical development of this idea shows that senescence is the inevitable consequence of selection acting to maximize reproduction over the entire lifespan.

Early vigor is a crucial component of fitness. The key to understanding the evolution of senescence is that reproduction early in life makes a much greater contribution to fitness than the equivalent amount of reproduction later in life. There are two reasons for this. Both can be explained in terms of saving money for your retirement, because the proliferation of a lineage is analogous to the growth of capital at compound interest. There are two options for you to consider.

- The first is to purchase an annuity, a fixed sum of money per year that becomes payable at a certain age. The later in life it begins, the less valuable it is at present. The reason is very simple: if it begins later, you are less likely to survive to collect it.

- The second is to make an investment that is to be liquidated at a fixed time in the future. The later in life it is made, the less valuable it is at present. The reason is that money laid down earlier will have accumulated more interest at any future date.

Thus, postponing an investment will reduce its value in proportion to the rate of mortality and the rate of interest. Transferring this to a biological context, reproduction early in life is more valuable than reproduction later in life in proportion to the rate of mortality and the rate of increase of the population. The direction of this effect is quite clear. To appreciate its significance, recall that lineages (like investments) grow exponentially through time. Early reproduction is not just more important than later reproduction. It is enormously more important. The age at first reproduction is generally the single most important component of overall fitness. Hence, any tendency to reproduce earlier in life will be strongly favored by selection.

The corollary of the importance of early reproduction is that selection will favor an increase in fecundity early in life, *even if this entails a catastrophic loss of vigor later in life.*

Selection for early reproduction reduces survival or fecundity later in life. The costs of reproduction that we have described so far are paid immediately, in terms of smaller offspring or an increased risk of death. It is easy to understand that similar costs may be deferred. A ewe that produces twins in a famine year will not only run a greater risk of dying, but may also run down reserves that would have supported her in future years. Hence, present reproduction may weaken survival or reproduction later in life.

Artificial selection for increased egg production in domestic fowl provides a practical example of this principle. It is desirable to increase the number of eggs laid by fowl, but the obvious procedure—choose the individuals laying the most eggs—is cumbersome because the birds live for three years or more, so progress would be slow. An alternative is to select birds that lay many eggs in the first few months of maturity, because this would speed up the process of selection. It is assumed that this high rate of egg-laying is maintained later in life. Where this scheme has been tried, it is very successful at first. An advance in the rate of laying early in life is accompanied by a corresponding advance in lifetime egg production. After a few generations, however, progress slows down and eventually stops altogether, so that the lifetime egg production of the selected stocks is no greater than that of their ancestors. The reason is that a greater rate of reproduction early in life has become accompanied by a reduction of fecundity later in life. This represents a *prospective* cost of reproduction: a cost of early vigor that is paid by older individuals.

You will probably ask why selection for greater reproductive effort early in life should first increase reproductive output later in life, and then, in later generations, reduce it. To understand this, you have to think like an evolutionary biologist. The ancestral flock of fowl would be genetically diverse, with some genotypes being generally more vigorous than others—living longer and reproducing more, in the particular conditions of the experiment. The less vigorous genotypes, with poor survival and fecundity, will tend to be eliminated in each generation. Because the combination of poor survival and low fecundity is constantly tending to be eliminated from the population, genotypes with greater rates of survival or higher fecundity, or both, will increase in frequency. The inevitable consequence is the evolution of a negative correlation between survival and fecundity, such that birds genetically predisposed to laying more eggs are more likely to die young. It is this evolved cost of reproduction that eventually sets a limit to further progress in increasing lifetime egg production.

10.4.5 Live fast = die young

If greater investment in reproduction early in life causes greater mortality later in life, then there should be two effective ways of extending normal lifespan: either reducing input, or restricting output.

Lifespan may be limited by the rate of living. The simplest theory of senescence is that it is caused by the wearing-out of tissues, and that the amount of wear depends on the amount of use.

Many physiological processes vary with body size: generally speaking, small animals work faster. Since they also die younger, this suggests that the metabolic expenditure per unit weight per lifespan is therefore approximately a constant: one gram of tissue performs the same amount of work (about 5 kJ, or 0.25 cal) in any organism. Hence, it is inferred, life ends when this fixed amount of work has been accomplished. This is the rate-of-living hypothesis. If it is correct, then lifespan can be extended by spreading this finite amount of work over a longer period of time, either by supplying less fuel or by consuming it more slowly. The rate-of-living hypothesis is certainly consistent with the effects of caloric restriction. Experiments on many organisms, including mice, have shown that individuals kept on a low (but nutritionally sufficient) diet tend to live longer than individuals who are given as much food as they can consume.

The rate-of-living hypothesis is supported by clk-1 *mutants in* C. elegans. The nematode worm *C. elegans* has been extensively used for studies of senescence. In particular, mutations in the *clk* ("clock") genes can dramatically extend lifespan. The normal lifespan of about 15 days is increased to 18–19 days in the single mutants *clk-1* and *clk-2*, and to 28 days in the *clk-1 clk-2* double mutant. This shows that lifespan is to some extent under genetic control, and will evolve in response to selection. The most obvious characteristic of these long-lived mutants is that they process food more slowly. The worms are cultured on bacterial slurry, and feed by pumping the bacteria through the gut. This pumping action is slowed down in *clk* mutants.

The rate-of-living hypothesis provides a physiological, not an evolutionary, mechanism to explain why older individuals become progressively enfeebled.

At the same time, it provides a basis for understanding how the life histories are likely to evolve.

10.4.6 Senescence may evolve directly through functional interference, or indirectly through mutation accumulation.

Early vigor may lead to exhaustion. Mutations that increase fecundity early in reproductive life have a large effect on fitness, and consequently will spread more quickly than mutations with similar effects later in life. Thus, selection will generally tend to increase early vigor. The converse is also true: selection will indirectly reduce vigor later in life, causing a senescent decline in rates of survival and fecundity. Senescence then evolves because the expression of prospective costs of reproduction is biased by the general weakening of selection with age.

This is compatible with the rate-of-living physiological theory of senescence, because a prospective cost of reproduction is necessarily entailed by the finite capacity of tissue to perform work. Increased metabolism early in life will incur a greater amount of damage and will thereby accelerate senescence and curtail lifespan.

Deleterious mutations may be expressed later in life. An alternative evolutionary theory of senescence is that the expression of deleterious mutations will increase with age. Suppose that deleterious loss-of-function mutations with variable ages of expression occur at many loci. The accumulation of deleterious mutations will be checked by selection at all ages, of course, but countervailing selection is more effective in younger than in older individuals. The equilibrium level of mutational load will therefore increase with age: deleterious mutations will accumulate to a greater extent in older age-classes.

Selection tends to reduce genetic variation, by eliminating less fit genotypes from the population. Stronger selection reduces variation to lower levels. If selection against deleterious mutations is stronger early in life and weaker later, then the genetic variance of survival and fecundity should increase with age. This has been confirmed experimentally in *Drosophila*.

The functional-interference and mutation-accumulation theories of the evolution of senescence are not mutually exclusive. There is experimental evidence for both processes, which may act together in most populations.

10.4.7 Selection for deferred reproduction causes the evolution of extended lifespan.

If the life history can evolve, it should be possible to extend lifespan permanently through selection. This can be done by using artificial selection (Chapter 12) to favor reproduction later in life. Only the offspring produced by older females are used to propagate the population. This annuls the usual advantage of early reproduction, and thereby creates selection favoring later reproduction and the postponement of senescence.

An experiment of this kind propagated a base population of *Drosophila* that had been maintained in the laboratory for over a hundred generations on the usual 14-day cycle. This was selected in two ways: one set of lines was transferred on the same schedule, while another set was transferred at longer and longer intervals—at first 28 days, then 35, 42, 56, and finally 70 days. Eggs were collected only from the oldest females, so that only females that survive into old age are able to reproduce at all in the latter treatment. The populations that were transferred at

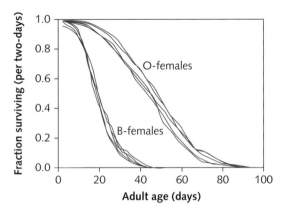

Figure 10.14 Experimental evolution of lifespan in *Drosophila*. Replicate selection lines were maintained with early (B) or late (O) age at last reproduction.

Rauser et al. 2009. Chapter 18 in *Experimental Evolution: Concepts, Methods, and Applications of Selection Experiments*, edited by Theodore Garland, Jr., and Michael R. Rose, Figure 18.2, p 560.

longer intervals evolved enhanced late-life fecundity and longer lifespan (Figure 10.14). At the same time, they evolved reduced fecundity early in life, reflecting the prospective cost of reproduction. Experiments like these show how the schedule of reproduction and the onset of senescence can be modified by selection, subject to the physiological constraints that limit the rate of living.

● CHAPTER SUMMARY

Multicellularity and differentiation give rise to complex individuals.

- The body develops by the proliferation and diversification of cell lineages.
  - *All multicellular organisms develop from single cells.*
  - *The bacterial cell wall obstructs the evolution of multicellularity.*
  - *The developing body is a diversifying clone of cells.*

- The fundamental division of labor in multicellular organisms is between somatic cells and germ cells.

- The germ line is often set aside early in development to protect it against invasion by selfish cell lineages.

Major clades have distinctive body plans.

- In most multicellular organisms, a range of cell types is produced by secondary somatic differentiation.

- Cell division is strictly regulated to avoid cancer.

- Each animal phylum has a distinctive body plan.

- Repeated structures give rise to modular organization.

- Life cycles often involve radical changes of form.
    - *Animals that live at several scales develop indirectly.*
    - *Vegetative and sexual phases of the life cycle may be separated.*
    - *Sexual organisms have a fundamental division between fusing and splitting phases.*
    - *The life cycle may involve a complex succession of forms linked by growth and reproduction.*

The evolution of tetrapod body plans is based on a group of multiply duplicated genes.

- Regulatory genes control the development of multicellular organisms.
    - *Gene expression is regulated by proteins that bind to DNA.*
    - *The homeodomain is the DNA-binding motif of many genes that regulate development.*
    - *Homeotic mutants reveal the identity of major genes involved in governing body plans.*

- The evolution of metazoan body plans involves modification of the genes regulating development.
    - *The Hox cluster evolved from simpler predecessors.*
    - *Bilateria have larger suites of Hox genes.*
    - *The Hox cluster is a gene family that has evolved by duplication and divergence.*

- The phylotypic stage of a group is the period of peak expression of the *Hox* genes.

The life history of an individual is its schedule of survival and reproduction.

- The amount of reproduction is limited by its cost.
    - *An island of sheep.*

- Reproductive allocation is optimized by selection.
    - *More means worse.*
    - *Reproduction can be risky.*

- Selection favors an optimal schedule of reproduction.
    - *All-or-nothing situations lead to the evolution of suicidal reproduction.*
    - *The life history will evolve when the intensity or the direction of selection varies with age.*

- Senescence is the correlated response to selection for early vigor.
    - *The rate of mortality increases with age.*
    - *Early vigor is a crucial component of fitness.*
    - *Selection for early reproduction reduces survival or fecundity later in life.*

- Live fast = die young
    - *Lifespan may be limited by the rate of living.*
    - *The rate-of-living hypothesis is supported by clk-1 mutants in C. elegans.*

- Senescence may evolve directly through functional interference, or indirectly through mutation accumulation.
    - *Early vigor may lead to exhaustion.*
    - *Deleterious mutations may be expressed later in life.*

- Selection for deferred reproduction causes the evolution of extended lifespan.

● FURTHER READING

Here are some pertinent further readings at the time of going to press. For relevant readings that have been released since publication, visit the book's Online Resource Centre at
www.oxfordtextbooks.co.uk/orc/bell_evolution/

Section 10.1   Kessin, R.H. 2000. Cooperation can be dangerous. *Nature* 408: 917–918.

Section 10.2   Bell, G. and Mooers, A.O. 1997. Size and complexity among multicellular organisms. *Biological Journal of the Linnean Society* 60: 345–363.

Bell, G. 1993. The comparative biology of the alternation of generations. *Lectures on Mathematics in the Life Sciences: Theories for the Evolution of Haploid-Diploid Life Cycles* 25: 1–26.

Nowell, P.C. 1976. The clonal evolution of tumor cell populations. *Science* 194: 23–28.

Martindale, M.Q. 2005. The evolution of metazoan axial properties. *Nature Reviews Genetics* 6: 917–927.

Section 10.3   Lemons, D. and McGinnis, W. 2006. Genomic evolution of Hox gene clusters. *Science* 313: 1918–1922.

Pick, L. and Heffer, A. 2012. Hox gene evolution: multiple mechanisms contributing to evolutionary novelties. *Annals of the New York Academy of Sciences* 1256: 15–32.

Slack, J.M.W., Holland, P.W.H. and Graham, C.F. 1993. The zootype and the phylotypic stage. *Nature* 361: 490–492.

Section 10.4   Bell, G. and Koufopanou, V. 1986. The cost of reproduction. *Oxford Surveys in Evolutionary Biology* 3: 83–131.

Godfray, H.C.J., Partridge, L. and Harvey, P.H. 1991. Clutch size. *Annual Review of Ecology and Systematics* 22: 409–429.

Reznick, D., Bryant, M. and Holmes, D. 2006. The evolution of senescence and post-reproductive lifespan in guppies (*Poecilia reticulata*). *PLoS Biology* 4: 0136–0143.

Ricklefs, R.E. 2008. The evolution of senescence from a comparative perspective. *Functional Ecology* 22: 379–392.

Wilson, A.J., Pemberton, J.M., Pilkington, J.G., Clutton-Brock, T.H. and Kruuk, L.E.B. 2008. Trading offspring size for number in a variable environment: selection on reproductive investment in female Soay sheep. *J Animal Ecology* 78: 354–364.

● QUESTIONS

1. Why have complex multicellular bodies evolved in eukaryotes and not in prokaryotes?

2. Why do the life cycles of multicellular organisms pass through a single-cell phase?

3. Why is the germ line of multicellular organisms segregated from somatic cell lineages early in the development of many groups?

4. How would you answer the preceding question from the point of view of a tree?

5. Discuss the idea that evolution inevitably leads to the appearance of more complex body plans.

6. A new drug has been developed that selectively kills cancer cells. Describe how it might affect the population dynamics of a growing tumor.

7. Describe the three basic body plans of Metazoa: sponge, cnidarian, and bilaterian. How does a simpler type of animal (such as a sponge) persist when a more complex type (such as a worm) has evolved?

8. Give as many examples of the modular organization of body plans as you can think of. Why should modular organization facilitate the radiation of groups with different body plans?

9. In the simplest life cycle a newborn would simply grow in size to become an adult. Why do complex cycles involving different kinds of individual evolve?

10. How has understanding the genetic basis of development helped us to understand better the evolution of body plans?

11. "The course of development reflects the course of phylogeny." Discuss this statement.

12. Why might females evolve to produce fewer offspring than the physiological maximum?

13. How would you expect a population to evolve when reproduction becomes less risky?

14. Explain why reproduction early in life has a much greater effect on fitness than the corresponding amount of reproduction later in life.

15. Define senescence. Is senescence inevitable in multicellular organisms?

You can find a fuller set of questions, which will be refreshed during the life of this edition, in the book's Online Resource Centre at **www.oxfordtextbooks.co.uk/orc/bell_evolution/**

11 The Dynamic Genome

The evolution of structure, development, and life history that you have read in the previous two chapters is based on genetic changes in populations, as more successful lineages replace the less successful. Thirty years ago, the nature of these genetic changes was only very poorly understood. The advances in knowledge that have been made since then have provided us with a new and powerful means of understanding evolutionary processes. Some of these have been used to describe the genetic basis for the evolution of novel body plans. We shall use others in tracing the genetic basis of how populations respond to selection, and how they adapt to new conditions of life (Chapters 13 and 14).

In this chapter, the focus is the genome itself. The genome is, after all, the source of all the hereditary information underlying the adaptation of organisms to their environment. The evolution of the genome is naturally one of the most pressing issues in evolutionary biology. Why are many genomes much larger than they need to be? How does the gain and loss of genes contribute to evolution? How do new genes come into being? Molecular genetics is beginning to provide answers to fundamental questions like these. The answers have turned out to depend on the Darwinian rules that govern genome dynamics. Some of them have been very unexpected indeed.

The success of modern genetics has been due in part to a relentless concentration on a very few species of model organisms, so that labs throughout the world can contribute to a cumulative increase in knowledge. These are by now very well understood. They are all undramatic: a gut bacterium, a small fly, a nematode worm, and a cress plant. Perhaps the least dramatic of all is a unicellular fungus that we all know: yeast.

Yeast is one of the most important domesticated organisms. People have used it to bake bread, to brew beer, and to make wine for thousands of year. More recently, it has been used in laboratories throughout the world to investigate cell biology and genetics, so that it is now better understood than any other organism. We shall use it to illustrate many topics in evolutionary genetics and experimental evolution.

Yeast was the first organism whose genome was completely sequenced, in 1996. Since then, many other genomes have been completely sequenced, including our own, and more are added each month. This flood of information is giving us new insight into evolutionary change at the level of whole genomes. One thing to bear in mind, however, is that it may be misleading to refer to (for example) "the human genome." There is no single human genome, because there are as many human genomes as there are humans. It is true that our genomes, taken collectively, are different from the genomes of chimpanzees, taken collectively. This is the consequence of evolutionary divergence and the origin of new genes. It is equally true that our genomes are variable. This is because of the continual flux of gene frequencies in populations. Divergence and flux are two of the leading features of the dynamic genome.

11.1 Evolving genomes diverge.

The familiar yeast of bakeries, vineyards and laboratories is called *Saccharomyces cerevisiae*. It was originally domesticated from a wild organism, of course. It can still be found in the wild, although the constant exchange between natural and humanized environments means that the properties of the wild isolates are difficult to interpret. However, it has a wild sister species that has never been domesticated, called *Saccharomyces paradoxus*, from which it diverged about 5–10 Mya. The two are very similar in most respects and can still mate together, although they are reluctant to do so, and most of the hybrid progeny fail to grow. This is typical of recently diverged sister species.

In Chapter 4, the relationships among organisms were illustrated by diagrams showing the branching and divergence of lineages. In diagrams like these, the succession of populations through time is indicated by a solid line, which forks at the time when the set of outbreeding populations representing the ancestral species becomes separated into two sets, each ancestral to one of the two extant species. The real situation is much more complicated than this, because each population actually consists of many lineages proliferating over time. This is drawn in Figure 11.1 as though the populations were asexual, or as though it depicted only a single gene. Outcrossing would make the diagram more complicated, but its basic features would remain the same.

11.1.1 Genotype lineages eventually correspond to species lineages.

At the time of the original divergence, both populations would contain lineages from the ancestral population, so that neither is monophyletic. To interpret the sequence of events, we can use the concept of the coalescent (Section 3.1.1): all individuals in both populations descend from the most recent common ancestor in the ancestral population. Two of the immediate descendants of the most recent common ancestor are thus the ancestors of the two divergent populations. We shall call these the last common

ancestor of each population. At the time of the split, each population may contain lineages descending from the last common ancestor of the other. Another way of putting this is to say that each population will contain some, but not all, of the descendants of its last common ancestor. Hence, one or both contain a mixture of lineages sharing an ancestry that postdates the most recent common ancestor.

As time goes on, however, the number of lineages they share will decrease, either by random loss, or because the separated populations experience different conditions of growth. If they remain separate (without exchanging genes) there will come a time when each descends uniquely from its last common ancestor. At this point they are truly monophyletic. Figure 11.1 illustrates the transition from an interbreeding population to two separate monophyletic lineages.

The lesson is that the early stages of divergence are genetically complex processes in which the diverging populations still share many lineages. It would be easy for them to re-unite, and in fact most probably do. The few that remain separate are on their way to becoming different species, and begin to accumulate fixed differences—alleles that are present in one but not in the other. Divergence has begun.

11.1.2 The rate of divergence can be estimated from molecular and geological data.

Domesticated and wild yeast are still so similar that corresponding genes (with some exceptions) are found on similar positions on the same chromosomes, and can be compared nucleotide by nucleotide. They are about 15% divergent; that is, the probability that homologous sites are occupied by different nucleotides is about 0.15. This is an important number. It reflects two processes. The first is selection: if the two species have become adapted to different ways of life, they will diverge genetically. This may happen very rapidly, but the rate will differ among genes according to their contribution to fitness in particular ways of life. The second is genetic drift (Section 6.1.1): being genetically isolated, random changes will cumulate

pop. A   pop. B

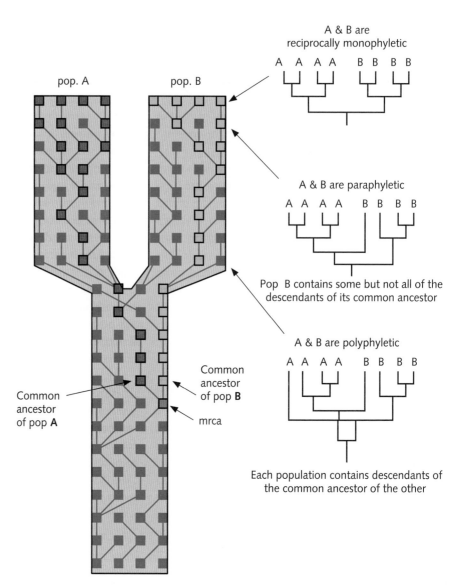

A & B are
reciprocally monophyletic

A A A A   B B B B

A & B are paraphyletic

A A A A   B B B B

Pop B contains some but not all of the
descendants of its common ancestor

A & B are polyphyletic

A A A A   B B B B

Each population contains descendants of
the common ancestor of the other

Common
ancestor
of pop **B**

Common
ancestor
of pop **A**

mrca

Figure 11.1 Lineage structure
in diverging populations. The
symbols represent asexual
individuals. It is only in the
final generation that the two
populations are different
monophyletic groups.

Image kindly supplied by Prof. C.
William Birky, Jr.

independently in the two populations. This may happen more slowly, and at a more or less constant rate.

Recall that each amino acid in a protein is specified by a particular triplet of nucleotides, according to a code that is partially redundant because the number of combinations of three nucleotides ($4^3 = 64$) is less than the 20 amino acids used in most proteins. Hence, some mutations will alter a nucleotide without altering the amino acid encoded by the triplet; these are called synonymous mutations. Others cause the substitution of a different amino acid into the protein encoded by the gene, and hence are called non-synonymous.

The rate of sequence divergence sets the underlying schedule of evolution. To understand how it can be calculated, suppose that we have identified two sister species and their ancestor. The (imaginary) sequence of part of a gene in the ancestor, comprising 12 nucleotides, is CAACGTGACGCA. Over the course of time, this diverges to give rise to different sequences in the two descendant taxa: CCACGTGACGCA and CAACGAGATGCA. This gives us the information necessary to calculate the fundamental rate of evolutionary divergence in terms of nucleotide substitutions.

There are in all three changes from the ancestral sequence. Hence, the overall sequence divergence is $d = 3/12 = 0.25$. This does not necessarily reflect the total number of changes that have occurred, however, because a nucleotide at any position might have mutated into a different nucleotide, which later mutated back to the original nucleotide. If so, we would detect no change, missing the double mutation that in fact took place. To account for this, we calculate the average amount of change that really took place, designated as K, given the observed amount of change. This is $K = -¾\ln(1 - 4d/3)$. (The proof of this lies beyond the scope of this book, but can be found in any textbook of population genetics.) Hence, $K = 0.3$. This is the best estimate of the amount of genetic change that occurred during the divergence of the two sister species from their ancestor.

To convert this into a rate, we need to know the length of time that has elapsed since the two sister lineages became separate, that is, reciprocally monophyletic. The most reliable information comes from a dated fossil record; suppose that this gives a time $T = 150$ My. The rate of sequence divergence is then $E = K/2T$. The factor of 2 is necessary because we are counting twice: both lineages have been diverging over the same length of time. Hence $E = 0.3/(2 \times 150\ \text{My}) = 10^{-9}$ substitutions per site per year. This is the fundamental rate that we seek. (Whether it is sufficient to account for adaptive evolution is explained below, in Section 11.1.4.)

We can go a little further by distinguishing between synonymous and non-synonymous (coding) changes. Two of the substitutions are synonymous, that do not change the encoded amino acid. Hence, the amount of synonymous change is $d_s = 2/12 = 0.17$, for which the corrected value is $K_s = -¾\ln(1 - 4 \times 0.17/3) = 0.19$. The third substitution is non-synonymous, encoding proline rather than glutamic acid in the protein expressed by the gene. Hence, the amount of coding substitutions is $d_n = 1/12 = 0.08$, with a corrected value of $K_n = 0.09$. This is the crucial rate of change that is associated with adaptive evolution.

The human–chimp difference is an example of recent genetic divergence. The lineages leading to modern humans and chimpanzees diverged in the Miocene, roughly 5 Mya (Chapter 5, Figure 5.3). The human and chimpanzee genomes have now been completely sequenced, showing that their overall divergence is about 1.5% (Chapter 5). Hence, the rate of divergence has been roughly $0.015/(5 \times 10^6) = 3 \times 10^{-9}$ per nucleotide per year. Using more precise estimates and correcting for multiple changes gives an estimate of about 1.3×10^{-9} per nucleotide per year. A more

natural timescale would use generations, so allowing for 10–30 y per generation gives a value of $1–4 \times 10^{-8}$ per nucleotide per generation. This is the probability that any given nucleotide will have become fixed for alternative states in the two lineages. Since our genomes are quite large (in excess of 10^9 nucleotides), this implies that several nucleotides—of the order of 10—become fixed for alternative states in each generation. Most of these will be at non-coding sites of one sort or another. Nevertheless, a good rule of thumb is that coding regions of genes undergo a change of state in diverging lineages every few generations. This is the slow but unceasing process by which lineages are driven apart.

This rate of divergence depends on how the sister lineages differ in ecology, population size and other factors, and so it will vary from case to case. For the mouse–rat divergence, for example, it is rather higher, at about 8×10^{-9} per nucleotide per year. Details apart, the important lesson of these molecular studies is that once lineages become isolated they begin to diverge, giving rise to great diversity within rather short periods of geological time.

The rates that have been calculated for eukaryotes are calibrated from the fossil record. It is more difficult to document the rate of divergence in groups that do not readily fossilize. In particular, there are no useful fossils of bacteria that we can use to estimate the rate of evolution of prokaryotes. However, some bacteria are obligate and highly specific endosymbionts of animals, for example the bacteria that aid the digestion of cellulose in insects such as termites and cockroaches. These specialized types cannot be older than their hosts. From the fossil record of insects, the rate of divergence of their endosymbiotic bacteria has been calculated at about $2–4 \times 10^{-10}$ per nucleotide per year. Given the greater proportion of coding DNA in bacterial genomes, this is not very different from the estimate for eukaryotes. Hence, the forces that push apart independent lineages act with similar strength in organisms of all kinds.

11.1.3 Synonymous substitutions are roughly clock-like, whereas non-synonymous substitutions are not.

The molecular clock can be used to estimate the date of divergence between clades from gene sequence data. Since $E = K/2T$ then $T = K/2E$, so that the time when two lineages first diverged can be estimated from molecular data alone, provided that E, the rate of sequence divergence, is a constant. (We still need one reliably dated calibration point so as to transform divergence from units of nucleotides into years.) For example, we can measure the genetic differences between groups of vertebrates and use these data to rank the dates at which they diverged—the branching order of sister taxa. The most recent common ancestor of mammalian carnivores and ruminants, for example, lived more recently than the most recent common ancestor of mammals and diapsids (reptiles and birds).

If we know the actual date at which any two sister taxa diverged from reliable fossil evidence, then we can calculate all other divergence dates—assuming that genetic divergence is strictly proportional to elapsed time. The divergence of people and birds, for example, must have been before 310 Mya (because synapsid and diapsid fossils date from the upper Carboniferous) but after 340 Mya (because modern amphibian groups date from the lower Carboniferous). Taking 310 Mya as a conservative estimate, the age of all other sister groups can be calculated. Considering the assumptions involved, the argument is remarkably successful: the dates obtained from fossils and the dates obtained from gene sequences of modern vertebrates are highly correlated. There are some inconsistencies. In particular, the fossil evidence demonstrates a rapid diversification of eutherian mammals immediately after the K/T boundary, whereas the molecular data suggest an earlier cryptic radiation (see Section 5.6). This might happen because the fossil record is poor, because genetically distinct lineages still look the same, or because the rate of genetic change varies over time.

Only neutral changes show the time. Neutral evolution happens when nucleotide substitutions cause no change in protein structure because of the redundancy of the genetic code. In principle, they provide an ideal clock that ticks at a constant speed set by the mutation rate, independently of environmental change. By contrast, beneficial mutations that alter protein structure and are fixed because they enhance adaptation to new conditions of growth will be more or less frequent according to the fluctuations of the environment. The contrast between the rates of neutral and adaptive evolution is illustrated for haplorrhine primates in Figure 11.2. The rate of synonymous substitution is rather uniform, being about $3–4 \times 10^{-8}$ per site per year. This suggests that the rate of neutral substitution depends only on the neutral mutation rate. The rate of non-synonymous substitution, on the other hand, is lower and much more variable from less than 10^{-9} to more than 3×10^{-9} per site per year. Neutral evolution will provide a more reliable estimate of elapsed time than adaptive evolution.

11.1.4 The observed rate of divergence is large enough to fuel adaptive evolution.

Conversely, we can take observed rates of divergence and ask whether they are capable of fuelling adaptive radiation. The rate of non-synonymous divergence is of the order of 10^{-9} per nucleotide per year. Let us assume that we are dealing with a relatively large and slowly-reproducing eukaryote with one generation per year, and that a representative protein is encoded by 1000 nucleotides. Then the rate of non-synonymous mutation is about 10^{-6} per locus per generation. Since the origin of eukaryotes there have been roughly 10^8 (annual) generations. Hence, each surviving lineage of eukaryotes will have experienced about $10^8 \times 10^{-6} = 100$ successful non-synonymous substitutions per locus and distantly-related eukaryotes will differ by about 200 amino acids per protein, or about two-thirds of the total. This provides ample room for adaptive evolution—and in more rapidly reproducing species this conclusion applies with even greater force.

11.2 Genes are modified at different rates.

An overall average rate of change is useful to give a rough date for the divergence of lineages and to establish the limits to adaptive change. Its usefulness is limited, however, by how this average rate of change itself varies over time. The variations of neutral and adaptive rates of change are illustrated for a range of mammalian genes in Figure 11.3.

*   Neutral change depends on demography. Neutral alleles will be exchanged more often by genetic

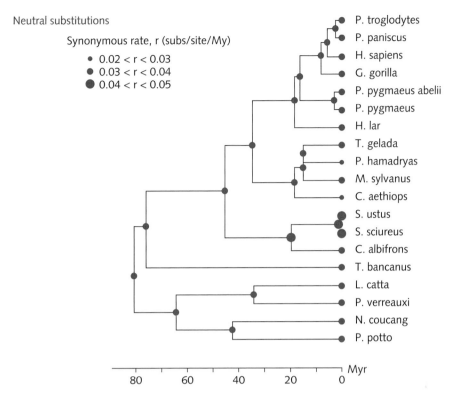

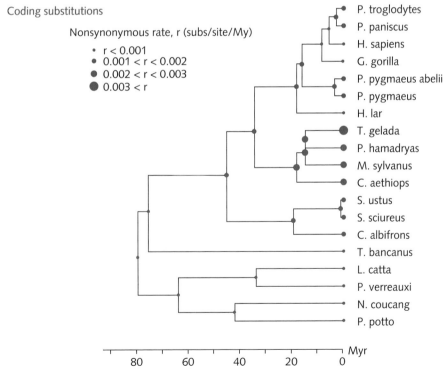

Figure 11.2 Rates of neutral and coding nucleotide substitutions in haplorrhine primates. The coding substitutions are less frequent but more variable among lineages.

Seo et al. 2004. Estimating absolute rates of synonymous and nonsynonymous nucleotide substitution in order to characterize natural selection and date species divergences. *Molecular Biology and Evolution* 21: 1201–1213, by permission of Oxford University Press.

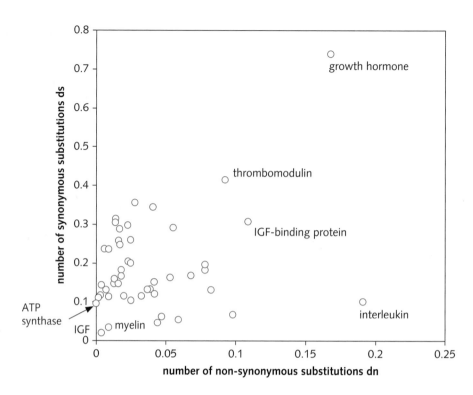

Figure 11.3 Rates of evolution of different genes in mammals.

Based on data from Yang and Nielson. 1998. *J. Mol. Evol.* 46: 409–418, Table 2.

drift in small populations, or in populations that fluctuate widely in abundance over time.

- Adaptive change depends on the environment. So long as conditions are constant, purifying selection (Section 14.1) will maintain the integrity of genes. Functionally different alleles will be exchanged by directional selection more often in populations that experience frequent environmental change.

As a general rule, genetic drift will lead to diversification whereas purifying selection will lead to conservation. The rate of synonymous nucleotide substitutions among mammals, for example is 4.6 $\times$ 10^{-9} per site per year, whereas it is only 0.9 $\times$ 10^{-9} per site per year for functional substitutions that lead to a change in protein structure. This shows how powerful purifying selection acts to restrict the genetic divergence that would otherwise occur.

11.2.1 Some genes are highly conserved.

Some genes and genetic elements, for example those involved in early development, do not respond either to demography or to environmental fluctuations. They change over millennia, but only very

slowly, and remain similar in distantly related species. These are genes whose protein products are central to the functioning of the cell and cannot be modified without severe impediment, regardless of conditions of growth. One example would be the genes that govern the assembly of ribosomes. This has given rise to the concept of a core genome: a small group of genes encoding proteins whose structure cannot be altered without severely disabling the organism. This group would be the genome of the ancestor of all cellular organisms. It can be defined by deletion. When genes are removed, one by one, from the genome of yeast, they can be put into two categories, with very few intermediate cases. In the first, the loss of the gene has little effect on growth, at least in comfortable laboratory conditions. In the second, the loss of the gene prevents growth completely.

It is this second category of genes which are lethal-when-deleted that provides candidates for the core genome. In yeast about 20% of the genome, amounting to about 1000 genes, falls into this category. Many of these, however, are not lethal-when-deleted in other species, so the core genome is actually much smaller than this. In bacteria, a recent survey identified a core genome of 144 genes,

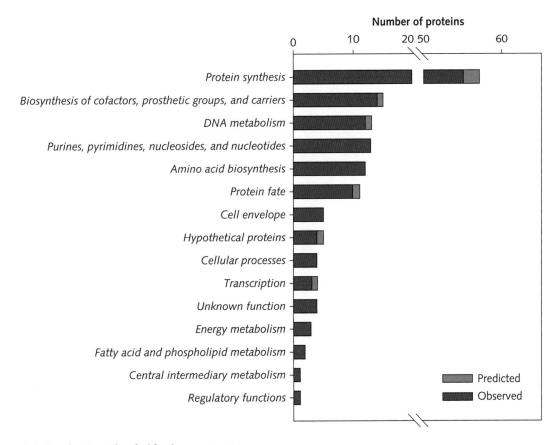

Figure 11.4 Gene functions identified for the core genome.

Reproduced from Callister et al. 2008. Comparative bacterial proteomics: analysis of the core genome concept. *PLoS One* 3 (2): e1542. doi:10.1371/journal. pone.0001542.

which are arranged by function in Figure 11.4. These occur in all species, are highly conserved, and are always lethal-when-deleted. They are candidates for the small set of genes assembled in the first cellular organisms.

11.2.2 Other genes evolve more rapidly.

Genes outside the core genome evolve more rapidly, are required in some conditions but not in others, may or may not be present in a species, and are usually viable-when-deleted. The most general and the most remarkable conclusion from gene deletion programs, in fact, is that the great majority of genes can be deleted with little appreciable effect on the rate of growth of the cell. Since the complete removal of a gene is the greatest possible effect of a mutation, this demonstrates conclusively that most mutations have only very slightly deleterious effects. The genome is therefore massively redundant: a loss of

function caused by the loss of one enzyme can often be compensated by redirecting the flow of material and energy through alternative pathways. On the one hand, this facilitates evolution because lineages that are deficient in a particular function do not become extinct immediately but persist until they may evolve an alternative. On the other hand, the redundancy of cell systems is itself an evolved character, because only lineages of cells able to survive damage to a particular metabolic pathway will persist.

Because most mutations are only mildly deleterious, a range of variant types will accumulate in the population over time. These are then available for selection when conditions should change. The rate of adaptive evolution varies widely over different kinds of protein, according to the frequency of change: for fibrinopeptides about 4×10^{-9} per site per year; for hemoglobin about 1×10^{-9} per site per year; and for cytochrome only 0.2×10^{-9} per site per year. This variation reflects the evolutionary difference between proteins like fibrinopeptides, which contribute to the continually shifting balance of power between species, and proteins like cytochromes, which are essential for core metabolic processes regardless of changes in the conditions of growth.

11.2.3 Genes decay when they are no longer selected.

Some sections of the genome look like genes but incorporate flaws that prevent them from producing a useful protein. There might be a small deletion at some point, for example, that shifts all downstream nucleotides out of frame, or a premature stop codon that abbreviates the transcript. These "pseudogenes" arise from normal genes that have suffered loss-of-function mutations. They often arise when a gene is duplicated, because one copy can then decay without lessening the original rate of expression. These defective copies are usually left untranscribed. Other pseudogenes arise by retroposition of part of an mRNA transcript; these are processed but yield only a non-functional transcript. In either case, they continue to be passively replicated while contributing nothing to the phenotype. Hence, any further mutations occurring in a pseudogene are neutral, and will continue to accumulate without check. Consequently, pseudogenes will diverge rapidly in sister lineages. In mammals, for example, a rough figure for the average divergence of pseudogenes is about 5×10^{-9} substitutions per nucleotide per year, whereas the rate for protein-coding genes is about 2×10^{-9} substitutions per nucleotide per year. Pseudogenes soon become genetic junk, littered with mutations.

Pseudogenes add to the metabolic burden of replicating the genome, and when they are removed or abbreviated the lineage benefits slightly. This consistent but weak selection gradually removes a pseudogene from the population altogether. In *Drosophila*, for example, the half-life of a pseudogene is about 10 My.

11.2.4 Mutations in regulatory genes can have far-reaching effects that are difficult to predict.

A phenotype will be expressed only if the genes that govern it are transcribed. Most genes in most cells are inactive most of the time, however, because if all were being transcribed simultaneously metabolism would be overwhelmed and specialization would be impossible. Transcription begins only when RNA polymerase binds upstream of a structural gene, and is controlled by allowing or preventing it from binding. This is mediated by a short section of DNA immediately upstream of the gene called the promoter, which at any given time may be either exposed, in which case transcription can proceed, or blocked by a particular kind of protein called a transcription factor, whose structure enables it to bind to the promoter. The gene that encodes the transcription factor, which may be situated anywhere in the genome, thereby regulates the expression of the structural gene, and through it the resulting phenotype.

The promoter region is an example of a cis-regulatory element; "cis" means that its position, close to the gene it controls, is essential for it to act. There are other kinds of cis-regulatory element that fine-tune gene expression by limiting it to a particular tissue or stage of development. There are also trans-acting elements producing diffusible molecules acting at a distance, such as micro-RNAs that can bind to mRNA transcripts. All these are described in any textbook of molecular developmental biology. As the mechanisms of gene regulation have been uncovered in recent years, their role in the evolution of phenotypes has begun to be studied.

Both structural and regulatory genes can be modified. Two structural genes govern the utilization of lactose by *E. coli*, a permease that transports it across the cell membrane and a β-galactosidase that cleaves it into glucose and galactose. These genes, arranged together with a third gene and transcribed in the same direction, evolve in a predictable manner in a lactose-limited chemostat (Section 13.3.7). They are regulated by two proteins, as shown in Figure 11.5.

The first is produced by a promoter gene *lacI*, situated upstream of the structural genes, which is produced constitutively and acts as a repressor by binding to a short operator sequence between the promoter and the structural genes. An isomer of lactose, allolactose, binds to the repressor protein and alters its shape so that it can no longer bind tightly to the operator.

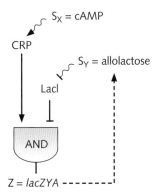

Figure 11.5 The regulation of lactose utilization in *E. coli.*

Reprinted by permission from Macmillan Publishers Ltd: Alon, U. Network motifs: theory and experimental approaches. *Nature Reviews Genetics* 8 (6). © 2007.

The second regulatory element is an activator, encoded by a distant gene, which binds to a site between *lacI* and the operator, provided that it has itself bound cyclic AMP. The level of cyclic AMP is inversely proportional to the intracellular concentration of glucose, so that the activator protein is bound only when glucose is lacking. The combination of the two proteins prevents expression of the structural genes unless lactose is present and glucose is absent. If both are absent, expressing the structural genes would waste material and energy. If both are present, it is more efficient to utilize glucose first, because it requires less processing to yield energy, and only when it has been exhausted to switch to lactose metabolism.

The flux through the lactose pathway is affected by mutations in the structural genes of the lac operon (Section 13.3.7), and evidently it may also be affected by mutations in the regulatory genes. For example, there is an intact regulatory system in the *lactis* strain of *Lactobacillus delbrueckii*, which is used for industrial milk fermentation. In the *bulgaricus* strain, however, the lac operon is expressed constitutively as the result of mutations that inactivate the repressor protein. These have spread through intense selection for rapid fermentation, which is why *bulgaricus* is now the preferred strain for making yogurt. There are many other examples of experimental evolution where the initial step in adaptation is a regulatory change, usually caused by a loss-of-function mutation leading to constitutive expression.

Regulatory networks are combinations of motifs. In the absence of glucose the lac operon has a simple regulatory system involving a single specific transcription factor. Other cases are more complex, but there seems to be a limited range of kinds of interaction involved in regulatory networks, some of which are illustrated in Figure 11.6.

- Feedback loops. The most common example of feedback is autorepression, whereby expression is limited by the gene product itself. A more complicated situation would involve an A protein activating a B locus whose product activates a C locus, whose product then represses (or activates) the A locus.

- Feed-forward loops. In a feed-forward loop, A activates B and C, in addition to B activating C. This is an efficient way of processing fluctuating signals, because it can be calibrated so that C is expressed only if the signal (such as the availability of a substrate) lasts long enough for both A and B to be activated.

- Single-input modules. A single transcription factor may uniquely regulate several structural genes or operons. This is efficient when the genes concerned all contribute to the same metabolic pathway.

- Dense overlapping regulons. A set of genes or operons may be regulated by a set of transcription factors, with each factor regulating several genes and each gene being regulated by several factors. The genes concerned are often functionally similar in some way, such as governing stress response or entry into cell division.

Hence, systems of gene regulation have a modular structure, in the sense that they can be dissected into a small number of kinds of subsystem. The evolution of structural genes is facilitated by the modular structure of proteins and genes (Section 6.3.2), and the evolution of regulatory systems may likewise be facilitated by the predominance of a few simple motifs that can be combined together in many ways.

New attributes can evolve through the modification of regulatory genes. The evolution of cis-regulatory elements may also be facilitated by the direct association between any modification of the element and its effect on the target gene, so that a new

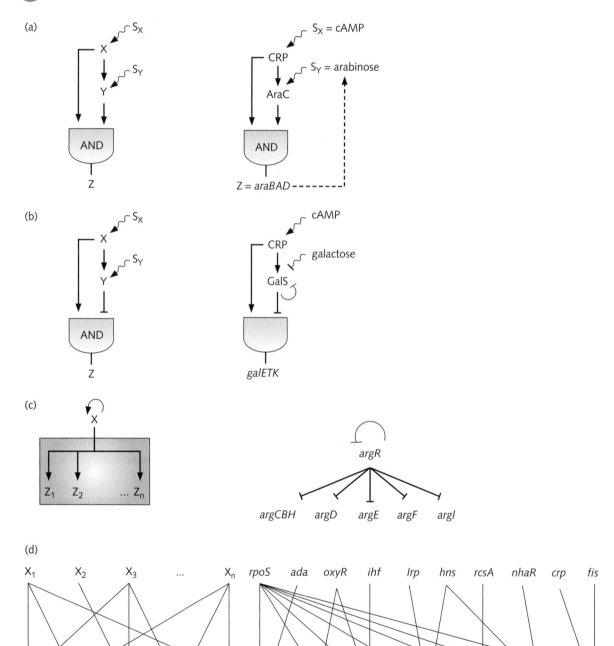

Figure 11.6 Recurrent motifs in gene regulation (left) with examples from *E. coli* (right). (a) Coherent feed-forward loop: X regulates Y, while Z is activated by both X and Y. The example is arabinose utilization. (b) Incoherent feed-forward loop: X activates Z but also activates Y, which represses Z. The example is galactose utilization. (c) Single-input module: X regulates a group of target genes and also regulates itself. The example is arginine biosynthesis. (d) Dense overlapping regulon: a set of regulators X$_n$ control in combination a set of target genes. The example is stress response.

Reprinted by permission from Macmillan Publishers Ltd: Alon, U. Network motifs: theory and experimental approaches. *Nature Reviews Genetics* 8 (6). © 2007.

phenotype is immediately expressed in heterozygotes. On the other hand, the phenotype may be so profoundly modified that it is unlikely to be viable, let alone advantageous. Bacteria have rather shallow regulatory networks, with only a few genes involved in any given pathway. Gene regulation in eukaryotes is more complicated. Genes that affect cell fate during development, in particular, are enmeshed in regulatory networks involving dozens of genes with long chains of interactions. Hence, a mutation in a regulatory gene may affect many other genes and thereby give rise to severely disturbed development. There is nevertheless the potential for major new attributes to evolve through the modification of regulatory genes, as illustrated by the *Hox* cluster (Section 10.3.2).

11.3 Eukaryote genomes have evolved novel features.

The simplest possible version of a genome—the whole set of genes replicated together—would be a solution of DNA in which each molecule was a separate gene. Something approaching this simplicity can be found in a few unusual cases, but otherwise genes are organized into larger or smaller groups because this increases the efficiency of replication. The way in which genes are arranged into groups constitutes the structure of the genome. There are two basic arrangements: a single circular molecule with a single bidirectional origin of replication, which most bacteria use, or a variable number of linear molecules with special arrangements for beginning and ending replication, which is the eukaryote state.

11.3.1 Bacteria have small, simple genomes.

Bacterial proteins are encoded by uninterrupted open reading frames, with neither introns nor extensive overlaps. These are the paradigm of the genome as a necklace of beads, one strung after another with little between.

Bacterial genomes consist of core, shell, and cloud. A very small fraction of the beads occurs in all necklaces; these constitute the core genome. They are highly conserved genes that encode the irreplaceable central processes of the cell. A much larger number occur in many but not most genomes, forming a shell supporting general-function pathways that are useful in many, but not all, ways of life. Finally, there is a very numerous cloud of genes with specialized functions, each found in only a very few lineages. These three categories of genes evolve in quite different ways. Core genes experience strong purifying selection

which preserves their integrity. Shell genes are also preserved by purifying selection but can also be modified quite readily through directional selection in response to changes in the conditions of growth. Cloud genes can often be exchanged freely among distantly related lineages, thereby providing the potential for rapid evolution of novel characteristics.

Bacterial genomes vary greatly in size. The very smallest genome would—in principle—comprise the core genes alone. The smallest cellular genome currently known is the skinny 0.2 Mb of *Carsonella*— but this is an obligate endosymbiont of insects which has transferred many of its metabolic demands to the genome of its host. Among free-living prokaryotes, the smallest genomes are about 1 Mb, as in the ubiquitous marine bacterium *Pelagibacter*. The gene density of bacterial genomes is about 1 gene per kb, so the core genome comprises somewhere between 200 and 1000 genes. Most bacteria have much larger genomes: the familiar *E. coli*, for example, has a genome of 4.6 Mb encoding about 4500 genes. Differences in genome size straightforwardly reflect differences in the number of genes: bigger genomes have proportionately more genes (Figure 11.7). The very largest bacteria known are *Epulopiscium*, which lives in the intestines of surgeonfish, and *Thiomargarita*, found in ocean sediments. Both are as large as big unicellular eukaryotes such as ciliates or dinoflagellates—they are visible to the naked eye—and have correspondingly large genomes consisting of 7000–8000 genes. Hence, there is a tenfold variation in the number of genes among bacteria, generated by an ongoing process of the acquisition of new genes and the loss of old ones in bacterial lineages.

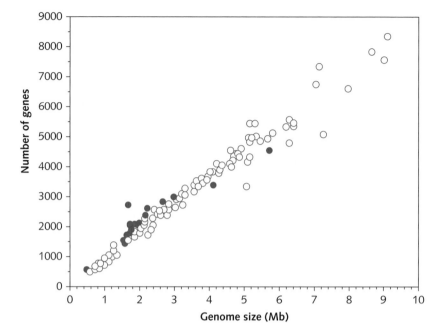

Figure 11.7 Variation in genome size among bacteria (open symbols) and archaea (solid symbols).

Reproduced from Gregory, T.R. and DeSalle, R. 2005. Comparative genomics in prokaryotes. In *The Evolution of the Genome*, edited by T.R. Gregory, 585–675. With permission from Elsevier.

11.3.2 Bacterial genomes are sculpted by horizontal gene transfer.

Among eukaryotes, species lineages are normally separate, and exchange genes very seldom, if at all. Bacteria are different. They do not form species in the same way, and very distantly related lineages may exchange genes.

Bacteria readily acquire new genes. Bacteria are infected by a great diversity of viruses—bacteriophages—which themselves originally evolved from bacterial genes. Once viral DNA has been injected into the bacterial cell it may be incorporated into the bacterial chromosome and replicated, if the host survives the infection. Other genes can be acquired from other cells—again by a process akin to infection—or simply by being taken up from the environment. Soil and water actually contain gene cassettes, complete with sequences facilitating their uptake, ready for incorporation into bacterial genomes (Section 6.2). Bacteria live in a sea of DNA. Consequently, bacterial genomes often include many newly-acquired elements. Pathogenic strains of *E. coli*, for example, contain hundreds of genes that are not present in standard non-pathogenic strains. These genes, which may account for 30% of the genome, have been acquired very recently from other lineages (see Section 6.2).

Bacterial evolution is often reticulate rather than branching. Asexual lineages of organisms like bacteria can be represented naturally by a branching diagram, the phylogenetic tree. This method of representing descent captures the history of lineages of dividing cells, but the pervasive exchange of genes between bacterial lineages connects these branches and turns the diagram into a network. If the rate of exchange is greater than the rate at which sister lineages diverge then bacterial evolution cannot be represented as a conventional phylogenetic tree. It is instead a web of lineages which intermittently exchange genes. Hence, bacterial genomes are often mosaics of genes acquired at different times from other lineages, often only distantly related. The rate of exchange varies among genes. Cloud genes are exchanged easily, shell genes less so, and core genes hardly at all. We can still draw the phylogenetic tree of bacteria, but it represents the ancestry of only a small cluster of highly conserved genes, the rest being readily exchanged between lineages.

11.3.3 Eukaryote genomes are sculpted by size and sex.

Eukaryote genes descend from genes in bacteria, although they are often extensively modified. The genes, genomes, and cells of eukaryotes, however, are quite different from their bacterial precursors.

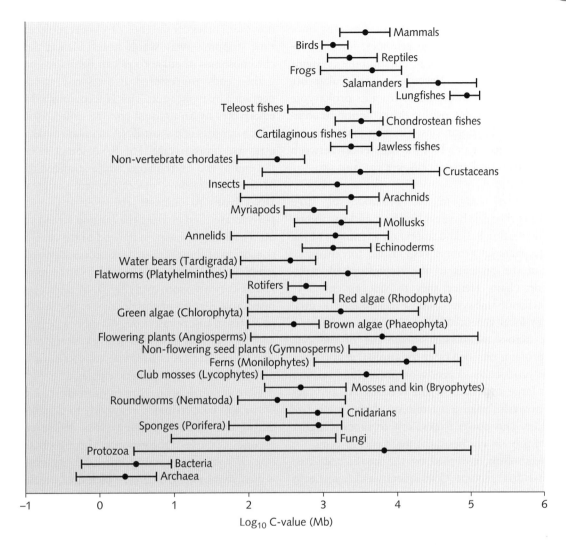

Figure 11.8 Genome size in bacteria, archaea, and eukaryotes.

Reprinted by permission from Macmillan Publishers Ltd: Gregory, T.R. Synergy between sequence and size in large-scale genomics. *Nature Reviews Genetics* 6 (9). © 2005.

- Eukaryote cells are much larger—on average, unicellular eukaryotes are about 10,000 times as large as bacteria.

- Eukaryote genomes are also much larger (Figure 11.8).

- Eukaryotes have compound cells which contain one or more genomes outside the nucleus—always a mitochondrial genome (secondarily transferred to the nucleus in some endoparasites) and often one or more others (see Section 8.3).

- The eukaryote genome consists of linear chromosomes with telomeres and centromeres, rather than a circular genome with a single origin of replication.

- The eukaryote genome usually contains a very large quota of non-coding DNA absent from bacterial genomes.

- The eukaryote genome is replicated by mitosis driven by a microtubular cytoskeleton.

- Eukaryote genes are mosaics of introns and exons, with spliceosomes to bridge the exons.

- Eukaryotes are sexual, with complete fusion of cells and nuclei, followed by the meiotic restitution division.

These differences are not as absolute as was once thought. Bacteria with large genomes, linear chromosomes and rudimentary microtubule systems have been discovered. Nevertheless, the eukaryote genome was a radical departure from its prokaryote precursors, and made possible the evolution of complex multicellular organisms. The two main reasons for this advance, which link the differences between prokaryote and eukaryote cells, are size and sex.

Endosymbiosis leads to the evolution of large cells. Mitochondria have evolved from endosymbiotic bacteria (Section 8.3) and still retain a remnant of the original bacterial genome responsible for encoding many of the proteins of the respiratory electron-transport chain. Hence, they provide an extensive internal membrane system whose membrane potential is under local genetic control. By contrast, bacteria respire across the plasma membrane of the cell, using proteins encoded by chromosomal genes. Since energy production depends on membrane surface area, a large amoeba or ciliate is thousands of times more powerful than a bacterium of the same size would be. This power is available to support higher rates of protein synthesis, the main energy drain for cell metabolism, governed by the small, dedicated mitochondrial genome. Consequently, eukaryote cells have the capacity to enlarge the nuclear genome while maintaining the same energy supply per gene. This is the crucial advance underlying the expansion of the genome from about 5000 genes in free-living bacteria to about 20,000 genes in multicellular eukaryotes.

Telomeres and centromeres are associated with mobile genetic elements. The large genome of eukaryotes has evolved in step with novel means of replicating and transcribing large quantities of DNA. In particular, the single circular chromosome of most bacteria has become fragmented into the multiple linear chromosomes of eukaryotes. The biochemistry of DNA replication, however, leads to a fundamental difficulty: when the terminal RNA primer (necessary to initiate synthesis) is removed, once the lagging DNA strand has been replicated, it leaves a gap of a few nucleotides. This does not matter in a circular molecule where continued rolling-circle replication fills in the gap. It does not matter much in a linear molecule either, to begin with. Continued replication,

however, chews off the ends of a linear molecule, over time stripping away whole genes. Hence, the linear chromosomes of eukaryotes can be maintained only if their ends are restored at the same rate they are degraded by replication. This is accomplished in most eukaryotes by an unusual enzyme called telomerase, which consists of a short RNA template molecule and a reverse transcriptase. The RNA template binds to the end of the overhanging single-stranded terminus of the chromosome, the reverse transcriptase adds DNA bases to elongate the overhang, and DNA polymerase then enables the lagging strand to catch up. The result is a cap on the end of the chromosome consisting of short repeated sequences supplied by the template RNA of telomerase which can be replaced as fast as they are worn away by DNA replication. The distinctive feature of the process is the involvement of reverse transcriptase, which is otherwise associated with viruses and mobile genetic elements. Indeed, in some cases, such as *Drosophila*, telomeres are directly maintained by mobile genetic elements.

Chromosomes which are not attached to a membrane will be distributed randomly to daughter cells during cell division, causing gross deficiencies and excesses of parts of the genome. The regular segregation of complete genomes into both daughter cells is mediated by centromeres, which nucleate microtubules to pull apart the two chromatids. Centromeres are diverse in structure, but often consist of a short central region flanked by regions containing repeated DNA sequences reminiscent of telomeres, and may be associated with mobile elements or their remains.

Gene density is much lower in eukaryotes than in prokaryotes. Eukaryote genomes vary widely in size, from about 10 to about 10,000 Mb, in contrast to the 1–10 Mb genomes of bacteria, as illustrated in Figure 11.8. A part of this difference is attributable to the roughly fourfold difference in gene number, but a much larger part to a hundred-fold difference in gene density: bacterial genomes contain about 1000 genes per Mb, whereas eukaryote genomes carry only about 10 genes per Mb. The bulk of many eukaryote genomes, including our own, consist of mobile genetic elements or the variety of repetitive sequences that they leave behind them (Section 1.5.2).

It has often been suggested that these elements are maintained by selection in the normal way because they have an important role in gene regulation and development. This is a most unlikely explanation. A regulatory element should have characteristically different insertion sites in different tissues which are the same for all individuals. Most mobile elements have the same insertion sites in different tissues which vary among individuals, exactly the opposite of the expected pattern. Some intragenic DNA may indeed have a straightforward function, but most of it does not contribute to the performance of the cell or the individual, and its prevalence therefore requires a special kind of evolutionary mechanism.

Eukaryote genes are interrupted by non-coding sequences. The low gene density of eukaryotes is also partly attributable to the low coding density of genes. The coding region of most eukaryote genes is broken up into a series of exons separated by non-coding introns. Consequently, transcription requires that the intron sequences be snipped out of the mRNA and the exon sequences connected up. Messages are spliced together in this way by the spliceosome, a complex structure of protein and RNA. About a quarter of the human genome consists of introns, and they present the same evolutionary problem as intergenic DNA. One solution is that they provide a real benefit. Exon shuffling sometimes results in new genes (Section 6.4.2), so introns provide a source of evolutionary flexibility. Although this is a consequence of the existence of introns, however, it is a very unlikely explanation of their evolution. Inserting non-coding sequences into a gene is likely to damage or destroy the transcript and will be eliminated by strong natural selection in the short term, without regard to any possible benefits in the long term. We must consider different explanations for the evolution and maintenance of non-coding intergenic and intragenic DNA. The most likely explanation is that they are genetic parasites that flourish in the sexual environment provided by eukaryote lineages.

11.4 Genetic elements may evolve cooperation or conflict.

Mitochondria and plastids are called "organelles," and it is natural to view their role in the cell as being analogous to the role of organs within the body. They are different, however, in one crucial respect: they possess their own autonomous genomes, which replicate independently of the nuclear genome. Even within the nuclear genome, mobile elements replicate independently of Mendelian genes. The eukaryote cell, therefore, houses a range of different kinds of replicators whose interactions can create conflicts that explain some of its most striking features.

11.4.1 Eukaryote genomes include several different kinds of replicators whose interests may differ.

All the genes of the nuclear genome are replicated at the same time, after a period of growth, and are evenly allocated to the daughter cells. By contrast, mitochondrial genomes are replicated during cell growth, more or less independently of the nuclear genome, and without any mechanism to ensure an even partition of genomes between mitochondria or mitochondria between cells. Deleterious mutations that compromise respiration arise quite often and will reduce the growth rate of the cell. They can easily be found by spreading yeast cells on solid agar growth medium containing glucose, letting them grow for a day, and then searching for unusually small colonies. These often have defective mitochondria that have suffered loss-of-function mutations in the mitochondrial genome. Because these lineages grow much more slowly than normal they should be eliminated from the population by purifying selection. Nevertheless, these "petite" types often occur at quite high frequencies – too high to be explained by mutation pressure alone. The reason is that defective mitochondrial genomes may be replicated quite successfully. Some kinds of mutation, indeed, increase the rate of replication, such as extensive deletions which reduce the size of the molecule. These mutations will tend to spread within the population of mitochondrial genomes of the lineage, replacing the intact genomes, as illustrated in Figure 11.9. Hence,

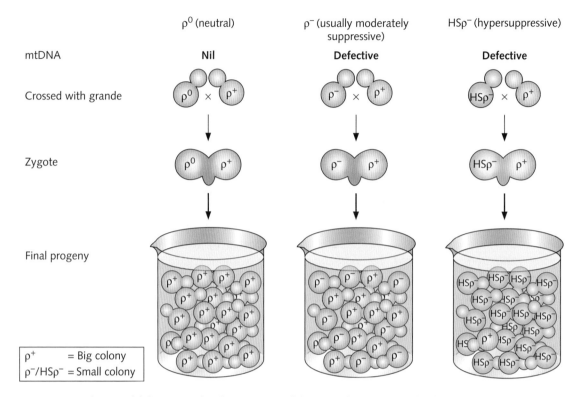

Figure 11.9 Over-replication of defective mitochondria in progeny of the crosses between normal and suppressive or hyper-suppressive yeast strains. When cells that lack mitochondria (p⁰) are crossed with normal (p⁺) cells, all their descendants have mitochondria. When mildly defective (p⁻) cells are crossed with normal cells, many of their descendants are "petite" cells with a defective phenotype. When severely defective (HSp⁻) cells are crossed with normal cells, almost all their descendants are HSp⁻ petites.

there are two opposed processes of selection acting on mitochondrial genomes.

- Selection among cells favors those with intact mitochondrial genomes and fully functional respiration.
- Selection among mitochondria may favor those with defective genomes, if the defect increases their rate of replication.

The outcome depends on the balance between these two processes. Crucially, selection at one level (among mitochondria, or mitochondrial genomes) can maintain defective "selfish" variants within the population, despite countervailing selection at another level (among cells, or nuclear genomes). This is the central principle for understanding the structure of eukaryote genomes. It is also relevant to issues of practical concern: many deadly genetic disorders are caused by mitochondrial defects, for example, whose spread is governed by the conflict of interest among replicators within the cell.

11.4.2 The replication system itself is vulnerable to genetic parasites.

Self-replicating systems can be exploited because they are not specific. The machinery that replicates the genetic material of the cell will replicate any DNA sequence it encounters. It could not be otherwise: no mechanism is known by which an enzyme could examine a DNA molecule, compute its phenotype, and having done so decide whether or not to replicate it. Within very broad limits, all sequences are replicated with the same efficiency and the same fidelity. A functional gene and its pseudogene version are treated equally, and only the minute metabolic penalty exacted by superfluous replication leads in the long term to the extinction of the pseudogene. A more fundamental consequence of this indifference is that it creates an opportunity for genomic parasites that utilize the replication machinery of another genome.

The nuclear and plastid genomes of eukaryotes harbor a host of parasitic genetic elements. A large proportion of eukaryote genomes consists of elements that rely on the non-specific property of DNA replication, while contributing nothing to the performance of the individual.

- Transposons encode their own excision and re-insertion, so they move from site to site in the genome. If a transposon moves from a region that has already been replicated to a region that is about to be replicated then it has twice the number of copies in one of the daughter genomes.

- Retrotransposons are transcribed as RNA and the transcript is then inserted as DNA at another site. Both daughter genomes bear twice the number of copies. In structure and life cycle, retroposons are similar to retroviruses, except that they do not leave the cell.

- LINEs (long interspersed nuclear elements) are broadly similar to retrotransposons, although they lack any detailed genetic similarity to retroviruses.

- Mobile introns are found in the genomes of chloroplasts and mitochondria; a copy spliced from an RNA transcript can be inserted into the homologous site.

This is by no means a complete catalogue. Evidently, the eukaryote genome provides a fertile environment for the evolution of a zoo of selfish genes.

11.4.3 Selfish elements compete for access to the germ line.

Selfish genetic elements can spread despite contributing nothing to the adaptedness of the individual, or even if they damage the individual, provided that they are preferentially transmitted to offspring. Hence, their activity in somatic tissue is irrelevant; they will spread only if they can somehow force their way into the germ line.

A selfish element will spread if it is directed preferentially into the germ line. B-chromosomes, or accessory chromosomes, are found in about 10% of all animals and plants. They resemble normal autosomes, but do not encode any essential functions; they are present in some individuals but not in others; they may vary in number among individuals that bear them; and individuals that lack them develop normally. They spread by exploiting the distinction between somatic and germ-line cells. They are transmitted through mitosis to all the tissues of the developing individual, but in many cases are known to become less frequent or numerous in differentiated somatic tissues; however, they accumulate in meristematic or germ-line cells that alone will give rise to gametes. This makes good sense from the point of view of a genomic parasite: in somatic cells it would only be a hindrance, injuring the development and activity of its host, and thereby retarding the host reproduction on which its own transmission depends. Once in the germ line, a B-chromosome can be further directed into progeny if it can navigate into the egg nucleus and avoid being eliminated in polar bodies, like the grasshopper chromosome shown in Figure 11.10.

Selfish elements can spread by destroying an allelic competitor in germ cells. Gamete-killers are elements that destroy gametes not carrying a copy of the element. The evolution of gamete-killers is more complicated than a gene which simply steers itself

Metaphase I in primary oocyte

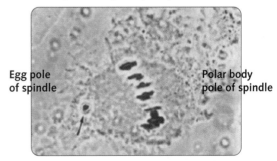

Figure 11.10 B-chromosomes in the grasshopper *Myrmeleotettix* migrate preferentially toward the egg pole, avoiding exclusion in the polar body.

Reproduced from Hewitt, G.M. 1976. Meiotic drive for B-chromosomes in the primary oocytes of *Myrmeleotettix maculatus* (Orthoptera: Acrididae). *Chromosoma* 56 (4). With kind permission from Springer Science and Business Media.

into the germ line, because an allele that destroys eggs or sperm bearing a different allele must be resistant to its own poison. The best-known example is the segregation-distorter (SD) element of *Drosophila*, which acts in heterozygotes to kill the sperm bearing the normal allele. A second locus encodes resistance to SD. Hence, a combination of the SD allele with the resistant allele produces sperm that overcome their competitors while not suppressing themselves.

Some selfish elements spread by killing cells that lack them. The ultimate example of selfish DNA is provided by systems in which an element is maintained because its absence kills the cell. The R1 plasmid of bacteria, whose behavior is sketched in Figure 11.11, is an example. It is a small circular molecule, replicated independently of the main genome of the cell, which bears two genes, *hok*

Cells infected by R1 survive because **hok** mRNA is bound by **sok** mRNA and thus remains untranslated.

There is no special machinery to distribute plasmid copies evenly between daughter cells at cell division. The plasmid is thereby sometimes lost simply by chance.

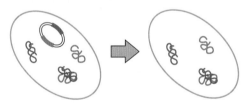

But although the plasmid may have been lost, some plasmid mRNA will inevitably remain...

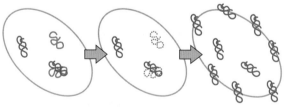

sok mRNA is much less stable than **hok**

hok is thereby expressed, killing the host cell

Figure 11.11 Cartoon of the poison–antidote system maintaining plasmid R1.

(for host killing) and *sok* (for suppression of host killing). The *hok* mRNA is translated into a small protein that collapses the proton gradient across the cell membrane, halting respiration and killing the cell. Cells infected by R1 nevertheless survive because *hok* mRNA is bound by *sok* mRNA and thereby remains untranslated. There is no special machinery to distribute plasmid copies evenly between daughter cells at cell division, so the plasmid is sometimes lost simply by chance. But although the plasmid may have been lost, some plasmid mRNA will inevitably remain. Because *sok* mRNA is much less stable than *hok*, *hok* mRNA is eventually unbound and then expressed, killing the host cell. Hence, the parasitic plasmid is maintained in an infected population because only cells that bear it survive.

Some selfish elements spread by destroying entire genomes. The paternal sex ratio (PSR) element of a small parasitic wasp, *Nasonia*, provides an even more extreme example of genetic ruthlessness. Figure 11.12 shows how it works as a B-chromosome transmitted through sperm. Fertilized eggs in wasps usually develop as diploid females, whereas unfertilized eggs develop into haploid males. The PSR B-chromosome, however, causes all other paternally derived chromosomes to degenerate. Consequently, all offspring develop as males—who will transmit the PSR element. The sex-determination system of the wasp enables the element to spread by destroying all the genes in whose company it is transmitted.

Selfish genes occasionally become domesticated. A successful selfish gene presents a paradox: having been fixed in the population, its selfish behavior is no longer an advantage. It will naturally proceed to decay, and eventually form part of the replicated but functionless DNA that forms such a large part of eukaryote genomes. Before this happens, however, its ability to manipulate the machinery of replication may be appropriated by the machinery itself. We have already seen two examples of this—bacterial plasmids carrying resistance genes and eukaryote telomeres based on transposons. The eukaryote genome demonstrates the Darwinian logic by which selfish elements spread at the expense of individual performance. It also demonstrates the Darwinian

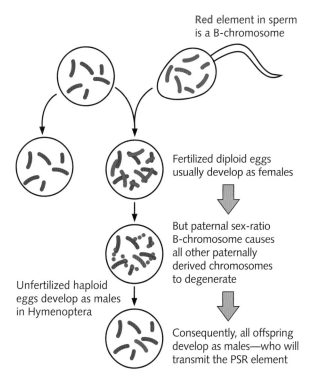

Red element in sperm
is a B-chromosome

Fertilized diploid eggs
usually develop as females

But paternal sex-ratio
B-chromosome causes
all other paternally
derived chromosomes
to degenerate

Unfertilized haploid
eggs develop as males
in Hymenoptera

Consequently, all offspring
develop as males—who will
transmit the PSR element

Figure 11.12 Transmission of paternal sex ratio (PSR) in *Nasonia*.

opportunism by which some of these elements can be conscripted to support fundamental features of sexual genomes.

11.4.4 Sex creates an opportunity for selfish genes to infect new lineages.

Any supplementary gene or genome in an asexual population is limited to the lineage in which it arises. It is akin to a vertically transmitted endosymbiont, such as a mitochondrion, which will persist only if it enhances the performance of its host. Bacterial plasmids are a common example. They compete with the main chromosome for the resources required for replication, and can spread only if they encode useful features such as antibiotic resistance—which they often do. It is only in sexual populations that harmful elements have an opportunity to spread. This is a consequence of gamete fusion: when a gamete bearing a selfish gene fuses with a gamete that does not, the zygote necessarily bears the gene, and may then transmit it to most or all of the gametes it eventually

produces. In this way, selfish genes in sexual populations can be transmitted to uninfected lineages, and in time become fixed despite damaging all the individuals which bear them.

Plasmids spread rapidly through bacterial populations when they encode conjugation. Some bacterial plasmids have the extraordinary ability to inject a copy into a neighboring cell. The infected cell forms a long narrow bridge through which part of the chromosome passes into the recipient cell, where it is recombined into its genome. Conjugative plasmids may bear useful genes; but even if they do not, they will spread by virtue of their ability to infect lineages that lack them (see Section 6.2).

The 2-micron plasmid of yeast spreads in outcrossing populations. The equivalent of a conjugative plasmid in eukaryotes is a selfish element that will only spread in sexual populations. Plasmids are rare in eukaryotes, but the 2-micron plasmid of yeast provides a clear example. This can be demonstrated by elegant laboratory experiments such as that illustrated in Figure 11.13. In asexual mitotic populations the frequency of the plasmid-bearing strains declines slowly because replicating the plasmid is a metabolic burden for the cell. If the population is self-fertilized, with gamete fusion occurring within the ascus, the plasmid does not increase appreciably in frequency. This is because it remains restricted to the single self-fertilized lineage in which it arose. When the ascus is disrupted before gamete fusion, however, the plasmid spreads rapidly through the population. This is because gametes from one lineage can now fuse with gametes from another, allowing the plasmid to spread from lineage to lineage through the population.

Homing endonuclease genes are parasites of sexual populations. Homing endonuclease genes, or HEGs, are particularly clear examples of selfish elements able to spread rapidly through outcrossed sexual populations. In a genome heterozygous for the HEG, an endonuclease is transcribed and cuts the homologous region in the sister chromosome at a recognition site specific to this HEG. The break is then repaired using the HEG sequence as template.

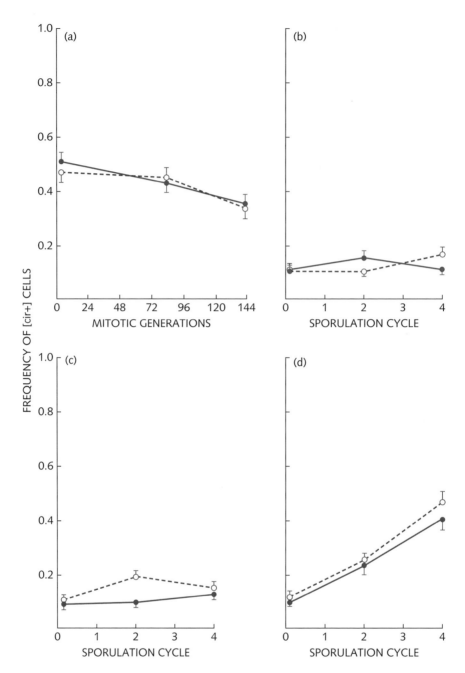

Figure 11.13 The spread of the 2-micron plasmid changes in experimental populations of yeast, from the paper by Futcher et al. (1988). The two lines in each figure are two different strains. (a) In asexual mitotic populations the frequency of the plasmid-bearing cir+ strains declines slowly. (b) If cultures pass through successive sexual cycles, with gamete fusion occurring within the ascus, the plasmid does not increase appreciably in frequency. (c) The same result is obtained if unfused vegetative cells are killed with ether during the sexual cycle. (d) When the ascus is disrupted before gamete fusion, enforcing a high rate of outcrossing, the plasmid spreads rapidly through the population.

Republished with permission of the Genetics Society of America from Futcher et al. 1988. Maintenance of the 2 micron circle plasmid of *Saccharomyces cerevisiae* by sexual transmission: an example of a selfish DNA. *Genetics* 118 (3); permission conveyed through Copyright Clearance Center, Inc.

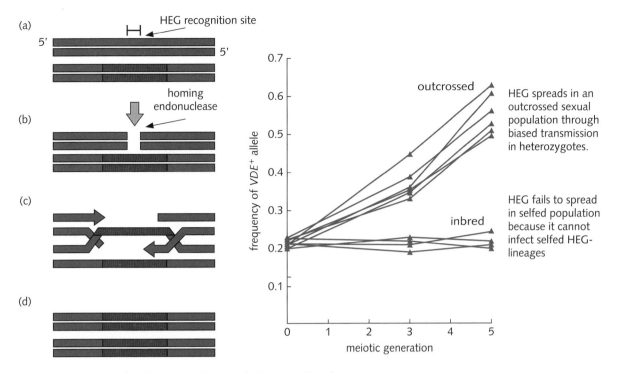

Figure 11.14 Genetic and evolutionary mechanism of a homing endonuclease gene.

Reproduced from Goddard, M.R. Outcrossed sex allows a selfish gene to invade yeast populations. *Proceedings B* 268 (1485) © 2001, The Royal Society.

This disrupts the gene, but it can be transcribed normally nevertheless because the HEG incorporates an intein, a protein able to splice across the gap that the HEG has created. The consequence, of course, is that the genome is now homozygous for the HEG and the individual retains normal viability.

The evolutionary implications of this biochemistry can be exposed in experiments where HEGs are introduced into experimental populations of yeast. The HEG fails to spread in self-fertilized populations because it cannot infect lineages that lack the element, whereas it spreads like wildfire through outcrossed populations by virtue of its ability to infect naïve lineages (Figure 11.14), becoming fixed within little more than a dozen generations or so.

Selfish genes can be used for population-genetic engineering. The properties of HEGs have suggested how they could be used to infect and eliminate a sexual species. The HEG is engineered to recognize and cut a site in the middle of an essential gene. The engineered element does not have an intein and

therefore fatally disrupts the gene. The knockout is recessive, however, and has no effect in heterozygotes. Finally, HEG expression is regulated by a meiosis-specific promoter. Now, imagine what would happen when the gene is then introduced into a population by releasing heterozygotes. These develop and mate normally because the knockout is recessive. Most of the gametes they produce are HEG+ so most of their progeny (instead of 50%) are heterozygotes. The gene thus spreads rapidly in an outcrossed population. Since it is present almost entirely in heterozygotes, its spread is at first imperceptible (except to molecular biologists). After a few generations, however, the heterozygotes become sufficiently frequent that many progeny are homozygote HEG+ and die, because the insertion is not compensated by an intein. This does not prevent the further spread of the element, which may eventually wipe out the population completely. This approach is currently being explored as a means of eliminating pests such as malarial mosquitos and thus relieving human populations in the tropics of a great burden of suffering and death.

● CHAPTER SUMMARY

Evolving genomes diverge.

- Genotype lineages eventually correspond to species lineages.

- The rate of divergence can be estimated from molecular and geological data.
  - *The human–chimp difference is an example of recent genetic divergence.*

- Synonymous substitutions are roughly clock-like, whereas non-synonymous substitutions are not.
  - *The molecular clock can be used to estimate the date of divergence between clades from gene sequence data.*
  - *Only neutral changes show the time.*

- The observed rate of divergence is large enough to fuel adaptive evolution.

Genes are modified at different rates.

- Some genes are highly conserved.

- Other genes evolve more rapidly.

- Genes decay when they are no longer selected.

- Mutations in regulatory genes can have far-reaching effects that are difficult to predict.
  - *Both structural and regulatory genes can be modified.*
  - *Regulatory networks are combinations of motifs.*
  - *New attributes can evolve through the modification of regulatory genes.*

Eukaryote genomes have evolved novel features.

- Bacteria have small, simple genomes.
  - *Bacterial genomes consist of core, shell, and cloud.*
  - *Bacterial genomes vary greatly in size.*

- Bacterial genomes are sculpted by horizontal gene transfer.
  - *Bacteria readily acquire new genes.*
  - *Bacterial evolution is often reticulate rather than branching.*

- Eukaryote genomes are sculpted by size and sex.
  - *Endosymbiosis leads to the evolution of large cells.*
  - *Telomeres and centromeres are associated with mobile genetic elements.*
  - *Gene density is much lower in eukaryotes than in prokaryotes.*
  - *Eukaryote genes are interrupted by non-coding sequences.*

Genetic elements may evolve cooperation or conflict.

- Eukaryote genomes include several different kinds of replicators whose interests may differ.

- The replication system itself is vulnerable to genetic parasites.
  - *Self-replicating systems can be exploited because they are not specific.*
  - *The nuclear and plastid genomes of eukaryotes harbor a host of parasitic genetic elements.*

- Selfish elements compete for access to the germ line.
  - *A selfish element will spread if it is directed preferentially into the germ line.*

– Selfish elements can spread by destroying an allelic competitor in germ cells.

– Some selfish elements spread by killing cells that lack them.

– Some selfish elements spread by destroying entire genomes.

– Selfish genes occasionally become domesticated.

• Sex creates an opportunity for selfish genes to infect new lineages.

– Plasmids spread rapidly through bacterial populations when they encode conjugation.

– The 2-micron plasmid of yeast spreads in outcrossing populations.

– Homing endonuclease genes are parasites of sexual populations.

– Selfish genes can be used for population-genetic engineering.

● FURTHER READING

Here are some pertinent further readings at the time of going to press. For relevant readings that have been released since publication, visit the book's Online Resource Centre at www.oxfordtextbooks.co.uk/orc/bell_evolution/

Section 11.1   Bromham, L. and Penny, D. 2003. The modern molecular clock. *Nature Reviews Genetics* 4: 216–224.

Section 11.2   Alon, U. 2007. Network motifs: theory and experimental approaches. *Nature Reviews Genetics* 8: 450–461.

Section 11.3   Lawrence, J.G. and Hendrickson, H. 2005. Genome evolution in bacteria: order beneath chaos. *Current Opinion in Microbiology* 8: 572–578.

Thomas, C.M. and Nielsen, K.M. 2005. Mechanisms of, and barriers to, horizontal gene transfer between bacteria. *Nature Reviews Microbiology* 3: 711–721.

Koonin, E.V. and Wolf, Y.I. 2010. Constraints and plasticity in genome and molecular-phenome evolution. *Nature Reviews Genetics* 11: 487–498.

Section 11.4   Werren, J.H. 2011. Selfish genetic elements, genetic conflict, and evolutionary innovation. *Proceedings of the National Academy of Sciences of the USA* 108: 10863–10870.

Burt, A. 2003. Site-specific selfish genes as tools for the control and genetic engineering of natural populations. *Proceedings of the Royal Society of London B* 270: 921–928.

● QUESTIONS

1. Explain the process of lineage sorting that occurs when two populations become permanently separated, from the time of their most recent common ancestor to the time when each has become monophyletic.

2. The two DNA sequences ATCGACTGCCACTAT and AACGATCGCCACCAT have diverged from the ancestral sequence ATCGATCGCCACTAC over a period of 180 My since their most recent common ancestor. Calculate the overall rate of substitution and the rates of synonymous and non-synonymous substitutions.

3. Discuss the merits and weaknesses of using divergence in DNA sequence to estimate the date of the most recent common ancestor of sister taxa.

4. "A billion years is not sufficient time for the genetic diversity of eukaryotes to evolve through random mutation." Evaluate this statement using estimated rates of DNA sequence divergence.

5. Describe the fate of a gene when all members of a population bear loss-of-function mutations. Compare it with the fate of a character state that is no longer functional, such as eyes in cave fish.

6. Compare and contrast the evolution of structural and regulatory genes.

7. Compare and contrast evolutionary patterns and processes in bacteria and eukaryotes.

8. How have the main features of the eukaryote cell evolved?

9. Explain how the structure of the eukaryote cell leads to conflicting processes of selection at different levels of organization, with particular reference to mitochondria.

10. Give examples of selfish genetic elements and explain how they evolve despite contributing nothing to the vigor of the individual.

11. Why do selfish genetic elements spread more readily in outcrossed sexual populations than in asexual or inbred populations?

12. Discuss the interpretation of genetic elements as beneficial or parasitic, with special reference to bacterial plasmids and eukaryote transposons and retrotransposons.

You can find a fuller set of questions, which will be refreshed during the life of this edition, in the book's Online Resource Centre at **www.oxfordtextbooks.co.uk/orc/bell_evolution/**

PART 5
Selection

Adaptation is the focus of evolutionary biology because it is the crucial feature of living organisms that requires a natural explanation. The mechanism that brings about adaptation is therefore the focus of evolutionary theory. This mechanism is natural selection. The operation of natural selection is modulated or impeded by other processes, such as inbreeding or randomness, but so far as we know natural selection alone can drive the evolution of complex, integrated bodies. We have learned about the properties of natural selection from three sources. The first is artificial selection, the deliberate choice of breeding stock by farmers and scientists. The second is experimental evolution, the study of adaptation in laboratory microcosms. Finally, the effects of natural selection have been measured in the wild, both in undisturbed populations and in populations affected by human activity. This part of the book is about natural selection as the mechanism of evolutionary adaptation.

Artificial Selection

We can study animals and plants to find out how natural selection produces evolutionary change. We may equally want not only to understand the world, but to change it. We can do this by using evolutionary principles as the basis of technologies to produce more desirable kinds of animals and plants. The most important of these involves artificial selection: the deliberate choice of certain individuals to propagate a line. This has been the mechanism for producing all the modern varieties of domestic animals and crop plants. The application of evolutionary principles to practical problems has been essential to the great expansion of agricultural productivity on which our modern civilization is based. At the same time, artificial selection is itself a powerful method for investigating these principles.

12.1 Artificial selection produces rapid and predictable change in the short term.

Attempts to produce new varieties of crop plants and domestic animals by selective breeding are evolutionary experiments on a large scale. They are bound to be complicated, time-consuming and expensive. The principles of artificial selection, however, can be worked out by experiments using model organisms in laboratory experiments. These have the usual advantages of being short-lived, numerous, and easily manipulated. We cannot use microbes, however, because the object of the experiments is to understand how the morphological or physiological characteristics of multicellular animals or plants can by modified by deliberate selection. For example, mice can be used to discover how characters such as the body weight, litter size or physical activity of a mammal respond to selection. This helps to guide longer-term experiments to improve cattle, swine, or sheep.

Taking this approach one step further, any animal could be used to investigate fundamental issues such as the effect of population size, genetic variation, or selection intensity on the outcome of artificial selection. The most extensive experiments have used the fruit fly *Drosophila*, which has been a workhorse for genetic studies since the early twentieth century. To show how these studies help us to understand the mechanism of evolution, we shall describe a careful and exact experiment that was carried out more than fifty years ago. It was designed to find out whether the outcome of selection can be deduced from a very simple theory that requires knowing only the amount of genetic variation in the population and the intensity of selection applied by the experimenters. The result was that it can—up to a point. Both the success and the failure of the simple theory tell us a great deal about the process of evolution.

12.1.1 The breeder's equation predicts the short-term response to selection.

The simplest question that we can ask is: how much improvement can be expected when selection is applied to a population over a single generation? The answer involves two kinds of quantity.

The first is how much variation there is in the population, with respect to some particular character of interest. If all individuals are exactly the same, selection cannot be applied at all. It is impossible to select for the number of heads or livers per individual, for example. Most characters, however, vary to some extent. Quantitative characters that vary on a continuous scale (such as the length of the tail of a mouse) or a nearly continuous scale (such as the number of vertebrae in the tail) almost always vary among individuals, sometimes very widely. Selection is then practicable, provided that character state can be scored. The experimenter applies selection by choosing exceptional individuals. Only these individuals are allowed to breed. These selected individuals have an average character state of S, which differs from the average of the population, A. This creates a selection differential, D, equal to the difference between the average of the selected individuals and the average of the population before selection: $D = S - A$.

Variation is sufficient for selection to occur, but it is not sufficient for evolution to occur. Some individuals may happen by chance to find more food than others, or remain free of disease, and thereby develop into exceptional adults. They are then recognized and selected by the experimenter. The offspring of these selected individuals, however, will develop in the same average conditions (amount of food supplied by the experimenter, or overall incidence of disease) as their parents. Consequently, they will have the same average character state as their parents. Selection has been applied, but has not caused any change in average character state in the following generation.

Hence, selection is effective only when the offspring of the selected individuals resemble their parents, with respect to the state of the character under selection. In a phrase, character state must be heritable for selection to be effective. Another way of expressing this is that variation among individuals must be genetic; that is, the difference among individuals, with respect to the state of the character being selected, must be caused by differences in their genotypes. For example, individuals may differ because they bear different alleles at a given locus. Hence, we conclude that selection can be effective when there is genetic variation in character state.

The effectiveness of selection therefore depends on the balance between two sources of variation: environmental (differences in conditions of growth experienced by individuals, independent of their genotype) and genetic (differences arising from the genotypes of individuals, independent of their conditions of growth). The fraction of the total amount of variation that is attributable to genetic differences among individuals is thus:

(genetic variation) / (genetic + environmental variation).

This quantity has a technical name: it is the *heritability* of a character. It is denotes by h^2, whose value varies from zero to one (the reason for using a square is purely historical). A low value (near zero) indicates that the variation of a character is mostly attributable to conditions of growth. In the offspring generation, individuals have similar character state regardless of their parents. A high value (near one) indicates that most variation is heritable: offspring resemble their parents, regardless of their conditions of growth.

Heritability is estimated by a breeding trial. Males and females are paired at random, and produce families of offspring. It is obvious that a highly heritable character will vary among families (because their parents are different), while offspring from the same family will be similar (because they share the same parents). This can be represented by plotting the average of the offspring against the average of their parents. The slope of this graph will be zero if offspring are the same, on average, regardless of their parents. It will be unity if offspring, on average, exactly resemble their parents. Almost all realistic cases will fall in between, of course. The slope of this graph, between zero and unity, is the heritability of the character, as illustrated in Figure 12.1.

The scheme of selection, combined with an estimate of heritability, gives us enough information to predict how the population will change over time. The response of the population, R, with respect to the character being selected, is the change in average character state from the parental generation to the offspring generation. This is the breeder's equation: $R = h^2D$. It has a straightforward interpretation, which is illustrated in Figure 12.1. Selection causes a difference, D, between the selected individuals and the parental population as a whole and a fraction h^2 of this difference is transmitted to the offspring of these selected individuals.

12.1.2 The Edinburgh bristle experiment illustrates the basic principles of artificial selection.

Under a low-power microscope, you can see that the body of a fruit fly is studded with long hairs, or

The selection differential

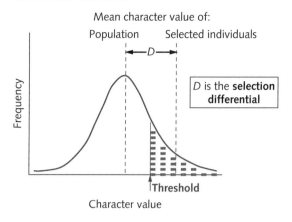

D is the **selection differential**

The heritability

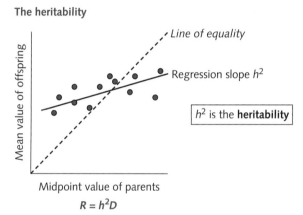

Line of equality

Regression slope h^2

h^2 is the **heritability**

$$R = h^2 D$$

Figure 12.1 The breeder's equation. The response to selection R is equal to the selection differential D depreciated by the heritability h^2.

bristles. These are not as simple as our body hair; rather, they are complex organs that enable the fly to sense its physical and chemical environment. They also offer an easy target for experiments in artificial selection: there are relatively few of them (about 40 or 50), so they are easily counted on anaesthetized flies, so that individuals with exceptionally many or exceptionally few bristles can be retained, to found the next generation, while all others are discarded. This provides us with a convenient and easily manipulated model system to investigate how average character value responds to artificial selection.

The change in bristle number over time, caused by a deliberate program of artificial selection, can then be used to predict how some useful character, such as the milk yield of cows or the egg production of poultry, could be improved by some proposed scheme of selection. This was the basis for a classical experiment designed to validate the theoretical basis of modification by selection, performed by a group of researchers at Edinburgh University half a century ago.

The first step was to set up a large population of flies with a good deal of variation in bristle number, so that the heritability of the character could be estimated from breeding trials. It was found that about half of all the variation among individuals was attributable to variation among families, so $h^2 = 0.5$. This is fairly representative of quantitative characters. It implies that when selection is applied, it is likely to cause quite rapid change in the average character state of the population.

The next phase was the selection experiment itself. Each selection line was propagated from 20 adult flies in each generation, but these individuals were chosen from different numbers of candidates. In one case, 25 flies were screened, and the 20 most extreme individuals chosen. This is rather weak selection: most of the candidates passed. Hence, the selection differential D is rather small. In another case, the 20 successful individuals were chosen by screening 100 flies. This is quite strong selection: most of the candidates failed, and the selection differential is correspondingly large. Other lines were intermediate. All were propagated for a few generations, while recording the mean bristle number of all the adults screened in every generation.

The outcome of selection confirms the general theory. We can now compare the actual modification of the experimental selection lines with the theoretical prediction. The lines that were more strongly selected (20/100 accepted) should have more bristles than the more weakly selected lines (20/25 accepted), when the individuals with the most bristles are chosen by the experimenter. Choosing the individuals with the fewest bristles should have the same outcome, in the opposite direction. The lines did all respond to selection within five generations, with mean bristle number increasing or decreasing; this was not unexpected, given the ease of scoring the character and its rather high heritability. Moreover, they responded in the expected way, with more intense selection causing a proportionately greater response (Figure 12.2).

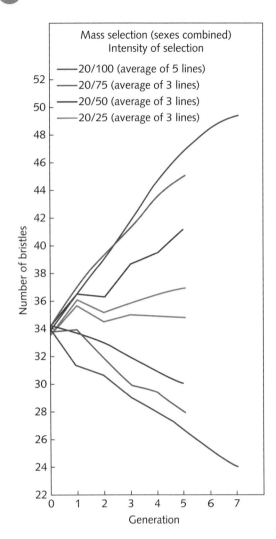

Figure 12.2 The Edinburgh bristle experiment. Response in relation to intensity of selection.

Reproduced with kind permission from Springer Science+Business Media: Clayton, G.A., et al. 1957. An experimental check on quantitative genetical theory. I. Short-term responses to selection. *Journal of Genetics* 55 (1).

With intense selection (20 flies selected from a sample of 100), there is a steep and regular response in the direction of selection; when selection is much weaker (20 flies selected from a sample of 25) there is little or no consistent response in the short term. Moreover, the actual number of bristles gained or lost was very close to that predicted by the breeder's equation (Figure 12.3).

Replicate selection lines tend to diverge. For each level of selection, several replicate selection lines were established, each from the same base population, and each selected by the same protocol. On average, they responded as expected. However, not all the replicate lines subjected to the same intensity of selection responded in exactly the same way. Some responded a little more than expected; others, a little less. Moreover, lines which started out by responding a little more rapidly than expected, or a little less rapidly, kept their rank relative to the others (Figure 12.4). This is caused by sampling error. The lines are set up as rather small random samples from the base population, and differ somewhat in their initial composition. This difference will subsequently tend to increase, as the lines are propagated, as biased samples, from generation to generation.

If the experimenter could scrutinize the genotypes of the flies directly, choosing only those individuals bearing the appropriate genes, these errors would not exist, or would be very slight. However, the experimenter must instead select phenotypes, and will often choose flies which express extreme phenotypes because they have obtained somewhat more or less food than average, which have developed from somewhat larger or smaller eggs, or which vary for any of a multitude of reasons attributable to the unique circumstances of their individual development. Consequently, any two selected samples will be somewhat different genetically, and because only genetic differences are transmitted, independent selection lines will tend to diverge. A single selection line thus represents a unique historical process that cannot be precisely repeated.

Relaxing selection causes reversion to the ancestral state. What will happen if we simply stop selecting, and transfer the lines by permitting all individuals to reproduce? This would re-create the original conditions of culture, where the flies reproduce without any direct human intervention. In these conditions, there is some optimal bristle number, which is reached by purifying selection acting against individuals with more or fewer bristles. Artificial selection forces the population away from this value. When it is relaxed, therefore, purifying selection of the same kind present in the ancestral population is the only source of selection. Consequently, the lines begin to converge on the ancestral state (Figure 12.4).

This shows how any domesticated stock is simultaneously affected by two sources of selection. The

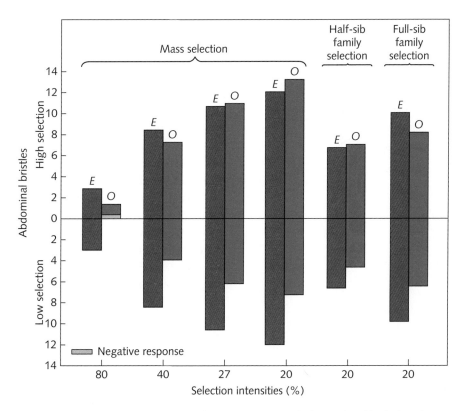

Figure 12.3 The Edinburgh bristle experiment. Comparison of the observed (*O*) with that predicted by the breeder's equation (*E*).

Reproduced with kind permission from Springer Science+Business Media: Clayton, G.A., et al. 1957. An experimental check on quantitative genetical theory. I. Short-term responses to selection. *Journal of Genetics* 55 (1).

first favors any kind that reproduces well in captivity. This is natural selection, even though it occurs in a humanized environment, such as a ploughed field, a cattle pasture or a laboratory culture vial. The second is the artificial selection applied directly by the experimenter. This runs contrary to natural selection, because it is deliberately used to produce a result that we desire, but that natural selection has not delivered. Hence, the response to artificial selection will be limited by the contrary operation of natural selection within the experimental lines.

Selected lines surpass their ancestor. In the short term of 3–4 generations, the experiment showed how we can predict the outcome of artificial selection from knowing how much heritable variation was originally available. What happens in the medium term of about 30–40 generations? The obvious answer is that the most extreme types present in the ancestral population will have been selected, and at this point no further improvement will be possible.

This was not what happened. Astonishingly, both upward and downward selection lines came to lie far outside the range of variation present in the ancestral population (Figure 12.5). The average bristle number was initially about 40, with the most extreme flies having about 50 bristles. In the high line, the average bristle number was about 90. No single fly in the original population could possibly have had 90 bristles. How could such individuals have arisen? There are two possibilities.

The first is that new mutations causing extreme phenotypes might have occurred. They could equally have arisen in the ancestral population, but would have been eliminated by natural selection. They spread because the individuals which bear them, however feeble they might be in other respects, have abnormally large numbers of bristles, and are therefore chosen by the experimenter.

The second is that alleles causing modest increases in bristle number are brought together by genetic recombination. Bristle number might be affected by

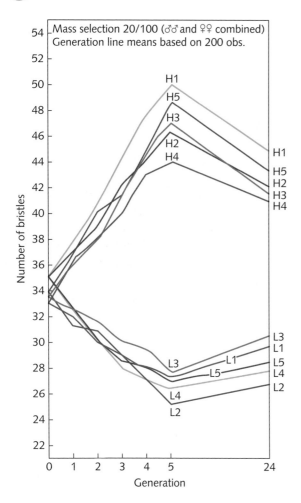

Figure 12.4 The Edinburgh bristle experiment. Divergence of replicate selection lines. Note that the lines tend to return towards the ancestral character value when selection is relaxed after generation 5.

Reproduced with kind permission from Springer Science+Business Media: Clayton, G.A., et al. 1957. An experimental check on quantitative genetical theory. I. Short-term responses to selection. *Journal of Genetics* 55 (1).

very many genes; let us consider just twelve, for simplicity, each of which has two alleles. One allele causes an increase in number and is designated +; the other allele, written as –, causes a decrease in number. A sequence of plus and minus alleles defines a genotype, for example – – + – + + + – – – + –. We can imagine that increased size is being favored by selection, and that the fitness of an individual is the count of + alleles in its genotype, 5 in this case. The population initially comprises a range of genotypes, such as

– + – – + – – – + – +
+ + – – – – – + – – + –

– + – – + – + – + – – +
+ + – + – – + + + – + –
– – + – + – – – + + – –

and so forth. The fittest genotype in this set is the second from last, with a score of 7. This would be the limit of sorting in an asexual population; this genotype would become fixed despite bearing the deleterious – allele at many loci. If the population is sexual, it can transcend this limit. Imagine a mating between the second-from-last genotype and the last. This would produce recombinant progeny with a range of phenotypes. Some would by chance inherit mostly – alleles, and would have very low fitness (the worst is – – – – – – – – + – – –, score 1). Others, however, would inherit mostly + alleles, and would have very high fitness (the best is + + + + + – + + + + + –, score 10). Recombination has the effect of making all the allelic diversity present in the initial population available to selection, which then drives the population beyond the original limits of variation. If the individual with a score of 10 mated with the first genotype on the list above, some of its progeny would have + alleles at all loci, and would have reached the theoretical limit of sorting in sexual populations.

The response to selection ceases when genetic variation is exhausted. Both of these processes contributed to forcing the population in the Edinburgh bristle experiment beyond the original limits of variation. After about 35–40 generations, however, no further progress could be achieved. The experimental populations had reached a plateau. What is it that limits the capacity for selection to produce change?

In general, selection will become ineffective once genetic variation has been exhausted. Eventually, the population will come to consist exclusively of a single extreme type, or a few extreme types, whether generated through mutation or recombination. Any residual variation among individuals is then purely environmental, and no further progress can be made. Many experiments end for this reason. A good example is the heroic attempt, lasting over 30 years, to evolve mice as big as rats. For the first 15 years or so, this was entirely successful. Body size increased steeply from an average of about 25 g to about 45 g. It then ceased to respond to selection, and for the next 15 years fluctuated irregularly without any

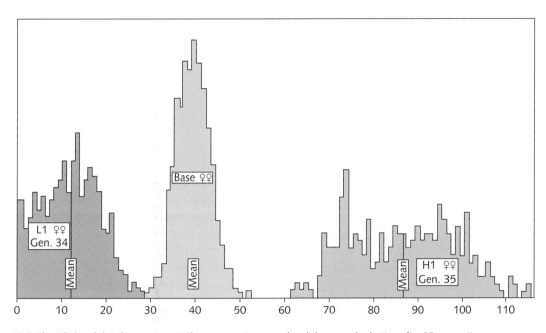

Figure 12.5 The Edinburgh bristle experiment. The response to upward and downward selection after 35 generations.

Reproduced with kind permission from Springer Science+Business Media: Clayton, G.A., et al. 1957. An experimental check on quantitative genetical theory. I. Short-term responses to selection. *Journal of Genetics* 55 (1).

definite trend, despite strong and continued selection. There was simply no genetic variation left in the experimental populations.

Artificial selection may be limited by countervailing natural selection. In the case of the Edinburgh bristle experiment, however, the limit to artificial selection was not set by the exhaustion of genetic variance. This was easily demonstrated by applying downward selection to the stalled upward selection lines. They readily responded, with a sharp decrease in bristle number. Clearly, they still contained abundant genetic variation. Why could this not be harnessed to increase bristle number still further?

In most organisms, this might have simply remained unresolved, but the genetic toolkit that has been developed for *Drosophila* made it possible to work out what had happened. An allele causing a large increase in bristle number had arisen by mutation. Heterozygous individuals, bearing a single copy of the allele, had many more bristles than average, and consequently were very likely to be selected. Homozygous individuals, however, bearing two copies, died early in development. This is an example of a recessive lethal allele. The consequence was that

the frequency of the allele was increased in every generation by artificial selection (heterozygotes were likely to be chosen) but reduced by natural selection (homozygotes always died). Because the homozygotes are produced by mating between the selected heterozygotes, the advance produced by artificial selection is cancelled by the regress caused by natural selection. This tension between artificial selection and natural selection imposed the limit on how far bristle number could be increased.

This is a special case of a common phenomenon in artificial selection. Experimental lines that evolve extreme phenotypes very often become abnormal and enfeebled in various ways. Further selection is then ineffective. The reason is usually the appearance of recessive alleles that enhance the selected phenotype in heterozygotes but cause severely defective phenotypes in homozygotes. There is an elegant evolutionary explanation for the prevalence of such alleles late in the history of selection lines. Some alleles will both depress the selected character and debilitate the individual. Such alleles will be rapidly lost. Others will enhance the selected phenotype and invigorate the individual. These will rapidly be fixed. Others will enhance the phenotype while debilitating the individual; these will be left segregating in the population, and are therefore likely to be instrumental in setting a limit to the progress made by artificial selection.

12.2 Domestication is applied artificial selection.

Our ancestors were hunter-gatherers who ate whatever wild produce they could catch or find. This phase of human history is almost past; it lingers only in the fish market. Since the agricultural revolution that began about 10,000 years ago, we have domesticated most of the animals and plants we rely on. There are very few of these. Of all the thousands of species of wild vertebrates, we rely on only a dozen or so for most of our wants—cattle, sheep, swine, chickens, and a few others. There are hundreds of thousands of species of plants, but we use only a very few of these for our food—the major crops such as rice, wheat, maize, and bananas. In the process of domestication, the species that we have chosen have been greatly modified from their wild ancestors. In each case, farmers have conducted extensive, long-term evolutionary experiments that have produced new kinds of animals and plants.

12.2.1 The process of domestication usually involves four phases that differ in how selection is applied to populations.

The first phase is the choice of organisms to domesticate. Planting annual grasses in an enclosure eases the task of gathering edible seeds. Planting oak trees for acorn production would not give the same benefit, because they take too long to mature. Consequently, wheat and barley were domesticated, whereas oaks were not. Social animals that follow a leader can often be induced to tolerate restraint, so cattle and dogs were domesticated whereas deer and badgers were not. (Cats are a mysterious exception to this rule.) Hence, domesticated animals and plants are not only a small sample of all the possible candidates, but also a highly unrepresentative sample.

The second phase is the proliferation of particular lineages in the early stages of domestication, without conscious human intervention. Imagine the early generations of wheat cultivation, for example.

The plants are enclosed, harvested, and part of the grain re-sown. This practice would create a novel environment to which the population would quickly become adapted. Successful genotypes would possess a suite of characters such as retaining the seed on the stalk (so that they could be efficiently harvested) and rapid germination (because dormant seed will not contribute to the crop). This phase is dominated by natural selection within a humanized environment.

The third phase is the deliberate selection by farmers of individuals with desirable attributes as the parents of the next generation. This phase requires a basic understanding of genetics (offspring resemble parents) and evolution (the next parent generation resembles the current offspring generation). It is readily applied to characteristics of individuals, such as palatability, docility, and ease of handling. Because it is applied by a conscious agent for a purpose, it is called artificial selection. It has been responsible for most of the qualitative modifications of crop plants and domestic animals in the first 10,000 years of agriculture.

The final phase is the systematic selection by scientists of attributes such as grain yield or milk production. Crops such as cabbage or celery are grown as individual plants, well-spaced out in rows. It is relatively easy for the farmer to recognize the largest or tastiest plants, and select them to perpetuate the crop. It is not as effective for field crops such as wheat, where it is the overall production of a population that is the target of selection. Indeed, it may be counterproductive: large wheat plants may be aggressive types that suppress the growth of their neighbors and thereby reduce overall seed yield. This phase requires substantial long-term investment in breeding trials on experimental farms where large plots of cultivars can be compared. It has been conducted systematically only in the last hundred years or so, since the establishment of publicly funded experimental stations in the late nineteenth century. These establishments have been responsible for much of the increase in agricultural productivity that has sustained industrial societies.

12.2.2 Domestication often involves a few regulatory genes of major effect.

Modern cereal grains such as barley, wheat, and maize were domesticated from wild ancestors by pre-historic farmers who chose particular kinds of plant to supply the seed corn for the next crop. By doing so, they often produced plants radically different from their ancestors. In many cases, they were identifying the phenotypic effects of a few major genes with a profound effect on the development of plant structures such as seeds and fruits.

The best-known set of major genes involved in domestication is that involved in the evolution of maize from wild teosinte in Mexico. Some affect the infructescence (cob) which has become modified like the ear of wheat. Thus, in teosinte each grain is protected by a tough covering consisting of a rachis segment and a glume, whereas in maize these are reduced in size and form the cob on which the naked grains are borne. The gene responsible is *teosinte glume architecture1* (*tga1*), which is thought to be a regulatory gene. Furthermore, the ears of teosinte disarticulate at maturity whereas those of maize remain intact and are easily gathered, and the size of the ear has been greatly increased. At the same time, there has also been a radical change in plant architecture: teosinte has long lateral branches bearing tassels (male flowers) whereas maize has short branches bearing ears. The gene primarily responsible for this change is *teosinte branched1* (*tb1*) which like a similar gene Q in wheat is a transcription factor.

12.2.3 Desirable features of domesticated animals and plants can be greatly altered by systematic artificial selection.

In the summer of 1896, agronomists at the University of Illinois measured the oil and protein content of 163 ears of a contemporary variety of maize called Burr's White. The most extreme individuals in either direction were chosen to found upward and downward selection lines for oil and protein content.

This experiment—begun before the rediscovery of Mendelian genetics—is still continuing, after more than 100 annual cycles of selection. The results (so far) are illustrated in Figure 12.6. Maize kernels usually contain about 10% protein and 5% oil. The upward selection lines now have about 30% protein and about 20% oil. The downward selection lines have about 5% protein, and so little oil that it cannot readily be measured. The selection lines, descending from a few individuals of a single cultivar, now span the entire range of variation known to occur in maize.

- The main outcome of the Illinois long-term corn experiment is similar to that of the Edinburgh bristle experiment: artificial selection can modify the features of a lineage far beyond the range of variation present in the ancestral population.

- The maize lines, like the fly lines, readily responded to reverse selection, with the upward lines rapidly evolving lower protein and oil content. Hence, they must still retain a considerable amount of genetic variation.

- Furthermore, the lines began to express deleterious side-effects after many generations of selection. The downward lines for both protein and oil, for example, had low germination success after 70 or so generations of selection. This is comparable with the morphological abnormalities observed in the fly lines, and likewise demonstrates a growing tension between artificial selection and natural selection.

This remarkable experiment illustrates the great modification that can be achieved by long-continued selection. It has taken much longer than the fly experiment, because flies reproduce twenty times faster than corn. Nevertheless, the main features of the two experiments are very similar. This shows how experiments on model organisms in the laboratory can be used to predict how the useful characters of domesticated animals and plants will respond to a planned program of artificial selection.

12.2.4 Crop yield has been greatly improved by systematic artificial selection.

The grain yield of cereals hardly changed in the first 10,000 years of domestication. Some characters evolve

(a)

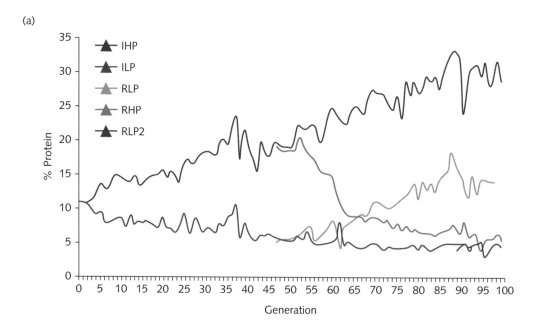

(b)

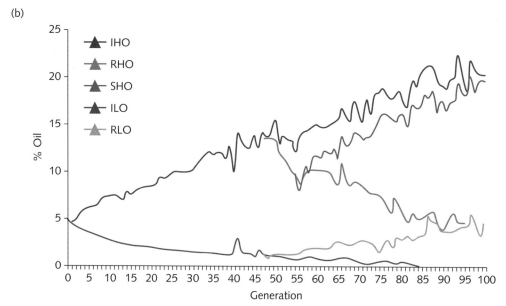

Figure 12.6 The first century of the Illinois long-term corn experiment. Note the response to reversing the direction of selection after 50 years. In the key labels, I Illinois (primary upward or downward selection), R reverse selection; H high, L low; P protein, O oil.

Reprinted from Moose et al. Maize selection passes the century mark: a unique resource for 21st century genomics. *Trends in Plant Science* 9 (7): 358–364 © (2004), with permission from Elsevier.

rapidly through natural selection when a cereal is brought into domestication. These include a tougher rachis to hold the seeds on the stem until threshing, and reduced seed dormancy. This is because subsistence farmers will sow the seed corn that has been preserved from the previous harvest. Lineages which readily shed their seeds, or fail to germinate promptly, will simply not be perpetuated by normal agricultural practice. The farmers are acting unconsciously as agents of natural selection. Other characters of individual plants, such as the size and palatability of seeds, are readily modified by unsystematic artificial selection. Seed yield, however, is a character of populations of plants, and is much more difficult to select.

In the last 60 years, systematic mass selection has been very effective in increasing yield. In cereal crops, the object of cultivation is to harvest as much grain as possible from a whole field. The best varieties can only be identified by comparing whole fields, or at least sizeable plots, each sown with a different variety. This was quite impracticable under prehistoric or medieval systems of cultivation. It became routine only when large-scale field trials were instituted by private corporations or publicly funded research stations in the 1950s, as mentioned above. The practical difficulties are formidable, not only because of the space needed for such experiments, but also because the yield of a variety might vary from year to year because of its response to climate, or might depend on the level of irrigation or fertilization. Nevertheless, the enterprise has been astonishingly successful. In the last 60 years, cereal yields have increased by an average of 100 kg per hectare per year. Figure 12.7 shows how, within living memory, the yield of cereal crops has become three or four times greater than had been accomplished in the previous 5000 years of agriculture.

The response to selection can be demonstrated by common-garden trials of stored obsolete cultivars. The improvement of modern crops cannot be conclusively demonstrated by comparing contemporary with historical yields, because other factors, such as an altered climate or improved harvesting techniques, might have made a contribution. The decisive observation is the common-garden experiment. Seeds of antique and modern varieties are sown at the same site, grown under the same conditions of cultivation, and their yield compared. This is possible because samples of seed from old varieties have been carefully stored. Common-garden experiments show that modern varieties are indeed more productive.

The contribution of selection can be evaluated by varying the conditions of growth. The common-garden experiment is not completely decisive, however, because yields are affected by agricultural management and inputs, as well as by the genotype of the crop. It is possible that modern cultivars are superior only under modern conditions of cultivation,

which often involve high inputs of water, fertilizers, and pesticides. This possibility can be investigated by a more complex type of experiment that involves growing both modern and obsolete cultivars under both new and old agricultural regimes. From the results of an experiment like that illustrated in Figure 12.8, the overall improvement in yield can be attributed to two causes: a genetic component caused by new varieties, and an environmental component caused by new conditions of growth. These two components are roughly equal in magnitude. Artificial selection has been responsible for roughly half the increase in yield that has been accomplished in the last 60 years.

The Green Revolution shows how selection can act at different levels. Modern agronomic techniques have been successful because the target of artificial selection has shifted from the individual plant (appropriate for cabbages) to the plant population (appropriate for wheat). The fundamental lesson to be learned is that selection can act at different levels. Varieties can be selected either for their performance as individuals or for their combined performance as populations. This insight led to the improvement of rice and other crops, often called the "Green Revolution." The new varieties that enabled rice yield to keep pace with the growing number of people were not, in general, large and vigorous plants. They were the opposite: short-stemmed and relatively feeble plants. This is because large and vigorous plants compete intensely with one another, and thereby stifle the production of the field as a whole. Short and feeble plants do not suppress the growth of their neighbors, so the field as a whole produces more. This demonstrates how selection can produce different kinds of outcome when it is applied at different levels: selecting individual performance, or selecting the combined performance of large groups of individuals.

Almost all desirable attributes of domestic animals and plants have been improved by systematic artificial selection. Almost every food that you can buy in the grocery store has been radically improved by systematic artificial selection. Milk yield in cattle, meat production in poultry, the palatability of beans, peas, and salads, and many other examples, have greatly advanced since your parents' or grandparents' days. Figure 12.9 shows that this improvement has often

(a)

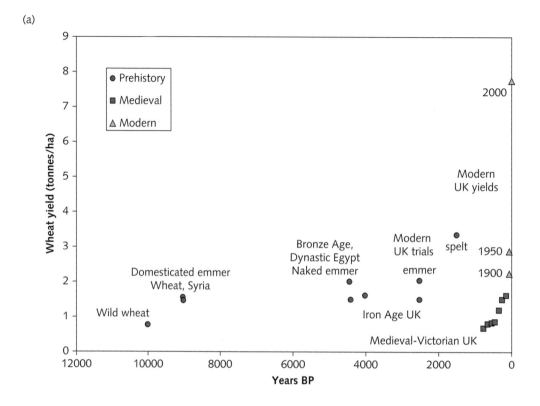

(b)

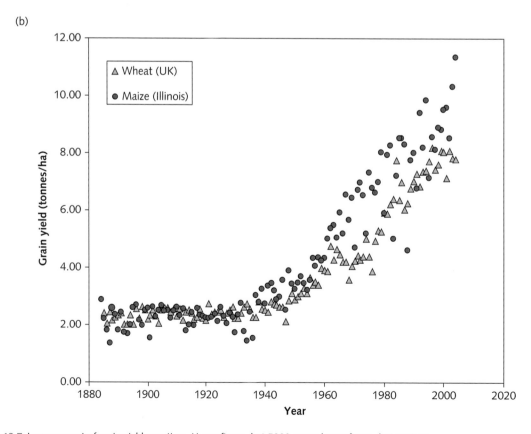

Figure 12.7 Improvement of grain yield over time. Upper figure, last 5000 years; lower figure, last century.

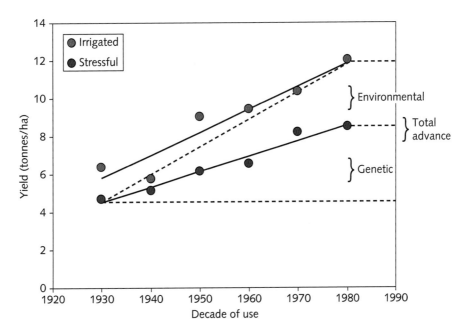

Figure 12.8 Yield of maize cultivars developed in past decades and grown together in a common garden. The use of irrigated and stressful conditions permits an estimate of the environmental and genetic contributions to improved yield.

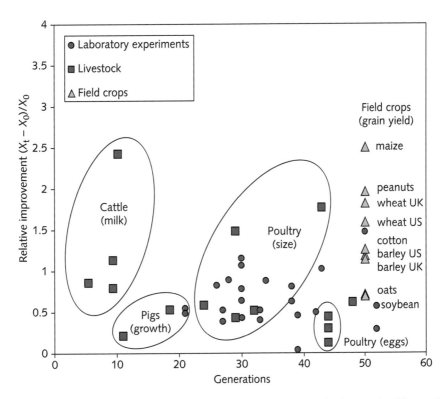

Figure 12.9 Improvement in domestic animals and crop plants through artificial selection. The data are the difference between the value after selection and the value before selection, as a fraction of the value before selection, after a given number of generations. A value of zero would indicate no improvement. Thus, poultry meat mass increased in one project by about 250% (plotted value about 1.5) after 30 generations of artificial selection.

been substantial. The variety, quality, and cheapness of our food are in large part due to the success of evolutionary technology.

There have been a few exceptions, where selection has been ineffective. Egg production in poultry has not changed much, for example, probably because there is a natural limit of one egg per day. This is an example of a physiological constraint that is difficult to modify by selection because there is little or no variation on which selection can act. A more mysterious exception is the speed of racehorses. A champion racehorse is one of the most valuable of animals, and a vast amount of money and effort is put into breeding winners. Nevertheless, intense selection has failed to increase the speed of winning horses over the past 50 years. Selection is usually effective, and is often astonishingly effective. But there is no guarantee that it will work, and sometimes it fails completely.

12.3 Artificial selection can produce extensive adaptive radiations.

Artificial selection, applied conscientiously over many generations to a single lineage of animals or plants, is capable of modifying its characteristics to extremes never seen in its ancestor. If it is applied in different directions to many lineages, it can produce a whole range of new types from a single ancestral stock. We have seen how natural selection has led to the diversification of types specialized for different ways of life in many groups. Divergent artificial selection has a similar outcome: a broad range of specialized types, all descending from the same common ancestor. We have already met one example of this, the range of vegetables produced from wild cabbage (Section 7.3.1). Dogs provide a more extreme example.

Dogs are much more variable and specialized than their ancestor. This point is evident to anyone who has pondered the variety of dogs to be seen every day in streets or fields. All modern breeds descend from the wolf, a large and rather variable canid that struck up an acquaintance with human bands towards the end of the last glacial advance, some ten thousand years ago. They were for long selected, no doubt largely unconsciously, for several useful characteristics, particularly for hunting and guarding, and have more recently been deliberately selected and inbred to create the vast range of modern breeds whose relationships are illustrated in Figure 12.10. In body size alone, the variation among domestic dogs not only exceeds that among wolves, but actually exceeds that among all members of the family Canidae—wolves, coyotes, jackals, wild dogs, and foxes.

The extent of behavioral modification has been even more extraordinary. The capture of prey by wild canids may involve a whole sequence of behaviors: flushing, tracking by scent, detection by sight, pursuit, either individually or in a pack, herding, crippling, killing, and finally carrying the prey back to be eaten. Domestic dogs have been selected to excel at one of these tasks, while often suppressing all others. Spaniels will flush game from low undergrowth; bloodhounds follow a scent trail; gazehounds such as salukis have exceptional visual acuity, while pointers will actually detect prey and then freeze without proceeding to pursue it; foxhounds and beagles hunt in packs; sheepdogs will herd flocks of sheep (and almost anything else) without attacking them; bulldogs were bred to grip their prey and hang on; mastiffs are fighting and killing dogs; retrievers will fetch dead or wounded prey without (in theory) damaging it; every component of the strategy of wild canids for capturing prey is represented by a specialist breed. Moreover, different breeds tackle different prey: deerhounds and otter hounds for large errant animals pursued in the open, the whole tribe of terriers for small game hiding in burrows and crevices. The exuberant diversity of dogs is a striking testimonial to the power of selection to direct adaptive change far beyond the limits of the original population within a few hundred generations.

Artificial selection provides an exemplar of how natural selection works. Anyone who has ever worked with natural populations will be familiar with two facts: the first is that they are always changing, and the second is that it is very difficult to identify the agents of change and to quantify their effects. To establish basic principles, it is always necessary to bring science indoors, where replication and control are possible.

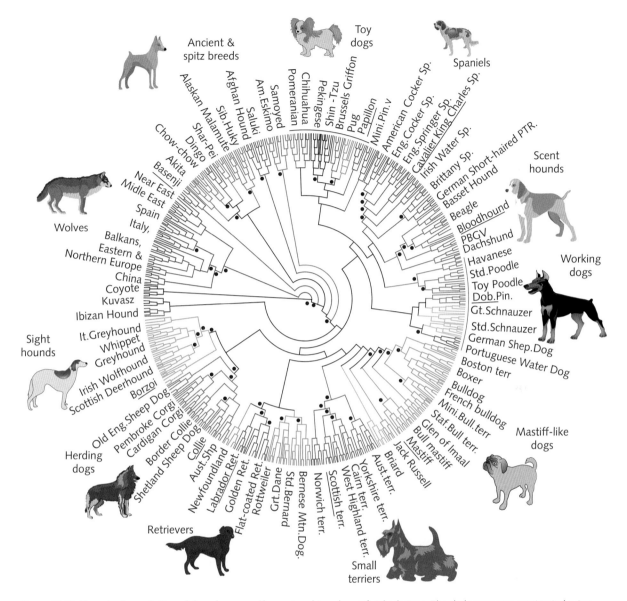

Figure 12.10 The adaptive radiation of dogs from a wolf ancestor, driven by artificial selection. The cladogram was constructed using variable loci throughout the genome. The dots mark particularly well-supported branches of the cladogram.

Reprinted by permission from Macmillan Publishers Ltd: VonHoldt et al. Genome-wide SNP and haplotype analyses reveal a rich history underlying dog domestication. *Nature* 464 (7290) © 2010.

Artificial selection is simple to understand because the experimenter provides the agent of selection and is free to vary how it acts. This has provided three crucial lessons.

- First, selection usually (but not always) drives adaptation in a predictable direction.
- Secondly, the response to selection initially depends on the quantity of variation in the population.
- Thirdly, the origin of new variation, by recombination or mutation, enables selection to drive the population far beyond the limits of variation in the ancestral population.

Artificial selection thereby provides an exemplar for evolutionary change: it shows clearly how evolution works, by consciously providing the agent of selection and controlling its effects. Natural selection is more difficult to study because it cannot be directly controlled and requires careful fieldwork over many years. Nevertheless, there are now many examples where selection and adaptation have been documented in laboratory and natural populations almost as precisely as they have been with domestic plants and animals on the farm. Natural selection in the laboratory and in the field is the subject of the next two chapters.

● CHAPTER SUMMARY

Artificial selection produces rapid and predictable change in the short term.

- The breeder's equation predicts the short-term response to selection.
- The Edinburgh bristle experiment illustrates the basic principles of artificial selection.
  - *The outcome of selection confirms the general theory.*
  - *Replicate selection lines tend to diverge.*
  - *Relaxing selection causes reversion to the ancestral state.*
  - *Selected lines surpass their ancestor.*
  - *The response to selection ceases when genetic variation is exhausted.*
  - *Artificial selection may be limited by countervailing natural selection.*

Domestication is applied artificial selection.

- The process of domestication usually involves four phases that differ in how selection is applied to populations.
  - *The first phase is the choice of organisms to domesticate.*
  - *The second phase is the proliferation of particular lineages in the early stages of domestication, without conscious human intervention.*
  - *The third phase is the deliberate selection by farmers of individuals with desirable attributes as the parents of the next generation.*
  - *The final phase is the systematic selection by scientists of attributes such as grain yield or milk production.*
- Domestication often involves a few regulatory genes of major effect.
- Desirable features of domesticated animals and plants can be greatly altered by systematic artificial selection.
- Crop yield has been greatly improved by systematic artificial selection.
  - *In the last 60 years, systematic mass selection has been very effective in increasing yield.*
  - *The response to selection can be demonstrated by common-garden trials of stored obsolete cultivars.*
  - *The contribution of selection can be evaluated by varying the conditions of growth.*
  - *The Green Revolution shows how selection can act at different levels.*
  - *Almost all desirable attributes of domestic animals and plants have been improved by systematic artificial selection.*

Artificial selection can produce extensive adaptive radiations.

  - *Dogs are much more variable and specialized than their ancestor.*
  - *Artificial selection provides an exemplar of how natural selection works.*

● FURTHER READING

Here are some pertinent further readings at the time of going to press. For relevant readings that have been released since publication, visit the book's Online Resource Centre at www.oxfordtextbooks.co.uk/orc/bell_evolution/

Section 12.1   Clayton, G.A., Morris, J.A. and Robertson, A. 1957. An experimental check on quantitative genetical theory. I. Short-term responses to selection. *Journal of Genetics* 55: 131–151.

Clayton, G.A. and Robertson, A. 1957. An experimental check on quantitative genetical theory. II. The long-term effects of selection. *Journal of Genetics* 55: 152–170.

Hill, W.G. and Caballero, A. 1992. Artificial selection experiments. *Annual Review of Ecology and Systematics* 23: 287–310.

Section 12.2   Moose, S.P., Dudley, J.W. and Rocheford, T.R. 2004. Maize selection passes the century mark: a unique resource for 21st century genomics. *Trends in Plant Science* 9: 358–364.

Section 12.3   vonHoldt, B.M., Pollinger, J.P., Lohmueller, K.E., et al. 2010. Genome-wide SNP and haplotype analyses reveal a rich history underlying dog domestication. *Nature* 464: 898–903.

● QUESTIONS

1. Explain how the short-term response to artificial selection depends on the heritability of a character and the manner in which individuals are selected. Why might this explanation fail in the long term?

2. Why do replicate selection lines tend to diverge in character state?

3. Why do selection lines tend to return to the ancestral character state if selection is relaxed?

4. Explain how artificial selection can shift the range of character states expressed by individuals in a population completely beyond the range of variation in the ancestral population.

5. Describe the main stages involved in the domestication of a species of plant.

6. What are the strengths and weaknesses of using laboratory experiments to understand and predict the outcome of artificial selection of livestock and crops?

7. Compare and contrast the results of selecting for (a) individual characteristics of vegetables such as cabbage, and (b) yield per acre of grain crops such as wheat.

8. Describe the adaptive radiation of dogs through artificial selection and compare it with the adaptive radiation of a group through natural selection.

You can find a fuller set of questions, which will be refreshed during the life of this edition, in the book's Online Resource Centre at www.oxfordtextbooks.co.uk/orc/bell_evolution/

13 Experimental Evolution

13.1 The laboratory microcosm is a time machine for evolution.

Many of the evolutionary processes that we have described extend over long periods of time—long, that is, on human timescales of a few decades. It would be futile to try to observe them on short timescales, of a few weeks or months, unless we can invent a time machine. Fortunately, this is quite easy: the time machine is a population of microbes, cultured in a small vial on the laboratory bench. A thimbleful of liquid will support a large population (say, about 10 million individuals), which reproduce every hour or so. With this resource, we can follow the evolution of large populations over thousands of generations in a few months. Such experiments can be used to probe evolutionary mechanisms in deep time. They have shifted evolutionary biology from a comparative science that interprets patterns of variation into an experimental science capable of documenting the smallest details of how evolution occurs.

To understand how an evolutionary experiment works, imagine yourself to be a bacterial cell inoculated into an utterly strange environment. Your ancestors have never before encountered such conditions: an uncomfortable temperature and unfamiliar food, for example. You will probably die, or reproduce so slowly that your lineage soon becomes extinct. If you are very lucky indeed, one of your offspring may bear a mutation that permits it to grow more rapidly. At an individual level, this is not very likely. But in the whole population there are so many individuals that it is very likely indeed—almost certain, in fact—that one of them will be lucky. Their lineage will spread through the population over time, supplanting their ancestors and increasing the level of adaptation.

There is a snag, however. The lineage will only spread as long as the population is capable of reproducing. In a small volume of culture medium, it will soon run out of resources. To continue the experiment, we need to create a population capable of unlimited growth. This population is a *microcosm*. As an imitation of the real world, it has its shortcomings. It is very small, and very simple. It does not pretend to have all the complexity of a complete natural ecosystem. Its simplicity is also its strength, however, because it allows us to study the basic mechanisms of evolution in a controlled laboratory setting.

The two basic ways of setting up a microcosm are sketched in Figure 13.1: both have been fertile methods for the experimental study of evolution.

- Batch culture. The simplest method is to transfer a small inoculum to a fresh vial of nutrient medium, as sketched in Figure 13.1 (a). Suppose that we transfer 1%; then each individual can divide 6 or 7 times before the population runs out of resources again. Hence, we can keep the population in a state of continued growth for 1000 generations in about three months, assuming the daily transfer that would be feasible for bacteria or unicellular eukaryotes such as yeast. The main advantage of serial transfer experiments using batch culture is that it is easy and inexpensive, so that it makes it possible to do evolution experiments in any kind of laboratory, from grade school to university. Its chief drawback is that the conditions of growth continually change, from an abundance of resources early in growth

(a)

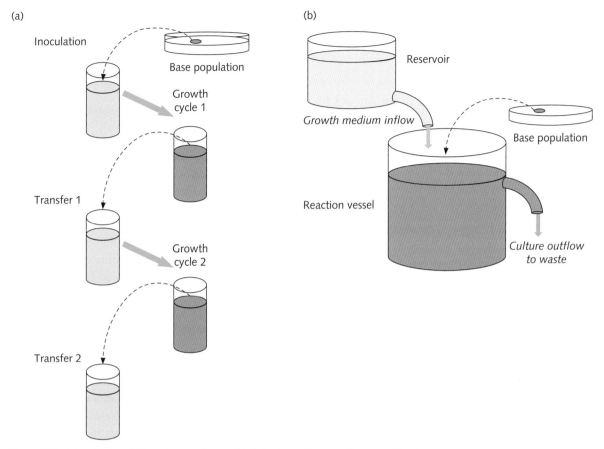

(b)

Figure 13.1 (a) A simple serial transfer experiment. (b) A simple continuous culture experiment.

to severe depletion just before transfer. This may or may not be a handicap, depending on the kind of experiment you intend to do.

• Continuous culture. A more sophisticated apparatus involves feeding nutrient medium continuously into the culture (Figure 13.1b). Naturally, the culture itself must be drained at the same rate. This is arranged by having the medium pumped into the culture vessel from a reservoir, while the vessel itself overflows to the waste. This machine is called a *chemostat*. After a brief initial period of adjustment, the inhabitants of the chemostat will grow at precisely the rate that is determined by the rate at which resources are being supplied. Hence, they will always experience perfectly uniform conditions of growth. The chemostat makes it possible to study adaptation to precisely defined environments. Its main drawback is that

it is difficult to maintain for long periods of time, because it is difficult to exclude foreign organisms. One anecdote to illustrate this is that beer is not made by continuous culture. It would be perfectly feasible to do this: put malted barley in the reservoir and yeast in the culture vessel, and draw beer from the overflow. In practice, industrial chemostats for brewing beer are sooner or later invaded by wild yeasts, which ruin the quality of the product. Hence, commercial beers are brewed by batch culture, because contaminants can be periodically removed by sterilizing the vessels.

In either case, a microbial culture seems to be a dull puddle of cloudy water. In fact, it is the scene of a continual drama in which dynasties arise, struggle, and prevail or perish. Its inhabitants are not, perhaps, as charismatic as dinosaurs or Darwin's finches. Nevertheless, they are likewise subject to natural selection, and evolve under the same laws. Moreover, they are much more convenient to study than large, long-dead, or remote animals and they provide us with first-hand evidence of the mechanism of evolution.

13.2 Deleterious mutation is balanced by purifying selection.

When bacteria reproduce by binary division each product is almost but not completely identical. The slight differences between them are caused by mutations—spontaneous changes in the structure of the DNA, most of which reduce the rate of replication. Mutations always occur and are generally deleterious (Sections 2.1 and 7.1). Moreover, they will tend to accumulate, because each genome propagates the mutations it has inherited while remaining vulnerable to new mutations. The growing population of a culture vial is not, therefore, a single clone, but rather a complex and shifting mixture of genotypes.

13.2.1 The genetic variance of fitness increases through mutation.

We can follow the accumulation of mutations over time by inoculating a chemostat with a small colony derived from a single cell and thereafter taking samples from the culture and measuring the variance of fitness. We do this by spreading the sample very thinly, so that individual cells are far apart, on solid medium made up of the chemostat medium solidified with agar. This sample is incubated for a short time and then inspected under the microscope to measure how many times each cell has divided. Thus, by extracting cells from the culture at intervals and immobilizing them on a two-dimensional agar surface we can find out how fitness varies in the three-dimensional culture itself. If we plot the variance of fitness on time, as in Figure 13.2, the intercept of the graph is the initial variance and the slope is the rate at which variance increases over time.

The initial variance is mainly attributable to chance or environment—cells may have been in different phases of the growth cycle when they were isolated, for example. The rate of increase is the additional variation which arises in unit time through mutation. The ratio of the two (rate of increase/initial value) is the amount of new genetic variation that arises in unit time as a fraction of all the non-genetic sources of variation. In experiments with a variety of organisms this quantity has a value of $1–5 \times 10^{-3}$ per generation. Selection will

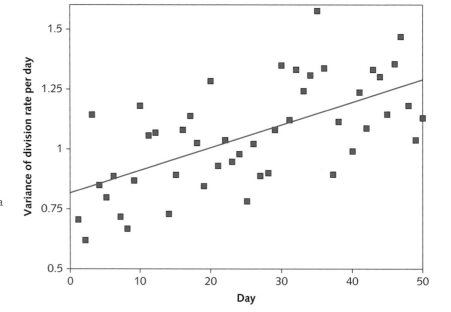

Figure 13.2 The increase in the variance of division rate over time in a laboratory culture of the green alga *Chlamydomonas*.

Reproduced from Goho and Bell. The ecology and genetics of fitness in *Chlamydomonas*. IX. The rate of accumulation of variation of fitness under selection. *Evolution* 54 (2) © 2007, John Wiley and Sons.

be effective when genetic sources of variation make up a substantial fraction of total variation (Section 14.1.1). In other words, even if we start with a single genotype, mutation will generate plenty of variation for selection to work on within a few hundred generations.

13.2.2 Isolate lines decay in the absence of effective selection.

The fate of a growing population in the absence of any source of genetic change except mutation can be investigated by setting up a batch culture experiment in which a single random cell alone survives to be transferred. This is done by spreading a sample from the culture on agar at the end of the growth cycle, incubating this sample for a short time and then using a single colony, chosen at random, to inoculate a fresh vial. Choosing a colony at random prevents natural selection from operating, so that the population evolves through mutation pressure alone.

Mean fitness declines over time in isolate culture. As the culture expands, each lineage descending from the founding cell will tend to accumulate mutations. Some will accumulate more than others, but by the end of the growth cycle almost all will bear more mutations than the founder. A cell chosen at random from one of these lineages will thus bear a greater or lesser number of additional mutations; and every one of these will necessarily become fixed in the population, because this cell is the sole founder of the next generation. Each successive growth cycle thus begins with a cell bearing more mutations than the founder of the previous cycle. Hence, over time the number of mutations tends to increase without limit and for this reason the average fitness of the population declines linearly over time.

Replicate isolate lines diverge over time. If two or more separate isolate lines are maintained, a similar process of mutation accumulation will occur in each of them. At the first transfer, however, a different lineage will be chosen to perpetuate each independent line. By chance, some will bear more mutations than others. The disparity between lines that is established in this way cumulates over time, increasing, on average, at each transfer. Hence, the variance of average fitness increases linearly over time.

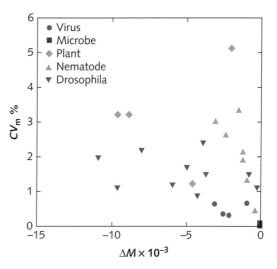

Figure 13.3 The outcome of mutation accumulation experiments using isolate lines, in terms of the standardized loss of mean vigor (ΔM) and increase in the coefficient of variation among lines (CV_m). "Vigor" includes fitness and characters thought to be closely related to fitness, such as flower number in plants.

Reproduced with permission from Halligan, D. L. and Keightley, P. D. (2009) 'Spontaneous Mutation Accumulation Studies in Evolutionary Genetics', *Annual Review of Ecology, Evolution, and Systematics*, Vol. 40: 151-172

The fate of isolate lines thus exemplifies two fundamental processes. The first is directional: the level of adaptedness will tend to decline over time when mutation is the only process governing the quantity of genetic variation in the population. The second is historical: replicate lines maintained in identical conditions will grow steadily more different over time. Figure 13.3 shows the extent of the deterioration and disparity of isolate lines from experiments using a variety of organisms. They show that the rate of loss of vigor can be up to 1% per generation, with an average value of about 0.3%, with the coefficient of variation among lines increasing by up to 3% per generation.

13.2.3 Diversity is maintained at equilibrium between mutation and selection.

In practice, of course, populations do not continually decay. This is because the accumulation of deleterious mutations is opposed by purifying selection.

Purifying selection opposes deleterious mutation. Let us think of what happens during the propagation of an isolate line in more detail. The single founding clone will immediately begin to diversify through mutation as it proliferates, forming a miniature phylogenetic tree whose branches are lineages that have acquired different mutations. The population just before transfer will contain many of these lineages, but not all. Some may acquire a lethal mutation and will instantly become extinct. Others may struggle under an unusually heavy burden of mutations so that there are few if any survivors at the end of the growth cycle. Hence, the single survivor is not chosen at random from the full spectrum of mutations that have occurred, but rather at random from the biased sample that has survived. This is an example of purifying selection: enfeebled types have been largely eliminated from the population. It is likely to be ineffective, because there is not enough time within a single growth cycle for the lineages with fewest mutations to replace all the others. Hence, the bulk of the population will consist of lineages bearing several mutations, and it is likely, purely by chance, that a cell from one of these will be chosen as the single survivor.

Purifying selection will become more effective when more cells are chosen to propagate the culture, because the sample will become more representative of the surviving population. When the inoculum consists of a few thousand cells each lineage will be represented roughly in proportion to its frequency in the population, so lineages with fewer mutations will consistently outnumber those with more, transfer after transfer. The tendency for fitness to decrease through mutation will then be opposed by a tendency for fitness to recover through purifying selection, because lineages with more mutations will become less frequent.

Deleterious mutations are not completely eliminated by selection. Purifying selection is not completely effective, because selection is necessarily subject to chance. An individual bearing a mutation that confers greater than average fitness is more likely to survive and reproduce, but this is only a tendency, never a guarantee. Conversely, an individual bearing a deleterious mutation is more likely than average to die without reproducing, but it may succeed nevertheless.

To be more precise, suppose that a mutation arises in a single individual that reduces its fitness by an amount s;

that is, the number of offspring it produces is $1 - s$, relative to an unaffected individual. The quantity s is called a selection coefficient. Clearly, this lineage will be eliminated from the population by selection. However, it will not be eliminated immediately. On average, it will persist for about $1/s$ generations, and while it does, it will contribute to the variability of the population.

What if $s = 0$? This is the special case of a neutral mutation, which has no effect on fitness and is therefore unaffected by selection. It may be lost quite soon, by chance, or it may slowly spread to high frequency. Because its dynamics are governed by the mutation rate alone, the probability that it will eventually become fixed—after a long period of fluctuating at intermediate frequencies—is simply equal to its mutation rate. Hence, any population is likely to contain large numbers of neutral alleles in the course of being fixed or lost.

Mutation and selection come into equilibrium. Mutation and selection will occur in any growing population, with mutation tending to increase the number of deleterious mutations per genome while selection acts to reduce it. Suppose that the rate of mutation at a particular site in the genome is u. The site can be a nucleotide or a gene or any other genetic element: in any case, a fraction $1 - u$ of offspring inherits exactly the parental state and a fraction u receive some altered state. The rate of selection at the same site is s: the fitness of offspring bearing a mutation is $1 - s$ relative to the unit fitness of the parental state. The frequency of the mutation at equilibrium is then just u/s. In other words, the opposed processes of selection and deleterious mutation lead to a dynamic equilibrium, at which the rate of mutation (which tends to increase the frequency of low-fitness alleles) is balanced by the rate of selection (which tends to remove them). Thus, purifying selection will not completely eliminate deleterious mutations from the population, but will rather prevent them from spreading to high frequency.

Asexual populations may have substantial levels of genetic variation. Consequently, a well-adapted asexual population will not usually consist of a single clone representing the best possible genotype for the environment it occupies. Instead, there will be a very large number of variants held at mutation–selection equilibrium. Each individual variant will be rare, but the overall pool of variation may be extensive. For example, in wild yeast populations, which usually reproduce asexually, genetic surveys have shown that any two random individuals differ at about 1 in every 1000 nucleotides, which is roughly equivalent to 1 nucleotide per gene. A first sight, this is a very low level of variation, but since it applies throughout the genome it implies that any two individuals will differ at many sites, unless they are closely related. Hence, the population contains a substantial stock of fluctuating variation. So long as the environment remains constant, this merely represents deleterious mutations, each held at low frequency by purifying selection. If the environment should change, however, some of these mutations may become beneficial, providing the population with the potential to respond to selection and adapt to the altered conditions of growth.

13.3 Beneficial mutation drives directional selection.

The effect of a mutation on fitness often depends on the environment that an organism lives in. Changes in some very ancient features of cells, such as ribosome synthesis, are usually lethal and are certainly very unlikely to be beneficial in any environment. There is a single allele which is superior in all conditions. Other features—such as carbohydrate metabolism, stress response, or secretion—are much more malleable. There may be many alleles, each superior in a particular range of conditions. Mutations in these genes are conditionally deleterious: they reduce fitness when the population is well adapted but may actually increase fitness when conditions change.

13.3.1 Permanent adaptation depends on the balance between variation and the rate at which conditions deteriorate.

The next experiment is to transfer a bacterial population from the environment to which it has become well adapted to a new and stressful environment where growth is slow. A few mutations that were previously deleterious become beneficial in these changed conditions, and they will tend to spread through the population under directional selection. How will the population change as a consequence?

Rapid and severe stress may cause extinction. If the stress is sufficiently severe, or if it is applied very rapidly, the population will simply become extinct. When a stream or a pond dries out, the fish simply die. They do not evolve into amphibians, because there is no variant in the population capable of surviving out of water. On a more prosaic level, bacterial infections are often cured by antibiotics. In this case, there may well be variants that are somewhat resistant to the antibiotic, and they will tend to increase in frequency through natural selection. Unless they are resistant enough, however, the population will become extinct anyway. It is not enough for the population to adapt—it has to adapt fast enough.

Adaptation may lead to evolutionary rescue. On the other hand, if the stress is less severe, or is applied only quite slowly, then the spread of resistant types through the population may rescue it from extinction. The population will at first decline, as susceptible types are eliminated, but will subsequently recover as resistant types spread to high frequency. The characteristic U-shaped time-course of evolutionary rescue has been demonstrated in the lab by exposing yeast populations to high concentrations of salt (Figure 13.4). It will happen whenever there is enough variation for selection to be effective, which is most likely in large populations. This point can be proven by setting up a range of populations from very large to very small and showing that evolutionary rescue fails to happen when the population falls below some threshold of size.

Evolutionary rescue can lead to tolerance of lethal conditions. A severe stress need not wipe out a population instantly. A stress that causes enough mortality for growth to fall below the replacement rate will produce a slower decline that eventually results in extinction many generations in the future.

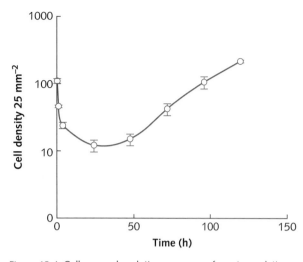

Figure 13.4 Collapse and evolutionary rescue of yeast populations exposed to high salt concentration in the laboratory.

Reproduced from Bell and Gonzalez. Evolutionary rescue can prevent extinction following environmental change. *Ecology Letters* 12 (9) © 2009 Blackwell Publishing Ltd/CNRS.

This extended period of time provides an opportunity for new genetic variation to appear by mutation or recombination. When this happens, rescue is not limited to the range of variation originally present, and populations may eventually become adapted to conditions that would have been lethal to their ancestors. This is most likely to happen in large populations that are declining slowly in abundance, such as bacteria exposed to inadequate doses of an antibiotic.

13.3.2 Beneficial mutations sweep through asexual populations.

Lethal stress lies at one extreme of the fluctuations that occur in all natural environments. Whether mild or severe, these events will shift the relative fitness of different types and thereby trigger an episode of natural selection. When we mimic them in the lab we can study the process by which directional selection improves adaptedness.

All the members of the current population descend from a single individual in the past. Each lineage in an asexual population is genetically isolated from every other lineage. They are not ecologically isolated, however—far from it, since each lineage will

compete with all others for resources. Consequently, there will be a continual flux of lineages, as the more successful replace the less successful because they are ecologically more efficient. In the end, only a single lineage will survive. This lineage must have descended from a single ancestral cell, which arose earlier in the history of the population. Hence, all individuals in the current population can trace their ancestry back to this cell, which represents the coalescent. (We have met the coalescent before (Section 3.1.1) in the context of phylogenetics.) This insight implies that only a very few individuals in the current population have any evolutionary future; they descend from the coalescent. All the others belong to the "living dead"—cells that are active and reproductive, but that will nevertheless make no genetic contribution to the population of the distant future.

Selective sweeps remove variation in asexual populations. How did this successful lineage arise? Its most recent common ancestor must have borne a beneficial mutation responsible for its ecological superiority. In a sexual population, this mutation would become dispersed among innumerable lineages through mating and recombination. In an asexual population, however, it will remain permanently linked with the rest of the genome in which it arose. This is why a single genome eventually replaces all others. The consequence is that the passage of a beneficial mutation sweeps out genetic variation by replacing the diversity of genotypes coexisting at equilibrium between mutation and purifying selection with a single clone. Once this clone has been fixed, of course, mutation will continue to operate and genetic variation will slowly be restored.

The passage of a beneficial mutation can be detected by marker genes. This is a valuable insight because it immediately provides us with a powerful and elegant technique for studying adaptation in asexual populations. Suppose that we mix two strains together, which are genetically identical except that one carries a marker. This marker is an allele that confers some easily distinguishable phenotype; the ones that are most commonly used provide resistance to an antibiotic, or require some essential nutrient, or change the appearance of colonies growing on agar. It is important that the marker has no effect, or only a very slight effect, on fitness in normal growth medium. The mixed population is cultured in a stressful environment to which it is at first poorly adapted, and transferred to fresh medium every day or so. The frequency of the marker is recorded at regular intervals, for example by spreading the culture onto agar plates containing antibiotic and counting the number of resistant and susceptible colonies.

For some time, this frequency will vary a little on one side or another of 50%, if marked and unmarked strains were at first equally frequent. After a little time, however, either the marked or the unmarked strain will suddenly spread rapidly through the population, until it reaches almost 100%. This is not because the marker allele itself is advantageous or deleterious: we chose it because it has almost no effect on fitness. It is because a beneficial mutation has arisen in one or the other strain, and has spread through the population. As it spreads, the marker state with which it associated necessarily spreads with it, because in an asexual population the beneficial mutation and the marker are not separated by recombination. The experimenter then isolates a few of the marked cells, by spreading the culture on agar and picking colonies, and confirms that they do indeed grow better than the ancestor in the medium where they have been cultured.

Mutants resistant to antibiotics or phage arise spontaneously in bacterial cultures, in the absence of antibiotic or phage, and tend to increase in frequency by recurrent mutation, provided that they do not impair fitness. They are readily scored by plating. While they are rare, however, any beneficial mutation is almost certain to arise in a normal, non-mutant individual. The passage of a series of beneficial mutations is therefore marked by a jagged rise and fall in frequency of these spontaneously arising markers, as illustrated in Figure 13.5. This process of "periodic selection" is an elegant demonstration of the cumulation of beneficial mutations during the process of adaptation.

These simple experiments give us a method for trapping the beneficial mutations responsible for adaptation to novel environments, measuring their effects, and identifying the genes responsible. It is worth emphasizing that it really is quite simple: a modestly equipped laboratory is easily capable of charting the course of evolution in microbial populations.

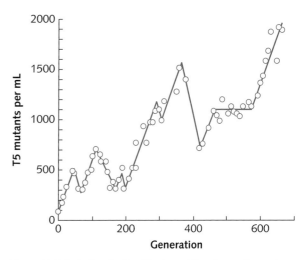

Figure 13.5 Periodic selection. The jagged line shows change in the frequency of mutants resistant to phage T5 in a chemostat population of *E. coli*.

Reproduced from Kubitschek, H.E. 1974. Operation of selection pressure on microbial populations. *Symp. Soc. Gen. Microbiol.* 24: 105–130 with permission from Cambridge University Press.

13.3.3 Substitution involves three phases of waiting, establishment, and passage.

The fixation of a single beneficial allele is the unit process of adaptive evolution. It can be broken down into three phases: the waiting time before the appearance of a beneficial mutation, the establishment time before an ultimately successful mutation begins to spread, and the passage time required for its eventual fixation.

The waiting time depends on the mutation supply rate. At the finest level of analysis, evolution depends on a supply of beneficial mutations. The crucial variable is the *number* of beneficial mutations that arise in each generation. If this is one or more, selection will be continuously active, and the rate of adaptation will be limited primarily by the rate at which these can be fixed. If it is less than one, then the population will lack selectable variation most of the time and the rate of adaptation will be limited primarily by the rate of beneficial mutation.

The mutation supply rate that provides the basis for adaptation in turn depends on three variables.

- The first is simply the fundamental rate of mutation per nucleotide, which can be estimated quite precisely.

- The second is the size of the population, because more mutations will occur in larger populations. This too can be estimated precisely.

- The third is the probability that a mutation will be beneficial. This is a more difficult calculation, because it depends on the state of the population with respect to the environment. If the environment has recently changed so as to severely reduce growth, the population is poorly adapted, and its fitness may be increased by many kinds of mutation. That is, the rate of beneficial mutation may be quite high in poorly adapted populations. On the other hand, a population that has experienced the same conditions for a long time is likely to be well adapted, the probability that a random mutation will increase adaptedness is very small, and the rate of beneficial mutation will be very low.

Clearly, the number of beneficial mutations that arise in each generation is not a fixed property of a species, or population. It will be large in very abundant species whose conditions of growth have recently changed. It will be much smaller in rare species living in a constant environment. In experimental evolution, we can calibrate the rate of adaptation by manipulating population size and the conditions of growth.

Adaptation halts when beneficial mutations are very rare. The most widespread and pervasive change in conditions of growth occurring at present is the steady rise of CO_2 concentration in the atmosphere, caused by the burning of fossil fuels. This is causing the warming of the Earth and the acidification of the oceans. Its most direct effect, however, is to increase the supply of inorganic carbon that photosynthetic organisms such as land plants, green algae, diatoms and cyanobacteria use for growth. They might be expected to adapt to this sustained shift in the conditions of growth. This does not happen: the evolutionary response to elevated CO_2 in the medium term of hundreds of generations has been studied in several species of phytoplankton with uniformly negative results. No specific adaptation to elevated CO_2 has yet been detected: there is no tendency, that is, for lines maintained in the laboratory for a thousand generations at 1000 ppm CO_2 to grow faster or to yield more at this concentration than lines maintained concurrently at today's ambient concentration of 400 ppm.

This is the most extensive and sustained failure of a selection experiment yet reported. Experiments that expose populations to stressful (but not lethal) conditions almost always observe adaptation in some lines at least. Improved conditions of growth, such as an ampler supply of carbon, may fail to elicit specific adaptation because the physiological systems involved, such as photosystems, are already optimally configured. Experimental evolution will make no progress when beneficial mutations are so rare that none are likely to occur within a few hundreds or thousands of generations.

Successful beneficial mutations only become established if they survive stochastic loss. Much more often, populations do adapt to changed conditions through beneficial mutations. Once a beneficial mutation has occurred, it will tend to spread through selection. This is only a tendency, however, and is by no means inevitable. Even in large populations, it is very likely that a particular beneficial mutation will arise in a single individual. This individual may die, or fail to reproduce, because of any of the multifarious hazards of mortality. Whether or not the lineage bearing the mutation becomes extinct while it still comprises only one or a few individuals depends on how great an advantage the mutation confers. If the advantage is only modest, the lineage is almost as likely to go extinct as a comparable lineage that lacks the mutation, so the mutation is likely to be extinguished before it has had the opportunity to spread. Conversely, a mutation that confers a very large advantage is more likely to survive.

As a rough rule of thumb, if a lineage bearing the mutation has fitness $1 + s$, relative to a lineage that lacks it, the probability that the mutation will eventually become fixed is about $2s$. Hence, the total waiting time until the appearance of a successful beneficial mutation is $1/2s$ times as long as the interval between successive beneficial mutations, most of which soon become extinct.

The passage time depends on the intensity of selection. A lineage bearing a favorable mutation that reaches a certain threshold—think in terms of about 50 individuals—becomes almost free of the risk of early extinction, and will steadily replace all others in the population. How long will this take?

The simplest case would involve two alleles, each mutating to the other at a rate of u per replication, one of which is initially deleterious with a fitness of $1 - s$ and therefore occurs at a frequency of u/s. When the environment changes, it becomes beneficial with a fitness of $1 + s$ and then proceeds to spread until the population is once again at equilibrium between mutation and selection. The time required for this to take place depends on:

• The intensity of selection: when selection is stronger (the advantage of the allele in the new conditions is large) a beneficial allele becomes fixed more rapidly.

• The initial frequency: when the allele is rarer (its disadvantage in the old conditions is large) it will take longer to spread.

For example, suppose the mutation rate is $u = 10^{-6}$ per locus and there is moderately strong selection with $s = 0.1$. The passage time is then about 230 generations (check this with the goldfish pond model, Section 2.2.1). Any other values may be substituted, of course, but a good rule of thumb is that environmental change that reverses the direction of selection will lead to the replacement of one allele by another in a few hundred generations.

Hence, we can divide the process of adaptation in an asexual population into three periods. The first is the waiting period before a beneficial mutation first arises. The second is the establishment period, during which several beneficial mutations arise before one escapes stochastic loss and begins to spread. The third is the passage period, in which a successful mutation that has broken through this barrier spreads and eventually becomes fixed. The total time that is required for the substitution of a beneficial mutation is the sum of the establishment time and the passage time. The overall rate of adaptation may be limited either by the establishment time or by the passage time, depending on the rate of supply of beneficial mutations and their effect on fitness.

13.3.4 The beneficial alleles responsible for adaptation often have a large effect on fitness.

There are two extreme accounts of adaptation. One is based on the frequency of mutations: most mutations have small effects, so adaptation is the consequence of the fixation of mutations at many loci, each of which has only a small effect on fitness. The other is based on the effect of mutations: beneficial mutations of small effect are likely to be lost soon after they appear, and only mutations of large effect are likely to spread.

The first allele to be fixed is expected to have a large effect on fitness. Once environmental conditions change, a wide range of mutations may become beneficial. Most will increase fitness only by a very small amount; only a very few will have a large positive effect. Beneficial mutations of large effect are rare. Nevertheless, most of the large number of mutations of small effect will quickly be lost, and those that remain will spread only slowly. Any mutation of large effect that appears while they are spreading is likely to become established and will then rapidly replace them. Provided that the mutation supply rate is high enough, therefore, the first beneficial mutation to contribute to adaptation is likely to have a large effect on fitness. The converse is also true: in small populations, mutations of small effect may become fixed before any large-effect mutation appears.

Adaptation is dominated by large-effect mutations in experimental populations of bacteria. This prediction can be tested by trapping beneficial mutations that have spread through the population by using fluctuations in the frequency of markers (Sections 13.3.2 and 14.3.2). This is easy to do when a large population is challenged by a severe stress. For example, we have cultured bacteria using the amino acid serine as the only source of carbon and energy. Most types are unable to grow at all, so those with any degree of proficiency have a large advantage—as much as twofold—and spread very rapidly. The fitness of the first beneficial mutations to become fixed is shown in Figure 13.6. Even in much more relaxed conditions—supplying glucose as the limiting resource, for example—the first types to be fixed are usually 10%–30% fitter than their ancestors. These greatly superior types simply override the crowd of beneficial mutations with more modest effects.

The rate of adaptation increases with population size. However, this conclusion depends on population size. If the population is large then the mutation supply

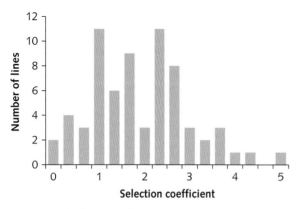

Figure 13.6 The frequency distribution of selection coefficients associated with the first successful beneficial mutant to spread in serine-limited cultures of *Pseudomonas*.

Reproduced from Barrett et al. 2006. Mutations of intermediate effect are responsible for adaptation in evolving *Pseudomonas fluorescens* populations. *Biology Letters* 2 (2): 236–238.

rate will be correspondingly large and beneficial mutations of large effect will dominate the process of adaptation because the waiting time is relatively short. If it is small, these are so rare that mutations of modest effect will arise and spread to fixation before any large-effect mutation has occurred—the waiting time is much longer than the passage time. Consequently, as population size increases the balance will shift from slow change via modest increments of fitness in small populations to more rapid change involving beneficial mutations of large effect in more numerous populations.

This phenomenon can be demonstrated experimentally by setting up microbial populations in vessels of different size, which will support populations from a few thousand to hundreds of millions of cells. The larger populations adapt faster, and the mutations responsible for adaptation have larger effects on fitness. In the world outside the laboratory, what this means is that the dynamics of adaptation may be quite different in low-abundance organisms such as mammals or cephalopods than in high-abundance organisms such as diatoms or ciliates.

13.3.5 Adaptation often leads to loss of ancestral function.

Microcosm experiments show very clearly that adaptation to novel conditions of life often evolves quickly, within a few hundreds or thousands of generations. Why, then, is the world not populated by a few extremely well-adapted kinds of microbe? The reason is that there is a cost of adaptation. Populations exposed to novel conditions become adapted to them, but at the cost of losing adaptedness to their original conditions of growth. The advance of fitness in new conditions is achieved at the expense of a debilitating regress in other conditions: adaptation is always conditional.

Regress may be caused by functional interference. Structures that become specialized for one use are thereby disqualified for others. Tools provide an obvious analogy: spoons and scalpels perform their respective functions very successfully, but you would not substitute the one for the other. You can discover a biological example by culturing algae or bacteria in a chemostat for a few weeks. The walls of the vessel will become coated with a film of cells, and clumps of cells may form on the bottom. These types evolve simply because by sticking to the walls or lurking in the depths they are less likely to be washed out in the overflow. Under normal conditions they are inefficient because without free movement their access to nutrients is restricted, but in the chemostat this is more than balanced by their lower mortality. The same phenomenon can be seen in natural bodies of water where microbial communities form layers on any available surface—the algae that make the slippery film on the rocks at the edge of the lake are less likely to be removed by the outflow stream than if they were freely floating.

Regress may be caused by mutational degradation. The second penalty of specialization is that it necessarily leads to the neglect of other ways of life. Tools that are long disused will rust or rot. Adaptations to a way of life that ceases to be experienced will deteriorate because loss-of-function mutations will be neutral and will therefore tend to accumulate in the population. It is not that any single mutation will spread through the population. Rather, all lineages will tend to incur damage to some of the genes responsible for adaptation to the original environment. This is why cave animals tend to lose pigmentation and sight.

The two sources of regress can be distinguished in experiments with many replicate selection lines by

(a)

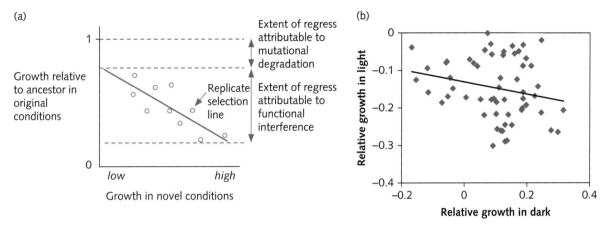

(b)

Figure 13.7 Causes of regress of selection lines when restored to their original conditions. (a) explains the concept. (b) is an example from an experiment with the green alga *Chlamydomonas,* using lines cultured for hundreds of generations in the dark.

Reproduced from Bell. Experimental evolution of heterotrophy in a green alga. *Evolution* 67 (2) © 2012 Graham Bell. *Evolution* © 2012 The Society for the Study of Evolution.

plotting performance in the ancestral environment against performance in the novel environment. The degree of functional interference will be greater in lines that have diverged more, leading to a negative correlation. The degree of mutational degradation, on the other hand, will be a function of time alone, so that all lines are expected to be equally impaired. This argument is illustrated in Figure 13.7 by an experiment in which lines of a green alga were cultured in the dark for hundreds of generations, after which many grew poorly in the light. There is a slight negative correlation, but most of the regress is attributable to mutational degradation.

13.3.6 The outcome of selection can be predicted within limits.

Our knowledge of cell biology will eventually progress to the point where the likely outcome of selection can be predicted from biochemical first principles. We already have two means of predicting the course of evolution.

The first is to calculate the consequences of modifying different components of a metabolic system. For example, lactose diffuses from the medium to the cell membrane through pores in the cell wall, is actively transported through the membrane by an enzyme (a permease), and then cleaved into glucose and galactose by another enzyme (beta-galactosidase). Since the fitness of the cell in a lactose-limited chemostat

is strictly proportional to the flux through this pathway, the consequences of modifying any part of the pathway can be calculated. Hence, if the properties of any mutant strain are measured (for example, the rate at which it can hydrolyze lactose) its fitness can be calculated, and its fate, in competition with the ancestral strain, can be predicted. Figure 13.8 shows that these predictions are very precise and successful.

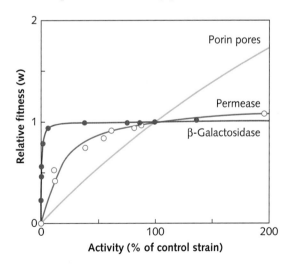

Figure 13.8 Prediction of fitness of mutants in the lactose utilization system of *E. coli.* The points are the observed activity and fitness of structural mutants in the permease and beta-galactosidase enzymes; the lines are the relationships predicted from biochemical principles.

Reprinted from Dykhuizen and Dean. Enzyme activity and fitness: evolution in solution. *Trends in Ecology and Evolution* 5 (8). © 1990, with permission from Elsevier.

The second method is to predict which genetic modifications will be selected in response to a given stress. This is much more difficult. In fact, it is not yet feasible to attempt them for a creature as complex as *E. coli*. A start has been made, however, with the simpler genomes of viruses. For example, most gene expression by phage T7 requires its RNA polymerase gene. If this is deleted, the equivalent gene from a related phage, T3, can be used, but is much less efficient. However, a single base-pair change (G → C) at a particular position (–11) in the T7 promoter produces a marked elevation in expression. Hence, we can predict that when the T3 gene is substituted for the T7 gene, the outcome of selection will be G → C mutations at –11 in T7 promoters. The experimental outcome was to some extent in agreement with this prediction: beneficial mutations at position –11 were often involved in adaptation. Sometimes other sites were involved, however; and at –11 the most frequent change was not G → C but rather G → A.

The stochastic, historical nature of evolution implies that there are limits to its predictability. We can reasonably expect to be able to predict the *themes* of adaptation to particular conditions, whereas the *variations* on these themes that evolve in a given selection line may be unpredictable, even in principle.

13.4 Long-term experiments document the main features of evolutionary change.

Bacterial evolution over 50,000 generations has been studied in the laboratory. Imagine the ideal evolution experiment. A population is cultured for thousands of generations in a novel environment; any increase in adaptedness is easily measured; the mutations responsible can be identified and characterized; and there is a complete fossil record of all intermediate stages. Hence, the course of evolution can be charted and its mechanisms unambiguously revealed. This may sound like an impossible ambition; yet, it is the staple of experimental microbial evolution, and has been realized many times.

In February 1988, Rich Lenski began a very simple evolution experiment at the State University of Michigan at East Lansing. He started with a single cell of the bacterium *E. coli*, which divided to form a colony. This was used to inoculate a dozen populations, which from this time on are kept separate. Each is maintained in batch culture in medium that has glucose as the only source of carbon and energy, with 1% of the population being transferred each day to fresh medium. This is a very unfamiliar environment for *E. coli*, which normally lives in the gut of mammals. Consequently, we expect the populations to evolve rapidly under strong selection. We can also expect them to evolve in full view, so to speak, since the genetics of *E. coli* has been studied for decades and its genome is thoroughly understood. We can use experiments like these both to document the course of evolution and to discover the succession of genetic changes that is responsible for adaptation.

An expansion of 100 times is equivalent to about 6–7 doublings, so that after a year of daily transfers a line has passed through about 2000 generations. Every 500 generations, each line is stored by freezing it at –80 °C. At the time of writing, the lines have passed through more than 50,000 generations, providing a replicated long-term process of adaptation with a complete fossil record. Simple though it is, this experiment gives us an unparalleled resource to study the detailed mechanics of adaptation through natural selection.

13.4.1 Adaptation can be measured precisely.

As expected, each line adapts to the novel conditions of growth it experiences, through the successive substitution of beneficial mutations. The degree of adaptation can be evaluated at any time by a competitive assay. The ancestral strain exists in two versions, which are genetically identical except that one bears the normal version of the gene encoding the enzyme responsible for the metabolism of the sugar arabinose (Ara^+), while the other bears a non-functional allele (Ara^-). The two can be distinguished by growing them on agar plates containing an indicator of metabolism, where Ara^+ cells form white colonies, whereas Ara^- cells form red colonies. Likewise, six of the experimental lines are Ara⁺ and six are Ara⁻. The line to be evaluated bears one marker (say, Ara⁺) and is mixed with the ancestral strain bearing the other marker (Ara⁻). This mixture is then cultured for a number of growth cycles, with the frequency of the markers being recorded (by spreading the cultures on indicator plates) at each transfer. If the experimental line has evolved, its fitness in the new conditions will be greater than that of the ancestral strain, and it will therefore increase in frequency in the mixture. The rate of increase in frequency, which is easily measured by counting colonies of different color, can then be used to calculate the amount by which the fitness of the evolved line has increased relative to its ancestor.

13.4.2 Adaptation increases rapidly then slows down.

Each line at first increases in fitness rapidly. Fitness increases rapidly at first in all the lines, but as time goes on the rate of increase slows down, as shown in Figure 13.9. Most of its adaptation is achieved within about 2000 generations; after 20,000 generations or more it is still responding to selection, but only very slowly. This is easy to interpret. The first mutations to be fixed are likely to be those which have a large effect on fitness, because they are least likely to be lost by chance. Once they have become fixed, only

mutations of lesser effect are available, and these will then be able to spread. As time goes by, the level of adaptation can be enhanced only by mutations of lesser and lesser effect. Consequently, the level of adaptation continues to rise, but more and more slowly, over time.

Populations remain well-adapted to a constant environment through purifying selection. This gives us an important insight into evolution: much of the time, selection is active not in producing change, but rather in preventing change. After the first few thousand generations, during which the population becomes well-adapted to novel conditions, selection will act mainly to eliminate deleterious mutations that reduce adaptedness. The initial phase of directional selection is followed by a second phase of purifying selection—until conditions change again.

Populations carry a large amount of potentially adaptive variation. Deleterious and neutral mutations do not contribute directly to adaptation, of course, so long as the environment remains constant. It is important to bear in mind, however, that "deleterious" and "neutral" are terms that refer to a particular set of environmental conditions. When conditions change, some mutations that were previously deleterious or neutral may become beneficial. These will permit the population to adapt rapidly to a new environment. A microbial culture that has long been maintained in the same culture medium may appear to be a uniform cloud of cells. This is misleading. In fact, it conceals a large quantity of fluctuating variation that is capable of responding rapidly to selection. It is not a uniform population, but rather a complex set of lineages in continual flux.

The hidden diversity and evolutionary potential of microbial populations can easily be demonstrated. Suppose that you culture a line, by serial transfer, on glucose-limited medium for a few hundred generations then spread it on an agar plate containing glucose as the only carbon source. Almost all the colonies will be similar in size, because the population has become adapted to metabolizing glucose, so almost all the cells grow at the same high rate.

Now, suppose that instead you spread the culture onto a plate containing a different limiting

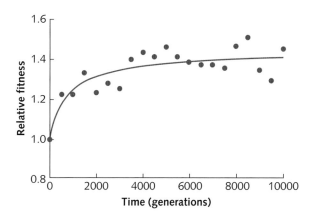

Figure 13.9 Trajectory of mean fitness in the first 10,000 generations of one of the long-term *E. coli* lines.

Reproduced from Lenski & Travisano. Dynamics of adaptation and diversification: a 10,000-generation experiment with bacterial populations. *PNAS* 91 (15) © 1994 National Academy of Sciences, U.S.A.

nutrient (such as maltose). The colonies that grow on this plate will vary considerably in size. This reveals the variability that was previously hidden in the population, in the form of mutations that are neutral, or only slightly deleterious, in the glucose environment but that vary in their capacity to grow on maltose.

13.4.3 Adaptation often involves repeatable genetic modifications.

With the advent of genomic technologies, it has become possible to discover the genetic changes that are responsible for adaptation. When these methods were applied to the long-term *E. coli* experiment, some consistent modifications were found.

* The most obvious was a loss of ability to metabolize the sugar D-ribose. This was caused by deletions in the ribose operon. The extent of the deletions differed among lines.

* A more subtle change was the modification, by point mutations, of four other genes. The same set of genes had been modified in most of the lines, but the precise mutations that had been selected differed among the lines.

These observations illustrate three general features of metabolic evolution in bacteria.

* First, many beneficial mutations involve a loss of function in one system (ribose metabolism) that causes a gain of function in another system (glucose metabolism).

* Secondly, the general targets of beneficial mutations (ribose operon, the four other genes) are often modified in most, or all, replicate selection lines, whereas the specific targets (extent of the deletion, particular sites within genes) are different in all cases. This can be expressed in terms of themes and variations: the general themes are repeatable, but the variations on these themes are idiosyncratic.

* Finally, the biochemical mechanism underlying the advantage of these beneficial mutations is often difficult to understand. It is not understood, in fact, for any of these examples. This is not because it cannot be discovered; no doubt it

will be. It is rather because our knowledge of how cells work is so poor that evolution is often more effective than any human engineer.

One of the major advances that is currently being made in experimental evolution is the use of the new potential to sequence whole genomes quickly and cheaply. In the next few years the rules governing the genetic changes underlying adaptation, so long sought, will be definitively established.

13.4.4 Evolutionary innovation may involve very rare events.

Adaptation to novel conditions is only possible if there are beneficial mutations that increase fitness. These may be very rare indeed, in which case a high level of adaptation will not evolve. This is why all organisms are imperfect. Very rare events are to be expected, however, if enough trials are allowed. If you play poker, for example, the probability of being dealt ♠10 ♠J ♠Q ♠K ♠A is very small (less than one in a million) if you play a hundred hands, but it is quite likely (about one chance in four) if you play a billion hands. You are hardly likely to play a billion hands of poker, but a bacterial lineage founded by a single cell inoculated into 10 mL of growth medium will go through a billion replications in 24 h, so that very rare events are not just likely, but almost certain to occur.

E. coli cannot utilize citrate as a carbon source. No known natural isolate can do so. Indeed, it is a characteristic that has been used to identify a particular bacterial strain as belonging to *E. coli*. Nevertheless, one of the long-term *E. coli* lines evolved the ability to metabolize citrate after about 30,000 generations in culture (Figure 13.10). We can calculate how many replications were screened by selection before this novel strain appeared.

* The lines are transferred daily by inoculating 1% of the grown culture into new medium.

* Hence, each growth cycle involves each cell in the inoculum going through an average of $\log_2 100 = 6.6$ doublings.

* Each doubling involves one episode of replication, so that the first generation is formed by one replication (giving rise to $N = 2$ cells), the second

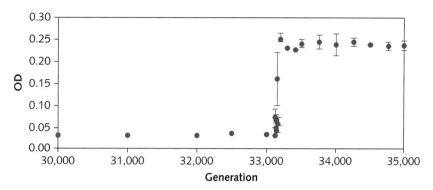

Figure 13.10 Evolution of a rare trait, citrate utilization, in the long-term *E. coli* experiment. OD is optical density, a measure of cell density; note how cultures tested in medium with citrate as sole carbon source increase sharply in density shortly after 33,000 generations.

Reproduced from Blount et al. Historical contingency and the evolution of a key innovation in an experimental population of *Escherichia coli*. PNAS 105 (23) © 2008 National Academy of Sciences, U.S.A.

by two (because each of the two cells replicates once, giving rise to $N = 4$ cells), the third by four (giving rise to $N = 8$ cells), and so forth.

- Hence, the total number of replications involved in producing 8 cells in three generations from a single cell is $1 + 2 + 4 = 7$. In general, the total number of replications involved in giving rise to G generations is then $2^G - 1$, or almost exactly $N = 2^G$ if G is large.

- A grown culture of *E. coli* contains (conservatively) about 10^9 cells/mL. The culture volume is 10 mL, so $N \approx 10^{10}$.

- The total number of transfers is about $30,000/6.6 \approx 4500$ (just over 12 years of daily transfers).

- Hence, the number of replications occurring in the line has been about 4500×10^{10}.

- Since there are 12 lines altogether, in any of which the evolution of citrate utilization might have occurred, the total number of replications screened $= 12 \times 4500 \times 10^{10} = 5.5 \times 10^{14}$, or 540,000,000,000,000 (540 trillion).

The evolution of a novel metabolic capacity might be a very rare event, but the opportunity for mutation is so extensive that it is, in fact, very likely to occur sooner or later. These figures are calculated for a laboratory experiment of modest size and short duration: the corresponding figures for natural populations living on much longer timescales would be much greater.

These calculations show how extremely rare events can contribute decisively to the evolution of novel adaptations. An event that occurs with a probability of (say) 10^{-12} per trial (one in a trillion) can be ignored for all practical purposes in most fields of science, or everyday life. It cannot be ignored in evolution, because selection will automatically tend to magnify the outcome of any one of the vast number of trials provided by exponential growth.

It should be borne in mind, however, that the opportunity for growth is much less in animals and plants than it is in bacteria. Bacterial cultures grown under optimal conditions take about 10 years to pass through 30,000 generations; insects and small plants might take 10,000 years; people take 1 My. In large, long-lived organisms, novel adaptations are only likely to emerge in the geological long term.

13.4.5 Chance, history, and necessity contribute to evolution in the microcosm.

Any episode of adaptation can be interpreted in terms of the contributions made by chance, history and necessity. For example, we have previously described how whales evolved from artiodactyl ancestors.

- Chance: the successful beneficial mutations required for some degree of initial adaptation to an aquatic way of life occurred in the first members of the lineage. They occurred in one artiodactyl lineage but not in others. They did not occur in perissodactyl lineages (odd-toed ungulates, such as horses and rhinoceroses). They did occur in other lineages, such as the carnivore lineage leading to seals.

- History: successive adaptations to an increasingly specialized aquatic way of life depended on those that had previously evolved. The tail fluke is horizontal because the ancestors of whales were galloping ungulates.

- Necessity: body form evolves predictably through the constraints encountered by any large swimming animal. The streamlined body of whales has evolved through natural selection because it minimizes drag.

In experimental evolution, we can use microbial populations to identify and to measure the contributions made by chance, history, and necessity to the process of adaptation. For our baseline, in which none of these forces operate, imagine a single population, founded by a single individual, which is cultured for a short time in the same conditions that its ancestors had experienced for many generations. We then plot the character state of the derived population on that of the founder. Since we do not expect any change to have occurred, this will be a single point lying exactly on the line of equality. Any more realistic situation will fail to conform to this ideal, for three reasons.

- Chance: we use several replicate populations, each founded by the same ancestor. Since each will experience a different succession of mutations, they will come to differ in some degree.

- History: we use populations founded from ancestors with different character states. This difference will tend to be retained over time, so the derived populations will differ from one another, to the same extent that their ancestors differed.

- Necessity: we culture the population in conditions different from those which its ancestor had long experienced. Consequently, selection will cause the character state of the derived population to differ from that of its ancestor.

To evaluate the contributions of each process, we set up an experiment in which several replicate lines are established from each of a number of ancestors,

and cultured in novel conditions. This enables us to partition the variation among the evolved lines into the effects of chance (they will tend to spread out), history (they will tend to retain the ancestral state), and necessity (they will tend to converge on the same state, exceeding the ancestor). The range of possible outcomes is shown in Figure 13.11(a).

The crucial experiment can be done by first culturing several independent lines of the bacterium *E. coli*, all derived from a single ancestral cell, in glucose-limited medium. This means that glucose is the only source of carbon and energy. After 2000 generations (about six months) the lines are transferred to medium in which maltose is the only carbon substrate. All the lines are now very well adapted to utilizing glucose, because any mutation that increases the metabolic flux of glucose will have been strongly selected. When such a mutation spreads through the population, it will drag with it any mutations that affect maltose utilization. These will be a more or less random selection of possible mutations, because none of them have any direct effect on fitness, so long as glucose is the only sugar available. Hence, the independent selection lines will vary greatly in their ability to utilize maltose, and this variation will be expressed when glucose is replaced by maltose as the limiting resource.

After a further 1000 generations of culture in maltose-limited medium, the response of the lines can be partitioned between history (the extent to which they have retained the character state of their ancestors), chance (the extent to which lines derived from the same ancestor have diverged), and necessity (the extent to which the lines have converged to a common character state) (Figure 13.11(b)).

The result depends on the character we measure. If we choose fitness (the ability of the derived lines to grow in maltose-limited medium, relative to their ancestor) then the contribution of necessity is overwhelming. Natural selection for enhanced growth in novel conditions almost completely obliterates the effects of history and chance.

Other characters do not necessarily follow this rule. For example, cell size is more strongly affected by history than by chance or necessity. This is because selection will not necessarily favor a different cell size in glucose-limited and maltose-limited media, so the ancestral state tends to be maintained.

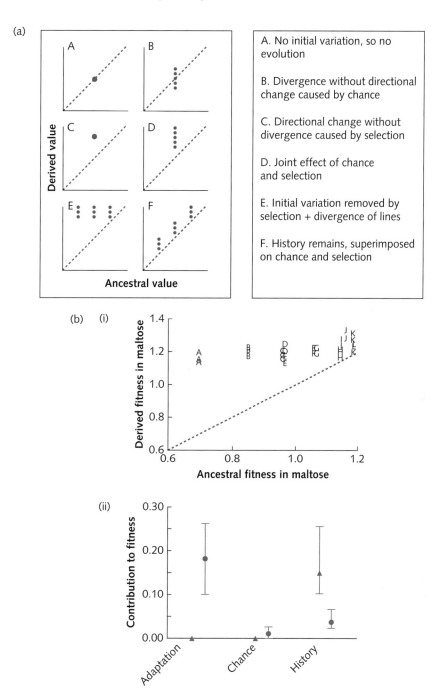

Figure 13.11 An experiment to partition the effects of history, chance, and necessity. (a) Experimental design. (b) The outcome of an experiment in which lines adapted to glucose are transferred to maltose. The upper panel (i) shows the response of replicate selection lines derived from 12 ancestors, labeled A to L. The lower panel (ii) shows how the response can be partitioned between the three sources of evolutionary change. Triangle symbols in the lower graph are ancestral lines; circle symbols are derived lines.

Reproduced from Travisano, M., et al. 1995. Experimental tests of the roles of adaptation, chance, and history in evolution. *Science* 267 (5194). Reprinted with permission from AAAS.

The lesson of this elegant experiment is that the fundamental processes of evolution can be studied and dissected in the laboratory. The interpretation of evolutionary change that normally occurs over very long periods of time can be validated by laboratory experiments carried out over a period of a few weeks or months. Experiments such as these do not require complicated or expensive technology. They are within the scope of advanced laboratory classes in high schools or community colleges.

13.5 Novel ways of life can evolve rapidly in the laboratory.

Populations of bacteria that evolve the ability to utilize glucose more efficiently provide a simple model system for studying the process of adaptation. It might be doubted that this experimental approach can be extended to the major shifts in ecology and morphology that characterize evolutionary change in the long term. For particular cases, this is undoubtedly true: no feasible experiment, for example, will ever retrace the origin of tetrapods from lobe-finned fish (Section 5.8). More generally, however, experimental evolution can show how radical shifts in the morphological and ecological attributes of organisms, such as autotrophy to heterotrophy, or unicellular to multicellular, can evolve over the relatively short term of laboratory experiments.

13.5.1 Green algae that normally grow in the light rapidly adapt to live in the dark.

Chlamydomonas is a unicellular green alga that has often been used to study the physiology and genetics of photosynthesis. It grows rapidly in minimal medium under bright light using atmospheric carbon dioxide as its only source of carbon. It will also grow, only much more slowly, in the dark, if it is provided with sodium acetate as a carbon source. (This is why it was chosen as a model organism to study photosynthesis: loss-of-function mutants that would be lethal in most photosynthetic organisms can be rescued by culturing them on acetate.) This provides an opportunity for the experimental evolutionist. Lines that are maintained for hundreds of generations in the dark evolve into efficient heterotrophs that are able to grow rapidly by using acetate. They are so superior to their ancestors that the normal means of measuring adaptation—by competition with the ancestor—do not work because the ancestor is so quickly overwhelmed. They have achieved a transition between two radically different ways of life, photoautotrophy and heterotrophy, within a few years in the laboratory.

13.5.2 Loss of ancestral function can lead to ecological separation.

Indeed, some lines have gone beyond that point. They are all impaired in some degree relative to their ancestor (Section 13.3.5 and Figure 13.7). In extreme cases, however, the ability to grow in the ancestral environment may be completely lost. Figure 13.12 shows that some lines from the same experiment have evolved efficient heterotrophic growth in the dark, using acetate as a carbon source, but are incapable of growth in the light, even if acetate is supplied. The most likely reason—yet to be confirmed empirically—is the fixation of loss-of-function mutations in the photosystem, coupled with sensitivity to the toxic by-products of photosynthesis. These lines are not merely efficient heterotrophs, but obligate heterotrophs, incapable of even surviving in their ancestral environment. In nature, these lines would never be found in the light, and would rapidly outcompete their ancestors in the dark. Hence, a relatively brief period of intense selection can result in the almost complete ecological separation of a derived population from its ancestor.

13.5.3 Freshwater algae rapidly adapt to marine conditions.

The laboratory species *of Chlamydomonas* lives in freshwater and soil. Wild-type strains are intolerant of salt and are severely impaired by more than about 6 g L^{-1} of NaCl. If the salt concentration is slowly increased by 1 g L^{-1} every two or three weeks, most lines become extinct, but a minority are rescued by

(a)

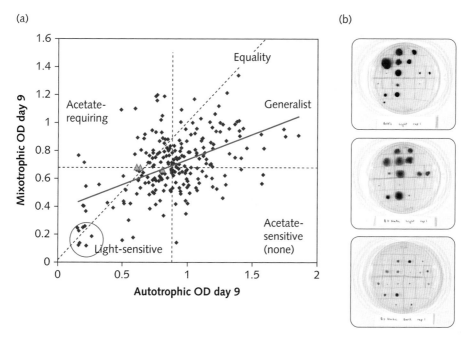

(b)

Figure 13.12 Adaptation to a novel environment can lead to complete ecological separation. These figures show the outcome of selection for heterotrophic growth in the dark, using acetate as a carbon source, in the green alga *Chlamydomonas*. There were 240 replicate selection lines. (a) The growth of the lines in the light, either by photosynthesis (using medium with no carbon source) or by a mixture of photosynthesis and heterotrophy (using medium supplemented with sodium acetate). Note the small group of lines at the lower left which are incapable of growth in the light, with or without supplementation. (b) The growth of some of these lines on agar plates. Top: light, without acetate; middle: light, with acetate; bottom: dark, with acetate. Reading from left to right and from top to bottom, colonies 1, 2, and 3 are ancestors, which grow well in the light and poorly in the dark. The remaining colonies are the experimental dark lines. Some derived lines (such as 4 and 5) can still grow well in the light. Others, such as 6 (at 2 o'clock) and 18 (at 3 o'clock), can no longer grow in the light at all, with or without acetate.

Reproduced from Bell. Experimental evolution of heterotrophy in a green alga. *Evolution* 67 (2) © 2012 Graham Bell. *Evolution* © 2012 The Society for the Study of Evolution.

natural selection and grow at salt concentrations that their ancestor could not tolerate. One by one, they too become extinct, but some lines nevertheless persist at very high concentrations. A few become capable of growing at 33 g L^{-1} NaCl and have evolved into fledgling marine organisms—indeed, they can be grown in seawater.

The likelihood of this physiological transition, with its wide ecological implications, is affected by the genetic system of the experimental population. Lines can be set up from one or many strains, and can be propagated vegetatively or sexually. A purely asexual vegetative line founded as a single clone can adapt only by fixing a succession of beneficial mutations. In contrast, a sexual line founded from many strains can generate a great deal of genetic variation by recombination. Most of the lines that survived at high concentrations of salt were sexual populations founded from a variety of ancestors, demonstrating that adaptation and evolutionary rescue are facilitated by an increased level of variability. There is a good deal of evidence from other systems that sex often accelerates adaptation (Section 15.2.3), showing how experiments can shed light on the evolution of fundamental characteristics of organisms.

13.5.4 Multicellularity can evolve rapidly in experimental populations of yeast.

Mild centrifugation of yeast cultures causes large cells, or groups of cells, to sediment rapidly out of suspension. These can be selected by discarding the supernatant and using the pellet to initiate the next growth cycle. Within a surprisingly short period of time—fewer than twenty transfers—multicellular aggregates have appeared in the cultures. They have

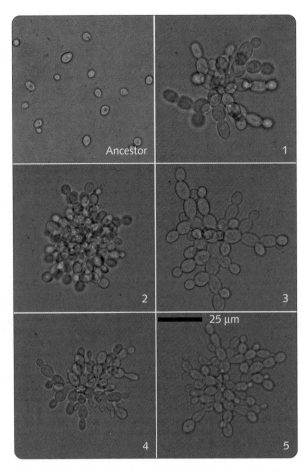

Figure 13.13 Multicellular "snowflake" individuals evolving from unicellular ancestors in experimental yeast populations.

Reproduced from Ratcliff et al. 2012. Experimental evolution of multicellularity. *Proceedings of the National Academy of Sciences of the USA* 105 (5): 1595–1600, with permission from *PNAS*.

a distinctive snowflake-like shape and seem to be distinct entities, because they reproduce by fragmentation, rather than merely by dissociating into individual cells (Figure 13.13). They are able to do this because cells within the snowflake die so as to cause it to separate into two smaller snowflakes that then proceed to grow before again dividing.

Sedimentation might be implicated in the evolution of multicellularity in yeasts or algae, although it is more likely to be a hindrance in most aquatic habitats. More plausible agents of selection for large size include protection from predators, and there is some evidence that the presence of predators can lead to the spread of multicellular forms of algae. The real significance of the yeast experiment, however, is

that even such a fundamental feature of body plans as multicellularity can evolve within a short span of time in the laboratory.

13.5.5 A new species of yeast has been deliberately evolved in the laboratory by hybridization.

Although hybridization usually acts as a brake on species formation (Section 7.3.5), it may actually lead to the formation of new species. The sunflower *Helianthus anomalus*, for example, which grows in arid regions of Arizona and Utah, originated as a hybrid between *H. annuus* and *H. petiolaris*, which are common species in western North America. This potential for stable hybrids to form new sexually isolated lineages has been exploited to create new species in the laboratory. *Saccharomyces cerevisiae* and *S. paradoxus* are two closely related yeasts that can occupy the same sites and mate readily, but 99% of hybrids die. The minority of viable F1 haploid progeny can be selfed to obtain fertile homozygotes. The F2 generation included viable strains that had high fertility when intercrossed but low fertility when back-crossed to either parent. Fertility dropped in F3 but recovered to very high levels in the F4. The inability of the F2 to mate successfully with either parent was associated with the replacement of some *paradoxus* chromosomes by *cerevisiae* chromosomes, suggesting that successful hybrids require particular mixtures of *paradoxus* and *cerevisiae* genomes. This remarkable experiment is the only successful attempt so far to create a new species in the laboratory, but I am sure that others will follow as speciation is gradually drawn into the research program in experimental evolution.

This chapter has shown that many of the fundamental processes that drive evolutionary change—even quite radical shifts to new ways of life—can be studied in real time using microbes in the laboratory. These microcosms enable us to study evolutionary mechanisms by experiment, just as physiological and genetic mechanisms are investigated by experiment. Evolution is normal science. Moreover, we can extend the experimental program to much more complex situations in which organisms evolve in natural environments. This is the subject of the next chapter of this book.

● CHAPTER SUMMARY

The laboratory microcosm is a time machine for evolution.

Deleterious mutation is balanced by purifying selection.

- The genetic variance of fitness increases through mutation.
- Isolate lines decay in the absence of effective selection.
  - *Mean fitness declines over time in isolate culture.*
  - *Replicate isolate lines diverge over time.*
- Diversity is maintained at equilibrium between mutation and selection.
  - *Purifying selection opposes deleterious mutation.*
  - *Deleterious mutations are not completely eliminated by selection.*
  - *Mutation and selection come into equilibrium.*
  - *Asexual populations may have substantial levels of genetic variation.*

Beneficial mutation drives directional selection.

- Permanent adaptation depends on the balance between variation and the rate at which conditions deteriorate.
  - *Rapid and severe stress may cause extinction.*
  - *Adaptation may lead to evolutionary rescue.*
  - *Evolutionary rescue can lead to tolerance of lethal conditions.*
- Beneficial mutations sweep through asexual populations.
  - *All the members of the current population descend from a single individual in the past.*
  - *Selective sweeps remove variation in asexual populations.*
  - *The passage of a beneficial mutation can be detected by marker genes.*
- Substitution involves three phases of waiting, establishment, and passage.
  - *The waiting time depends on the mutation supply rate.*
  - *Adaptation halts when beneficial mutations are very rare.*
  - *Successful beneficial mutations only become established if they survive stochastic loss.*
  - *The passage time depends on the intensity of selection.*
- The beneficial alleles responsible for adaptation often have a large effect on fitness.
  - *The first allele to be fixed is expected to have a large effect on fitness.*
  - *Adaptation is dominated by large-effect mutations in experimental populations of bacteria.*
  - *The rate of adaptation increases with population size.*
- Adaptation often leads to loss of ancestral function.
  - *Regress may be caused by functional interference.*
  - *Regress may be caused by mutational degradation.*
- The outcome of selection can be predicted within limits.

Long-term experiments document the main features of evolutionary change.

- *Bacterial evolution over 50,000 generations has been studied in the laboratory.*
- Adaptation can be measured precisely.
- Adaptation increases rapidly then slows down.

- *Each line at first increases in fitness rapidly.*
- *Populations remain well-adapted to a constant environment through purifying selection.*
- *Populations carry a large amount of potentially adaptive variation.*

- Adaptation often involves repeatable genetic modifications.

- Evolutionary innovation may involve very rare events.

- Chance, history, and necessity contribute to evolution in the microcosm.

Novel ways of life can evolve rapidly in the laboratory.

- Green algae that normally grow in the light rapidly adapt to live in the dark.

- Loss of ancestral function can lead to ecological separation.

- Freshwater algae rapidly adapt to marine conditions.

- Multicellularity can evolve rapidly in experimental populations of yeast.

- A new species of yeast has been deliberately evolved in the laboratory by hybridization.

● FURTHER READING

Here are some pertinent further readings at the time of going to press. For relevant readings that have been released since publication, visit the book's Online Resource Centre at **www.oxfordtextbooks.co.uk/orc/bell_evolution/**

Section 13.1 Dykhuizen, D. E. 1990. Experimental studies of natural selection in bacteria. *Annual Review of Ecology and Systematics* 21: 373–398.

Elena, S.F. and Lenski, R.E. 2003. Evolution experiments with microorganisms: the dynamics and genetic bases of adaptation. *Nature Reviews Genetics* 4: 457–469.

Buckling, A., Maclean, R. C., Brockhurst, M.A. and Colegrave, N. 2009. The Beagle in a bottle. *Nature* 457: 824–829.

Kawecki, T.J., Lenski, R.E., Ebert, D., Hollis, B., Olivieri, I. and Whitlock, M.C. 2012. Experimental evolution. *Trends in Ecology and Evolution* 27: 547–560.

Bell, G. 2013. Responses to selection: experimental populations. In *Princeton Guide to Evolution*, edited by Jonathan B. Losos et al., 230–237. Princeton University Press, Princeton, NJ.

Section 13.2 Halligan, D.L. and Keightley, P.D. 2009. Spontaneous mutation accumulation studies in evolutionary genetics. *Annual Review of Ecology, Evolution and Systematics* 40: 151–172.

Section 13.3 Bell, G. 2012. Evolutionary rescue and the limits of adaptation. *Philosophical Transactions of the Royal Society of London B* 368: 20120080.

Dykhuizen, D.E. and Dean, A.M. 1990. Enzyme activity and fitness: evolution in solution. *Trends in Ecology and Evolution* 5: 257–262.

Section 13.4 Lenski, R.E., and Travisano, M. 1994. Dynamics of adaptation and diversification: a 10,000-generation experiment with bacterial populations. Proceedings of the National Academy of Sciences, USA, 91: 6808–6814.

Woods, R., Schneider, D., Winkworth, C.L., Riley, M.A. and Lenski, R.E. 2006. Tests of parallel molecular evolution in a long-term experiment with *Escherichia coli*. *Proceedings of the National Academy of Sciences of the USA* 103: 9107–9112.

Blount, Z.D., Borland, C.Z. and Lenski, R.E. 2008. Historical contingency and the evolution of a key innovation in an experimental population of *Escherichia coli*. *Proceedings of the National Academy of Sciences of the USA* 105: 7899–7906.

Travisano, M., Mongold, J.A., Bennett, A.F. and Lenski, R.E. 1995. Experimental tests of the roles of adaptation, chance, and history in evolution. *Science* 267: 87–90.

Section 13.5   Bell, G. 2012. Experimental evolution of heterotrophy in a green alga. *Evolution* 67: 468–476.

Ratcliff, W.C., Denison, R.F., Borrello, M. and Travisano, M. 2012. Experimental evolution of multicellularity. *Proceedings of the National Academy of Sciences of the USA* 109: 1595–1600.

Greig, D., Louis, E.J., Borts, R.H. and Travisano, M. 2002. Hybrid speciation in experimental populations of yeast. *Science* 298: 1773–1775.

QUESTIONS

1. Discuss the merits and drawbacks of using batch culture (serial transfer) or continuous culture (chemostats) to study evolution in microbes.

2. Describe and explain what is likely to happen when a culture is propagated by transfer of a single randomly chosen cell at the end of each growth cycle. How would events differ if two random cells were chosen? Or if a thousand random cells were chosen?

3. Give an account of the dynamics of genetic diversity in a chemostat population of bacteria.

4. Explain the concept of evolutionary rescue and predict how the frequency of rescue will be affected by population size and the rate of environmental deterioration. Are there any other conditions that might affect the frequency of rescue?

5. Explain the concept of periodic selection and describe how it can be used to study adaptation in asexual populations. Would you expect similar dynamics in a sexual population?

6. With the aid of a diagram, describe the substitution of a beneficial mutation in terms of waiting time, establishment time, and passage time.

7. Adaptation in experimental populations often involves only a few mutations of large effect. Why is this? In what circumstances would adaptation be more likely to be based on many mutations, each of small effect?

8. Explain why populations adapting to a novel environment may exhibit reduced fitness in their ancestral environment.

9. "The course of evolutionary change may be unpredictable, even in principle." Discuss.

10. Write an essay about the contribution of the long-term *E. coli* experiment conducted by Rich Lenski to our understanding of evolutionary patterns and processes.

11. Design an experiment to evaluate how random, historical and selective processes contribute to the changes that evolve in a population exposed to novel conditions of growth.

12. "Experimental evolution may tell us something about minor modifications of organisms, but it will never be useful in understanding how novel characteristics evolve." Discuss.

You can find a fuller set of questions, which will be refreshed during the life of this edition, in the book's Online Resource Centre at www.oxfordtextbooks.co.uk/orc/bell_evolution/

14 Selection in Natural Populations

Think of a small patch of countryside, somewhere near where you live. It looks something like the sketch-map in Figure 14.1. A meadow by the edge of the woods is drained by a small stream that runs out of a slough and widens at one point into a pond. One upland portion has been fenced off and used as sheep pasture. There is an old mine working, long since abandoned and surrounded with a fence. It is a mundane piece of land such as might be found anywhere. It still retains features of an ancient landscape, although it has more recently been modified by human occupation, through agriculture and industrial activity. Walking across it on a summer's day you might imagine that the grass and the trees, the insects and the birds, have been there almost forever. But you are really seeing only a brief episode from a constantly changing scene. It has the appearance of permanence only because of the brevity of human life. A few hundred years ago, the meadow was a forest. A few hundred years hence, the pond will have been filled in by the sediment from the small stream that feeds it. All the myriads of plants and animals that you see around you will have departed, their place taken by others. Nature never stands still.

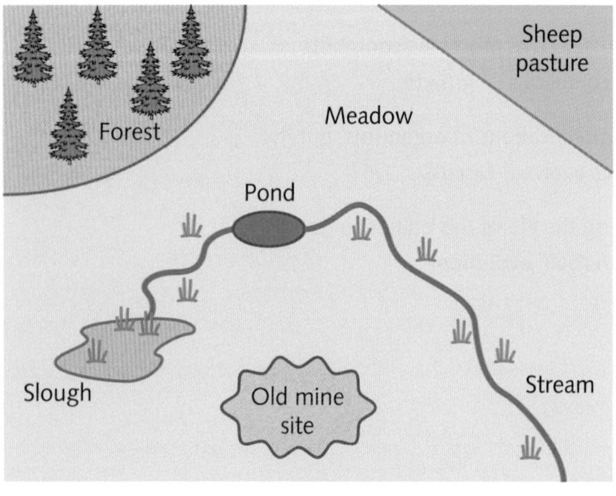

Look again at the grass that covers the meadow. It seems from a distance to be a uniform unvarying carpet of green, growing in the same way year after year. Nothing could be more deceptive. A closer look shows that leaves and seeds are being consumed by insects and fungi. Below ground, the roots of different species are engaged in a continual struggle for access to scarce resources in the soil. Far from being a peaceful and harmonious scene, the meadow is a battleground where no plant can survive unless it has beaten off the attacks of innumerable enemies.

Looking even closer, we can recognize some of the actors in this drama. *Anthoxanthum* is a grass that grows on the drier slopes of the meadow. The plants are being consumed by snails called *Cepaea*. Both are European species, in fact, that have only recently been introduced to North America. At the fringes of the woodlot, chickadees are foraging in the trees; one has just caught a moth. Below the birds, a field mouse, *Peromyscus*, slips into cover among the leaf litter. Growing down by the bank of the stream is the native touch-me-not, *Impatiens*, with its conspicuous orange flowers. Male sticklebacks (*Gasterosteus*) with bright red throats are building their nests in the shallows of the pond. All of these plants and animals are engaged in a ceaseless struggle for existence in which inferior lineages will be eliminated, while superior lineages will flourish.

Figure 14.1 Somewhere near here.

Grass, snails, mice, birds, moths, flowers, and fish will all be continually screened by natural selection. Consequently, we can study evolution in action, far from the laboratory or the farm, by closely observing how adaptation is maintained and extended in familiar organisms like these.

14.1 Purifying natural selection maintains adaptation.

Each of the species in this scene may be currently in one of two situations.

- It is well adapted to its conditions of life. These conditions have remained the same for a long period of time, so almost all the mutations that would increase fitness have been fixed in the population. Hence, almost any mutation that now occurs will be deleterious. The effect of selection is almost always to remove mutations, because the individuals which bear them are inferior to the average of the population. This is purifying selection.

- It is poorly adapted to its conditions of life. The environment has recently changed, and there has not yet been sufficient time for most potentially beneficial mutations to appear and become fixed. Most mutations will be deleterious, and will tend to be removed by purifying selection. Nevertheless, some will be beneficial, and will tend to spread through the population. This is directional selection.

In practice, of course, most species will fall between these two extremes: they are neither poorly adapted nor perfectly adapted. Nevertheless, the distinction between purifying selection, which maintains adaptedness by removing deleterious mutations, and directional selection, which extends adaptedness by favoring beneficial mutations, is clear and useful.

14.1.1 Purifying selection acts contrary to mutation and immigration.

Nature never stands still. But suppose that nature did stand still, that every place remained the same, and that every population were at present perfectly adapted to local conditions. This state of affairs could not last for long.

- The genomes of the next generation will suffer mutations. If the present generation is taken to be perfectly adapted, then the offspring are necessarily less well adapted. Mutation will reduce the mean fitness of the population.

- Some of the individuals growing in the site will be immigrants from nearby sites. If each local population is perfectly adapted to its own site, it must be less than perfectly adapted to other sites. Consequently, immigrants will have lower fitness than the offspring of residents. Immigration will reduce the mean fitness of the population.

Our imaginary perfect world cannot remain so, even for a single generation. Perfect adaptedness will be immediately degraded by mutation and immigration. It will not be degraded indefinitely, however, because poorly adapted individuals will often fail to survive or reproduce. Mutation and immigration are therefore opposed by purifying selection.

14.1.2 The intensity of purifying selection is equal to the genetic variation of fitness.

The effect of mutation and immigration is to introduce alleles into the local population that reduce the fitness of the individuals that bear them, relative to the average of the residents in the previous generation. In other words, mutation and immigration increase the genetic variance of fitness.

The local reduction in average fitness is countered by purifying selection, which in every generation restores the previous level of adaptedness. It acts by removing individuals whose fitness is lower than average, or in other words by reducing the genetic variation of fitness.

If the rate of mutation or immigration is too high, the frequency of deleterious alleles will increase without limit, and the local population will become extinct. In most realistic situations, however, they will be effectively opposed by purifying selection, and the population will persist.

Consequently, most situations represent a balance between the continual deterioration caused by the introduction of deleterious alleles and the continual amelioration caused by their removal. The deterioration attributable to mutation and immigration is equivalent to an increase in the genetic variance of fitness. Selection acts on this variance so as to restore the previous level of adaptedness. It follows that the intensity of natural selection is equal to the genetic variance of fitness.

This is an important principle, because it summarizes the normal effect of natural selection in maintaining a constant level of adaptedness. It is sometimes called "the fundamental theorem of natural selection."

14.1.3 Adaptedness is continually degraded and continually restored.

The fundamental theorem is useful because it provides a way of measuring the strength of purifying selection in natural populations. It is not practicable to score the number of deleterious mutations borne by individuals and then relate this to their production of offspring. It is practicable, however, to measure how many offspring are produced by individuals belonging to different families growing up in the same place. Individuals belonging to the same family are genetically similar, so the variation of fitness within families is largely caused by environmental differences—whether a seed falls on a dry patch or a wet patch of soil, for example. Conversely, average differences among families are largely genetic. The difficulty is that we can seldom recognize individuals as belonging to the same family, once offspring have dispersed away from their parents. The solution is to bring individuals into captivity, construct families by controlled breeding, mark the individuals, and then release them back into the place they were taken from. This is a field experiment. It is a powerful way of studying the dynamics of selection in natural populations.

Purifying selection increases fitness by a few percent per generation in sedges and birds. The touch-me-not, *Impatiens*, grows along watercourses and in damp places in the meadow. It is named for its seed capsules, which burst open explosively at the slightest touch when they are ripe. Some colleagues and I collected seeds from a population of *Impatiens* growing along a small stream on a wooded hillside. We germinated these seeds in a greenhouse and grew the plants to maturity—they are annuals, so this could be done in the following summer. These individuals were then crossed by artificial pollination to make families of known parentage. We germinated these seeds, and then planted the seedlings back into the site from which their grandparents had been collected. These seedlings then grew up into adult plants—or died before reproducing—and we carefully followed the fate of every one. At the end of our experiment, we had two pieces of information. The first is the aggregate variation of fitness among all the seedlings we had transplanted. The second was how fitness varied within and among families. We made three main discoveries.

- The first was that most of the seed produced by our experimental plants came from a very few individuals. About 30% of all the seed was produced by 1% of the original seedlings. This was largely because most of the seedlings died before developing into adults, a very common feature of natural populations of plants.

- Secondly, most of the variation in survival was environmental. There was in hindsight a clear gradient in seedling survival and adult reproduction from one side of our experimental plot to the other. This was visible only in hindsight; it did not coincide with any obvious change in conditions. The plants responded to some environmental gradient that we could not discern.

- Thirdly, there was a small residual amount of variation among families that was caused by genetic differences, independently of the conditions of growth. This amounted to about 3% of the total amount of variation in fitness.

What this experiment demonstrated is that fitness is degraded by about 3% in every generation in this population of *Impatiens*. In other words, offspring are on average about 97% as fit as their parents. The balance is restored by selection, which must raise fitness by about 3% in every generation.

The great tit, *Parus major*, is a small insectivorous bird that has been intensively studied in Britain and the Netherlands. It is similar in appearance and habits to the black-capped chickadee, *Poecile atricapillus*, of northern North America, a member of the same family of birds. In natural conditions it nests in holes which it readily forsakes for nest-boxes when these are made available. About a thousand nest-boxes have been set out each year in Wytham Wood, near Oxford, since the 1960s, and almost all the 100–500 resident tits use them. This allows the nestlings to be banded, so that each individual can be recognized and subsequently recorded for the rest of its life. Each adult bird will produce offspring, some of which will survive to be recruited into the breeding population a generation hence: the fitness of an individual is the number of recruits it produces during its lifetime. Forty or fifty years of continuous effort—many generations of students!—then provides pedigrees from which fitness can be followed generation after generation. These pedigrees tell us the number of recruits produced by

each bird and the amount of variation among individuals. They also show how individuals are related to one another, so we can calculate how much of the variation is genetic. From the fundamental theorem, this is equivalent to the amount by which fitness is increased in every generation by purifying selection. For males, this is about 2.5%—a very similar rate to *Impatiens*, despite the great differences between birds and annual plants. The estimate for females is much less, about 0.3%. There is often more variation in fitness among males than among females, for reasons that we shall explore in Chapter 15.

The mundane operation of natural selection is not dramatic. Indeed, it can be clearly perceived only by careful and laborious field experiments. It is nevertheless very important, because it has a general and universal application: all populations everywhere are continually exposed to purifying selection that maintains their level of adaptedness.

14.2 Directional natural selection drives adaptation to changing conditions.

Our homely landscape of meadow and stream stands, let us suppose, somewhere in Yorkshire or upper New York State. As recently as 10,000 years ago it would have been an arctic wilderness, locked in ice and without any living thing. As the climate warmed, tundra vegetation grew at the edge of the retreating ice sheet. Animals and plants slowly spread northward to colonize the newly exposed land and create soil. A forest sprang up.

Much later, human bands filtered in to hunt in the forest and clear patches of land for cultivation. Then newcomers arrived to clear the forest, drain the land, and parcel it out for agriculture. Other events—further west, in North America—might make this unprofitable, when forest and marsh would again begin to encroach. With or without our intervention, the landscape is never still. It shifts continually, partly in response to human activity, partly because of the incessant rhythm of natural change. Consequently, natural selection does not operate for long in a single fixed direction. Whether we look at a humanized landscape of field and woodlot, or at a completely pristine landscape where humans have never been, we shall find that populations seldom

remain well-adapted to the conditions of the day, but are instead continually buffeted by change.

14.2.1 New ecological opportunities provide the opportunity for new types to evolve.

The sticklebacks living in the pond are small fish we have met before (Section 7.3.4). They are heavily armored, with bony plates protecting their thorax, and long sharp spines that can be erected and locked into place on the pelvic girdle and along their back. This armor protects them from predatory fish: the plates prevent their bodies being crushed by the grip of a predator, and the spines deter attack. Sticklebacks live in the coastal zone of northern oceans. During interglacial periods, the retreat of the ice opens up streams running into the ocean, and sticklebacks move into these streams because they are able to tolerate fully freshwater conditions.

Freshwater conditions favor reduced armor in sticklebacks. Streams are very different from the marine littoral. There are fewer predators, because few fish are capable of making the transition from marine to freshwater conditions. There are also different predators; for example, dragonfly larvae do not live in the sea but are common in freshwater, where they attack juvenile sticklebacks. The armor that is an effective adaptation against predators in the sea is no longer necessary in freshwater—indeed, it may even be a disadvantage, as grappling predators like dragonfly larvae may actually use the pelvic spines to hold on to their prey. Consequently, individuals with reduced armor will be favored, and this leads to the loss of thoracic plates and pelvic spines in stream populations. The skeletons of marine and stream forms are illustrated in Figure 14.2.

Adaptation to freshwater is repeatable and predictable. The retreat of the ice opened up streams along all the coasts of the northern hemisphere, from British Columbia to Siberia. Stream populations adapt in a similar fashion, by evolving reduced body armor, throughout this vast range. This is not the result of migration between freshwater drainages. Lightly-armored forms have evolved independently on many different occasions. In this case, evolution follows a predictable course.

Adaptation may occur through loss-of-function mutations. The mutations responsible for the characteristic morphology of freshwater populations of sticklebacks occur mainly in two genes. Armor development is controlled largely by the *Ectdysoplasin (Eda)* gene. The product of this gene is a signal molecule that is required for normal scale development in other fish and for the development of ectodermal structures such as hair and teeth in mammals. The large pelvic spines of marine

sticklebacks are also reduced or completely lost in freshwater populations. Pelvic spine reduction is a Mendelian character involving a single regulatory mutation in *Pitx1*, a gene whose homologue is necessary for normal hindlimb development in mice. The mutations involved are loss-of-function mutations, because they result in the loss of a feature present in the ancestor. The loss of a feature specialized for life in the ancestral environment is often the first step in adaptation to novel conditions.

14.2.2 Novel environmental stresses lead to rapid evolution.

Insect populations evolve in polluted cities. About the middle of the nineteenth century, black varieties of several moths, formerly rarities, began to spread to a remarkable extent in the industrial regions of midland and northern England, and a little later in similar areas of the USA. The canonical example is the peppered moth, *Biston betularia*, in which a marked increase in melanic pigmentation is caused by a single mutation. Their spread followed the appearance of clouds of coal smoke above the newly-industrialized towns, smoke that was washed down by the rain to blacken walls and tree-trunks with soot. It is easy with hindsight to appreciate that the pepper-and-salt markings of the type originally common blended with the lichen-covered tree-trunks of the unpolluted countryside, whereas the black wings and abdomen of the melanic variant were inconspicuous when the moth was resting on the soot-encrusted bark of urban trees (Figure 14.3, upper panel). Visual predators such as birds would detect the pepper-and-salt variety more easily against a blackened background and tended to attack them first, creating selection that favored the melanics and caused their spread. Their effect on the frequency of melanics during a single episode of selection was demonstrated by following the fate of a relatively small number of moths, melanic and pepper-and-salt released into natural populations. In the countryside, where lichens still grew thickly on tree-trunks, the pepper-and-salt variant increased in frequency over the course of a few days; in polluted areas, it was the melanics that increased (Figure 14.3, lower panel). A single episode of selection, therefore, showed that melanic variants tended to spread in the novel environment furnished by sooty towns.

Marine form
Fully armored

Stream form
Lightly armored

Figure 14.2 Marine and freshwater forms of the three-spined stickleback, *Gasterosteus aculeatus*. The fish have been cleared and stained with alizarin to show the skeleton.

Photo courtesy of Nicholas Ellis and Craig Miller, UC Berkeley.

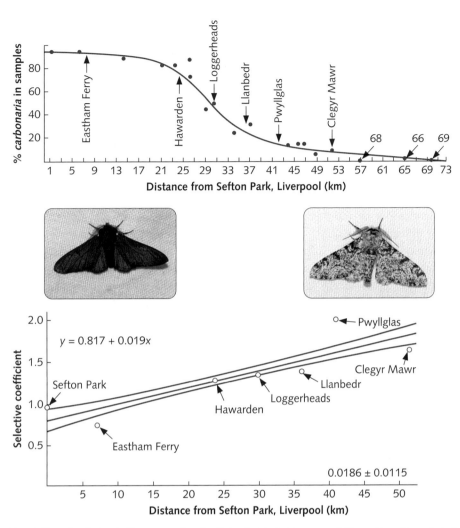

Figure 14.3 Rapid evolution of melanism in the peppered moth, *Biston betularia*. Upper panel. The pattern: the melanic form is abundant in the industrial zone of Liverpool, but is replaced by the normal pepper-and-salt form in rural areas of Wales. Lower panel. The process: experiments show that the normal form is less fit than the melanic form in industrial areas but more fit in rural areas. The y-axis is 1 + s, where s is the selection coefficient in favor of the normal form; a value of 1 indicates that melanic and normal forms have equal fitness.

Graphs reproduced from Bishop, J.A. 1972. An experimental study of the cline of industrial melanism in *Biston betularia* (L.) (Lepidoptera) between urban Liverpool and rural North Wales. *Journal of Animal Ecology* 41 (1): 209–243, with permission from Wiley. Photographs courtesy of Olaf Leillinger. These files are licensed under the Creative Commons Attribution-Share Alike 2.5 Generic license.

The selection caused by novel stresses can be very strong. Melanic variants of the peppered moth were first recorded in 1849. In 1875 they were still listed in catalogues as curiosities, so they were presumably still rather infrequent. By the mid-1880s, however, collectors were noticing that in some areas they were more common than the original pepper-and-salt type, and by 1898 they had reached a frequency of 98% in the Manchester–Liverpool area. Thus, within 50 generations—the moth has a single generation per year—the melanic phenotype had increased in frequency from about 1% to about 95%. Since in this case melanism is caused by a single dominant allele, the frequency of the melanic allele must have increased more than a hundredfold during this period, from about 0.005 to about 0.775. This implies intense selection, with a selection coefficient $s = 0.2$ or so, meaning that melanic individuals were 20% more likely to survive than normal individuals.

Selection is reversed when the stress is alleviated. Selection favoring the melanic form in the city reverses in rural areas, and this is reflected in the gradually decreasing frequency of melanic individuals from the city into the countryside. In the 1950s the effects of smoke pollution on human health became so clear that the use of smoke-producing fuels, such as domestic coal fires, was greatly restricted by legislation. Urban trees became gradually cleaner, and their trunks became covered again with lichens. This reduced and eventually reversed selection acting on the pigmentation of the moths. Consequently, the frequency of melanics began to decrease, and nowadays they are once again as rare as they were 150 years ago.

14.3 Local selection leads to divergent specialization.

The opening of freshwater lakes by deglaciation and the pollution of landscapes by industrialization are examples of environmental variation on the grand scale. Organisms may also become adapted to the smaller scale patchwork of meadow and woodlot, pond and stream.

14.3.1 Plants adapt to local soil conditions.

Local adaptation can occur over very short distances. Buttercups (*Ranunculus*) grow in both the meadow and the forest, where conditions are very different. When specimens were transplanted between adjacent grassland and woodland sites the residents consistently produced more stolons and leaves, showing that they had become locally adapted (Figure 14.4). Interestingly, the effect was asymmetrical, woodland incomers to the grassland being much more severely handicapped than grassland incomers to the woodland: the relative performance of incomers was only about 15% of residents in the grassland but more than 75% in the woodland. This may have been because the grassland population may have evolved from the woodland population when the wood was cleared, hardly more than a few decades previously. The experiment would then have censused both current selection and the historical effects of past selection, the grassland population still being adapted in some degree to woodland conditions.

Local adaptation can be demonstrated by long-term field experiments. In 1856 an old hay meadow of uniform appearance in the grounds of the Rothamsted Experimental Station at Harpenden, England, was divided into twenty plots that received different fertilizer treatments in order to determine their effects on the production of hay. The experiment is still continuing, and despite some changes in treatment and incomplete data recording it provides a 150-year record of biological changes attributable to specific environmental manipulations.

Many plots have received high inputs of specific nutrients, such as calcium and phosphate, that can influence growth and yield. The vegetative yield of the grasses such as *Anthoxanthum* is much greater in plots fertilized with calcium and phosphate, whereas other nutrients, such as magnesium, have no effect. This is an environmental effect, caused by the differences in soil nutrient concentrations between the plots, but there are also genetic changes in the populations. This can be demonstrated by extracting clones from the plots—easy to do, because most grasses proliferate vegetatively, and can be broken up into individual ramets that can be planted separately—and growing them under standard conditions in the greenhouse.

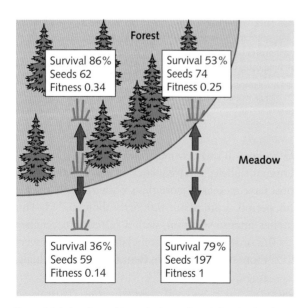

Figure 14.4 A reciprocal transplant experiment showing local adaptation in a buttercup, *Ranunculus*.

For example, ramets from limed plots are more responsive to calcium than ramets of the same species from unlimed plots: that is, when calcium concentration is manipulated in the greenhouse, the plants from the limed plots grow better at high concentrations, whereas the plants from unlimed plots grow better at low concentrations, as shown in Figure 14.5.

During the few hundred ramet generations since the beginning of the experiment, *Anthoxanthum* has adapted to the nutrient treatments that most affect its growth. The time course of adaptation is unknown, but populations on newly limed plots evolve heritable differences in yield and morphological characters within a few years. These results show that the intense natural selection that can be imposed on laboratory populations also acts to drive adaptation in open populations in the field.

14.3.2 Field mice adapt to living on beaches.

Mice become locally adapted to their background. Mice in the genus *Peromyscus* are common in grassland and woodland throughout North America. Similar species occur in other parts of the world. Most have dark brown fur over most of the body, changing to gray on the belly. They are inconspicuous creatures against their normal background of grass, dead leaves, or earth. Nevertheless, the color of their fur varies from place to place, and in exceptional

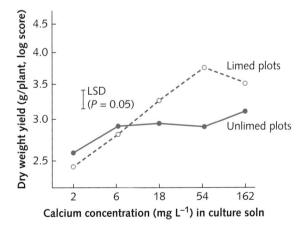

Figure 14.5 Rapid adaptation to liming of the soil by the grass *Anthoxanthum* in the Rothamsted park grass experiment.

Reproduced from Davies, H.I. and Snaydon, R.W. 1973. Physiological differences among populations of *Anthoxanthum odoratum* L. on the park grass experiment, Rothamsted. I. Response to calcium. *Journal of Applied Ecology* 10: 33–45, with permission from Wiley.

conditions the mice have a quite different appearance. In particular, some populations from the Gulf coast of Florida have very pale fur on the back and head. They are living on white sand beaches where dark mice would be very conspicuous and easily detected by visual predators such as owls and hawks. Conversely, pale individuals would be more conspicuous against a dark background. Hence, selection by visual predators is thought to maintain the variation in coloration among populations of field mice from the beaches of the coast to the dark, loamy soils further inland.

Adaptation is based on beneficial mutations of large effect. Fur pigmentation in mice is governed in part by the melanocortin gene *Mc1r*, which is regulated by another gene, *Agouti*. The Gulf coast beach populations bear a point mutation in *Mc1r* that tends to cause the development of pale fur. This only happens, however, if there is an increase in the expression of *Agouti*. Hence, the evolution of these populations is caused primarily by selection acting on just two genetic changes: a point mutation in the structural gene *Mc1r* and a derived over-expression of the regulatory gene *Agouti*.

Populations at different sites become independently adapted. Pale mice are also found on the Atlantic coast beaches of Florida. The molecular phylogeny of *Peromyscus* shows that dark and pale populations are not two separate clades. Instead, the Gulf and Atlantic coast populations have independently evolved pale coloration, as the consequence of selection for similar phenotypes, as shown in Figure 14.6. Their similarity is therefore an example of parallel evolution. Because they have evolved independently, the similar appearance of the beach populations on the two coasts need not necessarily imply that they have similar genotypes, and in fact they do not. The particular point mutation in *Mc1r* that is fixed in the Gulf coast populations does not occur in the Atlantic coast populations, where there are instead several other alleles of this gene. Hence, a particular agent of selection produces repeatable phenotypic changes, but these may evolve along different genetic pathways.

14.3.3 Plants adapt to polluted soils.

Landscapes that have been recently transformed by human activities often present very hostile conditions

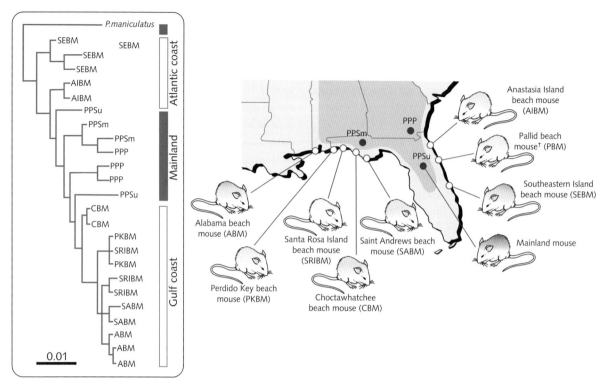

Figure 14.6 Independent evolution of pale coloration in beach mice. On the left is a phylogeny based on mitochondrial and nuclear DNA sequences that shows the independent derivation of the beach phenotype from inland populations on the Gulf and Atlantic coasts, as pictured on the right.

Steiner, C. et al. 2009. The genetic basis of phenotypic convergence in beach mice: similar pigment patterns but different genes. *Molecular Biology and Evolution* 26: 35–45, by permission of Oxford University Press.

for plant growth. Where the soil is polluted, in particular, plant populations must become adapted to a powerful and specific stress in order to persist. The humanized landscape provides dozens of unintentional experiments that test the limits of adaptation to novel stresses.

Grasses on old mine workings are resistant to heavy metals. The landscape of north Wales, the western promontory of Britain, has been humanized for thousands of years. A great deal is pasture grazed by sheep, with a mixture of grasses and low herbs, including *Agrostis*. The underlying igneous rocks, however, contain substantial deposits of heavy metals such as copper, zinc, and lead. These were being mined when the Romans came to Britain, but the pace of exploitation increased sharply when the Industrial Revolution created new markets. The deposits were relatively small and had been worked out by the late nineteenth century, leaving a fenced patch of rubble and spoil surrounding each

old mineshaft. This patch is at first sterile; nothing is able to grow on it, because high concentrations of heavy metals are leached from the low-grade ore that had been discarded there into the soil, creating a bare circle in the surrounding sheep pasture. After a while, a few plants from the pasture succeed in growing on the old mine area, which eventually became tufted with grasses such as *Agrostis* and herbs such as the bladder campion, *Silene*.

Adaptation may occur through gain-of-function mutations. By taking the progeny of plants growing on pasture or mine, it can be shown that the mine plants had evolved a heritable resistance to high levels of heavy metals, and were able to grow fairly well at concentrations that were lethal to the pasture plants. There is often an abrupt increase in tolerance to heavy metal pollution precisely at the boundary of the mine, as illustrated in Figure 14.7. In most cases resistance is conferred by a few major genes of large effect. The physiological mechanisms of resistance are not very

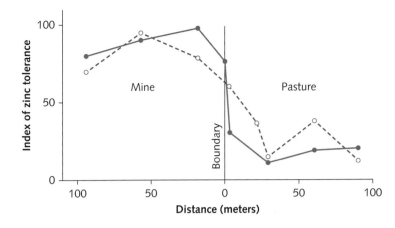

Figure 14.7 Precise local adaptation to heavy metal pollution. There is a sharp increase in zinc tolerance on different sides of the stone wall marking the boundary between mine and pasture. The two lines refer to two grasses, *Anthoxanthum* and *Agrostis*.

Reprinted from Macnair, M. Heavy metal tolerance in plants: a model evolutionary system. *Trends in Ecology and Evolution* 2 (12) 354–359. © 1987, with permission from Elsevier.

well understood, but they include binding the metal by specific proteins called metallothioneins, increased level of active efflux across the plasma membrane, and locking up the toxic metal in vacuoles. Similar genes are often involved in different species. These are gain-of-function mutations, conferring a feature lacking in the ancestor.

Adaptation to polluted sites can occur very rapidly. Resistance to heavy metals evolves quite rapidly: it has been little more than a century since the mines were abandoned, but resistant populations can appear on spoil-heaps within fifty years or so. It has even been reported that linear populations of plants resistant to zinc, a meter or so in width, appear beneath galvanized-iron fences within a decade of their construction, or in the small area beneath electricity transmission pylons within about 20 years.

Adaptation occurs in some species but not in others. The ancestral population of *Agrostis* was growing in a pasture together with many species of grass, to all appearances comparable. Some of these also evolve tolerance to heavy metals in polluted sites; others, however, never do. This is because alleles conferring tolerance exist at low frequency in some species but not in others, in the absence of pollution caused by mining. It may well be that in the long term rare beneficial mutations would enable all species to become adapted to the toxic soils. In the short term, however, the idiosyncratic properties of species mean that some species are able to adapt effectively whereas others are not. This is an example of how constraints arising from ancestry may hinder

the operation of natural selection and prevent the evolution of well-adapted phenotypes.

14.3.4 Local adaptation is balanced by selection and immigration.

Local adaptation is degraded by immigration. In a uniform environment, adaptation is continually degraded by mutation. Purifying selection prevents the accumulation of deleterious mutations, because less heavily loaded individuals are more likely to survive and reproduce. Where the state of the environment varies from site to site, selection tends to cause adaptation to local conditions, but it is opposed not only by mutation but also by immigration. This is because immigrants arriving at a given site may have been born in a different kind of site. Hence, they are likely to bear alleles that are well adapted to the site where they were born, but not to the site where they are growing up. This will consistently tend to reduce mean fitness, in much the same way as mutation.

For example, young field mice move about 250 m away from where they were born, before becoming adults and reproducing. This will tend to smear out any genetic differences between populations, and other things being equal would make the whole species genetically homogeneous. The extent of local adaptation will therefore depend on the balance between selection and immigration. High rates of immigration will prevent local selection from being effective unless it is correspondingly strong.

Distinctive local adaptation requires intense local selection. The erosion of adaptation by immigration was very clearly demonstrated by a small copper mine situated in a narrow glaciated valley down which the prevailing wind is funneled. A sketch-map of the mine and the pattern of tolerance in upwind and downwind populations of *Agrostis* are shown in Figure 14.8. At the upwind end of the mine, where pollen and seeds are blown from the pasture population on to the mine, an abrupt increase in copper tolerance occurred over less

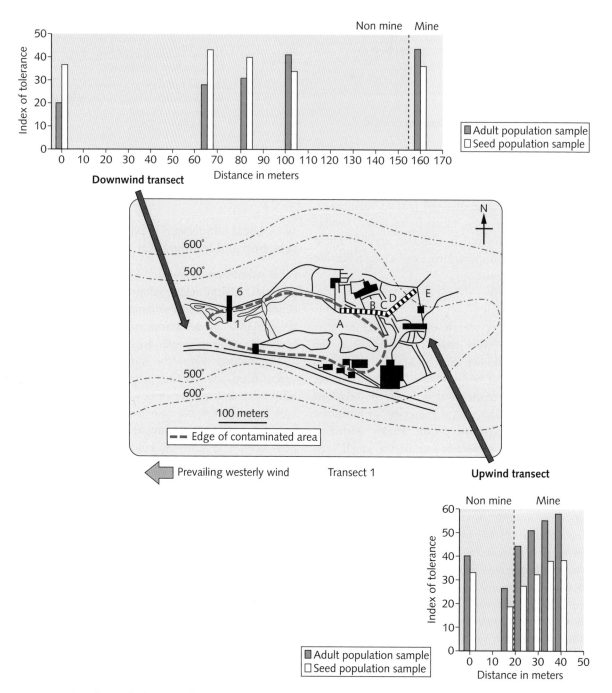

Figure 14.8 Gene flow and adaptation. The sketch-map shows the outline of a small copper mine and associated buildings in a steep-sided valley in Wales. The histograms show the degree of copper tolerance in adults and seedlings at the upwind and downwind boundaries of the mine. (Note that the upwind transect is ordered in the opposite direction to the map.)

Reprinted by permission from Macmillan Publishers Ltd: McNeilly, T. Evolution in closely adjacent plant populations III. *Agrostis tenuis* on a small copper mine. *Heredity*. © 1968.

than 20 m at the boundary between pasture and mine. Adult plants from the mine were much more tolerant than seedlings, grown from the seed that the same adult plants produced in the field; the difference between adults and seedlings represents the extent to which adaptation is broken down in every sexual generation by pollen from the pasture, and is therefore a minimal estimate of the selection that must be acting in every sexual generation to restore adaptation. In this case, the intense selection caused by the toxic soil is sufficient to keep mine and pasture populations distinct.

At the downwind end of the site, the wind blows pollen and seeds from mine to pasture. The evolution of tolerance on the mine involves a cost of adaptation: the growth rate of tolerant plants in uncontaminated soil is less than that of non-tolerant plants, so that mine plants growing on the pasture are only 70–90% as fit as the resident individuals. There is thus rather strong selection against mine plants on the pasture (if this were not the case, of course, metal-tolerant plants would often be common even in uncontaminated pasture), although it is not nearly as strong as selection against pasture plants on the mine. Consequently, there is only a gradual decline in the frequency of tolerant adults, over a distance of about 200 m, as one passes from mine to pasture at the downwind end of the site. As before, seedlings are less well-adapted than adults, showing that local adaptation tends to be restored in every generation, but less effectively because selection is weaker and gene flow is stronger.

14.4 Diversifying selection maintains extensive variation.

The coloration of mice and moths are simple examples of how distinct kinds of individual can be maintained within a species because the direction of selection is different in distinct kinds of site. Snails such as *Cepaea* are much more variable in appearance than mice or moths. Their shells vary in color from yellow through pink to dark brown, and they may be plain or streaked with one or more dark bands. Color and pattern do not vary continuously, but can be arranged into a large but definite number of combinations—yellow banded, brown unbanded, and so forth. This is because color and pattern are determined by a few linked genes whose alleles are responsible for the variation we observe. Hence, we can estimate gene frequencies from a survey of phenotypes.

In a handful of snails there may be no two that look exactly the same, yet all belong to the same species. How can this variation persist when purifying selection or directional selection continually eliminates poorly adapted types?

14.4.1 Polymorphism in snails is maintained by selective predation.

Snails are hunted by predatory birds. Snails have many enemies. Their shell protects them from some, by making them too tough to chew and too big to swallow. Some birds, such as thrushes and blackbirds, have found how to overcome this defense: they carry the snail in their beak to a nearby large stone, the anvil stone, where they hammer it until the shell breaks, then consume the body. The birds are territorial: each patrols a certain parcel of land, and uses a particular stone within its territory to kill its snails. Biologists can make use of this habit to discover the diet of the birds, because the smashed remains of their victims are scattered around their customary stone. The birds are often one of the main sources of mortality for adult snails, so they might in principle be agents of selection that influence the composition of the snail population within their territory.

Visual predation favors cryptic snails. The selective effect of bird predation can be evaluated by comparing the shells found around the anvil stone to the local population hunted by the resident thrush. The birds usually find snails that are a poor match to their background. In sheep-grazed turf, yellow shells are inconspicuous whereas brown shells stand out and are easily detected. Against a background of sodden fall leaves the reverse is true: dark brown shells are hidden whereas yellow shells are easily seen. In the stripy background of sun and shade in the bottom of a ditch, banding breaks up the outline of the shell and makes it more difficult to see.

The selective effect of visual predation can be evaluated by comparing the frequencies of different kinds of shell discovered by the birds, at the anvil stone, with those discovered by biologists making a careful search. Indeed, the discrepancy between the general population of snails living in an area and the sample of snails taken by the birds can be used to calculate the intensity of local selection. In one small slough, for example, the population contained 47% of snails with effectively unbanded shells, which have no bands that can be seen by looking down on the animal from above, as a bird would see it. The shells scattered round the stones where the snails were broken by the birds, on the other hand, included 56% of unbanded individuals. The proportionate difference between these two frequencies is $(47 - 56)/47 = -0.19$. This is the selection coefficient acting against effectively unbanded types in the slough. It constitutes rather

strong selection: unbanded individuals are almost 20% more likely to be killed by the birds.

Local selection leads to regional diversity. The snail population at any particular site has some combination of patterns. Another site—even a nearby site, a few hundred meters away—may have a quite different combination. Similar kinds of site, however, tend to have similar populations of snails. Most of the snails living in the sheep-grazed pasture with a short pale covering of grass have yellow unbanded shells. In a nearby ditch with long coarse grass the shells are yellow but banded. The meadow backs on woodland, where most of the shells are brown and unbanded. The pattern of variation in mixed English countryside is illustrated in Figure 14.9. The differences in color and pattern between different kinds of habitat are largely created by visual predators that tend to choose the types that are most conspicuous at each site, leaving the less conspicuous types behind.

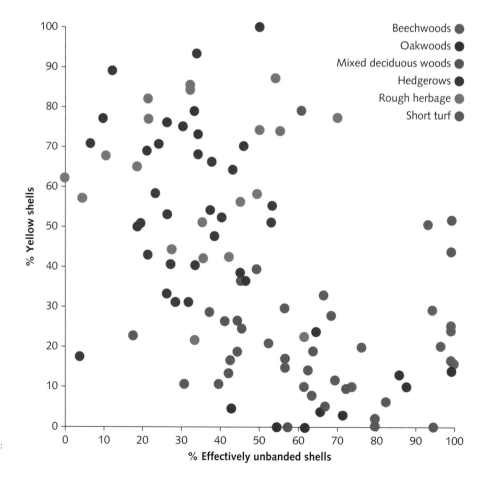

Figure 14.9 The distribution of shell color and pattern in relation to habitat among populations of the snail *Cepaea nemoralis* near Oxford.

Republished with permission of the Genetics Society of America, from Cain & Sheppard. Natural selection in *Cepaea*. *Genetics* 23: 99 © 1969.

The overall regional variation in appearance is thus composed of many different sites where there is a much more restricted range of variation.

14.4.2 Many agents of selection may act on a character.

Birds may kill many snails at some sites, but they are by no means the only sources of mortality. Individuals may be eaten by shrews, or killed by parasites, or starve to death, or freeze in winter, for example. All of these agents of mortality may also be agents of selection if their effect depends on the appearance of the shell.

Shell pattern varies with climate. For example, there is a broad tendency for the frequencies of pale types to increase towards the south in the European distribution of *Cepaea*. In Britain, Germany, and northern France, most populations have a fairly high frequency (often more than 50%) of brown shells. In southern France, Spain, Italy, and the Balkans, brown shells are much less frequent (often less than 10%) and populations consist mainly of yellow individuals. One explanation of this pattern is that snails with dark shells will heat up more rapidly, which will be advantageous in cool, cloudy regions but may result in heat stress in warmer countries.

Populations of *Cepaea* living in steep-sided valleys in Croatia vary in banding pattern, from fully-pigmented individuals with dark bands to paler individuals with faint bands. Populations on the floor of the valley are largely or entirely dark-banded, whereas those living high on the slopes have a high frequency of faint-banded individuals. This is probably also caused by climatic selection. The valleys act as frost hollows, with cold air flowing downslope to occupy the valley floor at night, creating a temperature difference of 15 °C or more between the floor and the upper slopes. The dark-banded individuals will warm up more quickly in the morning, and will be able to begin foraging earlier.

Shell pattern is also affected by chance and history. *Cepaea* also lives on downland in southern England. This is a landscape of low, rounded hills without a great deal of obvious ecological diversity. The pattern of variation is quite different from that found in a fragmented landscape of woodland and meadow. A particular type—brown unbanded, say—may be common over a large area, only to give way to a quite different type, such as yellow banded, over a short distance, without any evident ecological discontinuity. This patchy distribution on a fairly large scale of several square kilometers may have nothing to do with local selection. A new colony of snails might be founded by a few individuals that happen to be similar in coloration; this subsequently expands to occupy a small and relatively uniform region, whose population then necessarily resembles its ancestors. This gives rise to a patchy distribution of shell colors that has nothing to do with local selection, but is generated instead by chance and history.

The research program that developed around *Cepaea* has provided many insights into the evolution of natural populations. It has provided us with clear examples of adaptation driven by known agents of selection acting on clearly visible variation with a simple genetic basis. It also reminds us, however, that natural environments are much more complex and much less apparent than the laboratory, so natural populations are exposed to manifold agents of change.

14.5 Fluctuating selection changes in magnitude and direction over time.

Long-term surveys of Cepaea *show that selection changes in magnitude and direction over time.* One downland population of snails was followed for over 20 years. Each year, snails were captured and marked with a number scratched on the shell with a diamond point. A second sample was then taken to record the number of marked and unmarked individuals found. If all types have equal rates of mortality then all will have equal ratios of marked to unmarked individuals, so any discrepancy in this ratio is caused by differences in mortality.

The site was originally chosen to study visual predation, as it was occupied by thrushes which killed a large proportion of the snails. However, the thrushes themselves were killed after the first year of the survey by an exceptionally harsh winter, and never returned. For the next 20 years there was no overall trend in the composition of the population. However, the relative survival of the color types was not constant. Rather, it fluctuated irregularly from year to year. The lack of any long-term trend in the frequency of any particular type was not because every year was the same, but rather because every year was unpredictably different.

Stickleback populations shift on timescales of decades and millennia. Stickleback populations have also been studied over many years, with a similar result: the magnitude and direction of selection acting on features of the skeleton and spines often shifts from one generation to the next (Figure 14.10). The spines provide effective protection against gape-limited predators, such as trout and diving birds, which are reluctant to attack fully armored adult fish. They do not provide any protection against grappling predators such as dragonfly larvae, however, and might even make the fish more vulnerable, because such predators can cling on to the spines. Changes in the direction of selection may be associated with shifts in feeding behavior between foraging for zooplankton in the open water of the lake, when the fish are exposed to predation by birds, and foraging for worms and snails in the shallows, when they are attacked by dragonfly larvae. Any slight ecological

change from one year to the next (such as higher spring runoff, for example) may have effects that cascade through the lake community (for example, by increasing dissolved nutrients, leading to more phytoplankton and thus more zooplankton) and affect the pattern of selection acting on the stickleback population in unexpected ways.

One lake population of sticklebacks that lived in the Miocene was preserved in diatomite rock derived from lake sediments covering a total span of about 110,000 years. The fossils show most of the external morphology, including armor and spines, and can be dated precisely because of the seasonal pulses in sedimentation that produce distinctive annual layers that are preserved in the shales that the sediments eventually turn into. Two characters shifted radically and abruptly during this period, representing a shift from weakly armored forms with a single dorsal spine to strongly armored forms with up to three spines. This suggests a brief incursion of predatory fish or birds that produced temporary but powerful selection for enhanced body armor. Other characters, such as fin-ray number, fluctuated with lower amplitude and often showed slight trends over the entire period, or over shorter periods. Thus, morphology is continually shifting on all timescales: relatively short timescales of less than 10^4 years (body armor), intermediate timescales (body size), and long timescales of more than 10^5 years (fin rays).

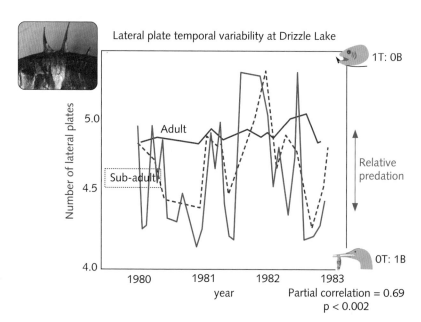

Figure 14.10 Variability of defensive armor in a population of sticklebacks. The variation in lateral plates is highly correlated with the balance of predation between trout (exclusively trout 1T:0B) and diving birds.(exclusively birds 0T:1B). Red indicates predation, blue indicates lateral plates (solid line, adult; dashed line, sub-adult)

Image courtesy of Dr. T. E. Reimchen.

14.6 Selection is commonplace and often strong.

In this chapter, we have described examples of natural selection involving familiar kinds of animals and plants living in an ordinary kind of place. There is probably somewhere like this within a few miles of where you are reading this book, and if you had the time you could yourself find out how selection works in a natural population. We have chosen these examples to emphasize that natural selection and adaptation are not unusual or exceptional processes that occur only in remote and exotic locations such as the Galapagos Islands. On the contrary: they are everyday processes that occur all the time, all around us, and are easily observed and studied.

Natural selection does occur in remote and exotic places, of course, and in fact some of the classical studies of selection have been carried out far from urban or agricultural centers. Some of them deserve a brief mention among more homely examples.

14.6.1 Natural selection is commonplace.

Mimicry in tropical insects is subject to strong selection acting through predators. Mimicry is a dramatic example of adaptation and has been studied since the earliest days of evolutionary biology. It is quite common among the insects of the meadow. Many wasps, hornets, and bumblebees, for example, are boldly patterned with black and yellow stripes. This makes them very conspicuous to potential predators such as birds or lizards, which avoid them because they can deliver a dangerous sting. Sharing the same color pattern thus provides a mutual benefit. On the other hand, some of the hover flies visiting flowers in the meadow also have conspicuous black and yellow stripes, although they are quite harmless. They are protected because they look like wasps and might be dangerous to attack.

The most spectacular examples of mimicry are found among tropical insects. South American butterflies in the genus *Heliconius*, for example, have conspicuous black and orange wings, and are unpalatable to birds because their bodies contain high levels of cyanide-containing compounds derived from their food plants. They are extremely variable, with many species having dozens of different color patterns. Each type is restricted to a particular region, however, and within that region only a single type is found. Even more strikingly, related species have similar galleries of color patterns, whose distributions correspond very closely.

Walking through the tropical forest in Peru, for example, you might observe that there are two species of *Heliconius*, *H. erato* and *H. melpomene*, which look very similar, both having a broad red or orange bar across the middle of the forewings and a white streak on the upper edge of the hindwings, the so-called "Postman" pattern. A few kilometers further on, both species are still common, but here all individuals have an orange patch at the root of the forewings, white markings in the middle of the forewings, and a series of narrow orange rays on the hindwings, forming the "Rayed" pattern. The reason for these abrupt, synchronized shifts is that a distinctive and conspicuous color pattern provides protection only when it is very common, so that local predators come to recognize it as a reliable signal of unpalatability. A rare pattern might not be recognized, so the insect would be likely to be attacked no matter how unpalatable it really was. Hence, species will come to resemble one another, and when different patterns of warning coloration evolve in different localities, the boundaries between them will become very sharply defined, because in any mixed site the most common type will have an advantage.

Natural selection acting on mimics can be studied by measuring the survival of individuals whose appearance has been altered, which shows that reducing mimetic resemblance by painting over color spots increases mortality. Alternatively, relative fitness can be measured by releasing individuals bearing some inconspicuous mark that enables the experimenter to recognize them in subsequent samples.

An experiment like this was carried out in northern Peru, in a part of the forest where there was a sharp boundary between an area dominated by Postman and an area dominated by Rayed. The results are shown in Figure 14.11. Mixtures of marked Postman and Rayed individuals were released at sites where the butterflies usually feed. The butterflies are thought

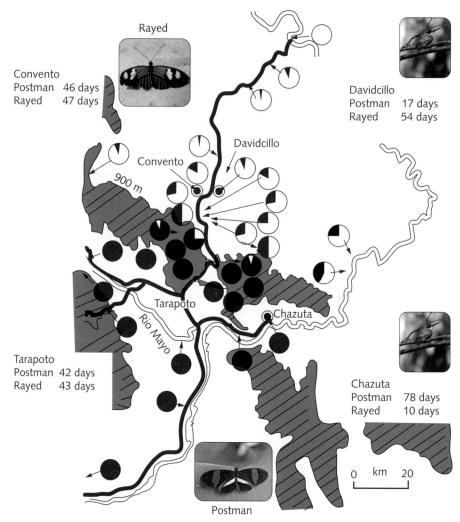

Figure 14.11 Survival of color morphs of *Heliconius* butterflies in relation to bird predation at experimental field sites in San Martin, Peru.

Reproduced from Mallet, J. and Barton, N.H. 1989. Strong natural selection in a warning-color hybrid zone. *Evolution* 43 (2): 421–431, with permission from Wiley.

to be attacked by jacamars, insect-eating birds that are common in many parts of the forest, although it is difficult to observe bird behavior directly in the field. At two release sites where there were few if any jacamars there was no detectable difference between the survivals of the two types. Jacamars were common at the two other sites. At one site, the resident populations of *Heliconius* were almost entirely Postman. Here, most of the recaptured marked individuals were Postman, showing that Postman has a higher rate of survival than Rayed. At the other site, the opposite happened: the resident population was Rayed and Rayed individuals survived better than Postman. Moreover, the differences in survival at both sites were considerable, showing that the boundary between the two types was maintained by

strong selection against the non-resident type, probably through visual predation by birds.

Beak form in Darwin's finches is modified by fluctuating selection acting through diet. The best-known example of an historical process of selection driven by a known selective agent is the change of beak shape in the large ground finch (Darwin's finch) *Geospiza fortis* on the island of Daphne Major in the Galapagos Islands (Figure 14.12). A prolonged drought in 1976–77 caused a change in the composition of the vegetation by favoring plants with large, tough-shelled seeds. These could be consumed only by finches with unusually large and powerful beaks, and between 1976 and 1978 beak depth increased. Heavy rain in 1983 reversed the trend in the vegetation by favoring

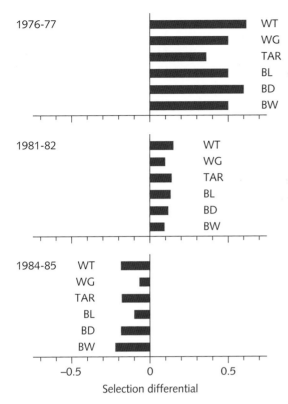

1976-77

WT
WG
TAR
BL
BD
BW

1981-82

WT
WG
TAR
BL
BD
BW

1984-85

WT
WG
TAR
BL
BD
BW

−0.5   0   0.5

Selection differential

Figure 14.12 Fluctuating natural selection of morphology in a population of Darwin's finch *Geospiza fortis*. Characters measured are: WT body weight; WG wing length; TAR tarsus length; BL beak length; BD beak depth; BW beak width.

Reproduced with permission from Gibbs, H.L. and Grant, P. R. (1987) 'Oscillating selection on Darwin's finches', *Nature* 327, 511 - 513.

plants with smaller, softer seeds that germinated more readily and thereby favored birds with smaller beaks that were more adept at processing them. Within a few years the response to reversed selection had more or less restored the status quo.

This study has become a classic example of selection in a pristine environment, the thoroughness of the fieldwork being buttressed by detailed knowledge of the ecology of the populations and the genetics of beak shape (the romantic location may also help). It is particularly noteworthy that selection is episodic and fluctuating, so that an opportune study would reveal strong natural selection, as indeed it did, whereas less fortunately scheduled surveys in (say) 1972 and 1992, however carefully executed, would have shown little if any change.

We may learn another lesson from these events, because the underlying genetic basis of changes in beak shape are now beginning to be understood.

Beak shape is modulated by *Bmp4*, whose product is a bone morphogen, which is strongly expressed early in the development of *Geospiza* species with deep beaks but not in those with long thin beaks. Thus, selection on this quantitative character may act primarily through alleles of a single gene to produce adaptation.

Mutations in the hemoglobin gene are selected because they confer resistance to malaria. Natural selection acts on all kinds of organisms, including humans. In many tropical regions, infectious diseases are a major cause of mortality. Malaria is one of the worst killers, causing about two million premature deaths every year. Most of these are caused by *Plasmodium falciparum*, the most malignant species of malaria parasite. The malaria parasite enters the bloodstream by a bite from an infected mosquito and destroys red blood cells, leading to a severe anemia. Several kinds of mutation confer some degree of resistance to malaria, including point mutations in the structural gene encoding hemoglobin that cause single amino acid changes in the molecule. The best known of these is HbS, in which the alteration of a single nucleotide (A to T) in the β-globin gene causes the seventh amino acid of the protein to be changed from glutamic acid to valine. This alteration makes it more difficult for the malaria parasite to enter the red blood cell. Consequently, individuals who carry one copy of the HbS mutation are more likely to survive infection by falciparum malaria. This creates strong selection for HbS in regions where falciparum malaria is endemic. As a result, the geographical distribution of the *HbS* allele coincides almost perfectly with the regions where falciparum malaria occurs.

Unfortunately, the allele has another effect: in homozygotes, the red blood cells are distorted into a curved, sickle-like shape (Section 6.1.1). These may jam in capillaries and obstruct the blood supply to organs, resulting in a potentially fatal anemia. Hence, there are two agents of selection acting on the *HbS* allele. It has a beneficial effect by protecting heterozygotes against severe malaria, whereas it is deleterious by causing anemia in homozygotes. Consequently, it will increase in fitness when it is rare, because almost all the individuals carrying it will be heterozygotes, but its spread will be curtailed when it becomes common, because mating between heterozygous parents will produce homozygous offspring.

Hundreds of examples of natural selection have been documented. We have given some examples of selection in natural populations, emphasizing evolutionary processes that have been described in the sort of places that are likely to be familiar to you, but also mentioning some less familiar situations. The examples that we have used are, of course, among the classical demonstrations of natural selection. But they are by no means exceptional. There are literally hundreds of studies that have documented the operation of selection in natural populations. There are so many that we can use them to make statistical generalizations about the frequency and intensity of natural selection.

14.6.2 Natural selection is often strong.

Studies of genetic polymorphism show that natural selection is commonplace and often strong. The work on *Cepaea* led to an explosion in studies of selection in natural populations. Until then, it had been assumed that selection would be so weak, even if it could be measured at all, that a field research program would be futile. After the demonstration that selection could be studied and measured like any other natural process, researchers flocked to the field.

The different colors and patterns of shells in *Cepaea* are an example of genetic polymorphism. Snail shells fall into a fairly small number of combinations of ground color and banding: yellow versus brown, banded versus unbanded, and so forth. Many similar examples in which a population consists of several discrete kinds of individual were used to measure relative fitness. The results were astonishing: studies of polymorphism where a selection coefficient is estimated for each discrete type yielded mean values of the selection coefficient $s = 0.33$ for undisturbed situations and $s = 0.30$ for situations that involved recent human disturbance.

The simplest way to summarize this result is that selection is seldom weak; about three-quarters of cases involve $s > 0.1$. For several reasons, this conclusion cannot be taken quite at face value. In the first place, many studies measured only one component of fitness, such as mortality, and if equally strong but countervailing selection acted on other components then net selection might be small. Secondly, an episode of strong selection might contribute

only a small fraction of overall variation in survival or fecundity and would then have little genetic effect. Finally, there may be a bias towards selecting, analyzing and publishing cases that are interesting because they are likely to involve strong selection. All these reservations being admitted, however, there is no doubt that field studies of genetic polymorphism have convincingly demonstrated that natural selection is easily detected in natural populations and is often found to be strong.

Studies of continuous variation likewise show that natural selection is commonplace and often strong. Much of the variation in natural populations does not fall into sharply defined categories; instead, it is continuously distributed (like body size) or nearly continuously distributed (like the number of vertebrae). This called for a different approach. Suppose that we plot the fitness of individuals expressing a given character as a function of character value. If the plot is flat with zero slope, individuals have the same fitness regardless of the value of the character. In this case, selection will have no effect on the mean value of the character. On the other hand, if individuals expressing different values of the character have different fitness, selection will tend to cause a change in the mean value of the character. Hence, the regression of fitness on character state (the "selection gradient") can be used to estimate the strength of selection on the character. The selection gradient for a single character is equivalent to the selection differential (Section 12.1.1, Figure 12.1).

For example, this approach was used to describe selection on a range of characters in *Impatiens pallida* by studying the fate of groups of seedlings emerging in a small area (about 0.1 ha) of forest floor. The first step was to measure a suite of characters and calculate their relation with viability, the first episode of selection. Since there were 24 groups of seedlings there were $7 \times 24 = 168$ opportunities for selection to occur, in 36 of which selection was detected in the form of a significant selection gradient. The surviving plants then flowered, and fecundity as a second component of fitness was then measured, yielding significant selection gradients in 47 cases. Thus, a random character in a random plot experienced selection through

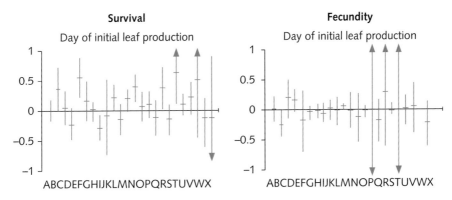

Figure 14.13 Local selection in *Impatiens*. The vertical axis is the slope of character value (day of initial leaf production) on fitness component (survival or fecundity), measured at 24 random locations (A–X) within a 30 m × 40 m study site.

Reproduced from Stewart, S.C. and Schoen, D.J. 1987. Pattern of phenotypic viability and fecundity selection in a natural population of *Impatiens pallida*. *Evolution* 41 (6): 1290–1301, with permission from Wiley.

differential viability in 21% of cases, and selection through differential fecundity in an additional 24% of cases.

In general, selection favored large plants with many leaves, although selection differed in strength and even in direction between sites separated by a few meters. This appeared to be caused by consistent differences in the conditions at nearby microsites: for example, seedlings that produced leaves later survived better in sunny sites, perhaps because their more precocious neighbors succumbed to water stress. This very detailed study, illustrated for one of the characters that were measured in Figure 14.13, showed that selection was frequent and often strong, but also varied substantially between sites within dispersal range.

For about a thousand similar cases involving many kinds of characters and organisms about a quarter of all estimates of the selection gradient were formally significant, with a mean value over all cases of 0.22. The same reservations—sampling bias, and so forth—apply to these studies as they do to studies of genetic polymorphism. Nevertheless, it is impossible to reconcile the dozens of studies reporting strong selection with the view that selection in natural populations is weak or infrequent. The very different data sets for selection coefficients acting on discrete characters and for selection gradients or selection intensities acting on continuous characters seem to agree quite well, and it seems to me inescapable that selection in open populations is commonplace and often rather strong.

14.6.3 Adaptation is often oligogenic.

Adaptation always proceeds through the spread of genotypes that have arisen through mutation or recombination. It might involve hundreds of loci where alleles each with a very small effect on character value are shifting in frequency; at the other extreme, it might involve one or two loci where alleles of major effect are segregating. Naturally, any intermediate between these two extremes is possible—and may often happen. Nevertheless, the second alternative seems often to be closer to the truth, at least in the early stages of adaptation. We have already seen this in previous chapters. The first steps in experimental evolution are often attributable to beneficial mutations of large effect. The fundamental changes responsible for the production of modern crop plants by artificial selection are caused by major genes. Even artificial selection on continuous characters is strongly affected by single genes of large effect. There is a good reason for this: genes with large effect are less likely to be lost by chance early in adaptation, and will thereby contribute disproportionately to the genetic make-up of the derived population.

Likewise, adaptation in natural populations is often, if not usually, based on the selection of alleles at one or a few loci. We have described many examples of this.

- The gracile types of stickleback that are successful in streams descend from marine ancestors by mutations in *Eda* and *Pitx1* genes.

- Industrial melanism is often attributable to dominant alleles at a single locus.

- Pale pelage in beach populations of mice is caused by an interaction between structural changes at the *Mc1r* locus and overexpression at the *Agouti* locus.

- Shell color pattern in *Cepaea* is governed primarily by a small number of linked genes.

- Mimic color patterns in butterflies are likewise attributable to a few tightly linked genes.

- Beak shape in Darwin's finches is strongly influenced by a bone morphogen gene.

- Hemoglobins resistant to malaria are encoded by point mutations in the structural gene.

This is very good news. It means that we shall often be able to work out precisely the genetic and physiological basis of adaptation by identifying the small set of genes that are largely responsible for it. It provides us with a basis for developing a predictive theory of adaptation that goes beyond phenotypes to the underlying genetic changes that are responding to selection.

This account of adaptation will not always be true. As adaptation proceeds, the supply of beneficial mutations of large effect will dwindle, and further progress can be made only through the substitution of a broader range of mutations of gradually diminishing effect. Before these smaller-effect mutations dominate the genetic architecture of adaptation, however, the environment may have changed again.

14.6.4 Strong selection causes rapid change.

Adaptation is the change in character state caused by natural selection in response to a change in the conditions of life. The simplest way of expressing the rate of evolution is to calculate the change from the initial state to the final state over a given period of time. This should be independent of the units in which the character is measured, so we use the ratio of final to initial state. This is best written in terms of logarithmic values, because of the exponential spread of favored types. Hence, the rate of change is (ln final state − ln initial state)/time, where ln denotes the natural (base e) logarithm. This expresses the proportionate change per unit time and is conventionally expressed per million years. The unit of change on this scale is the darwin (Dar), indicating an e-fold (e ≈ 2.72) change over 1 My.

In contemporary samples it is often more appropriate to use the kilodarwin (kDar) as the proportionate change per thousand years. A rate of 1 kDar thus corresponds to somewhat more than a doubling of character value in 1000 years.

The rate of evolutionary change can be measured in contemporary populations. The shore crab *Carcinas maenas* is a European species that was introduced accidentally to the eastern coast of North America in the nineteenth century and subsequently spread northwards from New England to Nova Scotia. It feeds on marine snails such as *Littorina* by crushing the shell and extracting the animal, a marine analogue of thrushes and *Cepaea*. How a crab handles a snail depends on the shape of the snail, because the adult shell preserves its juvenile whorls. In squat, low-spired shells the thin-walled juvenile whorls are protected by the thicker adult whorl, whereas the juvenile whorls remain exposed and vulnerable in tall, high-spired shells.

We can compare museum specimens collected before invasion by *Carcinas* with contemporary material from the same localities, which shows that there has been a trend from high-spired shells in the 1870s to low-spired shells in the 1980s, as illustrated in Figure 14.14, consistent with elevated rates of predation by crabs. By expressing shell shape geometrically, the rate of evolution can be calculated as 8.2 kDar. This is probably an under-estimate. Museum collections dating from 1915 suggest that most of this change may have taken place in the first 40 years after the appearance of *Carcinas*, in which case the time-averaged value is much less than the rate experienced by the population during the period over which most selection occurred.

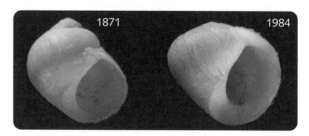

Figure 14.14 Rapid evolution of a marine snail caused by a new predator. These are two representative specimens of *Littorina* from Maine collected about a century apart.

Image kindly supplied by Robin Hadlock Seeley, Ph.D. Cornell University.

Evolution often happens rapidly on short timescales. The rate of change that is caused by natural selection acting on contemporary populations can be calculated for sticklebacks, melanic moths, Darwin's finches, marine snails, and many other examples. A survey of the literature has shown that we now have about 2000 estimates of the rate of evolution in natural populations. The median value is about 1 kDar. This cannot be taken quite at face value, because examples of rapid change are more likely to be reported than situations in which little change could be observed. Nevertheless, it has been very firmly established that rapid evolutionary change is by no means unusual. Natural populations are often exposed to strong selective agents and are rapidly changed as a result.

The magnitude and direction of selection frequently change over time. The rate of change in character value can be calculated equally well in living organisms or in fossils. When the two are put together, a striking and unexpected generalization emerges: the amount of change is almost independent of the length of time over which observations are made. For example, suppose that you were measuring the shapes of marine snails over a hundred years, or ten thousand years, or a million years. The amount of change you would record would be about the same in any case. Over a hundred years, this represents very rapid change; over a million years, very slow change. How can we reconcile these results? The answer is that selection often changes both in magnitude and in direction from generation to generation. Consequently, average character values change little more over millennia than they do in the course of a few years.

We can summarize briefly what we have learned of natural selection by studying populations in the field. Selection is commonplace and often strong; adaptation is rapid and often oligogenic; the magnitude and direction of selection often change over time. Selection can be studied like any other natural process, by measuring how populations evolve in response to stress, whether caused by nature or by human activities. The meadow, the forest, and the stream are not fixed unchanging backgrounds; instead, they are the crucibles of evolution. The animals and plants that live in them are not fixed unchanging populations, but rather adapt continuously as their conditions of life change.

● CHAPTER SUMMARY

Purifying natural selection maintains adaptation.

- Purifying selection acts contrary to mutation and immigration.
- The intensity of purifying selection is equal to the genetic variation of fitness.
- Adaptedness is continually degraded and continually restored.
  - *Purifying selection increases fitness by a few percent per generation in sedges and birds.*

Directional natural selection drives adaptation to changing conditions.

- New ecological opportunities provide the opportunity for new types to evolve.
  - *Freshwater conditions favor reduced armor in sticklebacks.*
  - *Adaptation to freshwater is repeatable and predictable.*
  - *Adaptation may occur through loss-of-function mutations.*
- Novel environmental stresses lead to rapid evolution.
  - *Insect populations evolve in polluted cities.*
  - *The selection caused by novel stresses can be very strong.*
  - *Selection is reversed when the stress is alleviated.*

Local selection leads to divergent specialization.

- Plants adapt to local soil conditions.
  - *Local adaptation can occur over very short distances.*
  - *Local adaptation can be demonstrated by long-term field experiments.*

- Field mice adapt to living on beaches.
  - *Mice become locally adapted to their background.*
  - *Adaptation is based on beneficial mutations of large effect.*
  - *Populations at different sites become independently adapted.*

- Plants adapt to polluted soils.
  - *Grasses on old mine workings are resistant to heavy metals.*
  - *Adaptation may occur through gain-of-function mutations.*
  - *Adaptation to polluted sites can occur very rapidly.*
  - *Adaptation occurs in some species but not in others.*

- Local adaptation is balanced by selection and immigration.
  - *Local adaptation is degraded by immigration.*
  - *Distinctive local adaptation requires intense local selection.*

Diversifying selection maintains extensive variation.

- Polymorphism in snails is maintained by selective predation.
  - *Snails are hunted by predatory birds.*
  - *Visual predation favors cryptic snails.*
  - *Local selection leads to regional diversity.*

- Many agents of selection may act on a character.
  - *Shell pattern varies with climate.*
  - *Shell pattern is also affected by chance and history.*

Fluctuating selection changes in magnitude and direction over time.

  - *Long-term surveys of* Cepaea *show that selection changes in magnitude and direction over time.*
  - *Stickleback populations shift on timescales of decades and millennia.*

Selection is commonplace and often strong.

- Natural selection is commonplace.
  - *Mimicry in tropical insects is subject to strong selection acting through predators.*
  - *Beak form in Darwin's finches is modified by fluctuating selection acting through diet.*
  - *Mutations in the hemoglobin gene are selected because they confer resistance to malaria.*
  - *Hundreds of examples of natural selection have been documented.*

- Natural selection is often strong.
  - *Studies of genetic polymorphism show that natural selection is commonplace and often strong.*
  - *Studies of continuous variation likewise show that natural selection is commonplace and often strong.*

- Adaptation is often oligogenic.

- Strong selection causes rapid change.
  - *Adaptation is the change in character state caused by natural selection in response to a change in the conditions of life.*

- *The rate of evolutionary change can be measured in contemporary populations.*
- *Evolution often happens rapidly on short timescales.*
- *The magnitude and direction of selection frequently change over time.*

● FURTHER READING

Here are some pertinent further readings at the time of going to press. For relevant readings that have been released since publication, visit the book's Online Resource Centre at
www.oxfordtextbooks.co.uk/orc/bell_evolution/

Section 14.1   Burt, A. 1995. The evolution of fitness. *Evolution* 49: 1–8.

Section 14.2   Shapiro, M.D., Marks, M.E. and Peichel, C.L. 2004. Genetic and developmental basis of evolutionary pelvic reduction in three-spine sticklebacks. *Nature* 428: 717–723.

Bishop, J.A. 1972. An experimental study of the cline of industrial melanism in *Biston betularia* (L.) (Lepidoptera) between urban Liverpool and rural North Wales. *Journal of Animal Ecology* 41: 209–243.

Section 14.3   Steiner, C.C., Rompler, H., Boettger, L.M., Schoneberg, T. and Hoekstra, H.E. 2009. The genetic basis of phenotypic convergence in beach mice: similar pigment patterns but different genes. *Molecular Biology and Evolution* 26: 35–45.

Bradshaw, A.D. 1991. Genostasis and the limits to evolution. *Philosophical Transactions of the Royal Society of London B* 333: 289–305.

Section 14.4   Jones, J.S., Leith, B. and Rawlings, P. 1977. Polymorphism in *Cepaea*: A problem with too many solutions? *Annual Review of Ecology and Systematics* 8: 109–143.

Section 14.5   Bell, G. 2010. Fluctuating selection: the perpetual renewal of adaptation in variable environments. *Philosophical Transactions of the Royal Society of London B* 365: 87–97.

Section 14.6   Joron, M. and Mallet, J.L.B. 1998. Diversity in mimicry: paradox or paradigm? *Trends in Ecology and Evolution* 13: 461–466.

Grant, P.R. and Grant, R.B. 2002. Unpredictable evolution in a 30-year study of Darwin's finches. *Science* 296: 707–711.

Williams, T.N. 2006. Human red blood cell polymorphisms and malaria. *Current Opinion in Microbiology* 9: 388–394.

● QUESTIONS

1. Distinguish between purifying selection and directional selection. What are the main causes of these processes?

2. Design a field experiment to measure the strength of purifying selection in a particular species of organism, such as a fish or a fern. What would you expect to find?

3. Describe an example of directional selection in a natural population. Can you identify cases of directional selection among organisms living near you?

4. Describe an example of local selection. Draw up plans to investigate local selection in a particular species of organism with respect to different kinds of habitat in which it is found, or with respect to alteration of its habitat by human activities.

5. Discuss how local adaptation depends on the balance between local selection and immigration. How might this balance contribute to limiting the range of a species?

6. Write an essay on the causes of variation in shell color and pattern in the polymorphic land snail *Cepaea*.

7. Draw up a catalogue of the agents of selection responsible for instances of rapid evolution in contemporary or near-contemporary populations, and describe how they have produced change. Do they have any features in common?

8. "Natural selection is commonplace and often strong." Evaluate the evidence for this statement and discuss how it might be affected by artefacts.

9. What is the evidence that adaptation is often caused by a few mutations of large effect, and in what circumstances would you expect this generalization to break down?

10. What is the evidence that the direction and intensity of selection often changes on short timescales? Be careful to take into account artefacts that might lead you erroneously to this conclusion.

11. How do studies of contemporary populations contribute to our understanding of long-term evolutionary processes?

12. Survey the collections in your local natural history museum. How might you use them to document evolutionary change?

You can find a fuller set of questions, which will be refreshed during the life of this edition, in the book's Online Resource Centre at **www.oxfordtextbooks.co.uk/orc/bell_evolution/**

PART 6
Interaction

Adaptation to some feature of the environment evolves through natural selection acting on variation among individuals. The situation becomes more complex when this feature consists of other individuals, or is modified by them. The operation of selection and the outcome of evolution then depend, not only on the attributes of individuals acting in isolation, but also on how these individuals interact with one another. There are two kinds of interaction: sexual and social. Sexual interactions govern mating success and lead to sexual selection. The outcome of sexual selection may be idiosyncratic or even bizarre features of bodies, often differing characteristically from the outcome of natural selection. Social interactions govern the success of individuals within groups, or the success of the groups themselves. Natural selection in a social context may lead to unexpected outcomes such as cooperation and altruism. When the social context involves individuals of other species, either mutualism or antagonism may evolve. This part of the book is about how the challenges and opportunities provided by neighbors modulate the operation of selection and thereby direct the evolution of sexual and social behavior.

Sexual Selection

We are so used to the sexual nature of animals and plants that we often fail to realize how odd it is. Like all other mammals, we are each male or female, and both must participate to produce offspring. Many other species, however, regularly reproduce with no appearance of either sex or gender. Sex is associated with reproduction in species like our own, but it is certainly not necessary for it. So why does it exist?

Gender is just as odd. The sticklebacks in the pond show how distinct freshwater types will evolve from marine ancestors, and furthermore how divergent natural selection will produce specialized limnetic and benthic phenotypes (Section 7.3.4). This is straightforward: but natural selection does not explain why, in most populations, the breeding males have vivid red throats. Why should they become so conspicuous to predators when drab coloration would protect them better?

The common touch-me-not *Impatiens*, a familiar annual plant growing in wet places in the forest, has been used to study natural selection in the field (Section 14.1.3). It is named for its ripe fruits, which explode when touched to scatter the seeds far and wide. How are these fruits formed? Most come from tiny inconspicuous green flowers which are self-pollinated. Others come from the very conspicuous large orange flowers that attract visits from bees and humming birds. How do large expensive flowers evolve if small cheap flowers would do the job as well?

The red throats of male sticklebacks and the orange flowers of touch-me-not are two examples of sexual ornaments. It would be difficult to explain how they evolve through natural selection because they are often either costly or risky or both. In fact they have not evolved through natural selection, but instead through the parallel process of sexual selection. That is the subject of this chapter.

15.1 The eukaryote life cycle is an alternation between vegetative and sexual cycles.

The simplest life cycle involves only growth and reproduction. An individual grows when it assimilates food in excess of its basal metabolic requirements, and reproduces when its size exceeds a certain limit. It may then produce another individual of nearly equal size from its accumulated reserves of tissue, or many smaller individuals, or somewhere between these extremes. Bacteria cycle endlessly between growth and reproduction (by binary fission), and have done so since they first evolved some 3500 Mya. Eukaryotes also grow and reproduce, like unicellular algae or hydras or water-fleas. In most eukaryotes, however, this simple vegetative cycle is interrupted sooner or later by a different kind of event. This involves a process during which two cells (never more) fuse and mingle their genomes, after which they divide in such a way that each daughter cell receives a mixture of the parental genomes. This is the sexual cycle.

15.1.1 The sexual cycle is a coupled process of sexual fusion and sexual restitution.

The eukaryote life cycle alternates between spore and gamete. In unicellular eukaryotes such as algae or yeasts there is a clear distinction between vegetative and sexual processes, as sketched in Figure 15.1. The

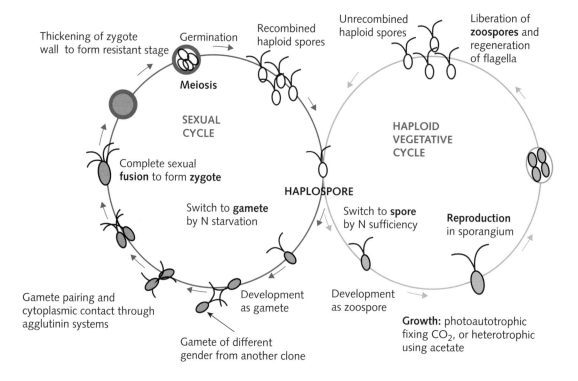

Figure 15.1 The life cycle of a unicellular green alga like *Chlamydomonas* consists of a sexual cycle and a vegetative cycle.

first is the growth of a cell, culminating in reproduction by division into two or more offspring cells, which then repeat the cycle through growth. This is the vegetative cycle, initiated by a cell specialized for growth, the spore. The second is the fusion of two cells, together with their genomes, to form a single cell, which subsequently divides into two or more cells with the original genomic complement. This is the sexual cycle, initiated by a cell specialized for fusion, the gamete.

The life cycle of animals and plants is more complicated because the growth and reproduction of the spore gives rise to a multicellular individual, within which gametes are formed in the germ line. Nevertheless the fundamental distinction remains between the spore as a vegetative cell that is specialized for growth and reproduction and the gamete as a sexual cell that is specialized for fusion.

Competition between spores and between gametes gives rise to different processes of selection. The spore, or the multicellular organism that the spore develops into, competes with others for resources in order to grow and reproduce. This leads to natural selection, and thereby to adaptation and adaptive radiation. This is the basis of almost all the examples of evolutionary change we have described so far. Gametes neither grow nor reproduce, and so they do not compete for resources. Instead, they fuse, and therefore compete for fusion partners. This leads to sexual selection through the spread of alleles which enhance the ability of gametes to fuse. Natural selection and sexual selection are fundamentally different kinds of process. Natural selection leads to forms adapted to the physical environment, such as the streamlined shape of large fast marine predators. Sexual selection leads to forms adapted to fuse with a partner; but it takes two to tango, and the partner may be differently inclined. The sexual cycle necessarily involves sexual interactions between individuals, and thereby generates novel and unexpected evolutionary outcomes.

15.1.2 Sex is primitive in eukaryotes.

Sex is such a complex process that it would not be surprising to find that it occurred only in a few unusual, highly derived lineages. On the contrary: sex is primitive and universally distributed, whereas the complete absence of sex is exceptional.

Sex is a shared derived character of eukaryotes. Sex is found in all the major groups of eukaryotes: animals and plants, ciliates and dinoflagellates, brown and red seaweeds all have sexual cycles. It is extremely unlikely that so complex a character should have evolved independently in all these lineages. Moreover, the sexual cycle of complete cell fusion followed by genomic restitution is never found in other organisms. Hence, we can be confident that the most recent common ancestor of modern eukaryotes was a sexual organism.

Most asexual lineages are recently derived from sexual ancestors. Some eukaryotes go through many generations without the intervention of any sexual process. Indeed, all unicellular eukaryotes must do so, because sex does not provide growth, so that a lineage of ciliates or yeasts may be propagated by simple division for hundreds or even thousands of generations. Multicellular eukaryotes such as lilies, duckweed or water fleas may also dispense with sex for long periods of time. They nevertheless maintain the potential for sexuality, and sooner or later each lineage passes through a sexual cycle. It is only in a few groups that sex has been lost entirely. There are completely asexual lizards, fish, and weevils, for example. All of these are groups consisting of only a few species, whose relatives are all sexual. It follows that they are groups descending from sexual ancestors in which sex has been lost. Moreover, they must be short-lived groups, or else they would be in the majority. Hence, the loss of sexuality must be associated with an increase in the risk of extinction.

Very few major groups of animals or plants are completely asexual. To find a large clade of exclusively asexual multicellular eukaryotes requires a microscope. There is a characteristic community of organisms that grows in fugitive waters: tiny rainwater pools on rocks or roofs, for example, or the water held in bromeliads and pitcher plants, or the very thin layer of water caught on the surface of mosses. It includes unicellular organisms such as ciliates, besides very small metazoans such as nematodes, tardigrades, gastrotrichs, and rotifers. These animals share several features that have evolved in response to their ephemeral habitats. Many can shrink, dry up, and cease all vital processes when the water evaporates, and yet continue to live; their mummy-like carcasses can survive for decades, then recover and resume active life within hours after water is restored. Most are self-fertilizing, or can produce eggs asexually, in order to maximize the rate of lineage expansion on the rare occasions when conditions are favorable.

One major group has no known sexual process whatsoever: no gametes, no males, and no meiosis. This is the group of bdelloid rotifers, tiny animals consisting of a few thousand cells that are immediately recognizable under the microscope by their looping, leech-like locomotion (see Section 7.1.1). You can find them by squeezing a pinch of sphagnum moss into a watch glass, or by steeping a few strands of dry hay in water for a day or two. There are several hundred species of bdelloids, but microscopists have sought in vain for two centuries to find any evidence of males, gametes, or sex (although some very recent genetic results suggest that some cryptic sexual process may nevertheless occur). They are the exception that proves the rule: every other major group of animals and plants contains sexual species.

15.2 Sex modulates the dynamics of evolution.

In many animals, reproduction is invariably associated with sex. This includes (for example) the great majority of insects, mollusks, echinoderms, fish, and mammals—such as us. Nevertheless, even in mammals, reproduction is not necessarily associated with sex. Dolly the

sheep proved that. It follows that there is some force that resists the replacement of sexual by asexual lineages. This force is not physiological—Dolly again—and must therefore be evolutionary. It is not straightforward to understand what it might be.

15.2.1 Sex is costly.

In an asexual population all individuals reproduce. Each produces, say, ten offspring. A population of 100 individuals will then produce 1000 offspring. Now consider a precisely comparable sexual population that is equivalent in all respects except that offspring develop only from a fertilized egg. This population comprises 50 females and 50 males. The 50 females will each produce ten offspring, and the total production will be therefore 500 offspring, or one-half of the production of the asexual population. The shortfall can be expressed in two ways.

- The cost of meiosis. Each partner, male or female, contributes only half their diploid set of genes in gametes to the zygote.
- The cost of males. Only the female half of the population is productive, the male half contributing nothing beyond their genes.

Both express the same argument: in a mixed population of sexual and asexual lineages, the asexual lineages will expand at twice the rate of the sexual lineages. Hence, the most basic Darwinian calculus leads us to the conclusion that sexual lineages will everywhere be replaced rapidly by ecologically equivalent asexual lineages.

This conclusion admittedly depends on two hidden assumptions. The first is that male gametes are much smaller than female gametes, so that the male parent makes only a negligible cytoplasmic contribution to the zygote. The second is that offspring are reared exclusively by mothers, so fathers make only a negligible physical contribution to the development of their offspring. These assumptions often fail: fusing gametes may be similar in size, and males may provide parental care. In the great majority of animals they hold: the female makes by far the greater cytoplasmic contribution to the zygote and provision to the developing offspring. In most cases, a sexual population should be rapidly replaced by an ecologically similar asexual population. This has not happened. There is therefore some countervailing process that restricts the proliferation of asexual lineages or reinforces the proliferation of sexual lineages. What is it?

15.2.2 Sexual progeny differ from their parents and among themselves.

In an asexual population lineages proliferate independently of one another. The population evolves through competition between lineages, some of which will spread at the expense of others. A sexual population is quite different because lineages are perpetually combined and recombined. They are combined by gamete fusion, when two haploid genomes fuse to form a single compound diploid genome. They are recombined by meiosis, when new haploid genomes are formed through crossing-over as mosaics of the haploid genomes contributed by the two parents.

There are two special cases for which sex has no genetic effect. First, the two fusing gametes might be genetically identical, in which case the products of meiotic recombination will also be identical with one another and with the two parental gametes. Secondly, there might be no recombination during meiosis, in which case the only products of meiosis will be the two unchanged gametic genomes. In either case, the outcome is the same as that of asexual reproduction through spores. Both of these cases occur in some organisms, in fact, but they are rare. In most organisms the fusing gametes are different and genetic recombination takes place during meiosis. The outcome is then very different from an asexual process. The branching lineages of an asexual population disappear, replaced by a network of genetic exchange. Moreover, the discrete and limited genetic repertoire of an asexual population is replaced by an almost endless diversity of combinations of alleles.

Sexual sibs differ among themselves. Offspring that are produced asexually are genetically identical, bar a very few novel mutations. By contrast, all of the offspring produced sexually by two unrelated parents are different, because each bears a different combination of parental alleles. They still resemble one another; everyone is more similar to their brothers and sisters than they are to a random stranger. Nevertheless, a sexually produced brood has a great deal of genetic variation that is almost completely absent from a clonal brood.

Sexual sibs differ from their parents. Sexually produced offspring vary because each bears a genome in which elements from both parental genomes are mingled. None will bear the same combination of alleles as either of their parents. Hence, any individual sexual offspring will be different from both its parents.

There are a few elements of the genome which escape genetic mingling. In most cases the mitochondria of the zygote, with their own separate but small genome, are contributed by the female parent alone. In species like ours, the Y-chromosome is transmitted by the male parent alone. The mitochondrial genome and the Y-chromosome are unusual examples of the asexual transfer of alleles within sexual lineages. For the great majority of genes, however, the fundamental genetic consequence of sex can be summarized very briefly: sex diversifies.

15.2.3 Sex facilitates adaptation.

Sex diversifies because it releases the potential variation that lies hidden in asexual populations. We have already described how the rate at which a population adapts to new and stressful conditions of growth is proportional to its level of genetic variation. Putting these together, sex is maintained in populations because it facilitates adaptation by providing a continual source of new genetic variation on which selection can act.

Sex allows beneficial mutations that arise independently to be combined in the same lineage. To understand more precisely how sex accelerates evolution, imagine an asexual population exposed to a new stress. It becomes adapted to this stress through the spread of alleles at two (or more) loci, each of which is individually advantageous. Both beneficial mutations may arise in the population at the same time, although any particular lineage will carry at most only one of them, given that the mutation rate is low. As the mutant lineages spread they will compete with one another, until a lineage bearing the mutation with the larger effect on fitness eventually replaces the other. The population is now partly adapted to the new environment. A second beneficial mutation will now arise in genomes which already bear the first, and in time this mutation will also be fixed. In other words, the population must evolve in series, with a stepwise accumulation of beneficial mutations.

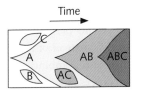

(a) Asexual: high rate of favorable mutaion

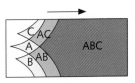

(b) Sexual: high rate of favorable mutaion

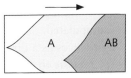

(c) Sexual or asexual: low rate of favorable mutaion

Figure 15.2 The pattern of adaptation in asexual and sexual populations in relation to the origin of new variation by mutation. Reproduced from Ridley, M. 2003. *Evolution*, 3rd ed., with permission from Wiley.

Adaptation follows a different course in sexual populations (Figure 15.2). While both beneficial mutations are present in the population, they can be combined into the same lineage by the fusion of gametes each bearing one of them, followed by their recombination into the same haploid genome during meiosis. This is likely to occur much more quickly than the fixation of the superior type. From then on, selection will drive the spread of individuals bearing both beneficial mutations. A sexual population evolves in parallel, so to speak, through the assembly of beneficial mutations into the same genome by gamete fusion and recombination. If more than two beneficial mutations contribute to adaptation, then the acceleration of evolution caused by sex is correspondingly greater. The crucial evolutionary consequence of the sexual cycle is that it rapidly generates lineages in which beneficial mutations that have arisen independently can be perpetuated in combination.

Sex is a more fundamental biological change than any of the other adaptations we have described. We have epitomized evolution as lineage dynamics. In asexual populations this simply involves the replacement of one lineage by another. In sexual populations a lineage has no permanent existence, and the population consists instead of temporary lineages that continually fuse and combine. Sex is an adaptation that alters the process of adaptation itself.

Sexual populations evolve faster. The acceleration of evolutionary change by sexuality has been demonstrated in some clever experiments with unicellular eukaryotes in the laboratory. The first point to be made is that large populations (consisting of many millions of individuals) adapt more rapidly when they are sexual. This has been demonstrated in experiments with yeast and algae, where the sexual cycle can be easily manipulated.

Yeast normally reproduces as a diploid, which on starvation undergoes meiosis to produce four haploid cells confined within a tough-walled structure called the ascus. These will normally fuse in pairs within the ascus when conditions improve, although outcrossing can be enforced by prematurely disrupting the ascus and mixing the haploid gametes. An asexual strain can be constructed by deleting two genes: *SPO11*, which encodes an endonuclease that initiates crossing-over by causing double-strand breaks, and *SPO13*, which affects the pairing of sister chromatids and is necessary for initiation of the second meiotic division. As the second division does not reduce chromosome number, the double

mutant produces non-recombinant diploid spores within the intact ascus. It can then be compared, under identical conditions, to an otherwise identical line in which meiotic reduction and recombination occur normally. When these lines are cultured in benign conditions there is little if any adaptation in either and no appreciable difference between them, so recombination has no effect on purifying selection. In harsh conditions (elevated temperature and salt) both types adapt, but the recombining lines adapt more rapidly: Figure 15.3 shows that the additional genetic variance generated by recombination increased the rate of adaptation by about 2.5% per sexual cycle.

Sexual populations evolve faster only if they are large. If the main effect of sex is to modulate the process of evolution then its success will depend on the availability of genetic variation. In a small uniform population sex will be ineffective because beneficial mutations are so rare that the first to arise is likely to become fixed before the next appears; it is only in a large population that several beneficial mutations will be available to be recombined into the same lineage. Hence, sex should accelerate evolution in large populations but not in small populations. This was demonstrated in experimental populations of the unicellular green alga *Chlamydomonas* exposed to high concentrations of bicarbonate.

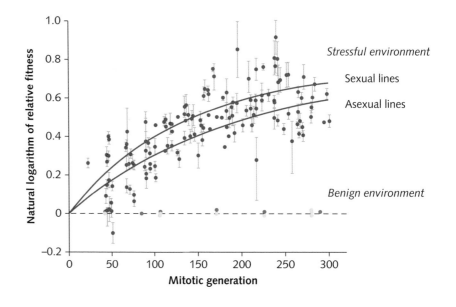

Figure 15.3 Adaptation to stressful conditions of growth in sexual and asexual yeast populations.

Reprinted by permission from Macmillan Publishers Ltd: Goddard et al. Sex increases the efficacy of natural selection in experimental yeast populations. *Nature* © 2005.

The populations were founded from clones and so had little if any genetic variation at first. Large sexual populations adapted more rapidly than large asexual populations, but in small populations sex had little effect. Hence, sex accelerates evolution only when the mutation supply rate is high enough for two or more beneficial mutations to be spreading simultaneously.

Sexual populations evolve faster only if they are variable. A different kind of experiment involves manipulating the potential genetic variation available at the outset. A sexual population will evolve more rapidly only if there is potential variation that can be released by selection. Sexual and asexual lines of yeast, created by genetic manipulation, can be constructed either as homozygotes or as heterozygotes. Sex is expected to affect the response to selection in a stressful environment only in the heterozygous lines. When these strains were mixed, the asexual strains at first increased in frequency, as has been found in similar experiments with *Chlamydomonas*. The fitness of sexual heterozygotes was particularly low, presumably because recombination tends to disrupt pre-existing beneficial combinations of alleles. Before the cultures could become fixed, however, the sexual strains rallied and subsequently eliminated

the asexuals in most cases for the heterozygous lines, while sexual and asexual strains were equally successful among the homozygous lines. The dynamics of these experiments, illustrated in Figure 15.4, supports a short-term advantage for sex generated through increasing the genetic variance of fitness.

The maintenance of sex requires continually renewed directional selection. These experiments show that sex accelerates evolution in populations exposed to novel stressful conditions. While this suggests an important evolutionary role for sex, it also gives rise to a serious difficulty. Sex will continue to be advantageous only if populations are continually exposed to new sources of directional selection. For this to be true, the environment must continually fluctuate over time, in such a way that new combinations of alleles have greater fitness than their predecessors every few generations. What agents could be responsible for a continual turnover of fitness? The most likely source of strong, specific, fluctuating selection experienced by a species is the antagonistic coevolution of its enemies, such as parasites and predators. We shall take up this theme in Chapter 17. For the rest of this chapter we shall take sex for granted, and explore some of its consequences.

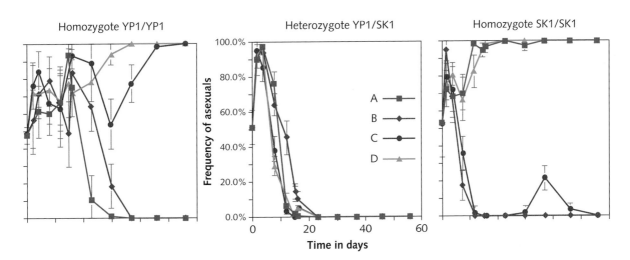

Figure 15.4 The fixation of sexual and asexual strains in homozygous and heterozygous yeast populations. A–D are four replicate lines for each type of population.

Greig, D., et al. The effect of sex on adaptation to high temperature in heterozygous and homozygous yeast. *Proceedings B* 265 (1400) © 1998, The Royal Society.

15.3 **Gametes are modified by sexual selection.**

The sexual cycle includes two events with contrary effects: the formation of one genome from two by gamete fusion, and the formation of two genomes from one by zygotic meiosis. These contrary effects have very different evolutionary histories.

Meiosis is highly conserved. You will have learned about the complex dance of the chromosomes during meiosis in elementary genetics courses. You may or may not have been told which organisms supplied the material. Some animals and plants have large and easily stained chromosomes that are convenient for illustrating the stages of meiosis—grasshoppers and lilies, for example. Apart from convenience, however, it does not really matter, because meiosis is much the same in all eukaryotes. There is a good deal of variation in detail (sperm are formed without crossing-over in *Drosophila*, for example), but the underlying process is remarkably similar. Some of the genes involved can even be recognized in bacteria, where they are responsible for DNA repair. Meiosis is an ancient innovation of eukaryotes whose essential features have been retained in all modern lineages.

Mating-type genes are idiomorphic and idiosyncratic. Sexual fusion is probably an even more ancient feature of eukaryotes, dating back to the origin of phagocytosis through the loss of a rigid cell wall. Far from being conserved, however, the means of fusion and the genes that govern them vary without limit. In unicellular algae and fungi fusion is usually regulated by mating-type genes: only gametes of different mating type may fuse. You might expect that there would be two mating-type alleles in each species and that related species would share similar alleles. This is far from being the case. In many algae and yeast alternative mating-type genes have different sequences, sizes and orientations, as illustrated in Figure 15.5. It is clear that they cannot have evolved from a recent common ancestor. Moreover, even closely-related species may have quite different mating genes. The genes that govern sexual fusion, then, evolve differently from genes involved in meiosis or vegetative function.

Hence, the two main events of the sexual cycle are strongly contrasted, with meiosis being conserved while fusion is idiosyncratic. The reason is that the constituent genomes of the zygote must collaborate

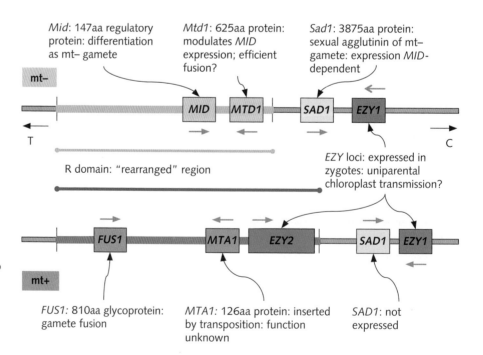

Figure 15.5 Cartoon of the mating type region in the two genders, mt+ and mt–, of *Chlamydomonas reinhardtii*. Many other elements are present. Arrows show the direction of transcription.

in order to go through meiosis successfully, whereas gametes compete with one another for fusion. It is the sexual competition of gametes that drives the idiosyncrasy of sexual systems.

15.3.1 Gender evolves to restrict gamete fusion.

The rule that governs sexual fusion is simple: gametes may fuse to form a viable zygote if they belong to the same species and to different gender. Hence, the concepts of "species" and "gender" are unique attributes of sexual lineages. We have already discussed species and species formation; sexual selection is about gender and its consequences.

Gender is primarily a property of gametes. To understand the biological nature of gender, it is useful to forget about animals and plants to begin with and think instead of a unicellular eukaryote that may develop either as a vegetative spore or as a sexual gamete. As a gamete it will fuse with any sexually compatible partner. In this case, gender is clearly an attribute of the individual cell. If the gamete has been produced by a multicellular organism such as a seaweed then nothing changes; it will still fuse only with some other compatible gamete, regardless of the form of its parent. Hence, "gender" is an attribute of gametes, not of the cells that give rise to them. (It is, of course, a rather unusual attribute because it can be defined only relative to other gametes of different gender.)

We commonly attribute gender to multicellular individuals (such as ourselves), but this takes its biological meaning from the gametes we produce. Bodies may nevertheless be endlessly modified for the better transmission of the gametes they produce, and this, indeed, is the central theme in the study of sexual selection in animals and plants.

Male and female evolve through a combination of natural and sexual selection. The original function of gender in unicellular organisms is to prevent mating between members of the same clone: it enforces a degree of outcrossing. It is not necessarily associated with any morphological difference between gametes. Nor is it necessarily bipolar: ciliate protozoans can have a dozen or so genders and some fungi have hundreds. Animals have just two: small motile male gametes, the spermatozoa, and much larger

non-motile female gametes, the ova. How did this fundamental male–female distinction evolve? The crucial idea is sketched in Figure 15.6.

Imagine a population which produces gametes of the same size regardless of gender. Many unicellular organisms such as phytoplankton and yeasts are like this, and their gametes are not much different in size from the vegetative cells. By contrast, the fertilized egg of a multicellular organism needs to grow into a larger individual, and therefore needs the reserves of material and energy to complete its development successfully. Hence, natural selection will favor the evolution of larger gametes because they will form larger and more successful zygotes when they fuse.

Now, suppose that in consequence the population evolves so as to produce very large gametes. A mutant type that produces very small gametes (simply through an increase in the number of divisions involved in gametogenesis) is then likely to spread because it produces many more gametes. The zygotes formed when one of these small gametes fuses with a large gamete is not as large as the zygote formed by the fusion of two large gametes, and the embryo that develops may not be as likely to develop successfully; but this drawback is more than compensated by the very large increase in the number of zygotes that the producer of small gametes produces. Hence, a population producing large gametes can be invaded by a novel type producing small gametes.

Conversely, a population that produced only small gametes could be invaded by a rare type producing large gametes, because these would give rise to much more successful zygotes. The evolution of male and female gametes thus arises from the balance between natural selection for producing fewer but larger gametes and sexual selection for producing more but smaller gametes.

15.3.2 There is intense competition for fusion among sperm.

The vast excess of sperm over ova implies that only a tiny majority will succeed in fusing. On the other hand, almost all ova will be fertilized in many cases. There is therefore a general and substantial imbalance in the forces of selection acting on sperm and ova. The ovum contributes most of the cytoplasm and stored resources to the zygote, and will therefore be modified

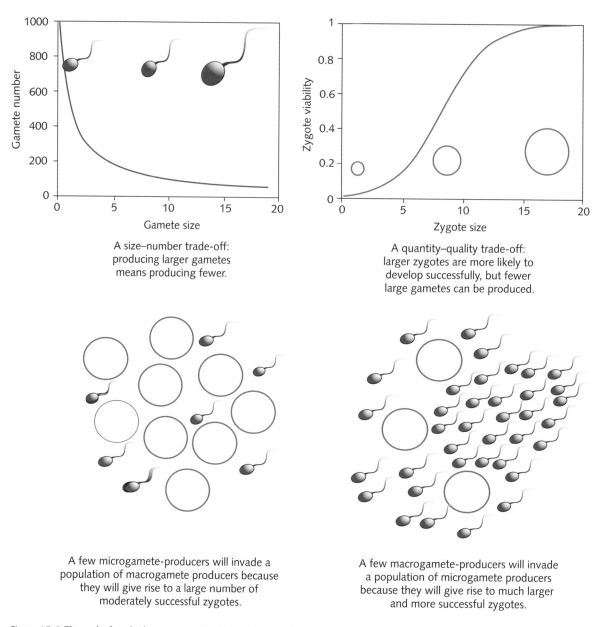

A size–number trade-off:
producing larger gametes
means producing fewer.

A quantity–quality trade-off:
larger zygotes are more likely to
develop successfully, but fewer
large gametes can be produced.

A few microgamete-producers will invade a
population of macrogamete producers because
they will give rise to a large number of
moderately successful zygotes.

A few macrogamete-producers will invade
a population of microgamete producers
because they will give rise to much larger
and more successful zygotes.

Figure 15.6 The male–female distinction evolves through the tension between sexual selection and natural selection.

primarily by natural selection acting through the vigor of the offspring into which this zygote develops. The sperm contributes very little to the zygote, but a successful sperm has succeeded in competition with millions of others. This competition shapes the characters of sperm through sexual selection.

Experimental sexual selection greatly increases the rate of fusion. Experimental evolution usually involves exposing populations to an environmental stress to which they adapt through natural selection. It is equally possible, however, to expose populations to a sexual stress by rewarding rapid mating and punishing the failure to mate. In the unicellular green alga *Chlamydomonas*, for example, mating is normally a response to nutrient starvation, which in nature signals the end of the growing season. In the laboratory, mating can be enforced by exposing cultures to chloroform, which kills the unfused vegetative cells but allows the zygotes to survive,

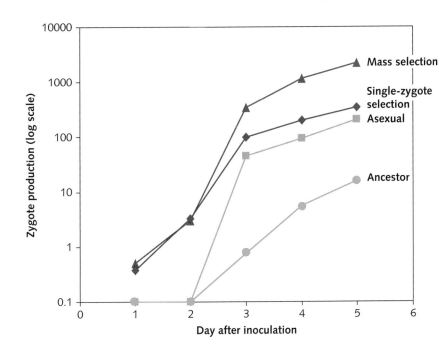

Figure 15.7 Experimental sexual selection. The number of zygotes produced in the period following the inoculation of a mating culture is greatly increased by sexual selection. Transferring large numbers of zygotes after each sexual episode (mass selection) is more effective than transferring a single zygote only. Note that the scale of the *y*-axis is logarithmic: the mass selection lines produce about a hundred times more zygotes than their ancestor.

because they develop a thick resistant cell wall. This amounts to very strong sexual selection, since the failure to mate quickly enough is lethal. After about a hundred sexual cycles, experimental populations mate very readily, even without the usual trigger of nutrient depletion. Indeed, they produce zygotes at 10–100 times the rate of unselected control populations, as shown in Figure 15.7. This shows that sexual characteristics can be modified by experimental sexual selection just as rapidly and thoroughly as vegetative characteristics can be modified by experimental natural selection.

Natural selection and sexual selection are fundamentally antagonistic. Enhanced sexual performance evolves at a price: experimental lines which mate rapidly grow slowly. Conversely, lines that are long propagated in purely vegetative cultures often lose the ability to mate. This is yet another example of evolutionary regression, whereby improvement in one characteristic is accompanied by the degradation of others, either through functional interference or through mutational degradation (Section 13.3.5). It is a particularly fundamental example, because it shows how adaptation in one phase of life (the sexual cycle) is countered by regress in the other (the vegetative cycle). This is a theme that dominates the evolution of sexual characteristics among individuals.

15.3.3 Males compete for access to females.

In many sessile marine organisms fertilization is necessarily external: eggs and sperm are shed into seawater and gamete fusion occurs outside bodies. Sexual selection then modifies the characteristics of gametes without much effect on individuals—individuals of seaweeds and sponges look much the same regardless of the gender of the gametes they produce. Indeed, many are hermaphrodites and produce both male and female gametes.

Motile males compete among one another. Motility changes everything, because individuals can then increase the success of their gametes by intimate association with another individual. This may involve no more than physical proximity when the gametes are shed: two individuals capable of movement will derive a mutual benefit from ensuring that their gametes are released in the same space at the same time in order to maximize the rate of fertilization. Frogs and salmon provide familiar examples: two individuals pair off, and when the female releases ova the male immediately releases sperm to fertilize the brood. The gross inequality in the size of sperm and ova, however, ensures that sexual selection will act differently on male and female individuals, because a male may be capable of fertilizing the ova of several females.

Hence, sexual competition will be more intense among males than among females. This will lead to a more extensive sexual modification of male characters, for example the massive hooked jaws of male salmon for tussling with rivals. It also leads to unexpected modifications of male development. Some male salmon become sexually mature at a much younger age and much smaller size than most. They cannot compete directly with the large full-grown males. What they can do, however, is to dash in quickly between a mating pair and fertilize some of the ova that the female has just emitted. In this case, sexual competition leads to the evolution of two distinct types of male: a large powerful fish able to drive away rivals, and a small agile fish that can exploit its success.

When gamete fusion requires copulation, the female sexual tract is an obstacle course for sperm. The close proximity of male and female at the point of gamete release is enhanced if the male ejaculates directly into the female sexual tract. The sperm must then make their way into the tract in order to fertilize the ova. This provides the female with the opportunity to select or reject incoming sperm, which can be diverted into different channels whereby it may be accepted, rejected, digested or expelled. Figure 15.8 illustrates

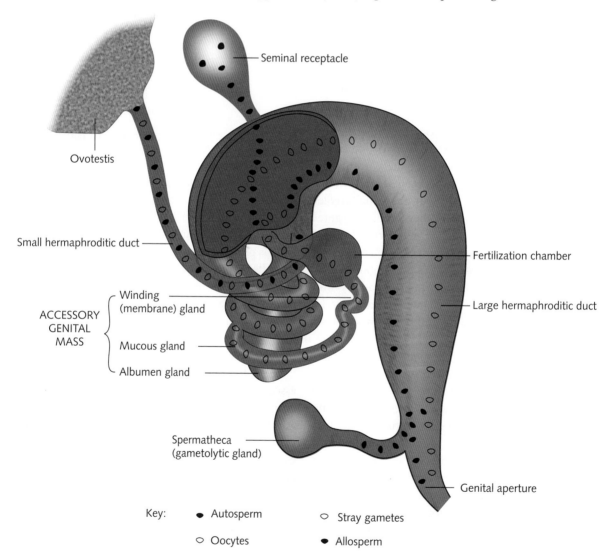

Figure 15.8 The pathways of male and female gametes through the reproductive tract of *Aplysia*, a marine mollusk. Self-ova, self-sperm, and foreign sperm follow different routes.

After Blankenship, unpublished.

the complex reproductive tract of a nudibranch (sea hare). How selective the female may be depends on the number of sexual partners she entertains. In strict monogamy there is no reason to prefer one sperm over another to fertilize an ovum, apart from an excess of deleterious mutations, if this can be detected. On the other hand, in promiscuous species sperm from the ejaculates of several males may be simultaneously present in the female tract and can be selectively screened. However, this adds another layer of complexity because the evolutionary interests of male and female are rarely identical, and any divergence of interest will give rise to sexual selection acting both on gametes and on the individuals which bear them.

Sperm can be highly modified for navigating the female tract. The selective screen of the female tract can drive the evolution of extraordinary sperm. For example, a mollusk called *Pseudopythina*—an unusual clam that grows on sea cucumbers—produces two kinds of sperm, which are illustrated in Figure 15.9. The first has a normal appearance with a small oval cell body and a long flagellum, whereas the second is a giant sperm that actually carries the normal sperm through the female tract. This illustrates two important processes. The first is the exaggeration that is often the consequence of sexual selection. The second, perhaps more fundamental, is the derogation of individual interest: the cell developing into the carrier sperm has no potential to fertilize, but rather assists others to do so. In other words, it commits sexual suicide to help the functional sperm it is related to. This is an extreme example of altruistic behavior, which is explained in the next chapter. There is one further twist: *Pseudopythina* is hermaphroditic. Sexual selection operates through the differential fertilizing abilities of gametes and does not necessarily require separate male and female individuals.

Male gametes respond to sexual selection in the laboratory. The normal sperm of *Pseudopythina* are normal from our point of view, and many metazoans, from sea urchins to frogs, do indeed have this kind of sperm. In other metazoans the sperm are highly modified. The fruit fly *Drosophila* is one of the most familiar of laboratory animals: its sperm are quite

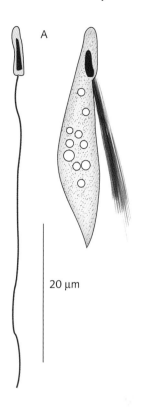

Figure 15.9 Normal sperm (left) and carrier sperm (right) of the marine bivalve *Pseudopythina*.

Reproduced from Lützen, J., et al. 2004. Morphology, structure of dimorphic sperm, and reproduction in the hermaphroditic commensal bivalve *Pseudopythina tsurumaru* (Galeommatoidea: Kellidae). *Journal of Morphology* 262: 407–420, with permission from Wiley.

conventional in morphology, but they are gigantic: including the flagellum, each is almost as long as the fly. The nematode worm *Caenorhabditis* is also a familiar laboratory model organism, but has unusual amoeboid sperm that crawl up the female reproductive tract. Sperm competition in these worms is normally minimal because they are self-fertilizing hermaphrodites and there is no selection for fertilizing ability beyond the bare minimum requirement that the eggs be fertilized. We can manipulate their development, however, so as to produce separate males and females. This necessarily leads to sexual selection among male gametes, which can be manipulated experimentally by restricting the access of males to females. Larger sperm are superior in competition because they crawl faster, but they are necessarily more expensive to produce. Hence, large size should evolve only when sperm from several different males often compete within the same female tract. This was confirmed by experiment: after 60 generations sperm volume had increased by about 20% in lines where sperm competition was more intense.

15.4 Males and females are divergently specialized through sexual selection.

In multicellular eukaryotes, individuals are attributed the gender or genders of the gametes they produce: male, female or hermaphrodite. This need not imply any substantial differentiation of structure or behavior. Sessile marine organisms such as sponges or seaweeds often broadcast their gametes directly into the water column, where they compete freely for fusion. Individuals cannot influence the fate of their gametes once they are shed and differ only in the structure of the gonads and their ducts. On the other hand, if individuals can influence the success of the gametes they bear then they may be profoundly modified by sexual selection.

15.4.1 Male and female function may be divided or united.

An individual may produce either male gametes exclusively, or only female gametes, or both, or either at different times. It is the economics of gamete transmission that determine how gender will become partitioned among individuals.

Large fixed costs favor the separation of male and female function. The economics of gender are to some extent similar to the economics of professions. To succeed as a lawyer, a plumber, or a musician a person must invest capital that confers an exclusive benefit: a violin cannot be used with its intended effect in stemming a leak or a wrench in convincing a court of law. The whole of this investment must be made before any return can be obtained; the violinist must purchase the whole instrument before even a single concert can be given. Transmitting male or female gametes may likewise entail an entire and exclusive investment. For example, the fertilized eggs of dogfish are provided with their external casing by the shell gland. This is a large and costly structure that does not contribute to male function but that is essential for female function, and which must be wholly developed before even a single successful egg can be produced. The marginal cost of eggs becomes less the more are made, of course, so material is economized and therefore offspring production increased by specializing as a female. Wherever costly structures are necessary in order to operate successfully as a female or a male the outcome is likely to be the evolution of separate female and male individuals.

Male and female function are united when costs can be shared. Conversely, there are other investments that may serve more than one function. In the economic sphere, education is an obvious example: the surgeon and the anesthetist, despite their different specializations, share a curriculum until an advanced stage of their training. The flower of insect-pollinated plants is a good example of a structure that serves both male and female function, for the insects that are attracted by its color or scent will both deposit pollen from other plants onto the stigma and take away pollen to fertilize neighbors. Hence, the flowers of insect-pollinated plants are often bisexual structures, and the individual plant is a hermaphrodite.

15.4.2 When gamete fusion involves courtship and copulation, male and female individuals can be strongly modified.

In multicellular organisms, individuals are the intermediaries whose behavior governs the fate of gametes. To put this in another way: the bodies of individuals become modified so as best to serve the sexual interests of gametes. This modification is very slight in organisms that broadcast their gametes into the surrounding medium, such as sponges or poplar trees. If gamete fusion requires the intimate association of two individuals, however, males and females may evolve very different sexual specializations. Sexual dimorphism varies from slight differences in size or appearance to bizarre extremes in which male and female seem to be quite different kinds of organism.

One principle underlies the whole range of outcomes: there is a fundamental asymmetry between the economics of male and female gametes.

Males are usually the more highly modified gender because they have greater variance of fitness. In many, but not all, animals and plants the following three generalizations are likely to hold.

- Fecundity is more strongly correlated with the number of matings among males than among females. Any male mating with a female will fertilize some fraction of her eggs on average. Hence, the number of offspring that a male produces will increase linearly with the number of matings he achieves. A female, on the other hand, is likely to produce about the same number of offspring regardless of how many or how few males she mates with, because even a single mating will provide enough sperm to fertilize all her eggs.

- There is a greater variance in the number of matings among males than among females. The average number of sexual partners is necessarily the same for males and females, but the variance is likely to be greater for males because males profit more from having more sexual partners. The most extreme case would be a population in which all the females were fertilized by a single male, all the other males being completely unsuccessful. The average number of offspring is the same for males and females (given an equal sex ratio), but the variance is much greater among the males.

- Hence, the variance of fecundity is greater among males because it depends on the number of matings they experience. Because females produce much larger gametes, they necessarily produce a limited crop the whole of which could be fertilized by a single male, with sperm to spare. Thus, the production of offspring by females is not usually limited by the supply of sperm. Offspring production by males, on the other hand, is limited by the supply of eggs, and is therefore increased by fertilizing more females.

These observations were first made in *Drosophila* by A.J. Bateman, and so are collectively known as Bateman's principle. They imply that sexual competition is stronger among males than among females, hence that sexual selection acts more strongly to modify male characteristics.

When the sexual biology changes, the principle changes. In some circumstances, however, sexual competition may be stronger among females. The same argument then applies, but with the genders reversed. In seahorses and pipefish, for example, the male broods the eggs and developing young in a brood pouch. In this case, males are the limiting reproductive resource, and we expect females to compete for access to them. Hence, females should be more strongly modified in these animals, and indeed they are often exceptional among fish in being more brightly colored than the males.

Males are generally small to increase the rate of mating. Males can be modified in many ways, as we shall see, but the most common outcome of sexual selection is the evolution of smaller size. In most motile organisms, males are smaller than females. In groups as diverse as fish, flies, water fleas, and even green algae the male is smaller than the female. This follows from the fundamental asymmetry between male and female gametes. Female individuals, bearing a crop of large expensive eggs, are likely to be more successful if they save energy by waiting for males to find them. Males bear cheap and numerous sperm which can be turned over rapidly but can be used only if females are located. Consequently, females in most animals are larger and more sedentary, whereas males are smaller and more active.

We have already mentioned the dwarf parasitic males of angler fish (Section 1.5.3). A similar pattern is found in other groups. In some motile animals such as rotifers, for example, the males are little more than motile testes. Even among algae, the males of *Volvox* (Section 8.1.1) are small colonies that hunt females by detecting pheromones, and in filamentous forms such as *Oedogonium* the dwarf male zoospores seek out sessile female cells. Regardless of phylogeny, the characteristics of male and female individuals generally reflect the duality of the gametes they bear.

Again, when the sexual biology changes, the application of the principle changes. Males can be highly modified in either direction, smaller or larger, depending on the agent of sexual selection: when males guard the offspring or fight for access to

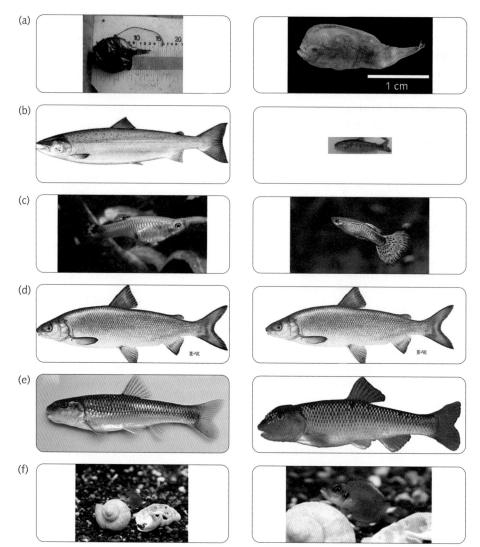

Figure 15.10 Sexual dimorphism in fish. Female on left, illustrated at same size; male on right, roughly to scale. (a) Angler fish, *Haplophryne*: dwarf parasitic male. (b) Atlantic salmon, *Salmo*: precocious small male. (c) Guppy, *Poecilia*: colorful male, smaller than female. (d) Whitefish, *Coregonus*: male similar to female. (e) River chub, *Nocomis*: larger male guards nest. (f) Cichlid, *Lamprolagus*: larger male gathers snail shells that female fits inside.

Atlantic salmon male courtesy of Ayrshire Rivers Trust. River chub male photograph by N. Burkhead & R. Jenkins, courtesy Virginia Department of Game and Inland Fisheries.

females, for example, they evolve larger size. Hence, male fish vary from dwarf parasites through precocious juveniles to large adults effective in protection or combat. A few examples from this range are illustrated in Figure 15.10.

15.4.3 Females may prefer certain kinds of male.

The asymmetry of male and female gametes may thereby lead to an asymmetry in the interests of male and female individuals. Male characteristics will usually evolve so as to maximize the quantity of mates (because sperm are cheap) whereas female characteristics will evolve so as to maximize the quality of mates (because eggs are expensive). Hence, females have the potential to choose among male suitors, accepting some and rejecting others. There is no doubt that females (or, more rarely, males) often do exercise a choice; female pigeons, guppies or fruit flies, for example, are usually courted by many males before accepting one. Why should they prefer one kind of male over others?

Female choice may be an innate sensory bias. In some cases, female preference may reflect nothing more than an accidental bias that originally evolved in a non-sexual context. For example, swordtails (*Xiphophorus*) are small stream-dwelling fish from Mexico. In some species the male bears a "sword," a long, tapering, brightly-colored process on the lower part of the tail fin. Females tend to prefer males with longer swords. However, when males of related species that lack swords are tested females prefer individuals to which an artificial sword has been attached. In this case, the preference shown by females seems to have evolved before the character expressed by males. Similarly, female guppies prefer males that display large orange spots, but they also tend to be attracted by orange spots in non-sexual contexts. It has been suggested that this reflects their preference for orange berries as food in wild populations. Hence, there may be an innate sensory bias that leads to the production of expensive orange pigments in males to attract females.

Females may receive a direct benefit from courtship. The clearest basis for choice is that females search for males who will contribute most to rearing their offspring. The fathead minnow, *Pimephales promelas*, for example, is a common stream fish of eastern North America in which the male builds a nest and guards the eggs against predators. Larger males usually secure the best sites—by excluding smaller rivals—and are more successful in attracting females to spawn in their nests. In many insects the male offers a food item as a "nuptial gift" to prospective partners. Females are more likely to mate with males offering gifts, or offering larger gifts, because they can use these resources for themselves or for their offspring.

Females may choose males with good genes that their offspring will inherit. Males contribute few if any resources to their offspring in many species, but they always contribute genes. Females that are able to identify genetically superior males—males bearing alleles that confer higher than average fitness in the current environment—can incorporate superior alleles into their offspring, thereby permanently enhancing the prospects of their lineage. Females are unable, of course, to scrutinize genes directly, and must instead infer genetic quality from the appearance and behavior of the male. The simplest test is vigor. Many courtship rituals involve lively and expensive displays by the male—for example, the wing-buzz and gymnastic dance performed by male fruit flies. Female preference for males able to perform such rituals successfully may evolve if the

offspring inherit their father's vigor. This argument has two weaknesses, however.

- Vigor may be the result of upbringing rather than environment: well-fed males, for example, may be more vigorous than starving competitors.

- If vigor is heritable, then the alleles responsible should have become fixed in the population through natural selection, removing genetic variation among males and thereby eliminating any basis for choice among females.

Hence, the two hurdles that a "good-genes" interpretation of female choice must surmount are to identify a perpetually renewed source of genetic variance in fitness among males and to explain how it can be exploited by females.

Females may choose males with lower mutational load. The most obvious renewable source of genetic variation is deleterious mutation. Random mutation will lead to variation among males, some of which have by chance received many mutations while others have received few. Males with fewer mutations will tend to be more vigorous, and females who mate with them will bear offspring which inherit fewer mutations than average and have a correspondingly greater chance of survival and reproduction. This can be studied experimentally by giving some females a choice of several males while others are given a single randomly chosen male to mate with. In experiments with fruit flies, females with a choice of males produce more vigorous larvae.

A more decisive test in species with external fertilization is to fertilize half the eggs of a female with sperm from a more vigorous male and half with sperm from a less vigorous male. This experiment was tried in gray tree frogs, where the male display is a loud call that attracts females. Males which called for longer were taken to be more vigorous than those with shorter calls. Tadpoles from eggs fertilized by more vigorous males generally grew and survived better than their half-sibs from eggs fertilized by less vigorous males. In short, females may enhance the fitness of their offspring by choosing to mate with more vigorous males.

Females may choose healthy males. A second possibility is that females select males carrying beneficial mutations. Although this may sound straightforward, the difficulty remains: any such mutations should have become fixed through directional selection, leaving the females with no variance among which to choose. Hence, we require a perpetually renewed source of directional selection. This could be provided by host–parasite coevolution. Females choose males who are resistant to contemporary pathogens. Their offspring will inherit any genetic source of resistance, and are likely to be healthy. In time, the parasites will evolve to overcome the resistance, so the source of variation among the hosts is continually renewed. Courtship rituals can then be interpreted as medical check-ups.

For example, female sticklebacks prefer to mate with males which have bright red throats. The red color is based on carotenoid pigments, which are expensive to deploy because they cannot be synthesized by vertebrates and must be obtained in the diet. Hence, bright coloration is a signal of health, because sick males infected with disease-causing parasites cannot obtain an adequate supply of carotenoids and have relatively drab throats. If parasite resistance is genetic, females mating with brightly-colored males can then transmit this resistance to their offspring. The vivid red crests and wattles of birds such as chickens and turkeys may have evolved in a similar way.

Females may choose compatible males. The situation is often more complicated, however, because parasite resistance may depend on a combination of alleles rather than on single unequivocally favorable alleles. The disease resistance of offspring is then enhanced if females choose to mate with compatible males. In vertebrates, the major histocompatibility (MHC) loci are an important source of resistance. They encode proteins able to bind small peptides produced by pathogens and present them on the cell surface, where they initiate an immune attack on the pathogen. MHC loci have thousands of alleles and can thus detect a very wide range of pathogens. Each individual has a small subset of these alleles, and an intermediate number of alleles seems to provide the optimum level of resistance. In natural populations of sticklebacks individuals have about 6 different alleles on average at MHC Class IIB loci. Levels of parasite infestation show that this is close to the optimal number of alleles. Hence, females will maximize the disease resistance of their offspring by mating with males bearing the complementary number of MHC alleles. This is a very onerous requirement, because it requires females both to detect their own internal state and to evaluate the MHC status of a potential mate. Nevertheless, they are capable of doing both.

Female choice has been analyzed through experiments with divided flow channels, where the female receives water from two tanks each holding a test male. Each male had few or many MHC alleles. Females who had few MHC alleles followed up the stream from a male with many alleles; conversely, females with many different MHC alleles preferred the stream from the male with few. Indeed, when the flow channels were "spiked" with MHC ligands, with no males present, the females preferred the stream that would have optimized the number of different MHC alleles in her offspring. Female sticklebacks are capable of choosing males with an optimum number of different MHC alleles.

15.4.4 Female choice causes sexual selection among males.

To this point, mate choice can be explained quite straightforwardly: alleles that affect female behavior will tend to spread if they lead to the production of more or better offspring through a preference for mating with a certain type of male. It is made more complicated because a female preference, once established, acts as a powerful selective force on males. Sexual selection among males will then favor character states that attract females. Female preference, however, is not based on male quality itself, but only on a perception of male quality. Males that promise more support than they intend to give, or are able to simulate vigor or health, will therefore be just as successful as genuinely superior rivals, and the alleles responsible for misleading courtship displays will spread because of their enhanced transmission through the males that bear them. In a manner of speaking, the characters of males that are expressed during courtship should be viewed not as signals but rather as advertisements. This leads in turn to selection among females for the ability to identify honest advertisements through shifts in perception which may turn out to favor new kinds of misleading display or character among males. Thus, the outcome of sexual competition within one gender will alter how sexual selection acts on the other gender. This reciprocal modification is the distinctive feature of sexual selection, and can lead to rapid, self-reinforcing and idiosyncratic evolutionary change.

Sexual advertisements are not always honest. Vigor and health usually provide reliable signals when they can be directly assessed. The red throat of male sticklebacks, for example, is expensive to produce, and only healthy males can do so; to this extent, it must be an honest advertisement. Where the male confers a direct benefit, especially a deferred benefit, the situation is less straightforward. Male sticklebacks provide parental care by building a nest and caring for the eggs and young. They must therefore partition their stored resources between courtship and parental care, and an expensive courtship display that leaves the male too enfeebled to provide adequate parental care would be deceptive. A similar dilemma is familiar in human economies, where any manufacturer producing a good must allocate a certain fraction of some fixed budget to improving its quality and the remaining fraction to advertisement.

Among plants, pollinators assist both male function by exporting pollen and female function by importing pollen. In many species, the sexual advertisement that attracts pollinators is the flower, while the direct benefit they receive is nectar. Both flowers and nectar are expensive to produce. The turtlehead (*Chelone glabra*) is a common herb of ditches and other wet places with large white flowers that are eagerly visited by bumble bees and other pollinators because of the large volume of nectar they can contain. The flower is almost completely closed, however, so that until an insect inserts its tongue it has no means of knowing whether any nectar is present. At this point, however, it has already deposited and received pollen. Consequently the plant can obtain the service without paying the reward, and if you survey turtleheads in your vicinity you will probably find that some of them secrete no nectar. These empty flowers are a false advertisement. If all the flowers were empty then the pollinators would stop visiting them, of course, so some plants can deceive pollinators only when others provide a true signal of the reward they offer.

Females sometimes prefer males with highly exaggerated traits. Sexual advertisements such as bright colors and conspicuous displays not only use up resources but may also expose males to greater risks of predation or starvation. If they were lethal then they could not evolve, of course, but in some cases males become so highly modified that their survival certainly seems to be jeopardized. Nevertheless, greatly exaggerated structures or behaviors often seem to be attractive to females. The peacock is the classical example.

In the African widow bird, *Euplectes*, the sexual benefits of exaggerated structures has been demonstrated experimentally. The male has extremely elongated tail feathers, about ten times as long as those of the female. The birds are territorial and polygynous, so that a successful male may sequester and mate with many more females than average. By cutting the tail-feathers, it is possible to create males with very short tails, with normal long tails (by re-attaching the cut section with glue), or with extremely long tails (by gluing a long piece onto a tail that had not been shortened much). The males with very short tails attracted fewer females than the normal birds, and those with extremely long tails attracted more females than did males of normal appearance (Figure 15.11). Highly exaggerated structures may, then, evolve because they are favored by sexual selection, despite their disadvantage under natural selection.

Highly exaggerated male traits evolve in promiscuous species. Birds-of-paradise are found mainly in the dense rainforest of New Guinea. The females are mostly inconspicuous russet-colored birds, whereas the males of many species have spectacular plumage and displays. The male Raggiana bird-of-paradise, for example, is a crow-sized bird with black breast feathers, a yellow collar, green throat, and a yellow crown; on its flanks it bears large, spectacular orange to red plumes; and it has two long thin tail feathers. The female is smaller and has none of these ornaments.

The courtship display opens with loud calls and wing-fluttering, followed by beating the wings together to produce a rhythmic thudding sound, and culminates in vigorous wing-beats, the erection of the flank plumes, and a swaying descent from the perch. A watching female may at this point move towards the male, who will puff up his breast feathers and rock violently from side to side before mounting her.

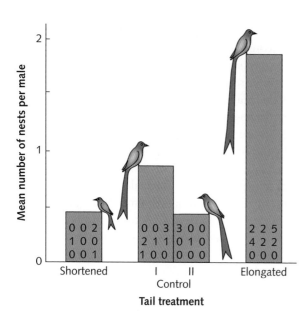

Figure 15.11 The effect of modifying a sexually selected character—tail length in widow birds—on sexual success. Numbers in boxes are numbers of females attracted by males with given phenotype.

Reprinted by permission from Macmillan Publishers Ltd: Andersson, M. Female choice selects for extreme tail length in a widowbird. *Nature* © 1982.

While all this is happening, the male is not alone; he is displaying in the company of several others, who compete for perches in a clump of trees. Females who are attracted to the site each choose one of these males to mate with, before retiring to lay eggs and rear the offspring entirely on their own. Hence, any male has the potential to fertilize many females, and in practice one or two males succeed in mating with most of the females who visit the group, many other males not mating at all. The mating system thus creates the potential for a great deal of variation in mating success among males.

Many other species, in fact, are also polygynous and highly dimorphic. In other birds-of-paradise, such as species of *Manucodia*, the males are monogamous and help to feed the nestlings; they are similar in size and appearance to the females. In accordance with Bateman's principle, therefore, males become more highly modified through sexual selection when the variance of mating success is much greater among males than among females.

Females may prefer fancy males because this enhances the sexual success of their sons. Highly exaggerated structures and displays, such as those of birds-of-paradise, may evolve through the reciprocal nature of sexual selection, if male traits become entrained with female preference. This can happen as follows.

- In the first place, females prefer males that express a character denoting high fitness, such as a red throat or crest as an advertisement of vigor and health.

- The alleles responsible for the expression of this character state are already favored by natural selection. They are now also favored by sexual selection. Alleles for female choice and male quality will rise to high frequency in the population.

- This creates selection for greater discrimination among females and for an exaggeration of sexual advertisements among males.

- Eventually the male character becomes so exaggerated that it is a liability that reduces survival. Other things being equal, natural selection would tend to suppress its expression. Nevertheless, it continues to evolve greater exaggeration because of the sexual advantage it confers.

- Eventually, the advantage of expressing the character through sexual selection and the corresponding disadvantage through natural selection come into balance, although at this point the males may have become very strongly modified.

The driving force in this process is the advantage gained by females who choose to mate with bizarre males through the sexual success of their sons, which follows from the high frequency of alleles governing this preference that has evolved in the population.

The evolution of bright coloration among male guppies illustrates several aspects of sexual selection. Guppies are highly dimorphic fish, as striking in their way as birds-of-paradise. The ease of breeding them and studying their behavior has made them popular among both hobbyists and scientists.

- They are highly promiscuous, with neither paternal care nor permanent pair bonds, and have brightly colored males and drab females.

- More brightly colored males, especially those with large orange spots, are more attractive to females. There may be some pre-existing sensory bias for orange spots.

- The trait is expensive: the most brightly colored males are 30% more likely than average to die before maturity, even in laboratory conditions. It is also risky: predators attack the most brightly colored males first.

- The color and position of spots are controlled by genes on the Y chromosome, whose expression is therefore limited to males.

- Artificial selection for enhanced orange coloration is effective, showing that female preference will result in heritable transmission of male characters to sons.

- Females from lines selected for orange color in males show a greater preference for orange, suggesting that female preference and male appearance can be entrained in evolution.

Hence, guppies exemplify within the same species many of the basic principles of sexual selection, which can be investigated by anyone with an aquarium and a notebook.

15.4.5 Males may compete directly for access to females in risky combats.

Sexual advertisements evolve to attract mates; likewise, sexual weapons can evolve to discourage competitors. Weaponry will be effective in sexual competition when there are resources that can be monopolized by stronger or better-armed males.

Male weapons can evolve to protect sexual resources. Fiddler crabs live above the tide marks on beaches and mudflats. Females have two small claws that they use to gather sediment from which they filter out edible particles. In males, one of these claws grows to such an enormous size that it may weigh as much as the rest of the body put together. It is used to defend breeding burrows. Females

inspect these burrows before choosing to remain in one, mate with the resident male and lay her eggs. She emerges a couple of weeks later to release the brood of larvae into the sea. Meantime, the male has guarded the burrow entrance, and may have built more burrows and mated with more females. His enlarged claw is used to guard these burrows and to evict smaller males from their burrows. It is a very effective weapon against other males, but it is also very expensive: enlarging one claw not only halves the number of feeding appendages but also creates a major metabolic burden. At the same time, it is necessarily a rather reliable indicator of vigor, and males attract females by waving their claws in the air as an advertisement of quality.

Male weapons can evolve to guard females. In protecting the nest site, male fiddler crabs necessarily guard females and prevent other males from mating with them. In herding animals like deer and cattle, males may sequester a group of females and guard them against other males simply in order to enjoy exclusive sexual access, without providing any resources beyond some relief from harassment. The earliest ungulates, such as *Eotragus* from the Miocene (18–20 Mya), bore short, simple horns or tusks. Similar weapons are borne by modern chevrotain and duiker, where they are used in dangerous dodge-and-stab contests. A great diversity of antlers and horns subsequently evolved and are exhibited by modern forms.

In some species, such as oryx and buffalo, the females bear horns that are usually smaller versions of those borne by males and are used against other females or predators. Larger and more complex structures are used in male combat involving ramming (as in buffalo and sheep) or wrestling (as in kudu and red deer). These have evolved as defensive structures that can be used to evaluate the strength of a rival without the need for lethal combat.

In red deer (*Cervus elephas*) the male weapons are branched antlers with pointed tines. Fighting is expensive, because they are large costly structures, and risky, because losers may be wounded. A fight is initiated when a challenger approaches a rival and roars loudly. The two will exchange roars for a while before approaching more closely and walking side by side. Both the roars and the parallel walk enable each male to weigh up its rival, and either may retreat at this point. If not, one will swing its head toward the other to invite combat and the two lock antlers, twisting and turning to gain the advantage of higher ground. Eventually one is pushed backwards, disengages and retreats; the winner may pursue its beaten rival for a while, and even gore it severely if it stumbles. Fighting ability is strongly correlated with mating success because the winner gains control of up to a dozen or so females.

Beetle horns used in male combat have evolved in parallel with bovid horns. There is a remarkable parallel between the horns and antlers of mammals and comparable structures borne by the males of scarab beetles. Scarabs (dung beetles) fight for access to females, usually inside tunnels they have dug, using their horns to overturn their rivals. As in ungulates, there is a great diversity of morphology among species, with the detailed shape of the horn being unique to each species.

15.5 Male and female attributes may be antagonistic.

Males generally compete with one another for access to females, while females compete with one another for access to resources. Thus, members of the same gender necessarily have conflicting interests. Conversely, individuals of unlike gender who mate together have a common interest in producing and raising offspring. Nevertheless, their interests may not be perfectly aligned, and this will have evolutionary consequences.

15.5.1 Males may damage mates to increase fertilization success.

It has long been known that sterile females live longer than fertile females in the fruit fly *Drosophila*. More generally, frequent mating reduces the lifespan of females. The effect is caused by accessory-gland proteins that are transferred in seminal fluid. These enhance the mating success of the male by inhibiting remating by the female and thus reducing competition with foreign sperm. The male is indifferent to the earlier death of the female because she will long before have laid all the eggs he has fertilized. Accessory-gland proteins are evolving very rapidly in the genus because the three-way conflict of interest between courting males, rival males and females drives a continual process of sexual selection.

In some cases mating can cause traumatic damage. In hermaphroditic flatworms, for example, sperm may be injected directly into the partner's body by a sharp-tipped intromittent organ. This manner of hypodermic impregnation can cause severe stab wounds.

15.5.2 Sexual conflict varies with mating system.

It may be in the interest of either partner to damage the other, if the likely outcome is greater overall reproductive success. This is only likely to be the case in species with promiscuous mating such that neither partner has a long-term interest in the wellbeing of the other. In monogamous species the sexual interests of male and female become entrained, because anything that damages the reproductive interests of one equally damages the interests of the other. The degree of sexual conflict will therefore vary with breeding system, being least for strictly monogamous species.

Sexual selection often drives the evolution of bizarre and idiosyncratic features. The extreme variation of sexual structures, whether designed to attract females or to subdue males, is a common outcome of the idiosyncratic course of evolution driven by sexual selection. Few structures that evolve through natural selection are as variable from species to species as the plumage of birds-of-paradise or the horns of scarab beetles. Few organisms are as odd as the world's smallest vertebrate, the male of the angler fish *Photocorynus*, which lives permanently embedded within the body of its mate. On a molecular level, few proteins evolve as fast as those associated with mating. Many of the most arresting features of organisms stem from the evolution of male and female and the irreconcilable conflicts that ensue.

● CHAPTER SUMMARY

The eukaryote life cycle is an alternation between vegetative and sexual cycles.

- The sexual cycle is a coupled process of sexual fusion and sexual restitution.
  - *The eukaryote life cycle alternates between spore and gamete.*
  - *Competition between spores and between gametes gives rise to different processes of selection.*
- Sex is primitive in eukaryotes.
  - *Sex is a shared derived character of eukaryotes.*
  - *Most asexual lineages are recently derived from sexual ancestors.*
  - *Very few major groups of animals or plants are completely asexual.*

Sex modulates the dynamics of evolution.

- Sex is costly.
- Sexual progeny differ from their parents and among themselves.
  - *Sexual sibs differ among themselves.*
  - *Sexual sibs differ from their parents.*
- Sex facilitates adaptation.
  - *Sex allows beneficial mutations that arise independently to be combined in the same lineage.*
  - *Sexual populations evolve faster.*
  - *Sexual populations evolve faster only if they are large.*
  - *Sexual populations evolve faster only if they are variable.*
  - *The maintenance of sex requires continually renewed directional selection.*

Gametes are modified by sexual selection.

  - *Meiosis is highly conserved.*
  - *Mating-type genes are idiomorphic and idiosyncratic.*
- Gender evolves to restrict gamete fusion.
  - *Gender is primarily a property of gametes.*
  - *Male and female evolve through a combination of natural and sexual selection.*
- There is intense competition for fusion among sperm.
  - *Experimental sexual selection greatly increases the rate of fusion.*
  - *Natural selection and sexual selection are fundamentally antagonistic.*
- Males compete for access to females.
  - *Motile males compete among one another.*
  - *When gamete fusion requires copulation, the female sexual tract is an obstacle course for sperm.*
  - *Sperm can be highly modified for navigating the female tract.*
  - *Male gametes respond to sexual selection in the laboratory.*

Males and females are divergently specialized through sexual selection.

- Male and female function may be divided or united.
  - *Large fixed costs favor the separation of male and female function.*
  - *Male and female function are united when costs can be shared.*

- When gamete fusion involves courtship and copulation, male and female individuals can be strongly modified.
    - *Males are usually the more highly modified gender because they have greater variance of fitness.*
    - *When the sexual biology changes, the principle changes.*
    - *Males are generally small to increase the rate of mating.*

- Females may prefer certain kinds of male.
    - *Female choice may be an innate sensory bias.*
    - *Females may receive a direct benefit from courtship.*
    - *Females may choose males with good genes that their offspring will inherit.*
    - *Females may choose males with lower mutational load.*
    - *Females may choose healthy males.*
    - *Females may choose compatible males.*

- Female choice causes sexual selection among males.
    - *Sexual advertisements are not always honest.*
    - *Females sometimes prefer males with highly exaggerated traits.*
    - *Highly exaggerated male traits evolve in promiscuous species.*
    - *Females may prefer fancy males because this enhances the sexual success of their sons.*
    - *The evolution of bright coloration among male guppies illustrates several aspects of sexual selection.*

- Males may compete directly for access to females in risky combats.
    - *Male weapons can evolve to protect sexual resources.*
    - *Male weapons can evolve to guard females.*
    - *Beetle horns used in male combat have evolved in parallel with bovid horns.*

Male and female attributes may be antagonistic.

- Males may damage mates to increase fertilization success.

- Sexual conflict varies with mating system.
    - *Sexual selection often drives the evolution of bizarre and idiosyncratic features.*

● FURTHER READING

Here are some pertinent further readings at the time of going to press. For relevant readings that have been released since publication, visit the book's Online Resource Centre at **www.oxfordtextbooks.co.uk/orc/bell_evolution/**

Section 15.1   Schurko, A.M., Neiman, M. and Logsdon, J.M. 2008. Signs of sex: what we know and how we know it. *Trends in Ecology and Evolution* 24: 208–217.

Section 15.2   Burt, A. 2000. Sex, recombination and the efficacy of selection: was Weismann right? *Evolution* 54: 337–351.

Goddard, M.R., Godfray, H.C. and Burt, A. 2005. Sex increases the efficacy of natural selection in experimental yeast populations. *Nature* 434: 571–573.

Section 15.3   Ramesh, M.A., Malik, S-B. and Logsdon, J.M. 2005. A phylogenomic inventory of meiotic genes: evidence for sex in *Giardia* and an early eukaryotic origin of meiosis. *Current Biology* 15: 185–191.

Parker, G.A. and Pizzari, T. 2010. Sperm competition and ejaculate economics. *Biological Reviews* 85: 897–934.

Section 15.4   Andersson, M. and Simmons, L.W. 2006. Sexual selection and mate choice. *Trends in Ecology and Evolution* 21: 296–302.

Clutton-Brock, T. 2007. Sexual selection in males and females. *Science* 318: 1882–1885.

Emlen, D.J. 2008. The evolution of animal weapons. *Annual Review of Ecology, Evolution and Systematics* 39: 387–413.

Section 15.5 Chapman, T., Amqvist, G., Bangham, J. and Rowe, L. 2003. Sexual conflict. *Trends in Ecology and Evolution* 18: 41–47.

Swanson, W.J. and Vacquier, V.D. 2002. The rapid evolution of reproductive proteins. *Nature Reviews Genetics* 3: 137–144.

● QUESTIONS

1. Describe how the eukaryote life cycle is divided into vegetative and sexual phases, and show how this leads to distinct processes of natural and sexual selection.

2. In what sense is sex a costly process and why does this raise difficulties for evolutionary theory?

3. How does sex modulate the process of adaptation? How is this affected by mating behavior (inbreeding versus outbreeding) and meiotic recombination?

4. Compare and contrast the evolution of genes regulating (a) meiosis, and (b) gamete fusion.

5. Account for the evolution of male and female gender.

6. Why are males usually more highly modified than females through sexual selection?

7. On what grounds might females evolve to prefer one type of male rather than another?

8. Bright and conspicuous sexual ornaments threaten the survival of the males that bear them. How do such exaggerated male characters evolve?

9. In what circumstances does sexual selection favor small size in males? Conversely, in what circumstances does sexual selection favor large size, weaponry, and aggression in males?

10. In what circumstances is competition more intense between (a) individuals of the same gender, and (b) individuals of different gender?

You can find a fuller set of questions, which will be refreshed during the life of this edition, in the book's Online Resource Centre at **www.oxfordtextbooks.co.uk/orc/bell_evolution/**

16 Cooperation and Conflict

Many features of organisms evolve because they enhance individual fitness in some defined situation. Tuna are streamlined because this reduces drag; snails are cryptically colored because this reduces predation. The external environment is taken to be a settled set of conditions to which a particular population adapts. This is a very useful simplification that is often roughly true, and it provides powerful explanations of most aspects of morphology and physiology.

There are other features of organisms, however, that cannot be readily explained in this way. The most interesting involve social behavior, and especially any behavior that involves a degree of helpfulness or cooperation among individuals. Unqualified exploitation or aggression, of course, needs no special explanation: it is easy to understand that selection will often favor the selfish pursuit of individual interest. It is more difficult to explain why individuals should help one another. It is much more difficult to explain why individuals should lay down their lives, without hope of return, in order that others should live. The purpose of this chapter is to show how natural selection can lead to the evolution of helpful behavior in certain kinds of society. We begin, as usual, with the simplest possible situation, involving bacterial cultures where social relations are of the simplest kind. Simple as they are, they show us how much more complicated kinds of society can be understood.

16.1 Neighbors are an important feature of the environment.

Social behavior occurs in societies, which we usually take to mean well-organized groups of complex organisms which interact with one another according to a system of rules. It does not seem likely at first sight that organisms as simple as bacteria could live in societies. A growing colony of bacteria, however, always has neighbors—other bacteria or yeasts or any other microbe. It will compete with them for resources, poison them with its metabolic products, or be inhibited by them in its turn. Its social relations are no doubt as simple as they can possibly be, but for exactly this reason we can use bacteria to chart the beginnings of social life.

A bacterial culture growing in liquid medium in a glass tube is just about the simplest society we can imagine. All cells require the same resources, which are dissolved in the medium and are thus equally available to all. There will be competition for these resources, of course, but this need not imply any sort of social interaction.

To express this more precisely, let us conduct a thought experiment in which we compare two culture systems. In the first place, the growing culture is maintained in a single tube and a certain number of cells are transferred at intervals to fresh medium. It will consist of a number of lineages, either because it was set up as a genetically diverse population or because mutations occur as it grows. As we have seen, however, the most rapidly growing lineage will eventually replace all others.

Now imagine instead that every lineage is cultured in a separate tube and each is again transferred at intervals, in such a way that the total number of cells transferred is held constant while each lineage contributes in proportion to the growth it has achieved. This will eventually lead to exactly the same result: only the tube occupied by the most rapidly growing lineage will contain any cells. Since the outcome is the same whether the lineages are mixed or kept separate, the process is completely asocial. Uniform competition does not constitute social behavior.

Conversely, we can define social behavior in terms of our thought experiment as any behavior that would cause a different outcome according to whether lineages are mixed or isolated. This can arise in bacteria when lineages use different resources, or use the same resource in different ways. It may seem a little eccentric to use bacteria to illustrate the basic principles of social behavior, but bear with us for the moment.

16.1.1 Diversity can be maintained by metabolic specialization.

There is a fundamental trade-off between the rate and yield of ATP production. The substrates dissolved in the growth medium are examples of public goods— resources that are freely available to all. When heterotrophic organisms degrade organic substrates to obtain energy, some of this energy is used to drive the reactions involved in degradation, while the rest is stored as high-energy phosphate bonds in ATP. If more of the energy is used to drive the reactions, they go faster, but the yield of ATP is less. Hence, when a substrate is degraded there is a fundamental trade-off between the rate of ATP production (moles of ATP per unit time) and the yield of ATP production (moles of ATP per mole of substrate). The two principal modes of deriving energy from organic compounds reflect this dichotomy.

- Fermentation uses an organic compound such as ethanol as an electron acceptor. It has a high rate of ATP production but a low yield (2 moles ATP/ mole glucose).

- Respiration uses an external electron acceptor, usually oxygen. It has a low rate of ATP production but high yield (32 moles ATP/mole glucose).

In short: fermentation is a rapid but profligate way of processing resources, whereas respiration is slow but frugal. Competition for freely available resources, such as sugars in solution, favors profligate fermenting types which convert them rapidly into new biomass, and in doing so removes them from the medium so that they cannot be used by more efficient respiring types. Hence, bacteria and yeasts growing on external nutrient sources evolve a fermentative lifestyle, whether or not oxygen is present. This is a wasteful use of resources, because most of the potential chemical energy remains unused in byproducts. Nevertheless, it will be favored by natural selection because any restraint in resource use is punished by a lower rate of growth. Selection does not necessarily maximize the yield or the efficiency of biological processes.

The situation is completely different when nutrients are internalized, like food items phagocytosed by a cell or swallowed by an animal. They then become private goods, the property of a single individual that cannot be appropriated by others. In this case they are usually processed efficiently by respiration, with a high yield of ATP.

Inefficient metabolism generates incompletely degraded substrates that can be used by others. Hence, a simple bacterial culture will quickly become dominated by fermenting types because they grow more rapidly, even though this leads to low yield. This is not the end of the story, however. Yield is reduced because the substrate is incompletely degraded. Intermediate products of metabolism such as ethanol and acetate are excreted into the medium by fermenters because they inhibit metabolism when they build up within the cell. These excreted substances can then be used as substrates by strains able to respire. When these types arise by mutation they will spread because they are able to scavenge the waste products of the fermenters. This is called "cross-feeding." It is a distinctively social process because its outcome depends on both types living together in the same culture. The outcome is a metabolically diverse community generated by the mutual invasibility of fermenters and scavengers, in the way sketched in Figure 16.1.

Cross-feeding often evolves in laboratory microcosms, even when these are founded by a single strain of bacteria. It is readily detected when fermenters and

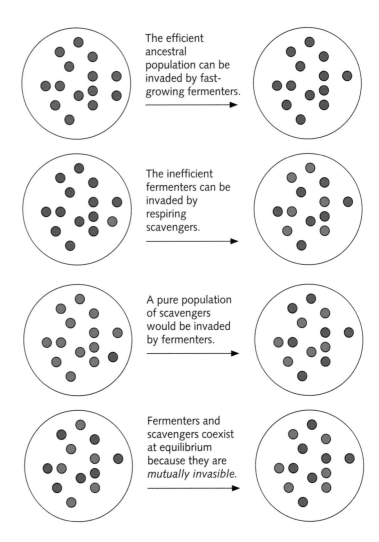

The efficient ancestral population can be invaded by fast-growing fermenters.

The inefficient fermenters can be invaded by respiring scavengers.

A pure population of scavengers would be invaded by fermenters.

Fermenters and scavengers coexist at equilibrium because they are *mutually invasible*.

Figure 16.1 Cross-feeding is maintained by mutual invasibility. The mutual invasibility of prodigal (fermenting) and frugal (respiring) lifestyles.

respirers form visibly different kinds of colony when they are spread onto solid medium. In the experiment illustrated in Figure 16.2, respirers formed large colonies whereas fermenters formed small colonies. Both will grow in normal nutrient medium. Both will also grow on spent medium – medium from which the cells are filtered out once growth is complete— provided that it is supplemented with glucose. Neither will grow on unsupplemented spent medium in which the same type has previously grown. The large-colony type, however, is capable of growing on spent medium in which the small-colony type has previously grown, proving that it is capable of scavenging the end-products of small-colony metabolism. The small-colony type, on the other hand, is not capable of growing on spent medium in which the large-colony type has previously grown. This simple experiment demonstrates how social interactions can

evolve within bacterial populations through metabolic specialization.

Given that fermenters have greater fitness than respirers when they are rare, while respirers likewise have greater fitness than fermenters when they are rare, there must exist some intermediate frequency at which the two types have equal fitness. Figure 16.3 shows that this frequency represents a stable equilibrium because:

- A small increase in the frequency of respirers reduces the supply of intermediate substrates, on which the respirers depend, so that fermenters now have greater fitness.

- A small increase in the frequency of fermenters lowers the supply of the primary resource and increases the supply of intermediate substrates, so that respirers have greater fitness.

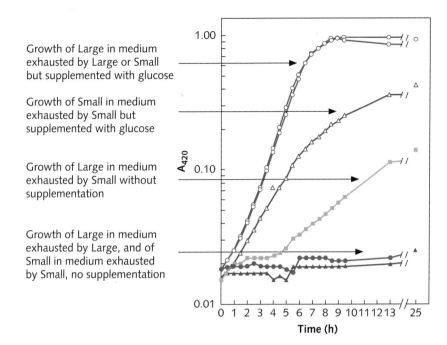

Growth of Large in medium exhausted by Large or Small but supplemented with glucose

Growth of Small in medium exhausted by Small but supplemented with glucose

Growth of Large in medium exhausted by Small without supplementation

Growth of Large in medium exhausted by Large, and of Small in medium exhausted by Small, no supplementation

Figure 16.2 Social interactions evolve in bacterial populations by cross-feeding. This experiment shows that a respiring type (forming "Large" colonies when spread on agar) can scavenge resources that are wastefully secreted by a fermenter ("Small"). Small did not grow in medium exhausted by Large, although this is not shown in the diagram.

Republished with permission of the Genetics Society of America, from Helling, R.B., et al. Evolution of *Escherichia coli* during growth in a constant environment. *Genetics* 116: 349–358 © 1987.

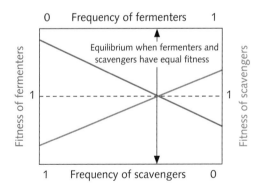

Figure 16.3 Mutual invasibility implies frequency-dependent selection.

In this way, the diversity of the population is actively maintained at some characteristic frequency for each type by negative frequency-dependent selection.

16.1.2 Diversity may evolve in structured environments.

The divergence of fermenters and respirers evolves in a well-stirred, homogeneous culture medium. In more realistic circumstances, the environment is likely to be patchy, with different places providing different conditions of growth. Spatial structure provides a further opportunity for diversification. For example, *Pseudomonas fluorescens* is a bacterium that lives on plant roots in the soil; it is a harmless relative of *Pseudomonas aeruginosa*, one of the agents that cause multiple sclerosis. If a vial of culture medium is inoculated with *P. fluorescens* it becomes cloudy within a day as the bacteria multiply. If the vial is left on the bench for three or four days, however, an odd thing happens: the vial becomes clearer, while a tough white mat covers its surface. This is the visible sign of a cryptic adaptive radiation within the vial.

Cells from different regions of the vial form colonies with characteristic shapes. If we take samples from different regions of the vial and spread them on agar, the colonies that grow have different shapes, as illustrated in Figure 16.4.

• Cells from the interior of the culture form smooth, glossy colonies that are similar to the ancestor with which the vial was inoculated. This is the smooth type.

Normal bacteria growing in broth

Mat formed at surface of medium

Pile of cells at bottom of vial

Smooth:
ancestral type, from interior of broth

Wrinkly spreader,
from surface mat

Fuzzy spreader,
from bottom of vial

Figure 16.4 Adaptive radiation of bacteria in an untended vial.

Reprinted by permission from Macmillan Publishers Ltd: Rainey, P.B. and Travisano, M. Adaptive radiation in a heterogeneous environment. *Nature* © 1998.

- Cells from the surface mat form flattened colonies with a pleated surface. These are called wrinkly spreaders.

The wrinkly spreader phenotype is caused by mutations in the *wss* operon that cause constitutive production of sticky cellulose-like fibrils that bind the cells together in a mat. The distinctive appearance of the colonies on agar is just a fortunate coincidence that can be used to study the dynamics of diversification within the culture.

Diversity requires environmental heterogeneity. A glass vial containing nutrient medium might seem to be as homogeneous as an environment could possibly be. Even in a freshly inoculated vial, however, there will be a gradient of oxygen from the surface downward, because diffusion is slow relative to the respiration of the culture. This gives an advantage to cells at the surface because they have a better oxygen supply. Hence, wrinkly spreaders arising by mutations in *wss* invade because they can maintain their position at the surface by sticking together to form a sheet of cells fixed to the glass wall of the vial. Moreover, the formation of the mat actually steepens the oxygen gradient because it inhibits oxygen

diffusion into the medium. The wrinkly spreaders pay for a more abundant supply of oxygen, however, in terms of reduced access to nutrients in the bulk of the medium, which are readily available to smooth cells living underneath the mat.

The coexistence of two distinct types thus depends on the existence of two distinct niches in the vial—an oxygen-rich but nutrient-poor region at the surface and a nutrient-rich but oxygen-poor region beneath. This is easily proven by repeating the experiment while shaking or stirring the vial. This prevents any oxygen gradient from forming, and the population remains uniform because wrinkly spreaders do not evolve.

As the culture ages, it continues to diversify. Several slightly different wrinkly spreader types appear, reflecting further specialization in different zones of the mat. A quite new type called the fuzzy spreader (its colonies have a characteristic blurred appearance on agar) also colonizes the surface, but does not form a mat and often falls to the bottom to form a loose pile of cells. It has a more complicated ecology than smooth or wrinkly spreader: it can invade cultures only when smooth is also present—that is, it will not invade a pure culture of wrinkly spreaders—and it is especially successful when the culture is attacked by a predatory bacterium. Thus, complex social interactions develop even in the simplest kind of bacterial culture and generate divergent selection resulting in the evolution of a range of specialized types.

Environmental heterogeneity leads to negative frequency-dependent selection. A pure culture of any type can be set up by picking a colony from agar and using it as the sole ancestor of a new population. Once grown, we can easily see whether it can be invaded by another type. Not surprisingly, wrinkly spreaders can invade a population of smooths, while smooths can invade a population of wrinkly spreaders. This mutual invasibility reflects frequency-dependent selection: either type has an advantage when rare, as shown in Figure 16.5. In this case, the advantage of rarity is generated by environmental heterogeneity because the two main niches in an unstirred vial are refuges for specialized types. The surface is a region where wrinkly spreaders predominate and cannot be displaced by smooths; in the bulk medium, smooths predominate and cannot be displaced by wrinkly spreaders. Each region necessarily constitutes a

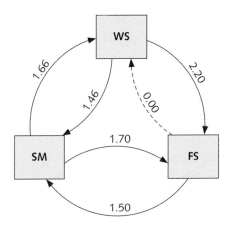

Figure 16.5 Mutual invasibility in the *Pseudomonas fluorescens* system. The figure on each arrow is the fitness of a rare invader (root of arrow) relative to the unit fitness of the resident type (tip of arrow).

Reprinted by permission from Macmillan Publishers Ltd: Rainey, P.B. and Travisano, M. Adaptive radiation in a heterogeneous environment. *Nature* © 1998.

certain fraction of the habitable space within the vial and thus guarantees the persistence of the type best fitted to live there. This is a very general rule: when the niches of a heterogeneous environment provide refuges for specialized types, frequency-dependent selection will act to preserve diversity. It is not necessary to visit distant locations to understand the basic principles of how biodiversity evolves: many of them are illustrated by a situation as mundane as an untended bacterial culture on a laboratory bench.

16.1.3 Social interactions maintain diversity through mutual invasibility.

Cross-feeding and surface growth are two examples of a general phenomenon. Imagine a pure culture of

respirers growing in nutrient medium. We introduce a few fermenters. These will spread because they have a higher rate of growth. Conversely, imagine a pure culture of fermenters to which we introduce a few respirers. These will also spread, because they can use a resource that the fermenters cannot. In other words, either type is able to invade a population dominated by the other. Hence, fermenters and scavengers coexist at equilibrium because they are mutually invasible. In a social context, natural selection does not necessarily, or even usually, lead to the fixation of a single best-adapted type. It is often the case that two or more differently specialized types are mutually invasible. In these circumstances, selection acts to preserve diversity.

We can extend this principle by considering the fitness of either type in relation to the composition of the population. At one extreme, consider a few fermenters introduced into a population of respirers. The fitness of the fermenters is greater than that of the respirers. In the converse case, when a few respirers are introduced into a population of fermenters, the fitness of respirers is greater—or, in other words, the fitness of the fermenters is less than that of the respirers. More generally, the fitness of fermenters decreases as their frequency increases. Hence, mutual invasibility implies frequency-dependent selection.

The lesson of these simple experiments with microbes is that when neighbors are an important environmental factor the fitness of any type will depend on the composition of the population. The fitness of a type will vary with its frequency, because similar individuals, using the same resources, compete with one another more intensely than they compete with dissimilar individuals, which use different resources.

16.2 Social interactions introduce a new aspect of adaptation.

Competition is a familiar aspect of everyday life, and its effect on the division of labor in human societies is obvious. If there are too many bakers or botanists in a town then their income will fall and prospective recruits will choose some other profession. Cooperation is no less familiar: we often help our neighbors, friends, and relatives, and in turn they help

us. In a biological context, however, there is a wide difference between the two. Competition is easy to understand, because lineages that fail to compete effectively are eliminated. Cooperation is more difficult, because it confers a benefit on a potential competitor. Hostility and aggression do not usually require special explanation, because selection will normally reward types that succeed in competition with others. It is much more difficult to explain why individuals should sometimes be helpful and trustworthy.

16.2.1 Individuals usually live in groups.

In a stirred culture, each bacterial cell interacts only indirectly with other cells, through the effect that each produces on the common medium. There are other kinds of organism that seem completely asocial, such as dinoflagellates or copepods in the open sea. By contrast, most animals live in groups and therefore interact directly with their neighbors. In a shoal of fish, a flock of birds, or a herd of animals, each individual will experience and respond to the behavior of a few others close to it, and will likewise influence them. But why do flocks or herds form in the first place? There are three reasons for this.

- The first is simply inability to disperse. Plants very often form patches of individuals of the same species because of their limited seed dispersal and lack of motility. Even bacteria are usually concentrated in soil pores or on particles suspended in water, so that each colony interacts mainly with a few neighboring colonies, rather than with the wider population.

- The second is that the environment is variable, with some patches that provide good conditions of growth and others that do not. Organisms able to disperse will quickly find the good patches and grow there.

- The third reason is that individuals form groups because they choose to do so, even though they are able to disperse freely.

16.2.2 There is often safety in numbers.

To see why there can be an advantage merely in belonging to a group, suppose that you are one of three fish living in a lake. There is a predator nearby—a pike, say—which you cannot see but which will sooner or later attack the closest prey. If there are two equally close, then one or the other is attacked with equal probability. The probability that you will be attacked is, of course, one in three. If you do not know where the predator lurks, how can you reduce the chance that you will be attacked? The answer

is, to move as close as you can to one of the other fish. The probability that you and your neighbor will be closer to the pike than the third, isolated, fish is one-half; and if so, the probability that you will be attacked is also one-half; so the overall probability that you will be attacked is $\frac{1}{2} \times \frac{1}{2} = \frac{1}{4}$. Just by moving close to a neighbor, you have increased your chance of survival from $\frac{2}{3}$ to $\frac{3}{4}$.

Precisely the same argument applies to the isolated third fish, which could now increase its chance of survival by joining the other two, forming a group of three. There is now a small shoal of fish. At this point, of course, the chance of being attacked is again one in three. Now this is the same as it was originally, with three isolated individuals. Nevertheless, the self-interest that led to the formation of the group will equally ensure its persistence, because leaving the group will entail an increase of risk, from $\frac{1}{3}$ to $\frac{1}{2}$. What applies to three will apply as well to a thousand, and a shoal of fish, a flock of birds or a herd of animals will be the result.

Just as prey may group to protect themselves against predators, so predators may group in order to overcome prey. A single wolf would be unable to bring down a moose, but a pack of wolves will readily do so. There may then be a mutual advantage for individuals to band together in order to pursue prey that an isolated individual would be unable to overcome.

16.2.3 Social interactions govern success in groups.

However they are formed, life in groups adds a new dimension to adaptation. Every individual must be able, of course, to withstand the physical stresses of life, must be able to gather enough food, and so forth. But they now need one other ability: they need to get along with the neighbors. Neighbors that are always disagreeing will naturally move apart, or destroy one another. Hostility and aggression do not usually require special explanation, because selection will normally reward types that succeed in competition with others. It is only if neighbors cooperate, however, that groups will persist. The central problem of social behavior, then, is why individuals should sometimes be helpful and trustworthy. More exactly: in what circumstances will natural selection favor types that are helpful and trustworthy? To focus the question, let us recognize four grades of behavior that are successively more onerous in terms of the burden that they place on the individual.

- **Sharing:** accommodating rather than attacking or exploiting partners.
- **Subsidizing:** contributing public goods that others use.
- **Cooperating:** trading honestly despite the risk of loss to cheats.
- **Sacrificing:** sacrificing personal fitness for the gain of others.

All of these cases incorporate a social dilemma: to provide a benefit to others at the expense of a cost to oneself. They differ in the ratio of benefit to cost. When the cost is slight, even a small benefit may be decisive, and the standard theory of natural selection can be readily extended to a social context. When the cost is death or sterility (equivalent, in evolutionary terms) then the theory has to be modified. The modification, however, serves to illustrate even more clearly the true nature of the Darwinian process.

16.2.4 Selection among groups may drive social evolution.

Cooperation is defined in relation to other individuals, and the nature and extent of cooperation are features of a group of individuals as a whole, rather than of any isolated individual. The vigor and stability of human societies depends on their laws and customs, and in non-human social organisms the fate of a group will likewise depend on how individuals interact with one another. Groups whose members cooperate with one another may be more productive or less likely to collapse than groups of selfish individuals. They are likely to be more productive when there is a food source that can be exploited more effectively by an organized group than by individuals; they are less likely to collapse when banding together enables a group to withstand attacks by enemies to which individuals are vulnerable. Cooperative groups will then expand at the expense of selfish groups, and as they expand the frequency of cooperation will increase in the community as a whole. This is a process of selection among groups, rather than among individuals.

Any individual trait that affects the viability or fecundity of an individual and that varies among individuals will be subject to individual natural selection, as in all the previous examples of natural selection in this book. Any social trait that affects the productivity of a group and that varies among groups will likewise be subject to group selection.

Selection may act at several levels of organization. Group selection is a process of selection acting at a level above that of the individual, usually among local populations. I have already described (Section 11.4) how selection can act at a level below that of the individual, among components of the cell, such as mitochondria, or among components of the genome, such as transposons. In colonial organisms such as corals and beech trees, where individuality itself is dubious, there will be a process of colony selection. Many characters will affect the characteristics of entities at several different levels of organization: mutations in respiratory genes, for example, might affect the growth of mitochondria, cells and colonies. In such cases, selection will operate at several levels simultaneously, and the outcome of selection will depend on the balance between them.

Group selection is weakened by dispersal. In territorial animals such as the Scottish red grouse some individuals are strong enough to defend feeding territories in the heather for their exclusive use, driving out any trespassers. These successful birds are able to appropriate the whole of the food supply in their territory and are likely to survive the harsh winter. Unsuccessful birds that do not acquire a territory must wander from place to place and will probably starve to death. Individual natural selection is likely to favor aggressive behavior that increases individual fitness, of course, but territoriality might also affect the fate of the population as a whole.

Suppose that every individual were able to roam at large; the area from which it could draw nutrition would necessarily be much smaller than a territory, and might well be inadequate to supply a sufficient diet, so that every bird would starve to death. Territoriality ensures that some birds at least will survive the winter, provided that other individuals are willing to forego the opportunity to acquire a territory, and are therefore willing to die to ensure the survival of the group as a whole. Territorial populations will expand at the expense of non-territorial populations, through the sacrifice of the non-territorial birds, and territoriality will spread through the whole grouse community through group selection.

So long as populations are completely isolated from one another, group selection is likely to occur and provides a plausible explanation for the evolution of the extreme degree of cooperation that this interpretation of territoriality entails. The weakness of the explanation is that group selection is effectively obstructed by dispersal between populations. Selfish birds who attempt to set up a territory without regard to the interests of the group as a whole will be more likely to survive and reproduce than cooperators who refrain from competition. Alleles that encode selfish behavior will therefore increase in frequency in each local population. When individuals move from one population to another they will from time to time inoculate highly cooperative populations with selfish alleles, ensuring that selfish behavior is always spreading in every local population. Even very low rates of dispersal—one or two immigrants arriving each generation into any given population—are sufficient to block the evolution of cooperation through group selection. This is why group selection is not generally considered to be a reasonable explanation of social traits such as territoriality.

Group selection is strengthened by aggregation. The situation is different if dispersal is selective, such that cooperative individuals tend to occur together in the same local population more often than would be expected by chance. This would ensure that most cooperative acts are directed towards other cooperative individuals, increasing the fitness of the alleles governing cooperative behavior. Cooperation can then evolve through group selection even in the face of very high rates of dispersal. It is difficult to imagine, however, how cooperative individuals could consistently and reliably recognize one another. It is not sufficient for cooperation to be associated with some arbitrary signal, because selfish individuals who copied the signal would then be likely to evolve. The difficulty of providing plausible general reasons that cooperating individuals should band together to the exclusion of selfish competitors is the main weakness of group selection as a general theory for the evolution of social behavior.

There is one common situation, however, in which neighbors are likely to resemble one another strongly: this is when the local population is a family group related by descent. Competition among family groups leads to kin selection (Section 16.6), which provides a convincing explanation for the evolution of highly cooperative societies.

16.3 Sharing can evolve when fighting is risky.

Hydractinia is a marine hydrozoan that forms colonies of zooids on gastropod shells. When two founding zooids settle on the same shell, they will proliferate until the edges of the two colonies come into contact. Colonies that have the same genotype at a locus that determines somatic compatibility will then fuse. If they are dissimilar, they will respond in one of two ways. The first is to lay down a fibrous matrix that forms a stable boundary at which growth by both colonies ceases. The two colonies then share the resource provided by the limited area of the shell. The second is to proliferate a stolon, into which nematocysts from nearby zooids migrate. These nematocysts are discharged into the foreign tissue of the neighbor, forming a necrotic zone of dying zooids. This continues until the neighboring colony is destroyed.

Some genotypes (which can be replicated by dividing the growing colony into portions) are always stoloniferous; some are never stoloniferous; others are able to switch from one behavior to the other. When a stoloniferous colony meets a matrix-forming colony, it invariably destroys it after a more or less prolonged struggle. When two stoloniferous colonies meet, one—the faster-growing colony—destroys the other. The stoloniferous colonies are certainly the stronger competitors, in the sense of repressing the growth of their neighbor more effectively. Yet both types coexist in *Hydractinia* populations, so aggression cannot always pay.

In what circumstances will individuals choose to share a resource rather than fighting for it?

16.3.1 Social interactions can be analyzed as games.

Imagine that two individuals simultaneously discover a good that could either be shared, or completely consumed by either of them. An example would be two raccoons finding a garbage bin. Each can choose to fight to defend the resource, or to share it without fighting. The success of either strategy (= set of rules that govern actions) depends on what the other individual decides to do.

- Fighting is a good idea if the other has decided to run away. If they have also decided to fight, however, you might get hurt.

- Sharing is a good idea if the other has also decided to share. If they have decided to fight, however, you will have to run away and end up with nothing.

Situations in which the consequences of an act depend on the corresponding act of a partner or rival can be treated as biological games. The simplest game is played between two kinds of individual—Hawks, who fight, and Doves, who share. An individual who finds a pile obtains the good—unless another individual finds it at the same time. There is then a contest with the following three rules.

(1) When two Doves meet, they share the good equally.

(2) When two Hawks meet, they fight. The winner gets the whole of the good; the loser is injured and gains no benefit.

(3) When Hawk meets Dove, the Dove runs away and the Hawk gets the whole good.

The outcome of each pairwise contest for goods of total value V can be expressed in the form of a pay-off matrix, like this:

		Given that opponent behaves as:	
		Dove	Hawk
Pay-off to self when behaving as:	Dove	$V/2$	0
	Hawk	V	$V/2 - C/2$

Here, C is the cost of losing a fight to another Hawk. Value and cost are both accounted as the increment of fitness resulting from the contest.

Given these rules, which of the two strategies will prevail? We cannot appeal to optimality, because the success of an individual employing either strategy depends on context—the behavior of the individual with which it is competing. Instead, we can only provide a solution that is conditional on the state of the population.

- **Pure Doves?** Suppose that the population does consist entirely of Doves. A few Hawk individuals then appear by mutation or immigration. They will be very successful, because almost all their contests will be with Doves, whom they beat every time. A population consisting entirely of Doves is therefore not a stable outcome of selection.

- **Pure Hawks?** Suppose instead that the population consists entirely of Hawks, until a few Doves appear. Almost all their contests will be with Hawks, and they will lose. Hence, a population consisting entirely of Hawks may be a stable outcome of selection.

16.3.2 The outcome of social evolution is an evolutionary stable state (ESS).

The criterion that we have used to determine whether a particular outcome is stable is *invasibility*. Given that a population consists predominantly of some type, will another type have greater fitness, and therefore tend to spread, when it is introduced at low frequency? If there is any such type, then the current population is not the end-point of evolution. To put this the other way round, if there is no other type that can invade the population, then it has reached an evolutionary end-point. This non-invasible state is called the evolutionarily stable state, or ESS for short. Formally: the ESS is the behavior such that if almost all individuals express that behavior, no other behavior increases fitness.

The ESS may be a pure strategy or a mixture of strategies. It is easy to understand that Dove is never an ESS, whereas Hawk may be. However, Hawk is not necessarily an ESS. If the population consists predominantly of Hawks, then Doves will lose most encounters, and will gain nothing from contested goods. Neither will they lose anything, however, because they immediately retreat from aggression. Most encounters in this population will be between Hawks, one of whom gains the good while the other is injured. If the injury reduces fitness to a greater extent than possession of the good increases fitness, then the expected outcome of combat is negative—on average, fitness will be reduced for either individual entering a combat. In this case, Doves will indeed tend to spread, since they lose less from contests. They will not entirely replace Hawks, because they will eventually become so numerous that many of the contests will partner a Hawk against a Dove,

which the Hawk will always win without penalty. Thus:

- Pure Dove is never an ESS.
- Pure Hawk is an ESS provided that the cost of fighting is not too great.
- When fighting is very costly ($C > V$), the ESS is a mixture of Hawk and Dove in some proportion, which can be calculated to be a frequency V/C of Hawk.

This explains how sharing goods—such as the sharing of gastropod shells by *Hydractinia* colonies—can evolve despite the presence of unconditionally aggressive types.

16.4 Subsidizing evolves when cheats can prosper.

A colony of a hydrozoan such as *Hydractinia* may obtain sole possession of the outside of a gastropod shell. A hermit crab might be sole possessor of the inside of the same shell. In either case, the shell is an example of a good that can be reserved for the use of a single individual which may be able to prevent others from using it. In this case, the evolutionary outcome depends on the balance between the cost and benefit of defense. Other goods are non-excludable and can be shared by any number of individuals. These lead to a different kind of evolutionary dynamic that depends on how goods are produced and consumed.

- The first distinction is the cost of production. Common goods such as air and sunlight are natural resources that exist without any effort by the individuals using them. Public goods are resources produced by some individuals that can be used by all, such as roads or radio stations. Universities are public goods.
- The second distinction is the effect of consumption. Goods that are consumed may or may not be depleted. Air is not depleted (to any meaningful extent) by breathing, whereas water may well be depleted by irrigation. Lecturing at universities is a non-depletable resource.

Putting these two categories together, we have a simple classification of non-excludable goods, with an example of each combination:

	Depletable	Non-depletable
Common goods	land	air
Public goods	streets	street lighting

The use of non-depletable goods leads to the simplest kind of natural selection, whereby the type with the greatest rate of increase supplants all others. When goods are depletable, the share taken by one will reduce the share available to another. Hence, there will be direct competition for depletable resources. This leads immediately to the fundamental social dilemma: individual interest will often conflict with the interest of the group as a whole.

16.4.1 Selfish use of depletable goods leads to the tragedy of the commons.

Competition between individuals for depletable resources may damage the population as a whole in two situations.

- A depletable common good such as a pasture could be managed by a community for maximal productivity. When it is available to all, however, it will be in the interest of each to add more livestock to the pasture, resulting in overgrazing and loss of productivity. This is the "tragedy of the commons."

- A public good such as public education supported by taxation could be supported by the whole community. When it is costly to all, however, it will be in the interest of each to reduce their contribution, resulting in the deterioration of the good. This is the "tragedy of the taxes."

In both cases there is a tension between community interests and individual interests. Everyone would benefit if each would refrain from using more than their share or contributing less than their share. Given that everyone else follows these rules, however, each will gain by using more or contributing less. Hence, everyone will do this, and the common good is depleted or degraded.

16.4.2 The production of public goods provides an opportunity for cheating.

A public good created by a group of individuals can be exploited by others who made no contribution to it—freeloaders, in a word. Microbial systems, despite their rudimentary social systems, provide very clear examples of freeloaders and show how they can spread and persist.

- Yeast hydrolyzes sucrose outside the cell with an invertase encoded by the *SUC* genes. The fitness of a group of cells would be maximized by some optimal rate of invertase production. However, cheats that produce less than this amount will have a greater pay-off and will thus tend to spread. Cheats are readily fabricated by deleting *SUC* genes. At low density cheats will have lower fitness than wild-type, because there are few neighbors for them to exploit. Hence, the fitness of cheats should increase with inoculum density, which was confirmed by experiment, as shown in Figure 16.6. This demonstrates the potential for cheating to evolve. In fact, *SUC* genes are exceptionally variable in structure and copy number, perhaps reflecting the instability of cooperation in natural conditions.

- Siderophores are an important public good for many bacteria. The greenish color of normal

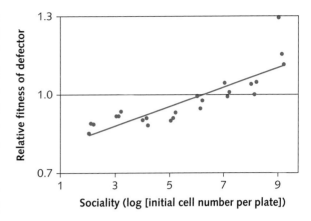

Figure 16.6 The fitness of yeast mutants unable to hydrolyze sucrose increases with inoculum density.

Greig and Travisano. The Prisoner's Dilemma and polymorphism in yeast SUC genes. *Proceedings B* 271 (Suppl 3) © 2004, The Royal Society.

Pseudomonas cultures is caused by the production of siderophores. These are iron-chelating molecules that are released by the cell. They bind ferric iron, and are then recaptured by the cell, which releases the iron and uses it. Any cell can recapture the siderophores made by another cell, however, without having to go to the trouble of making them itself. These types can be recognized by their pale, non-fluorescent coloration; they arise frequently in laboratory culture. The normal strain is thus a helpful type producing a public good that can be exploited by a freeloader.

Hence, the production of public goods often leads to a situation in which freeloaders are maintained by social interactions. How can we explain this in evolutionary terms?

16.4.3 The Producer–Scrounger game shows how free-riders are maintained.

Imagine a group of individuals, such as a flock of birds, searching for food so that interactions may occur between several individuals simultaneously. There are two kinds of individual, Producer and Scrounger. The Producers search actively, whereas the Scroungers loiter about until food has been discovered by a Producer, whereupon they all rush to share it. Before they can get there, the Producer has consumed the "finder's share," but must share the rest with the Scroungers. This is the Producer–Scrounger game. In this case the Scroungers exploit the Producers and lose by interacting with other Scroungers, although the gains and losses are less dramatic than in the Hawk–Dove game.

Unlike the Hawk–Dove game, the Producer–Scrounger game is played out in everyday situations. If you look carefully at a flock of pigeons or starlings foraging on the streets of your home town you will often be able to identify Producers and Scroungers. For example, the Carib grackle, *Quiscalus lugubris*, is a generalist group-foraging bird that feeds on discrete food items in urban areas of Barbados. Groups normally include both Producers and Scroungers.

Individuals are not bound to either strategy, and often change from one to the other.

At any one time, the frequency of Scroungers differs among groups, and in groups with more Scroungers the pay-off to this strategy is less, showing the negative frequency-dependent selection that is necessary to maintain both kinds of behavior in the population. Moreover, the balance of advantage between producing and scrounging can be altered by making food easier or more difficult to exploit once it is found, thereby altering the finder's share: when scrounging was made less profitable the frequency of the strategy fell. Hence, the predictions of the game-theoretic models can be checked by manipulating the pay-off matrix.

16.5 Cooperating can evolve through repeated encounters.

The Hawk–Dove game shows how selection does not necessarily favor unconditionally aggressive behavior. The Producer–Scrounger game shows how some individuals may produce public goods even though they may be exploited. Social behavior often goes further than this, however, and involves active cooperation. This must involve some degree of trusting other individuals, despite the fact that their interests are not the same as yours. How can this evolve through natural selection?

16.5.1 Costly helpful behavior can evolve only when there are repeated encounters between individuals.

Imagine that an individual may confer some benefit on another, at some slight cost to itself. For example, it might give warning of a predator. If the other individual is a stranger who will never be encountered again, then costly helpful behavior reduces fitness and will not evolve. If the other individual is a friend, however, whom you will often meet again, then it may well be worth warning them, because in the future they can return the favor by warning you. Friends will help one another, to their mutual benefit, because they can expect the favor to be returned. This begs a question, however: why should the favor be returned?

The answer is that the return of a favor can be reinforced by punishment. If you are not warned by your neighbor then you will in turn refrain from warning them, so punishing them for their selfishness. When they perceive this they may mend their ways and a mutually profitable warning system will then develop. Repeated encounters between two individuals may lead to mutually helpful behavior because either can punish the other for failing to reciprocate their assistance. Small birds will often cooperate to mob a predator, for example, in order to drive it away. This behavior is quite dangerous, so it's safer to let others do it—but they may not be keen to let you off the hook. Pied flycatchers, *Ficedula*, will mob model owls placed outside their nest-boxes. Most birds will join in a mobbing initiated by neighbors who had themselves previously mobbed the owl. However, they will not assist neighbors who had previously denied assistance to them. Hence, the mobbing behavior seems to be an example of reciprocity maintained by punishing uncooperative individuals.

16.5.2 Trading goods and services requires reciprocal cooperation.

When one kind of good is traded for another, how can a fair exchange—one acceptable to both parties—be guaranteed? Imagine that two societies live on either bank of a river. Farmers live on one side of the river: they grow corn. Smiths live on the other side: they make tools. The farmers need tools, and are willing to give corn in exchange; the smiths need corn, and are willing to give tools in exchange. Every week the farmers place a box full of corn on the island in the river. They remove the box put there by the smiths. The smiths put a box of tools, and take the corn. The outcome is that both parties gain, because each has received a benefit greater than the costs they have incurred.

A particularly avaricious farmer reasons that putting an empty box on the island would enable his community to gain the tools without paying for them. This reasoning is correct. His opposite number among the smiths, however, has come to the same conclusion. Both parties then find an empty box awaiting them, and trade comes to a halt, impoverishing both parties. The situation that leads to this dismal conclusion is the pay-off matrix of the Trader's Dilemma, illustrated by Figure 16.7.

The game is defined by: $C > A > D > B$. (It is better known as the Prisoner's Dilemma, but trading seems a more natural and general analogy.)

Cheats often prosper. It is obvious that one ESS of the Trader's Dilemma game is "always cheat." This can be demonstrated by simple evolution experiments with microbes. Viruses inside a cell synthesize proteins responsible for replication and encapsidation. When the cell is infected by several viruses, a cheat can utilize the proteins made by others while producing less than its fair share. Serial transfer of phage φ6 in these conditions resulted in the evolution of a strain that had high fitness in co-infected cells. When the ancestral and evolved strains compete, the pay-offs in terms of phage particles per host cell relative to the ancestor are:

		Your neighbor:	
		ancestor	evolved
Yourself	ancestor	A = 1	B = 0.65
	evolved	C = 1.99	D = 0.83

Hence, the selfish evolved strain becomes fixed. The helpful ancestor cannot invade because it loses to the selfish strain when it is rare; hence, the selfish strain has greater fitness at any frequency. Note how such social interactions reverse the normal logic of natural selection. Adaptation to the physical environment always causes an increase in fitness. Evolution in the social environment, where the agents of selection are neighbors which may themselves evolve, may result, as here, in a *decrease* in fitness.

Extreme kinds of cheat are the "defective interfering particles" that often evolve in virus populations. They have lost most or all protein-coding genes, and are incapable of replicating in the absence of intact virus. Hence, $D = 0$. This is not the canonical game: it leads to the coexistence of defective and intact virus.

Reciprocal cooperation succeeds because it creates the social environment in which it prospers. If you interact repeatedly with the same individual—unlike virus in a bacterial cell—it pays to cooperate, provided that you can trust them. But how do you know? You might know them personally, and know from long experience that they are trustworthy. You might even analyze their behavior in past encounters, and conclude that they can be relied on to cooperate. But then, being a rational individual, you cheat so as to take advantage of your knowledge. Is there any strategy that will make cooperation an unbeatable strategy for both players?

One powerful way of answering this is to conduct a selection experiment among possible strategies: a kind of evolutionary tournament. R. Axelrod (an economist) and W.D. Hamilton (an evolutionary biologist) invited a very wide range of people to submit

Your Neighbour:

	Play fair	Cheat
Play fair	**A = Award** for honesty with a dishonest neighbour	**B = Booby prize** for cooperating
Cheat	**C = Cheat** Gets something for nothing	**D = Deadlock** of dishonest traders

Yourself:

Figure 16.7 The pay-off matrix of the Trader's Dilemma.

computer programs which generated responses in the Trader's Dilemma when matched against a rival program. There was no limit on the complexity of the programs. The initial "population" was set up with an equal number of copies of each program, and programs were paired at random to compete against one another in a series of encounters. In the next "generation" the number of copies of each program was proportional to its payoffs in the previous generation. This was continued for 1000 generations, to enable the most successful kinds of strategy to emerge. Thus, the issue was decided by an unusual kind of selection experiment.

Many programs performed poorly and were rapidly lost from the population. A few did well and increased in frequency. The range of trajectories is illustrated in Figure 16.8. Remarkably, all except one of the strategies that increased in frequency at least briefly were "nice" strategies—that is, they are never the first to cheat. The only "nasty" strategy in the top 15 was: "be nice to begin with, but cheat on move 37 and thereafter cheat with increasing frequency if you find that you can get away with it." Why do nice strategies do so well? It is not because they are somehow able to exploit more aggressive strategies; on the contrary, they can draw at best and usually lose. Rather, it is because they modify the social environment so as to create conditions in which they prosper. Aggressive strategies perform poorly against one another because they fall into deadlock without ever realizing the benefits of cooperation. Cooperative strategies realize large gains when they meet, and if they take sensible precautions do not lose too badly when they are paired with an aggressive partner. Hence, the social environment becomes gentler and cooperative strategies prosper.

The winner was the simplest strategy (submitted by Anatol Rapoport of the University of Toronto): Tit-for-Tat. Here it is:

- On the first move, cooperate.
- On every subsequent move, do what your opponent did on the previous move.

This has three attributes that seem to be responsible for its success.

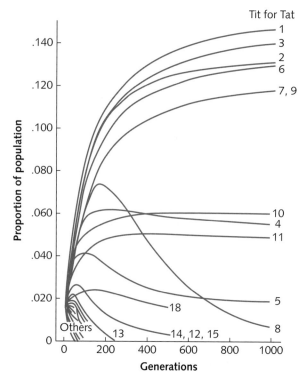

Figure 16.8 The Axelrod–Hamilton tournament. The lines show the trajectories of frequencies of programs in the tournament. Program 6 is "be nice to begin with, but cheat on move 37 and thereafter cheat with increasing frequency if you find that you can get away with it."

Reproduced from Axelrod, R. More effective choice in the Prisoner's Dilemma. *Journal of Conflict Resolution* 24 (3): 379–403 © 1980. Reprinted by permission of SAGE Publications.

- **It is nice:** it will never cheat first.
- **It is easily provoked:** a single attempt to cheat is instantly punished.
- **It quickly forgives:** if the opponent resumes cooperation, all past offenses are forgotten.

Tit-for-Tat is an example of reciprocity reinforced with punishment. It is tempting to speculate—many have—that this sounds like a promising basis for ethical behavior. From the point of view of this book, a more important point is that implementing the strategy does not need a human intellect: such a simple program could be followed by a bacterium. Some of the basic principles of social behavior, then, require no consciousness or even mental ability but apply to all organisms.

16.6 Altruism evolves only in family groups.

Ants, termites, and mole rats all form societies in which most individuals desist from reproducing in order to assist a small elite. This is a stage beyond cooperation. Cooperative behavior can evolve through reciprocity and retribution, but it does so because reciprocity confers an indirect benefit on the donor. Altruism is costly behavior that confers no benefit whether direct or indirect, current or deferred, on the donor. It cannot be explained by selection among individuals, which always rewards individual viability and fecundity. However, it can be explained by selection among families. When the family is the unit of selection, the process is called kin selection.

16.6.1 There is a profitable division of labor among cells in multicellular organisms.

The somatic cells of multicellular organisms are extreme examples of altruism: they forego any possibility of reproducing in order to enhance the reproductive potential of the germ cells. The flagellated somatic cells on the surface of *Volvox* (Section 8.1.1), for example, are responsible for motility and keep a rather large organism close to the surface—and the light. This makes it possible for the germ cells to grow so fast that the reproductive output of the colony as a whole is greater than an equivalent number of unicells could achieve. It is easy to see, in this case, how selection favors the evolution of a somatic caste of cells, because they are genetically identical to the germ cells. An allele directing the soma–germ distinction would spread because it would proliferate faster than an alternative allele directing all cells to reproduce.

In slime molds some of the aggregating cells will form the stalk of the colony and thereby enhance the success of those that develop as spores (Section 10.1.3). This situation is less clear than growth in *Volvox*, since the slime mold is formed by the aggregation of independent trophic amoebas and the cells are not necessarily identical. It is reasonable to suppose that most are identical, since dispersal in soil is probably a slow process. Hence, the evolution of the soma–germ distinction depends on the relatedness of the cells making up the body of an individual. If they

are indeed identical, then a separate caste of somatic cells will evolve provided they perform a function that elevates the reproduction of the individual above that which could be achieved by the same number of autonomous unicells. If they are merely cells drawn at random from the population then a somatic caste will never evolve, because an allele directing its development would not be systematically enhancing the proliferation of copies of itself. Evidently, there must be some intermediate state—some low frequency of unrelated cells within an otherwise homogeneous individual—where the advantage of multicellularity just balances that of independence.

16.6.2 There is a profitable division of labor among individuals in eusocial organisms.

The same principle that applies to cells in multicellular organisms applies to individuals in colonial organisms—provided that they are strictly asexual. The ecological circumstances that provide an advantage for separating vegetative and reproductive functions are quite different, of course. The most common involve a large external resource that is more profitably exploited by a group than by isolated individuals (like a farm) or a shelter that can be defended effectively only by a group (like a fort). The origin of eusociality thus requires a sufficiently large benefit to recipients of belonging to a colony, which could arise in either of two ways.

- Foraging. A colony can exploit a large external resource, while providing extensive care to helpless young. Examples include foraging red ants and bees seeking nectar. The original caste would be foragers, to provision the colony.

- Defense. A colony can defend a large, valuable resource against possible invaders. An example would be a termite nest in a dead tree. The first caste to evolve would be soldiers, to defend the nest.

Social insects such as ants and termites build large colonies with a marked division of labor, in which different castes are responsible for foraging, defense, and other duties, in order to support the few individuals who reproduce. Even in vertebrates there may be a simple division of labor between reproductive and sterile individuals. Meerkats and naked mole rats form large communities that are dominated by a single breeding pair. The others help to feed their young. It is only when this pair dies that subordinate individuals have a good chance of reproducing.

16.6.3 An allele that helps other copies of that allele to reproduce will tend to spread.

To understand how altruism evolves, suppose that a certain allele affects individual fitness such that one copy is transmitted to the next generation successfully, whereas an alternative allele encodes behavior that results in early death, before reproduction but also aids another individual bearing the same allele to produce three offspring. From the point of view of the gene, the first allele produced one copy per individual. The second produced $3/2 = 1.5$ copies. Any such allele will tend to spread, with a selection coefficient of $3/2 - 1 = 0.5$. Hence, helpful behavior will be selected, provided that the benefit (to copies of the helpful allele) is greater than the cost.

In asexual organisms, clonal offspring are (almost) certain to bear a copy of the same allele. In sexual organisms, however, offspring have only a 50% chance of bearing a copy of the same allele. A normal individual will produce two surviving offspring, each with 50% chance of bearing a copy of the allele it bears, so that the expected number of copies is $0.5 \times 2 = 1$, as before. A rare allele governing helpful behavior in one full sib helping another, at the expense of sacrificing its chance to replicate directly, would produce $(0.5 \times 3)/2 = 0.75$ copies of itself per individual. The helpful allele would spread only if it produces more than 1 copy per individual. Hence, it must assist its full sib to produce at least 4 offspring, if the cost to itself is to produce none.

Extending this argument, the benefit conferred on kin must exceed the cost to the individual, by a factor expressing the probability that the relative bear a copy of the helpful allele. In sexual organisms, this factor is equal to relatedness, leaving aside complications arising from asymmetrically inherited elements such as mitochondria.

> Relatedness r = probability that the focal individual bears a copy of the same allele.

This is normally but not necessarily the coefficient of relatedness by descent, because normally (but not necessarily) only relatives will have a greater probability of bearing the helpful allele than a random member of the population. For example:

Full sib	relatedness $r = 0.5$
Parent–offspring	relatedness $r = 0.5$
Grandparent–grandchild	relatedness $r = 0.25$
First cousins	relatedness $r = 0.125$

Relatedness can be used to partition the increment of fitness associated with a particular pattern of behavior into two components. The first is the direct effect of the behavior on the production of offspring by a focal individual, which corresponds to the normal definition of fitness in a non-social situation. The second is the indirect effect of the behavior on the production of offspring by other individuals, discounted by their relatedness to the focal individual. We can then define inclusive fitness as the sum of direct and discounted indirect effects, as illustrated in Figure 16.9. Inclusive fitness defined in this way is the currency of kin selection, giving us a straightforward way of extending the normal process of natural selection to the interpretation of social evolution.

Altruism will tend to evolve when a benefit conferred on relatives, discounted by their relatedness, exceeds the cost to the giver. To summarize this argument, let us define:

B	the benefit received by the recipient of an altruistic act (in units of fitness increment)
C	the cost of the act to the actor (ditto)
r	the relatedness of actor and recipient.

Altruism will then spread under natural selection if and only if $rB > C$. This is called "Hamilton's rule" after the evolutionary biologist who first cast it in this quantitative form.

The evolution of altruism depends only on the probability that two interacting individuals bear copies of the same "helpful" allele. There is a popular misconception that kin selection favors helpfulness because relatives share many of their genes. This is not the case. Kin selection favors helpfulness because relatives are more likely to bear copies of the allele that governs helpful behavior.

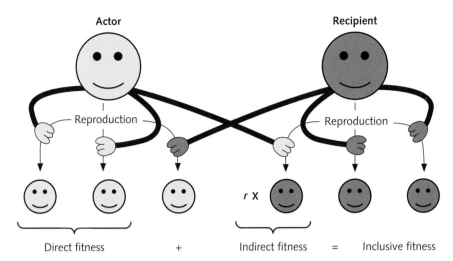

Figure 16.9 Inclusive fitness is the sum of the direct and discounted indirect effects of a pattern of behavior.

Reprinted from West, S.A., et al. Evolutionary explanations for cooperation. *Current Biology* 17 (16): R661–R672. © 2007, with permission from Elsevier.

16.6.4 Social insect colonies are family groups.

In social hymenopterans such as honey bees the colony descends from a single queen, so that all the workers are sisters. This high degree of relatedness provides the basis for kin selection. The same is true for other highly social animals such as termites and mole rat. The general principle is very simple: when families live together and act as a group, the most successful families will transmit their attributes disproportionately to the next generation, even if most family members only assist others, rather than reproducing themselves. Selection among families—kin selection—accounts for the evolution of altruistic behavior.

A large hymenopteran colony is a complex and highly integrated society that often involves highly altruistic behavior. The sting of honey bees, for example, is a barbed structure that sticks in the enemy—and thereby eviscerates and kills the bee. A stinging bee commits suicide for the benefit of her kin. However, there is a darker side. The sting of the queen bee is a curved, unbarbed structure that is used solely for killing her rivals, that is, her sisters. Thus, the integrity of the colony is preserved both by highly altruistic and by highly selfish actions.

Moreover, any highly organized society remains vulnerable to invasion by selfish types, such as workers who lay eggs themselves rather than rearing the offspring of the queen. In a small colony the queen might dominate by force, but in a large colony this is clearly impossible. The social contract must then be enforced by the workers themselves, who have a common interest in preventing selfish workers from reproducing. In animal societies, as in human societies, cooperation may be reinforced by punishment.

We have seen in this chapter how social behavior evolves through direct benefits, indirect benefits, and benefits to relatives:

- **Grouping**: obtaining direct individual benefits by being a member of a group. Individual natural selection accounts for this.
- **Sharing**: accommodating rather than attacking or exploiting partners. The direct benefit obtained by exploiting partners may be too risky.
- **Subsidizing**: contributing public goods that others use. A certain proportion of free riders will be tolerated.
- **Cooperating**: trading honestly despite the risk of loss to cheats. Partners that cannot readily separate have entrained fitness. Partners that can separate may receive deferred benefits by reciprocity.
- **Sacrificing**: sacrificing personal fitness for the gain of others. Helping relatives may evolve if it increases the fitness of the family as a whole.

Members of the same species live together in societies, with tighter or looser social bonds, and cooperate to a greater or lesser degree. Individuals of any species also interact with members of other species, with much greater ferocity or even greater intimacy than with members of their own. This is the subject of the next chapter.

● CHAPTER SUMMARY

Neighbors are an important feature of the environment.

- Diversity can be maintained by metabolic specialization.
  - *There is a fundamental trade-off between the rate and yield of ATP production.*
  - *Inefficient metabolism generates incompletely degraded substrates that can be used by others.*

- Diversity may evolve in structured environments.
  - *Cells from different regions of the vial form colonies with characteristic shapes.*
  - *Diversity requires environmental heterogeneity.*
  - *Environmental heterogeneity leads to negative frequency-dependent selection.*

- Social interactions maintain diversity through mutual invasibility.

Social interactions introduce a new aspect of adaptation.

- Individuals usually live in groups.

- There is often safety in numbers.

- Social interactions govern success in groups.

- Selection among groups may drive social evolution.
  - *Selection may act at several levels of organization.*
  - *Group selection is weakened by dispersal.*
  - *Group selection is strengthened by aggregation.*

Sharing can evolve when fighting is risky.

- Social interactions can be analyzed as games.

- The outcome of social evolution is an evolutionary stable state (ESS).
  - *The ESS may be a pure strategy or a mixture of strategies.*

Subsidizing evolves when cheats can prosper.

- Selfish use of depletable goods leads to the tragedy of the commons.

- The production of public goods provides an opportunity for cheating.

- The Producer–Scrounger game shows how free-riders are maintained.

Cooperating can evolve through repeated encounters.

- Costly helpful behavior can evolve only when there are repeated encounters between individuals.

- Trading goods and services requires reciprocal cooperation.
  - *Cheats often prosper.*
  - *Reciprocal cooperation succeeds because it creates the social environment in which it prospers.*

Altruism evolves only in family groups.

- There is a profitable division of labor among cells in multicellular organisms.

- There is a profitable division of labor among individuals in eusocial organisms.

- An allele that helps other copies of that allele to reproduce will tend to spread.

- Social insect colonies are family groups.

● FURTHER READING

Here are some pertinent further readings at the time of going to press. For relevant readings that have been released since publication, visit the book's Online Resource Centre at **www.oxfordtextbooks.co.uk/orc/bell_evolution/**

Section 16.1   Kassen, R. 2002. The experimental evolution of specialists, generalists, and the mainte-nance of diversity. *Journal of Evolutionary Biology* 15: 173–190.

Pfeiffer, T., Schuster, S. and Bonhoeffer, S. 2001. Cooperation and competition in the evolution of ATP-producing pathways. *Science* 292: 504–507.

Rainey, P. B. and Travisano, M. 1998. Adaptive radiation in a heterogeneous environment. *Nature* 394: 69–72.

Section 16.2   Lehmann, L., Keller, L., West, S. and Roze, D. 2007. Group selection and kin selection: two concepts but one process. *Proceedings of the National Academy of Sciences of the USA* 104: 6736–6739.

Section 16.3   Nowak, M. and Sigmund, K. 2004. Evolutionary dynamics of biological games. *Science* 303: 793–799.

Section 16.4   Morand-Ferron, J., Giraldeau, L-A. and Lefebvre, L. 2007. Wild Carib grackles play a producer–scrounger game. *Behavioral Ecology* 18: 916–921.

Section 16.5   Axelrod, R. and Hamilton, W.D. 1981. The evolution of cooperation. *Science* 211: 1390–1396.

Greig, D. and Travisano, M. 2004. The Prisoner's Dilemma and polymorphism in yeast *SUC* genes. *Proceedings of the Royal Society of London B* 271: S25–S26.

Section 16.6   West, S.A., Griffin, A.S. and Gardner, A. 2007. Evolutionary explanations for cooperation. *Current Biology* 17: R661–R672.

Nowak, M. 2006. Five rules for the evolution of cooperation. *Science* 314: 1560–1563.

● QUESTIONS

1. Describe the social dynamics of a mixture of fermenting and respiring genotypes of bacteria in a chemostat.

2. Describe the social dynamics of stirred and unstirred cultures of *Pseudomonas fluorescens* maintained by serial transfer.

3. How is genetic diversity maintained by environmental heterogeneity?

4. What are the benefits of living together with other individuals as a member of a group?

5. Explain how selection can act among groups as well as among individuals. Compare the efficacy of individual selection and group selection.

6. Give an account of the Hawk–Dove game and explain how it helps to understand the evolution of sharing.

7. Give an account of the Producer–Scrounger game and explain how it helps to understand how the production of public goods can evolve.

8. Give an account of the Trader's Dilemma game and explain how it helps to understand the evolution of honesty.

9. Describe how kin selection can lead to the evolution of altruistic behavior. Distinguish between direct and indirect effects, and show how indirect effects discounted by relatedness lead to a definition of inclusive fitness.

10. Compare and contrast group selection and kin selection in the context of helpful behavior.

You can find a fuller set of questions, which will be refreshed during the life of this edition, in the book's Online Resource Centre at **www.oxfordtextbooks.co.uk/orc/bell_evolution/**

17 Symbiosis and Struggle

Animals and plants do not live in isolation; they do not even live in communities isolated from others; each species lives as one thread in a living fabric of hundreds or hundreds of thousands of other species. This simple fact has the profound implication that evolutionary processes are likely to be dominated, not by physical factors, but rather by the multifarious interactions among species.

Natural communities consist of partners, enemies, and rivals. A bee passes from flower to flower, in the forest or meadow or your garden, collecting pollen or nectar. How has this behavior evolved? In the simplest terms, more efficient foragers will replace less efficient foragers within the population of bees. In fact—as we have seen in the previous chapter—the situation is more complicated because bees are organized into family groups with complex social behavior. But even this greatly understates the true complexity of the situation. Bees and flowers are trading services, the flowers providing food and the bees providing sex. Their relationship is helpful, to a point, but each will gain by exploiting the other, if it can: the flower will produce as little nectar as possible and the bee will make as few visits as possible, each to gain as much as it can for as little as it can give.

Moreover, each is engaged in other kinds of relationship. The plant depends on soil fungi growing around its roots for a reliable supply of nitrogen; it competes for nitrogen and other resources with neighboring plants; its leaves are eaten by caterpillars. The bee is competing for nectar with other kinds of bee as well as with flies and wasps; it runs the risk of being attacked by crab spiders that lie concealed within the flower; and its brood may be destroyed by parasitic fungi.

These friendships and animosities beget others in a cascade of ecological effects. The fungus that destroys the bee's brood reduces pollinator service to the plant, whose root fungi thereby suffer. By eating the plant, the caterpillar deprives the bee of food and thereby its fungal pathogen of a healthy host. Every species is embedded in a complex web of partners, enemies and rivals that will shape how it evolves.

Permanence ensures cooperation. The basic principle that governs how species interact is the same principle that governs how individuals of the same species behave towards one another. If the physical relation between two individuals of different species is close and cannot easily be changed or dissolved, each will tend to evolve to promote the other's wellbeing. They will evolve as mutualists. If they readily separate, on the other hand, one will evolve to exploit the other as ruthlessly as possible. They will evolve as parasite and host. The evolution of social relations between two species depends on the extent to which the interests of a couple are entrained.

17.1 Partners are more or less permanently associated.

Cooperation between individuals of the same species is surprising because they have similar attributes and competing interests. Individuals belonging to different species – especially if they belong to remotely related groups—are likely to have very different attributes and may scarcely compete at all. It is difficult to imagine, for example, what a mouse and a mushroom would compete for, or how any kind of evolutionary change would be driven by a shift in the relative abundances of mice and mushrooms. There is more opportunity for fruitful collaboration when partners differ, and the most extreme examples of cooperation involve quite different kinds of organism.

17.1.1 Cooperation between species is based on reciprocation.

When individuals can recognize one another, cooperation can be based on a reciprocal exchange of services. This benefits both partners, but of course is vulnerable to cheating. Cooperation will evolve only when cheating can be detected and punished.

Repeated interactions between individuals of different species are involved in cleaner–client symbioses. Cleaner fish (usually small wrasses or gobies) enter the mouths of much larger predatory fish, such as groupers, to remove dead tissue and ectoparasites. Groupers would normally eat small fish, but readily tolerate their cleaners. Indeed, they will visit the sites occupied by cleaners and solicit their attention. Crocodile birds likewise act as animated toothpicks for crocodiles basking on sandbanks, mouths agape.

Reciprocity is policed by punishment. These are clearly mutually beneficial relationships, but just as clearly they could break down: the grouper, for example, could make a quick profit by eating the cleaner, or the cleaner could take advantage of its privileged position by taking a bite out of its client. In fact, it sometimes does break down. An affronted client will then have several possible courses of action: it may punish the erring cleaner by chasing and threatening it or it may simply choose another cleaner in future. A cleaner that is rejected by most clients because it has previously cheated them—or has been observed by them to cheat others—will lose its livelihood. Reciprocal benefits can be established by custom when partners can recognize one another and have the ability to punish transgressors.

17.1.2 When two partners cannot easily separate, their fitnesses become entrained.

Cleaner fish and crocodile birds have a loose relationship with individual clients that can readily be dissolved. The bird or the crocodile may move away or die; another bird or crocodile will take its place. Some partnerships are more permanent, simply because the partners are physically unable to separate. Many plants, for example, are associated with soil fungi that invest their roots or even penetrate root cells and establish themselves within the living plant. Neither plant nor fungus is able to move apart. They are partners for life. In a situation like this, where two partners cannot easily separate, any fitness increment gained by one is transmitted in part to the other. Hence it is in the interests of either to support the other, because their fitnesses are entrained. Long-lasting interactions may evolve to benefit both partners because the fitnesses of two partners become mutually entrained when the partnership cannot easily be dissolved.

Long-lasting partnerships may become indissoluble. If a partnership is likely to last for a lifetime then both partners will gain by maximizing their joint productivity. They will gain even if they delegate vital functions that an isolated individual would need to survive. Green hydra, for example, are freshwater hydrozoans that capture small crustaceans with their tentacles. Despite being animals, they are green because they carry unicellular green algae in their bodies. The algae donate fixed carbon obtained by photosynthesis to the hydra, which reciprocates by providing nitrogen obtained from its prey. This is a very close partnership, but if food is scarce the hydra can end it simply by digesting the algae.

The acoel flatworm *Convoluta* is a more extreme case. It, too, is green, because its cells are filled with an endosymbiotic alga, *Tetraselmis*. In the adult worm, however, the pharynx and gut degenerate, leaving the animal completely dependent on photosynthesis. At this point the partnership has become indissoluble, at least for one of the two partners.

Vertical transmission leads to partners, horizontal transmission to enemies. Offspring may inherit their parents' partners, or the offspring of their parents' partners, for example if neither can disperse far. The relationship between partners may thus be continued beyond the lifetime of a single individual into future generations. This extends the concept of cooperation from individuals to lineages, and leads to a general principle that governs how intimate relations between organisms will evolve.

- If partners can readily separate and later acquire new partners then they will be selected to exploit one another because their fitnesses are uncoupled.

- This does not apply if the new partners are related to the old partners, because the genes governing the interaction will continue to be associated, and hence their fitness will be coupled, in proportion to the relatedness of successive partners.

The most straightforward reason that lineages should continue to be propagated in parallel is that one partner is transmitted through the germplasm of the other. In the extreme case, the partners live as a single compound individual. For example, lichens are a compound organism formed from algal and fungal partners. Many reproduce by means of soredia, small clumps of algal cells wrapped in fungal hyphae. Hence, both partners in the new lichen will be closely related to the partners in the parental lichen. The eukaryote cell is an even more intimate partnership, with chloroplast and mitochondrion descending from prokaryotic endosymbionts.

The converse is equally true: when lineages can readily separate, they will evolve as enemies because their reproduction is unlinked. This is the subject of the next section.

17.2 Enemies stimulate perpetual coevolution.

Every species lives in close association with many others. Mutualists such as hydras and green algae stand at one extreme; at the other there are parasites that kill or castrate their hosts. All manner of intermediate conditions can be found in nature. The fish lice (actually isopods) attached to salmon, for example, are external parasites living on body fluids, and they may severely damage the fish; the barnacles attached to whales, on the other hand, are making use of a convenient feeding platform without damaging their host beyond some slight increase in drag. In all cases, either is part of the environment for the other—the whale's skin serving the same function for the barnacle as a rock or a ship's hull, for example.

17.2.1 Other organisms act as agents of selection.

Individuals often differ in their ability to exploit any given environmental factor, and other organisms are no exception. When a population is in close contact over a long period of time with members of another species, therefore, it is likely to evolve adaptations that enable it to deal with them effectively.

Individuals vary in response to enemies. Infectious diseases often involve highly specific and intense interactions between the pathogen and its host. In most cases, host individuals vary in susceptibility to a particular disease. In human populations, for example, there is a great deal of genetic variation in susceptibility to leprosy and gastric ulcers, rather less to polio and smallpox, and only low levels of variation in susceptibility to mumps and rubella. Interestingly, the lowest levels of variation are found among diseases that have been successfully controlled by vaccines, for which there is an effective, long-lasting immune response. Our diseases likewise vary, as we learn from the seasonal shift in influenza serotypes each year requiring a new inoculation to keep pace with shifts in the virus population. Hence, both host and pathogen have the potential to adapt to one another.

Serial passage causes the rapid evolution of virulence. A graphic example of specific adaptation to another species is provided by infecting animals with bacteria, harvesting the bacteria after growth and then re-inoculating new host individuals taken from the same non-evolving ancestral stock. This continued serial passage is a special kind of selection experiment that has often been used in medical research, for example in the development of vaccines.

Serial transfer in a novel host is almost invariably accompanied by an increase in virulence, which often increases very rapidly to high levels. For example, Figure 17.1 shows how ten passages of *Salmonella typhimurium* through naïve mice increased mortality from less than 10% to more than 90%. The host's body provides a huge stock of resources which the pathogen is at first poorly adapted to exploit; types that exploit it more effectively will be selected, and increased virulence—damage to the host—is the consequence of their spread. Very high levels of virulence can evolve because the parasites are transmitted by the experimenter and are thus indifferent to the health of their host.

Host populations adapt to resident pathogens. Naturally, the converse is also true: hosts can likewise evolve resistance to repeated attacks by pathogens. The Composite Cross barley populations were set up forty or fifty years ago in California by wide crosses among many varieties of barley. They have been monitored at intervals ever since for various agronomic characters, including disease resistance. They are attacked by a variety of pathogenic fungi, including *Rhynchosporium*, which causes scald disease—large white blotches on the leaves that reduce photosynthesis and can cause a reduction in yield of several percent. Both host and pathogen are variable: different barley genotypes are resistant to different races of the fungus. Most of the varieties used to set up the initial hybrid populations were highly susceptible to most or all races of scald. After 45 generations of propagation, however, the frequency of resistance to most races had increased substantially, as shown in Figure 17.2. Moreover, many plants bore genes at different loci conferring resistance to several races of scald. An originally rather susceptible population, then, that is exposed to pathogens under more or less natural conditions evolved higher levels of disease resistance, through selection sorting both the initial variation and the variation arising subsequently from recombination.

17.2.2 Counter-adaptation leads to complex evolutionary dynamics.

As agents of selection, however, other organisms are fundamentally dissimilar to physical factors of the environment. A fish may well evolve resistance to higher temperature or lower pH, for example, but that is the end of the episode: the temperature or pH of the lake will be unaffected by anything the fish do. If it evolves resistance to a trematode parasite, on the other hand, that is only the beginning of the matter, because by the same token the trematode may subsequently evolve enhanced virulence when attacking the fish.

Costs of adaptation set limits to parasite virulence and host resistance. This implies that hosts will be selective agents driving the evolution of virulence in parasites, while parasites will in turn drive the evolution of resistance in hosts. If there are no other sources of selection then there will be a simple outcome—all hosts will become resistant and all parasites will become virulent. This is unlikely to happen in practice, however, because both resistance and virulence are likely to reduce fitness in the absence of enemies.

This principle has often been documented for bacteriophage, which attach to bacterial cells by binding to "phage receptors" expressed on the surface of the cell. These receptors are molecules (such as proteins or lipopolysaccharides) or structures (such as flagella or pili); for example, the receptor for phage φ6 is

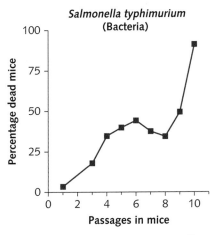

Figure 17.1 Serial passage increases the virulence of bacteria in hosts.

Graph redrawn from Ebert, D. 1998. Experimental evolution of parasites. *Science* 282 (5393): 1432–1436. Reprinted with permission from AAAS. Original data from Sutherland, I.A., et al. 1996. *Exp. Parasitol.* 83: 125.

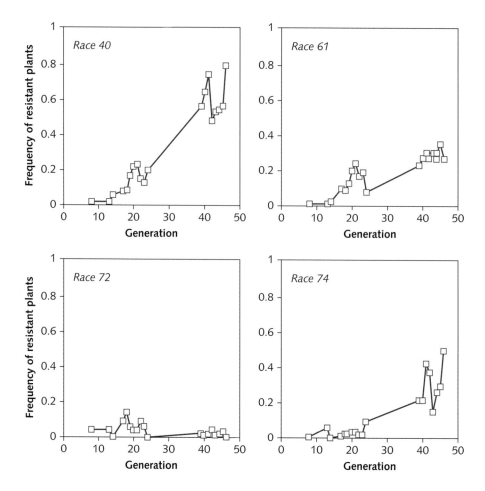

Figure 17.2 Evolution of resistance to races (strains) of scald, *Rhynchosporium secalis*, in barley Composite Cross II over 45 generations.

Based on data from Webster et al. 1998. *Phytopathology* 76: 661–668, Table 4.

lipopolysaccharide exposed on the outer membrane of its bacterial host *Pseudomonas*. Binding to phage is not the normal function of such molecules, of course—they are involved in sensing the environment, transporting substances across the cell surface, and protecting the cell. The evolution of resistance to phage often involves modifying or covering up or completely deleting these receptors. They cannot then function normally, and as a result resistant cells grow more slowly than the original susceptible cells when there is no phage present, as shown in Figure 17.3.

The same rule applies to parasites: types that are very successful in attacking some kinds of host are likely to be unable to infect others. For example, *Melampsora lini* is a fungal pathogen of wild flax, *Linum marginale*, where it causes rust disease on the leaves. There is a range of resistant genotypes in the plant, and a corresponding range of virulence types in the fungus. Each fungal race produces a certain

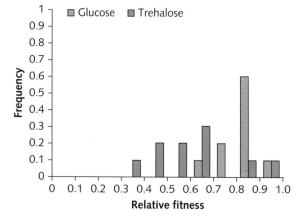

Figure 17.3 Cost of resistance to phage in experimental populations of bacteria. The ancestral susceptible line has unit fitness in the absence of phage; all evolved resistant lines have lower fitness. Glucose and trehalose are carbon sources in growth media.

Reproduced from Bohannan, B.J.M. and Lenski, R.E. 2000. Linking genetic change to community evolution: insights from studies of bacteria and bacteriophage. *Ecology Letters* 3 (4): 362–377, with permission from Wiley.

number of spores when it reproduces on one of the host types it is able to infect. As a general rule, the fungal genotypes that are able to infect a greater number of host types tend to have lower spore production on each. Hence, there are costs of adaptation, arising from the physiology of infection, that limit the virulence of parasites and the resistance of hosts.

Coevolution of enemies may lead to coupled oscillations in genotype frequency. With this in mind, suppose that there are two genotypes of host (H1 and H2) and two genotypes of parasite (P1 and P2). P1 can infect H1, but not H2, whereas P2 can infect H2, but not H1. This is the "allele-for-allele" system illustrated in Figure 17.4. How will such a system evolve?

- a population of H1 hosts will drive the spread of P1 in the pathogen population (the community is then mostly H1/P1);
- which will result in the spread of H2 in the parasite population (H2/P1);

- which will select for P2 pathogens (H2/P2);
- which will select for H1 hosts (H1/P2);
- which will favor P1 parasites (H1/P1);

and we are back where we started. The genetic coupling of parasite and host leads to an endless cycle of genetic change.

Planting very large areas of land with genetically uniform crops provides the ideal conditions for disease epidemics, and can lead to coupled genetic cycles driven by the natural selection experienced by the pathogens and the artificial selection applied to the crops. An example is shown in Figure 17.5. Barley cultivars carrying a new gene *Mla12* for resistance to powdery mildew began to be used on a large scale in the late 1960s, occupying about a quarter of the acreage by the end of the decade. At this point they began to lose their effectiveness as the corresponding virulence genes *Va12* spread in the fungal population, and were largely withdrawn from cultivation. Nevertheless, varieties bearing *Mla12* were still being used in breeding programs, where they were necessarily selected in the presence of virulent populations of mildew. The outcome was the re-emergence of *Mla12*, combined in new ways with other sources of resistance.

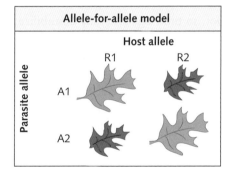

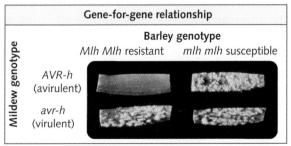

Figure 17.4 Two gene systems that may govern parasite virulence and host resistance. Top: The gene-for-gene system which governs parasite virulence and host resistance in many plant-parasite interactions. Bottom: An allele-for-allele system which has been used in several evolutionary models although few biological examples are known.

Reproduced from Brown, J.K.M. 2003. Little else but parasites. *Science* 299: 1680–1681.

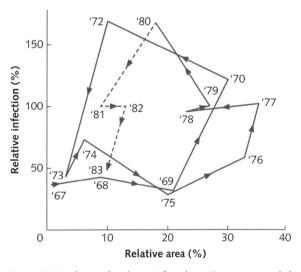

Figure 17.5 Evolution of virulence in fungal parasites causes coupled oscillations in the use of cultivars of crop plants. The line traces out in time the history of utilization and infection of barley varieties resistant to powdery mildew in the UK between 1967 and 1983.

Reproduced from Wolfe and Caten. 1987. *Populations of Plant Pathogens: Their Dynamics and Genetics.* Blackwell Scientific Publications, with permission from Wiley.

These new cultivars were planted extensively during the 1970s, occupying about a third of the acreage by 1977. However, their resistance to mildew was eroded year by year, until by the end of the decade it had been overtaken by the evolution of virulent strains of mildew. Naturally, these cultivars were in turn largely abandoned, and a third range developed. The coupled evolution of virulence in the pathogen and resistance in the host thus drives a cyclical process in which the response of fungal strains carrying *Va12* virulence first checks and then reverses the spread of barley varieties carrying *Mla12* resistance.

Coevolution of enemies may lead to an arms race. Alternatively, suppose that P2 can infect both H1 and H2, whereas P1 can infect only H1, as before. H2 will then prosper, but only so long as P2 is not around. P2, on the other hand, will do well regardless of the state of the host population. (This possibility, illustrated in Figure 17.4 as the "gene-for-gene system," is a more realistic description, in fact, of most of the parasite–host systems in crop plants whose genetics have been worked out.) This will evolve to a steady state at H2/ P2, because H2 has an advantage over H1 provided there are any P1 parasites present, while P2 has an advantage over P1 provided there are any H2 hosts present. This leads to strong selection among H2 hosts, however, for any variant resistant to P2. In practice, this will often lead to perpetual directional selection fueled by an arms race between parasite and host.

Crop plants provide some clear examples. In Iowa during the 1940s, for example, almost the whole acreage of oats was occupied by Richland and related cultivars. These were highly susceptible to oat stem rust (*Puccinia graminis*) and rapidly disappeared from cultivation after its appearance in 1942. They were replaced by the hybrid Victoria cultivars, which were resistant to stem rust races 1, 2, and 5, but susceptible to races 8 and 10. These latter races, rare in 1942, rapidly increased in frequency. At the same time, an epidemic of crown rust (*Helminthosporium victoriae*) began to develop. With the rise of these new pathogens, Victoria was forced out of cultivation, being replaced by Bond derivatives. These were resistant to races 8 and 10, but susceptible to race 7, which thereupon began to spread. At the same time, races of crown rust virulent on Bond also appeared.

This continual coevolutionary arms race is a very common outcome of planting very large acreages to uniform, or very similar, crops: at the time when these events occurred, the lifespan of a new oat cultivar was only about five years; wheat cultivars lasted fifteen years in Canada, but only five in Mexico; barley varieties are often overwhelmed in three or four years.

17.2.3 Enemies foster continual evolution in laboratory and field populations.

Parasites naturally tend to produce frequency-dependent selection in hosts, because rare host types will have a systematic advantage. The converse is equally true: rare parasite races that can attack the prevalent host types will spread rapidly. In both cases, however, the advantage is gained only after a certain time lag. When a host possessing some specific resistance increases in abundance it creates an opportunity for parasites that can circumvent its defenses, but there is inevitably a gap between the spread of the resistant host and the subsequent spread of the newly virulent parasite. In short: the interaction of enemies creates frequency-dependent selection that is lagged in time. When a reaction is delayed it tends to overshoot, so time lags generally destabilize equilibria. The interactions between enemies thereby lead to complex evolutionary dynamics.

Bacteria and phage coevolve in the laboratory. The crucial point is how the mean fitness of a population is related to the state of an enemy. In the laboratory, this can be studied using a bacteria–phage system, because both agents can be stored during a selection experiment in which bacteria and phage are propagated together and subsequently used to evaluate how their interaction changes over time. For example, the bacterial population from a given point in the experiment can be cultured with phage from the same time point, or from a previous time, or from a future time. We expect that the population will be better adapted to past phage than to future phage, with contemporary phage in between. Conversely, the phage from a given time can be tested against bacteria from past, contemporary or future generations. We expect them likewise to be better adapted to the past than to the future. This pattern of delayed response has been tracked in the clever experiment illustrated in Figure 17.6, where it fuels coevolutionary change between *Pseudomonas* and its parasite φ2 over hundreds of generations.

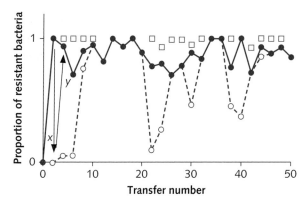

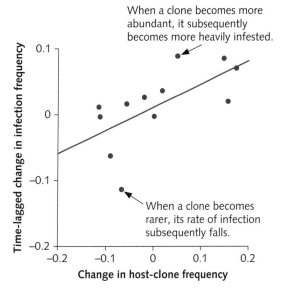

Figure 17.6 Time-lagged evolution of resistance to phage in experimental populations of bacteria. (a) Open squares: proportion of current bacteria resistant to ancestral phage. (b) Solid circles and lines: proportion of current bacteria resistant to current phage. (c) Open circles and broken lines: proportion of current bacteria resistant to future phage (two generations on). Evolution of phage virulence is shown by difference between (b) and (c) (for example, arrow x). Subsequent evolution of bacterial resistance is shown by difference between (c) and (b) (for example, arrow y).

Figure 17.7 Time-lagged response of infection by *Microphallus* to host genotype frequency in *Potamopyrgus*. The graph related the change in infection of clones of the snail between years y+1 and y+2 to changes in its frequency between years y and y+1.

Snails and trematodes coevolve in the field. It is much more difficult to find out how enemies interact in natural populations, where intensive long-term research programs are needed to work out the dynamics of parasite–host evolution. This has been achieved for the parthenogenetic snail *Potamopyrgus* and its trematode parasite *Microphallus* in New Zealand lakes. Studies extended over many years have shown that clones of the snail become more vulnerable after they have become common, and that chronically rare clones are consistently under-infected, as shown in Figure 17.7.

Host–pathogen interactions are often involved in rapid evolutionary change. The evidence from laboratory and field suggests that genes involved in the interaction between enemies should evolve much faster than others. The evidence from comparative genomics bears this out. In particular, the innate and acquired immune systems both bear witness to the pervasive selection pressures applied by short-lived pathogens. The ability to recognize billions of antigens and to produce the appropriate antibodies shows that multicellular individuals are continually

being attacked by a wide range of pathogenic microbes and viruses. Moreover, parasites may evade the immune response by similar means, such as the ability of *Plasmodium* (malarial parasite) to alter the antigenic properties of infected red blood cells (Section 2.5.3).

One extensive survey of nearly 4000 homologous sequences found that 9 of the 17 gene groups where directional selection seemed to be acting encoded surface antigens of pathogens. Genes encoding immuno-globulins and other defense-related proteins are very often highly variable, pointing to selection for rare alleles likely to confer resistance to common patho-gen species or genotypes. The pronounced differences between species show that selection is often excep-tionally rapid. These lines of evidence are consistent with the frequent occurrence of powerful selection driven by pathogens and capable of causing both diversification and rapid change.

Host–parasite coevolution drives perpetual evolution. The Red Queen famously remarked to Alice that in her country it was necessary to run very fast just to stay in the same place. Mutually antagonistic organisms live in the same country. It is a country with a Darwinian environment in which adaptation is merely provisional, and evolution a continual process of adaptation and counter-adaptation that would continue even if the physical environment were no more changeable than a chemostat. From day to day, the living world evolves continually as competitors, partners, rivals, and enemies shape one another's destiny. Evolution is not limited to being an exceptional response to some genetic innovation or environmental emergency, but is rather the usual condition of life on Earth.

● CHAPTER SUMMARY

Natural communities consist of partners, enemies, and rivals.

Permanence ensures cooperation.

Partners are more or less permanently associated.

- Cooperation between species is based on reciprocation.
  - *Repeated interactions between individuals of different species are involved in cleaner–client symbioses.*
  - *Reciprocity is policed by punishment.*
- When two partners cannot easily separate, their fitnesses become entrained.
  - *Long-lasting partnerships may become indissoluble.*
  - *Vertical transmission leads to partners, horizontal transmission to enemies.*

Enemies stimulate perpetual coevolution.

- Other organisms act as agents of selection.
  - *Individuals vary in response to enemies.*
  - *Serial passage causes the rapid evolution of virulence.*
  - *Host populations adapt to resident pathogens.*
- Counter-adaptation leads to complex evolutionary dynamics.
  - *Costs of adaptation set limits to parasite virulence and host resistance.*
  - *Coevolution of enemies may lead to coupled oscillations in genotype frequency.*
  - *Coevolution of enemies may lead to an arms race.*
- Enemies foster continual evolution in laboratory and field populations.
  - *Bacteria and phage coevolve in the laboratory.*
  - *Snails and trematodes coevolve in the field.*
  - *Host–pathogen interactions are often involved in rapid evolutionary change.*
  - *Host–parasite coevolution drives perpetual evolution.*

● FURTHER READING

Here are some pertinent further readings at the time of going to press. For relevant readings that have been released since publication, visit the book's Online Resource Centre at **www.oxfordtextbooks.co.uk/orc/bell_evolution/**

Section 17.1   Saikkonen, K., Wali, P. and Faeth, S.H. 2004. Evolution of endophyte-plant symbioses. *Trends in Plant Science* 9: 275–280.

Section 17.2   Ebert, D. 1998. Experimental evolution of parasites. *Science* 282: 1432–1435.

Bohannan, B.J.M. and Lenski, R.E. 2000. Linking genetic change to community evolution: insights from studies of bacteria and bacteriophage. *Ecology Letters* 3: 362–367.

Lively, C.M. and Dybdahl, M.F. 2000. Parasite adaptation to locally common host genotypes. *Nature* 405: 679–681.

● QUESTIONS

1. Describe, with example, the circumstances in which two species are likely to evolve a mutually beneficial relationship.

2. Differentiate between the conditions favoring the evolution of mutualism and those favoring the evolution of antagonism between two species.

3. Describe and explain the outcome of serial-passage experiments in which bacteria are incubated in a series of host individuals.

4. What sets limits to the virulence and host range that will evolve in a pathogen population?

5. Explain how the coevolution of host and parasite can lead to coupled genetic oscillations.

6. Explain how the coevolution of host and parasite can lead to an arms race.

7. Design an experiment to investigate the coupled evolutionary dynamics of host and parasite in the laboratory. What would you predict to be the outcome of the experiment?

8. Would you expect the evolutionary dynamics of predators and prey to be similar to those of parasites and hosts? In what respects might they be different?

You can find a fuller set of questions, which will be refreshed during the life of this edition, in the book's Online Resource Centre at **www.oxfordtextbooks.co.uk/orc/bell_evolution/**

GLOSSARY

I have tried to avoid jargon as much as possible in this book, but it may be helpful to collect definitions of some terms specific to evolutionary biology that have been used in the text.

Altruism Altruism is costly behavior that benefits another but confers no benefit whether direct or indirect, current or deferred, on the donor.

Ancestral constraint The failure to evolve a character state because of the lack of appropriate genetic variation.

Artificial selection The deliberate choice of breeding stock by farmers and scientists.

Balancing selection Selection acting in such a way as to preserve two or more distinct types within a population.

Clade A monophyletic group; an individual and all its descendants.

Cladogenesis Lineage splitting: the divergence of two independent lineages from a single ancestral lineage.

Cladogram A phylogenetic tree inferred from cladistic analysis.

Cladistics, cladistic analysis The method of inferring ancestry from the possession of shared derived character states.

Cline Continuous change in character state in a particular geographical direction.

Coalescent The individual which lived in the past from whom all individuals living now descend. Also applied to genes, species, or any non-recombining entity.

Coevolution The coupled evolution of two or more interacting species.

Continuous culture A technique to ensure perpetual growth of a microbial culture by adding fresh growth medium continuously at a given rate to a vessel containing the culture that overflows at the same rate.

Convergent evolution The independent evolution of a similar character state in two lineages: the convergence of non-homologous characters in distantly-related organisms. Similar to parallel evolution.

Cross-feeding The evolution of types (usually of bacteria) able to subsist on the metabolites excreted by others.

Cultivar A variety that has been deliberately produced by artificial selection.

Darwin A unit of the rate of evolutionary change (Dar), indicating an e-fold ($e \approx 2.72$) change over 1 My.

Directional selection Selection causing the spread of beneficial types.

Disruptive selection Selection acting simultaneously in different directions = Diversifying selection.

Diversifying selection The maintenance of variation as the consequence of selection acting simultaneously in different directions, for example in different habitats.

Drift (genetic drift) The stochastic fluctuation of allele frequencies over time.

Ecotype A variety with a restricted ecological distribution.

Endosymbiosis Serial endosymbiosis is the theory that eukaryote lineages have evolved through the sequential acquisition and subsequent domestication of bacteria as mitochondria, plastids, and other organelles.

Establishment period The period of time during which several beneficial mutations may arise in an asexual population before one escapes stochastic loss and begins to spread.

Evolutionarily stable state (ESS) The ESS is the behavior such that if almost all individuals express that behavior, no other behavior increases fitness.

Fitness The rate of proliferation of a defined category of individuals or lineages. Relative fitness is the rate relative to another lineage, or to the average of other lineages.

Fluctuating selection Variation in the direction or magnitude of selection over time.

Freeloader An individual which exploits a public good without contributing to it.

Frequency-dependent selection Selection operating on types whose fitness varies consistently with their frequency in the population.

Functional interference The loss of function with respect to an activity caused by the gain of function with respect to another.

Group selection Variation in the rate of proliferation of entire populations.

Hitch-hiking The spread of a gene through the effect it produces at a linked locus is called "hitch-hiking."

Homology, homologous Characters that are similar among species because they are inherited from the common ancestor of the group are said to be homologous.

Implementing innovation An implementing innovation is a character that, once established in an ancestral lineage, may itself become divergently specialized in descendant lineages.

Inclusive fitness The sum of direct effects on the fitness of a focal individual, contributed by itself, and indirect effects, contributed by relatives and discounted by their relatedness.

Innovation (as used here) A shared ancestral character that facilitates the adaptive radiation of a clade.

Invasibility The liability of a population to be invaded by rare types.

Isolate line An experimental line perpetuated by choosing a single individual from time to time as the sole ancestor of succeeding generations.

Kin selection Selection among families.

Lineage A chain of ancestors and descendants that propagates itself through time.

Lineage backwards-in-time An individual together with all its ancestors.

Lineage forwards-in-time An individual together with all its descendants.

Monophyletic A monophyletic group comprises an individual and all its descendants.

Morph A type that is distinguished by a single genetic difference, or a few simple differences, which may involve cryptic characters such as alleles at enzyme-coding loci.

Most recent common ancestor The lineage that is ancestral to all the members of a clade and to no other taxa.

Mutation supply rate The number of mutations arising in a population per generation.

Mutational degradation The loss of function expressed by a line that has evolved in a novel environment when transferred to its ancestral environment.

My Million years.

Mya Million years ago.

Natural selection The tendency for a type with a rate of proliferation greater than average to increase in frequency.

Neutral Without effect on fitness.

Optimality The theory that a feature (a structure, or behavior, or schedule) has evolved towards the best compromise between conflicting constraints.

Parallel evolution The independent evolution of a similar character state in two lineages. The convergence of clearly homologous characters in closely-related organisms. Similar to convergent evolution.

Parsimony (in this context) The principle that, in evaluating phylogenetic hypotheses, those involving fewer changes in character state should be preferred.

Passage period The period of time during which a successful mutation that escaped stochastic loss in an asexual population spreads and eventually becomes fixed.

Periodic selection The passage of a series of beneficial mutations through an asexual population.

Phylogenetic tree, phylogeny The branching diagram representing an hypothesis about the ancestry of a set of taxa.

Phylotypic stage The point during development at which the embryos of a monophyletic group most strongly resemble one another.

Polymorphism A character that takes two or more discrete states in a population is said to be polymorphic. If this variation is caused by alternative alleles it constitutes genetic polymorphism.

Potentiating innovation A potentiating innovation is a character that, once established in an ancestral lineage, enables other characters to diverge in descendant lineages.

Primitive The character state of the common ancestor of a clade.

Purifying selection Selection acting to eliminate deleterious mutations.

Radiation, adaptive radiation The divergence of character state among the descendants of a common ancestor.

Race A variety with a restricted geographical distribution.

Reciprocity Rewarding an individual whose acts have benefitted you.

Red Queen The theory of perpetual coevolution driven by reciprocal selection among antagonists.

Rescue Evolutionary rescue is the recovery of a population that has been subjected to a severe stress through the spread of resistant types.

Selection coefficient The fitness of one type relative to another.

Selection differential The difference between the average value of a character after an episode of selection and its average value before selection within a population. It is equal to the regression of relative fitness on character state.

Selection gradient The partial regression of relative fitness on character state; the regression of relative fitness on character state when the states of other characters (belonging to some pre-defined set of characters) are held constant.

Selfish Behavior that benefits the entity (such as a selfish gene) while damaging the collectivity (such as a cell or individual).

Serial passage The practice of infecting animals with pathogens, harvesting the pathogens after growth, and then re-inoculating new host individuals.

Serial transfer A technique to ensure perpetual growth of a microbial culture by regular transfer of a small sample to fresh growth medium.

Sexual conflict Variation in the direction of selection with gender.

Sexual selection The tendency for a type with a rate of sexual fusion greater than average to increase in frequency.

Sibla taxon (a made-up word, from "shortest inferred branch length from ancestor") The taxon that is currently regarded as being most closely related to (having the shortest branch length separating it from) the true ancestor of a clade is the sibla taxon of that clade.

Sister species (taxa). Two species (taxa) whose most recent common ancestor is not the ancestor of any other species (taxon) in a given set of species (taxa).

Species (ecological definition) A strongly marked variety.

Species (sexual definition) A species is the set of lineages which exchange genes among themselves while being isolated from all other lineages.

Stabilizing selection Selection acting to maintain the current average state of the population = Purifying selection.

Subspecies A race so strongly and consistently differentiated from others that it is given a formal taxonomic name.

Sweep A selective sweep is the replacement of all lineages in an asexual population by a single lineage following a stress. Also applied to an allele in a sexual population.

Synonymous A nucleotide substitution that does not affect the amino acid sequence of a protein, and hence is usually assumed to be neutral; other substitutions are non-synonymous.

Taxon (pl taxa) Any group of species.

Variety The most general term for a distinct kind that is not recognized as a species.

Vestigial structure The highly reduced, non-functional remnant of a structure no longer preserved by purifying selection.

Waiting period The period of time before a beneficial mutation first arises in an asexual population.

EVOLUTION: INDEX OF ORGANISMS

Echinodermata, echinoderm 121, 162, 270, 276
Echinoidea 121
Ectoprocta 125
edrioasteroid 10, 160
eel 13
electric eel 118
electric fish 270
Electrophorus 118
elephant 106, 264
elephantfish 116
emu 110, 268
Engladina 218
Ensatina eschscholtzii 203
Enteropneusta 121
eocrinoid 10
Eoherpeton 152
Eotragus 425
Epibulus insidiator 256
Epulopiscium 321
Equisetophyta 31
Ericineinae 107
Eryops 153
Euarchontoglires 106
Eudorina 228
euglenid 130, 242
Eukaryota, eukaryote 129, 166, 240, 321, 322, 323, 324, 325
euphorb 271
Euplectes 423
eurypterid 221
Eusthenopteron 154
Eutheria, eutherian 20, 105, 146
Excavata 130

fathead minnow 421
fern 131, 155, 240, 293
Ficedula 442
fiddler crab 425
fin whale 75
fire-bellied toad 210
fish 82, 270
fish lice 452
flamingo 110
flatworm 124, 255, 272, 426
Flaveria 232
 anomala 234
 bidentata 234
 pringlei 232
 ramosissima 234
 vaginata 234
flax 454
flowering plant 265, 275

fly 201, 292; *see also* Diptera
flying dragon 262
flying lemur 109
flying phalanger 109
flying squirrel 109
Folivora 108
Foraminifera, foraminiferan 131, 166
fowl 110
fox 350
freshwater shrimp 266
frog 81, 113
frogmouth 110
Fungi 125, 128, 154, 165, 287, 413, 451, 453

Gadus morhua 221
Gallinula chloropus 219
Gammaridae 266
gannet 110
gar 116, 156, 271
Gasterosteiformes 13
Gasterosteus 378 see also Stickleback
Gasterosteus aculeatus 208 see also Three-spined stickleback
gastropod 222
Gastrotricha, gastrotrich 124, 162
Geospiza fortis 394
giant salamander 112
Giardia 240
gibbon 103
Gila monster 72
gingko 9, 131, 220
giraffe 279
Gnathostomata, gnathostome 117, 157, 253
Gnathostomulida 124
gnetophyte 131
goby 451
golden mole 106, 107
goldentops 232
goldfish 42
Goliath beetle 269
Gonium 228
Gorilla:
 berengei 102
 gorilla 103
Graptolithina, graptolite 160, 221, 222
grass 378
grasshopper 327
grayling 303
great tit 301, 380 see also *Parus major*
greater glider 109
grebe 110

green alga 372, 451 *see also* Chlorophyta
green hydra 451
grey tree frog 421
grey whale 75
groundhog 109
Grypania 167
guppy 29, 117, 301, 303, 421, 424
Gymnophiona 113
Gymnospermae 131
Gymnotidae, gymnotid 116, 118, 271

Hadrocodium 147
hadrosaur 271
hagfish 120, 160
Haikouella 160
hawksbill turtle 253
hedgehog 106, 270
Helianthus:
 annuus 374
 anomalus 374
 petiolaris 374
Heliconius erato 393
Heliconius Melpomene 393
helicoplacoid 10, 160
Hemichordata, hemichordate 24, 121, 162
hermit crab 440
herring 202
herring gull 204
Heterodontus 157
Heterokonta 130
heteropod 278
Heterostraci 158
hexactinellid 164
Hexapoda 124
hippopotamus 14
Holocephali 118
holostean 269
Holothuroidea 122
Hominoidea 143
Homo 139
 erectus 139
 ergaster 140
 habilis 142
 heidelbergensis 140
 neanderthalis 140
 sapiens 99, 139
 see also Human
Homolozoa 160
honey bee 447
honey possum 109
honeycreeper 271

EVOLUTION OF LIFE: SUBJECT INDEX